T0344492

A FIRST COURSE IN RANDOM MATRIX THEORY

The real world is perceived and broken down as data, models and algorithms in the eyes of physicists and engineers. Data is noisy by nature and classical statistical tools have so far been successful in dealing with relatively smaller levels of randomness. The recent emergence of Big Data and the required computing power to analyze them have rendered classical tools outdated and insufficient. Tools such as random matrix theory and the study of large sample covariance matrices can efficiently process these big datasets and help make sense of modern, deep learning algorithms. Presenting an introductory calculus course for random matrices, the book focuses on modern concepts in matrix theory, generalizing the standard concept of probabilistic independence to non-commuting random variables. Concretely worked out examples and applications to financial engineering and portfolio construction make this unique book an essential tool for physicists, engineers, data analysts and economists.

MARC POTTERS is Chief Investment Officer of CFM, an investment firm based in Paris. Marc maintains strong links with academia and, as an expert in random matrix theory, he has taught at UCLA and Sorbonne University.

JEAN-PHILIPPE BOUCHAUD is a pioneer in econophysics. His research includes random matrix theory, statistics of price formation, stock market fluctuations, and agent-based models for financial markets and macroeconomics. His previous books include *Theory of Financial Risk and Derivative Pricing* (Cambridge University Press, 2003) and *Trades, Quotes and Prices* (Cambridge University Press, 2018), and he has been the recipient of several prestigious, international awards.

A FIRST COURSE IN RANDOM MATRIX THEORY

for Physicists, Engineers and Data Scientists

MARC POTTERS
Capital Fund Management, Paris

JEAN-PHILIPPE BOUCHAUD
Capital Fund Management, Paris

CAMBRIDGE
UNIVERSITY PRESS

University Printing House, Cambridge CB2 8BS, United Kingdom

One Liberty Plaza, 20th Floor, New York, NY 10006, USA

477 Williamstown Road, Port Melbourne, VIC 3207, Australia

314–321, 3rd Floor, Plot 3, Splendor Forum, Jasola District Centre, New Delhi – 110025, India

79 Anson Road, #06–04/06, Singapore 079906

Cambridge University Press is part of the University of Cambridge.

It furthers the University's mission by disseminating knowledge in the pursuit of education, learning, and research at the highest international levels of excellence.

www.cambridge.org
Information on this title: www.cambridge.org/9781108488082
DOI: 10.1017/9781108768900

First published 2021

A catalogue record for this publication is available from the British Library.

Library of Congress Cataloging-in-Publication Data
Names: Potters, Marc, 1969– author. | Bouchaud, Jean-Philippe, 1962– author.
Title: A first course in random matrix theory : for physicists, engineers and data scientists / Marc Potters, Jean-Philippe Bouchaud.
Description: Cambridge ; New York, NY : Cambridge University Press, 2021. | Includes bibliographical references and index.
Identifiers: LCCN 2020022793 (print) | LCCN 2020022794 (ebook) | ISBN 9781108488082 (hardback) | ISBN 9781108768900 (epub)
Subjects: LCSH: Random matrices.
Classification: LCC QA196.5 .P68 2021 (print) | LCC QA196.5 (ebook) | DDC 512.9/434–dc23
LC record available at https://lccn.loc.gov/2020022793
LC ebook record available at https://lccn.loc.gov/2020022794

ISBN 978-1-108-48808-2 Hardback

Additional resources for this title at www.cambridge.org/potters

Contents

Preface

Physicists have always approached the world through data and models inspired by this data. They build models from data and confront their models with the data generated by new experiments or observations. But real data is by nature noisy; until recently, classical statistical tools have been successful in dealing with this randomness. The recent emergence of very large datasets, together with the computing power to analyze them, has created a situation where not only the number of data points is large but also the number of studied variables. Classical statistical tools are inadequate to tackle this situation, called the large dimension limit (or the Kolmogorov limit). Random matrix theory, and in particular the study of large sample covariance matrices, can help make sense of these big datasets, and is in fact also becoming a useful tool to understand deep learning. Random matrix theory is also linked to many modern problems in statistical physics such as the spectral theory of random graphs, interaction matrices of spin-glasses, non-intersecting random walks, many-body localization, compressed sensing and many more.

This book can be considered as one more book on random matrix theory. But our aim was to keep it purposely introductory and informal. As an analogy, high school seniors and college freshmen are typically taught both calculus and analysis. In analysis one learns how to make rigorous proofs, define a limit and a derivative. At the same time in calculus one can learn about computing complicated derivatives, multi-dimensional integrals and solving differential equations relying only on intuitive definitions (with precise rules) of these concepts. This book proposes a "calculus" course for random matrices, based in particular on the relatively new concept of "freeness", that generalizes the standard concept of probabilistic independence to non-commuting random variables.

Rather than make statements about the most general case, concepts are defined with some strong hypothesis (e.g. Gaussian entries, real symmetric matrices) in order to simplify the computations and favor understanding. Precise notions of norm, topology, convergence, exact domain of application are left out, again to favor intuition over rigor. There are many good, mathematically rigorous books on the subject (see references below) and the hope is that our book will allow the interested reader to read them guided by his/her newly built intuition.

Readership

The book was initially conceived as a textbook for a graduate level standard 30 hours course in random matrix theory, for physicists or applied mathematicians, given by one of us (MP) during a sabbatical at UCLA in 2017–2018. As the book evolved many new developments, special topics and applications have been included. Lecturers can then customize their course offering by complementing the first few essential chapters with their own choice of chapters or sections from the rest of the book.

Another group of potential readers are seasoned researchers analyzing large datasets who have heard that random matrix theory may help them distinguish signal from noise in singular value decompositions or eigenvalues of sample covariance matrices. They have heard of the Marčenko–Pastur distribution but do not know how to extend it to more realistic settings where they might have non-Gaussian noise, true outliers, temporal (sample) correlations, etc. They need formulas to compute null hypothesis and so forth. They want to understand where these formulas come from intuitively without requiring full precise mathematical proofs.

The reader is assumed to have a background in undergraduate mathematics taught in science and engineering: linear algebra, complex variables and probability theory. Important results from probability theory are recalled in the book (addition of independent variables, law of large numbers and central limit theorem, etc.) while stochastic calculus and Bayesian estimation are not assumed to be known. Familiarity with physics approximation techniques (Taylor expansions, saddle point approximations) is helpful.

How to Read This Book

We have tried to make the book accessible for readers of different levels of expertise. The bulk of the text is hopefully readable by graduate students, with most calculations laid out in detail. We provide exercises in most chapters, which should allow the reader to check that he or she has understood the main concepts. We also tried to illustrate the book with as many figures as possible, because we strongly believe (as physicists) that pictures considerably help forming an intuition about the issue at stake.

More technical issues, directed at experts in RMT or statistical physics, are signaled by the use of a different, smaller font and an extra margin space. Chapters 3, 6, 7 and 13 are marked with a star, meaning that they are not essential for beginners and they can be skipped at first reading.

At the end of each chapter, we give a non-exhaustive list of references, some general and others more technical and specialized, which direct the reader to more in-depth information related to the subject treated in the chapter.

Other Books on Related Subjects

Books for Mathematicians

There are many good recent books on random matrix theory and free probabilities written by and for mathematicians: Blower [2009], Anderson et al. [2010], Bai and Silverstein

[2010], Pastur and Scherbina [2010], Tao [2012], Erdős and Yau [2017], Mingo and Speicher [2017]. These books are often too technical for the intended readership of the present book. We nevertheless rely on these books to extract some relevant material for our purpose.

Books for Engineers

Communications engineers and now financial engineers have become big users of random matrix theory and there are at least two books specifically geared towards them. The style of engineering books is closer to the style of the present book and these books are quite readable for a physics audience.

There is the short book by Tulino and Verdú [2004]; it gets straight to the point and gives many useful formulas from free probabilities to compute the spectrum of random matrices. The first part of this book covers some of the topics covered here, but many other subjects more related to statistical physics and to financial applications are absent.

Part I of Couillet and Debbah [2011] has a greater overlap with the present book. Again about half the topics covered here are not present in that book (e.g. Dyson Brownian motion, replica trick, low-rank HCIZ and the estimation of covariance matrices).

Books for Physicists

Physicists interested in random matrix theory fall into two broad categories:

Mathematical physicists and high-energy physicists use it to study fundamental quantum interactions, from Wigner's distribution of nuclear energy level spacing to models of quantum gravity using matrix models of triangulated surfaces.

Statistical physicists encounter random matrices in the interaction matrices of spin-glasses, in the study of non-intersecting random walks, in the spectral analysis of large random graphs, in the theory of Anderson localization and many-body localization, and finally in the study of sample covariance matrices from large datasets. This book focuses primarily on statistical physics and data analysis applications.

The classical book by Mehta [2004] is at the crossroads of these different approaches, whereas Forrester [2010], Brézin and Hikami [2016] and Eynard et al. [2006] are examples of books written by mathematical physicists. Livan et al. [2018] is an introductory book geared towards statistical physicists. That book is very similar in spirit to ours. The topics covered do not overlap entirely; for example Livan et al. do not cover the Dyson Brownian motion, the HCIZ integral, the problem of eigenvector overlaps, sample covariance matrices with general true covariance, free multiplication, etc.

We should also mention the handbook by Akemann et al. [2011] and the Les Houches summer school proceedings [Schehr et al., 2017], in which we co-authored a chapter on financial applications of RMT. That book covers a very wide range of topics. It is a useful complement to this book but too advanced for most of the intended readers.

Finally, the present book has some overlap with a review article written with Joel Bun [Bun et al., 2017].

Acknowledgments

The two of us want to warmly thank the research team at CFM with whom we have had many illuminating discussions on these topics over the years, and in particular Jean-Yves Audibert, Florent Benaych-Georges, Raphael Bénichou, Rémy Chicheportiche, Stefano Ciliberti, Sungmin Hwang, Vipin Kerala Varma, Laurent Laloux, Eric Lebigot, Thomas Madaule, Iacopo Mastromatteo, Pierre-Alain Reigneron, Adam Rej, Jacopo Rocchi, Emmanuel Sérié, Konstantin Tikhonov, Bence Toth and Dario Vallamaina.

We also want to thank our academic colleagues for numerous, very instructive interactions, collaborations and comments, including Gerard Ben Arous, Michael Benzaquen, Giulio Biroli, Edouard Brézin, Zdzisław Burda, Benoît Collins, David Dean, Bertrand Eynard, Yan Fyodorov, Thomas Guhr, Alice Guionnet, Antti Knowles, Reimer Kuehn, Pierre Le Doussal, Fabrizio Lillo, Satya Majumdar, Marc Mézard, Giorgio Parisi, Sandrine Péché, Marco Tarzia, Matthieu Wyart, Francesco Zamponi and Tony Zee.

We want to thank some of our students and post-docs for their invaluable contribution to some of the topics covered in this book, in particular Romain Allez, Joel Bun, Tristan Gautié and Pierre Mergny. We also thank Pierre-Philippe Crépin, Théo Dessertaine, Tristan Gautié, Armine Karami and José Moran for carefully reading the manuscript.

Finally, Marc Potters wants to thank Fan Yang, who typed up the original hand-written notes. He also wants to thank Andrea Bertozzi, Stanley Osher and Terrence Tao, who welcomed him for a year at UCLA. During that year he had many fruitful discussions with members and visitors of the UCLA mathematics department and with participants of the IPAM long program in quantitative linear algebra, including Alice Guionnet, Horng-Tzer Yau, Jun Yin and more particularly with Nicholas Cook, David Jekel, Dimitri Shlyakhtenko and Nikhil Srivastava.

Bibliographical Notes

Here is a list of books on random matrix theory that we have found useful.

- Books for mathematicians
  - G. Blower. *Random Matrices: High Dimensional Phenomena*. Cambridge University Press, Cambridge, 2009,
  - G. W. Anderson, A. Guionnet, and O. Zeitouni. *An Introduction to Random Matrices*. Cambridge University Press, Cambridge, 2010,
  - Z. Bai and J. W. Silverstein. *Spectral Analysis of Large Dimensional Random Matrices*. Springer-Verlag, New York, 2010,
  - L. Pastur and M. Scherbina. *Eigenvalue Distribution of Large Random Matrices*. American Mathematical Society, Providence, Rhode Island, 2010,
  - T. Tao. *Topics in Random Matrix Theory*. American Mathematical Society, Providence, Rhode Island, 2012,
  - L. Erdős and H.-T. Yau. *A Dynamical Approach to Random Matrix Theory*. American Mathematical Society, Providence, Rhode Island, 2017,

  - J. A. Mingo and R. Speicher. *Free Probability and Random Matrices*. Springer, New York, 2017.
- Books for physicists and mathematical physicists
  - M. L. Mehta. *Random Matrices*. Academic Press, San Diego, 3rd edition, 2004,
  - B. Eynard, T. Kimura, and S. Ribault. Random matrices. *preprint arXiv:1510.04430*, 2006,
  - P. J. Forrester. *Log Gases and Random Matrices*. Princeton University Press, Princeton, NJ, 2010,
  - G. Akemann, J. Baik, and P. D. Francesco. *The Oxford Handbook of Random Matrix Theory*. Oxford University Press, Oxford, 2011,
  - E. Brézin and S. Hikami. *Random Matrix Theory with an External Source*. Springer, New York, 2016,
  - G. Schehr, A. Altland, Y. V. Fyodorov, N. O'Connell, and L. F. Cugliandolo, editors. *Stochastic Processes and Random Matrices*, Les Houches Summer School, 2017. Oxford University Press, Oxford,
  - G. Livan, M. Novaes, and P. Vivo. *Introduction to Random Matrices: Theory and Practice*. Springer, New York, 2018.
- More "applied" books
  - A. M. Tulino and S. Verdú. *Random Matrix Theory and Wireless Communications*. Now publishers, Hanover, Mass., 2004,
  - R. Couillet and M. Debbah. *Random Matrix Methods for Wireless Communications*. Cambridge University Press, Cambridge, 2011.
- Our own review paper on the subject, with significant overlap with this book
  - J. Bun, J.-P. Bouchaud, and M. Potters. Cleaning large correlation matrices: Tools from random matrix theory. *Physics Reports*, 666:1–109, 2017.

Symbols

Abbreviations

BBP: Baik, Ben Arous, Péché
CDF: Cumulative Distribution Function
CLT: Central Limit Theorem
DBM: Dyson Brownian Motion
EMA: Exponential Moving Average
FID: Free Identically Distributed
GOE: Gaussian Orthogonal Ensemble
GSE: Gaussian Symplectic Ensemble
GUE: Gaussian Unitary Ensemble
HCIZ: Harish-Chandra–Itzykson–Zuber
IID: Independent Identically Distributed
LLN: Law of Large Numbers
MAP: Maximum *A Posteriori*
MAVE: Mean Absolute Value
MMSE: Minimum Mean Square Error
PCA: Principal Component Analysis
PDE: Partial Differential Equation
PDF: Probability Distribution Function
RIE: Rotationally Invariant Estimator
RMT: Random Matrix Theory
SCM: Sample Covariance Matrix
SDE: Stochastic Differential Equation
SVD: Singular Value Decomposition

Conventions

0^+: infinitesimal positive quantity
$\mathbf{1}$: identity matrix

$\sim$:	scales as, of the order of, also, for random variables, drawn from
$\approx$:	approximately equal to (mathematically or numerically)
$\propto$:	proportional to
$:=$:	equal by definition to
$\equiv$:	identically equal to
$\boxplus$:	free sum
$\boxtimes$:	free product
$\mathbb{E}[.]$:	mathematical expectation
$\mathbb{V}[.]$:	mathematical variance
$\langle . \rangle$:	empirical average
$[x]$:	dimension of x
i:	$\sqrt{-1}$
Re:	real part
Im:	imaginary part
$⨍$:	principal value integral
$\sqrt[\oplus]{\cdot}$:	special square root, Eq. (4.56)
$\mathbf{A}^T$:	matrix transpose
supp(ρ):	domain where $\rho(\cdot)$ is non-zero

Note: most of the time $f(t)$ means that t is a continuous variable, and f_t means that t is discrete.

Roman Symbols

A:	generic constant, or generic free variable
$\mathbf{A}$:	generic matrix
a:	generic coefficient, as in the gamma distribution, Eq. (4.17), or in the free log-normal, Eq. (16.15)
B:	generic free variable
$\mathbf{B}$:	generic matrix
b:	generic coefficient, as in the gamma distribution, Eq. (4.17), or in the free log-normal, Eq. (16.15)
C:	generic coefficient
C_k:	Catalan numbers
$\mathbf{C}$:	often population, or "true" covariance matrix, sometimes $\mathbf{C} = \mathbf{A} + \mathbf{B}$
$\mathcal{C}$:	total investable capital or cross-covariance matrix
c_k:	cumulant of order k
d:	distance between eigenvalues
dB:	Wiener noise
$\mathbf{E}$:	sample, or empirical matrix; matrix corrupted by noise
$\mathcal{E}$:	error
$\mathbf{e}$:	normalized vector of 1's $\mathbf{e} = (1, 1, \ldots, 1)^T / \sqrt{N}$

$\mathbf{e}_1$:	unit vector in the first direction $\mathbf{e}_1 = (1,0,\ldots,0)^T$
F:	free energy
$F(\cdot)$:	generic function
$F_\beta(\cdot)$:	Tracy–Widom distributions
$F_K(\cdot,\cdot,\cdot\cdot)$:	K-point correlation function of eigenvalues
$\mathbf{F}$:	dual of the sample covariance matrix, see Section 4.1.1
$\mathbf{F}(\mathbf{x})$:	force field
$\mathcal{F}$:	generalized force in the Fokker–Planck equation
$\mathcal{F}_n$:	replica free energy
$f(\cdot)$:	generic function
$\mathbf{G}(\cdot)$:	resolvent matrix
$\mathcal{G}$:	return target of a portfolio
$g_N(\cdot)$:	normalized trace of the resolvent
$\mathbf{g}$:	vector of expected gains
$\mathfrak{g}(\cdot)$:	generic Stieltjes transform
$\mathfrak{g}_{\mathbf{A}}(\cdot)$:	Stieltjes transform of the spectrum of $\mathbf{A}$
$\mathfrak{g}_{\mathbf{X}}(\cdot)$:	Stieltjes transform of a Wigner matrix
$\mathbf{H}$:	rectangular $N \times T$ matrix with random entries
$H(\cdot)$:	log of the generating function of a random variable, $H(k) := \log \mathbb{E}[e^{ikX}]$
$H_{\mathbf{X}}(\cdot)$:	log of the generating function (rank-1 HCIZ) of a random matrix
$\hat{H}$:	annealed version of $H_{\mathbf{X}}(t)$
H_n:	Hermite polynomial (probabilists' convention)
$\mathcal{H}$:	Hamiltonian of a spin system
h:	auxiliary function
$\mathfrak{h}(\cdot)$:	real part of the Stieltjes transform on the real axis (equal to the Hilbert transform times π)
I:	generic integral
$I(\mathbf{A},\mathbf{B})$:	HCIZ integral
$I_t(\mathbf{A})$:	HCIZ integral when $\mathbf{B}$ is of rank-1 with eigenvalue t
$\mathcal{I}$:	interval
i, j, k, ℓ:	generic integer indices
$\mathbf{J}$:	Jacobi matrices, or interaction matrix in spin-glasses
K:	number of terms in a sum or a product, or number of blocks in a cross-validation experiment
$K(\cdot)$:	kernel
$K_\eta(\cdot)$:	Cauchy kernel, of width η
$\mathbf{K}$:	Toeplitz matrix
L:	length of an interval containing eigenvalues
$L_n^{(\alpha)}$:	generalized Laguerre polynomial
$L_\mu^{C,\beta}$:	Lévy stable laws
$\mathbf{L}$:	lower triangular matrix

$\mathcal{L}$: log-likelihood
M: generic matrix
$\mathbf{M}_p$: inverse-Wishart matrix with coefficient p
m_k: moment of order k
N: size of the matrix, number of variables
$\mathcal{N}(\mu, \sigma^2)$: Gaussian distribution with mean μ and variance σ^2
$\mathcal{N}(\boldsymbol{\mu}, \mathbf{C})$: multivariate Gaussian distribution with mean $\boldsymbol{\mu}$ and covariance **C**
n: number of eigenvalues in some interval, or number of replicas
$O(N)$: orthogonal group in N dimensions
O: generic orthogonal matrix
$P(\cdot)$: generic probability distribution function defined by its argument: $P(x)$ is a short-hand for the probability density of variable x
$P_\gamma(\cdot)$: gamma distribution
$P_>(\cdot)$: complementary cumulative distribution function
$P_0(\cdot)$: prior distribution in a Bayesian framework
$P_i(t)$: probability to be in state i at time t
P_n: Legendre polynomial
$P(x|y)$: conditional distribution of x knowing y
$P_n^{(a,b)}$: Jacobi polynomial
P: rank-1 (projector) matrix
$\mathbb{P}[X]$: probability of event X
p: variance of the inverse-Wishart distribution, or quantile value
$p(x)$: generic polynomial of x
$p(\mathbf{A})$: generic matrix polynomial of **A**
$p_N(\cdot)$: generic monic orthogonal polynomial
$p(y,t|x)$: propagator of Brownian motion, with initial position x at $t = 0$
$Q(\cdot)$: generic polynomial
$Q_N(\cdot)$: (expected) characteristic polynomial
q: ratio of size of matrix N to size of sample T: $q = N/T$
q^*: effective size ratio $q^* = N/T^*$
$q(\mathbf{A})$: generic matrix polynomial of **A**
$q_N(\cdot)$: normalized characteristic polynomial
R: circle radius
$R_{\mathbf{A}}(\cdot)$: R-transform of the spectrum of **A**
$\mathcal{R}$: portfolio risk, or error
r: signal-to-noise ratio; also an auxiliary variable in the spin-glass section
$r_{i,t}$: return of asset i at time t
$S_{\mathbf{A}}(\cdot)$: S-transform of the spectrum of **A**
S: diagonal matrix of singular values
s: generic singular value
T: size of the sample

T^*:	effective size of the sample, accounting for correlations
T_n:	Chebyshev polynomials of the first kind
$\mathbf{T}$:	matrix T-transform
$\mathfrak{t}_{\mathbf{A}}(\cdot)$:	T-transform of the spectrum of $\mathbf{A}$
$U(N)$:	unitary group in N dimensions
U_n:	Chebyshev polynomials of the second kind
$\mathbf{U}$:	generic rotation matrix defining an eigenbasis
$\mathbf{u}$:	eigenvector (often of the population matrix $\mathbf{E}$)
$V(\cdot)$:	generic potential
$\mathbf{V}$:	generic rotation matrix defining an eigenbasis
$\mathcal{V}$:	generalized potential, or variogram
$v(x,t)$:	velocity field in Matytsin's formalism
$\mathbf{v}$:	eigenvector (often of the sample matrix $\mathbf{C}$)
$W(\cdot)$:	Fokker–Planck auxiliary function
$W_n(\pi,\sigma)$:	Wiengarten coefficient
$\mathbf{W}$:	white Wishart matrix, sometimes generic multiplicative noise matrix
$\mathbf{W}_q$:	white Wishart matrix of parameter q
w:	generic weight
$\mathbf{X}$:	Wigner matrix, sometimes generic additive noise matrix
x:	generic variable
$\mathbf{x}$:	generic vector
$\hat{x}$:	estimator of x
y_k:	data points
$\mathbf{y}$:	generic vector
Z:	normalization, or partition function
Z_n:	Kesten variable
z:	generic complex variable
$\mathfrak{z}(\cdot)$:	functional inverse of the Stieltjes function $\mathfrak{g}(\cdot)$

Greek Symbols

α:	generic scale factor
β:	effective inverse "temperature" for the Coulomb gas, defining the symmetry class of the matrix ensemble ($\beta = 1$ for orthogonal, $\beta = 2$ for unitary, $\beta = 4$ for symplectic)
β_i:	exposure of asset i to the common (market) factor
$\boldsymbol{\beta}$:	vector of β_i's
Γ:	gamma function
Γ_N:	multivariate gamma function
γ:	generic parameter or generic quantity
γ_c:	inverse correlation time

Δ:	Jacobian matrix, or discrete Laplacian
$\Delta(\mathbf{x})$:	Vandermonde determinant $:= \prod_{i<j}(x_j - x_i)$
δ:	small increment or small distance
$\delta(\cdot)$:	Dirac δ-function
ϵ:	generic small quantity $\epsilon \ll 1$
ϵ_i:	deviation of eigenvalue i from its most likely position
ε:	generic IID noise
ζ:	ridge regression parameter
$\zeta(\cdot)$:	inverse of the T-function $\mathrm{t}(\cdot)$
η:	small quantity, $1/N \ll \eta \ll 1$
θ:	edge exponent
$\theta, \theta_{k\ell}$:	angle and generalized angles
$\Theta(\cdot)$:	Heaviside function, $\Theta(x < 0) = 0$, $\Theta(x > 0) = 1$
κ_k:	free cumulant of order k
Λ:	diagonal matrix
λ:	eigenvalue, often those of the sample (or "empirical") matrix
λ_1:	largest eigenvalue of a given matrix
λ_+:	upper edge of spectrum
λ_-:	lower edge of spectrum
μ:	eigenvalue, often those of the population (or "true") matrix
π:	portfolio weights, or partition
$\Pi_N(\cdot)$:	auxiliary polynomial
$\Pi(\cdot)$:	limit polynomial
$\Pi(x,t)$:	pressure field in Matytsin's formalism
Ξ:	Rotationally Invariant Estimator (RIE)
ξ_i:	RIE eigenvalues
$\xi(\cdot)$:	RIE shrinkage function
ρ:	correlation coefficient
$\rho(\cdot)$:	eigenvalue distribution
$\rho(x,t)$:	density field in Matytsin's formalism
ϱ:	square overlap between two vectors
σ:	volatility, or root-mean-square
$\tau(\cdot)$:	normalized trace $1/N\,\mathrm{Tr}(\cdot)$, or free expectation operator
$\tau_R(\cdot)$:	normalized trace, further averaged over rotation matrices appearing inside the trace operator
τ:	time lag
τ_c:	correlation time
Υ:	volume in matrix space
$\Phi(\cdot)$:	auxiliary function
$\Phi(\lambda, \mu)$:	scaled squared overlap between the eigenvectors of the sample (λ) and population (μ) matrices

ϕ:	angle
$\phi(\cdot)$:	effective two-body potential in Matytsin's formalism
$\varphi(\cdot)$:	generating function, or Fourier transform of a probability distribution
$\Psi(\lambda, \lambda')$:	scaled squared overlap between the eigenvectors of two sample matrices
$\Psi(\cdot), \psi(\cdot)$:	auxiliary functions
$\boldsymbol{\psi}$:	auxiliary integration vector
Ω:	generic rotation matrix
$\omega(\cdot)$:	generic eigenvalue field in a large deviation formalism (cf. Section 5.5)

Part I

Classical Random Matrix Theory

1
Deterministic Matrices

Matrices appear in all corners of science, from mathematics to physics, computer science, biology, economics and quantitative finance. In fact, before Schrodinger's equation, quantum mechanics was formulated by Heisenberg in terms of what he called "Matrix Mechanics". In many cases, the matrices that appear are deterministic, and their properties are encapsulated in their eigenvalues and eigenvectors. This first chapter gives several elementary results in linear algebra, in particular concerning eigenvalues. These results will be extremely useful in the rest of the book where we will deal with random matrices, and in particular the statistical properties of their eigenvalues and eigenvectors.

1.1 Matrices, Eigenvalues and Singular Values

1.1.1 Some Problems Where Matrices Appear

Let us give three examples motivating the study of matrices, and the different forms that those can take.

Dynamical System

Consider a generic *dynamical system* describing the time evolution of a certain N-dimensional vector $\mathbf{x}(t)$, for example the three-dimensional position of a point in space. Let us write the equation of motion as

$$\frac{\mathrm{d}\mathbf{x}}{\mathrm{d}t} = \mathbf{F}(\mathbf{x}), \tag{1.1}$$

where $\mathbf{F}(\mathbf{x})$ is an arbitrary vector field. Equilibrium points $\mathbf{x}^*$ are such that $\mathbf{F}(\mathbf{x}^*) = 0$. Consider now small deviations from equilibrium, i.e. $\mathbf{x} = \mathbf{x}^* + \epsilon\mathbf{y}$ where $\epsilon \ll 1$. To first order in ϵ, the dynamics becomes linear, and given by

$$\frac{\mathrm{d}\mathbf{y}}{\mathrm{d}t} = \mathbf{A}\mathbf{y}, \tag{1.2}$$

where $\mathbf{A}$ is a matrix whose elements are given by $\mathbf{A}_{ij} = \partial_j F_i(\mathbf{x}^*)$, where i, j are indices that run from 1 to N. When $\mathbf{F}$ can itself be written as the gradient of some potential V, i.e. $F_i = -\partial_i V(\mathbf{x})$, the matrix $\mathbf{A}$ becomes symmetric, i.e. $\mathbf{A}_{ij} = \mathbf{A}_{ji} = -\partial_{ij} V$. But this is not always the case; in general the linearized dynamics is described by a matrix $\mathbf{A}$ without any particular property – except that it is a square $N \times N$ array of real numbers.

Master Equation

Another standard setting is the so-called Master equation for the evolution of probabilities. Call $i = 1, \ldots, N$ the different possible states of a system and $P_i(t)$ the probability to find the system in state i at time t. When memory effects can be neglected, the dynamics is called Markovian and the evolution of $P_i(t)$ is described by the following discrete time equation:

$$P_i(t+1) = \sum_{j=1}^{N} \mathbf{A}_{ij} P_j(t), \tag{1.3}$$

meaning that the system has a probability $\mathbf{A}_{ij}$ to jump from state j to state i between t and $t+1$. Note that all elements of $\mathbf{A}$ are positive; furthermore, since all jump possibilities must be exhausted, one must have, for each j, $\sum_i \mathbf{A}_{ij} = 1$. This ensures that $\sum_i P_i(t) = 1$ at all times, since

$$\sum_{i=1}^{N} P_i(t+1) = \sum_{i=1}^{N}\sum_{j=1}^{N} \mathbf{A}_{ij} P_j(t) = \sum_{j=1}^{N}\sum_{i=1}^{N} \mathbf{A}_{ij} P_j(t) = \sum_{j=1}^{N} P_j(t) = 1. \tag{1.4}$$

Matrices such that all elements are positive and such that the sum over all rows is equal to unity for each column are called *stochastic matrices*. In matrix form, Eq. (1.3) reads $\mathbf{P}(t+1) = \mathbf{A}\mathbf{P}(t)$, leading to $\mathbf{P}(t) = \mathbf{A}^t\mathbf{P}(0)$, i.e. $\mathbf{A}$ raised to the t-th power applied to the initial distribution.

Covariance Matrices

As a third important example, let us consider random, N-dimensional real vectors $\mathbf{X}$, with some given multivariate distribution $P(\mathbf{X})$. The covariance matrix $\mathbf{C}$ of the $\mathbf{X}$'s is defined as

$$C_{ij} = \mathbb{E}[X_i X_j] - \mathbb{E}[X_i]\mathbb{E}[X_j], \tag{1.5}$$

where $\mathbb{E}$ means that we are averaging over the distribution $P(\mathbf{X})$. Clearly, the matrix $\mathbf{C}$ is real and symmetric. It is also positive semi-definite, in the sense that for any vector $\mathbf{x}$,

$$\mathbf{x}^T\mathbf{C}\mathbf{x} \geq 0. \tag{1.6}$$

If it were not the case, it would be possible to find a linear combination of the vectors $\mathbf{X}$ with a negative variance, which is obviously impossible.

The three examples above are all such that the corresponding matrices are $N \times N$ square matrices. Examples where matrices are rectangular also abound. For example, one could consider two sets of random real vectors: $\mathbf{X}$ of dimension N_1 and $\mathbf{Y}$ of dimension N_2. The *cross-covariance* matrix defined as

$$C_{ia} = \mathbb{E}[X_i Y_a] - \mathbb{E}[X_i]\mathbb{E}[Y_a]; \qquad i = 1, \ldots, N_1; \qquad a = 1, \ldots, N_2, \tag{1.7}$$

is an $N_1 \times N_2$ matrix that describes the correlations between the two sets of vectors.

1.1.2 Eigenvalues and Eigenvectors

One learns a great deal about matrices by studying their eigenvalues and eigenvectors. For a square matrix $\mathbf{A}$ a pair of scalar and non-zero vector $(\lambda, \mathbf{v})$ satisfying

$$\mathbf{A}\mathbf{v} = \lambda \mathbf{v} \tag{1.8}$$

is called an eigenvalue–eigenvector pair.

Trivially if $\mathbf{v}$ is an eigenvector $\alpha\mathbf{v}$ is also an eigenvector when α is a non-zero real number. Sometimes multiple non-collinear eigenvectors share the same eigenvalue; we say that this eigenvalue is degenerate and has multiplicity equal to the dimension of the vector space spanned by its eigenvectors.

If Eq. (1.8) is true, it implies that the equation $(\mathbf{A} - \lambda\mathbf{1})\mathbf{v} = 0$ has non-trivial solutions, which requires that $\det(\lambda\mathbf{1} - \mathbf{A}) = 0$. The eigenvalues λ are thus the roots of the so-called characteristic polynomial of the matrix $\mathbf{A}$, obtained by expanding $\det(\lambda\mathbf{1} - \mathbf{A})$. Clearly, this polynomial[1] is of order N and therefore has at most N different roots, which correspond to the (possibly complex) eigenvalues of $\mathbf{A}$. Note that the characteristic polynomial of $\mathbf{A}^T$ coincides with the characteristic polynomial of $\mathbf{A}$, so the eigenvalues of $\mathbf{A}$ and $\mathbf{A}^T$ are identical.

Now, let $\lambda_1, \lambda_2, \ldots, \lambda_N$ be the N eigenvalues of $\mathbf{A}$ with $\mathbf{v}_1, \mathbf{v}_2, \ldots, \mathbf{v}_N$ the corresponding eigenvectors. We define Λ as the $N \times N$ diagonal matrix with λ_i on the diagonal, and $\mathbf{V}$ as the $N \times N$ matrix whose jth column is $\mathbf{v}_j$, i.e. $V_{ij} = (\mathbf{v}_j)_i$ is the ith component of $\mathbf{v}_j$. Then, by definition,

$$\mathbf{A}\mathbf{V} = \mathbf{V}\Lambda, \tag{1.9}$$

since once expanded, this reads

$$\sum_k \mathbf{A}_{ik} V_{kj} = V_{ij}\lambda_j, \tag{1.10}$$

or $\mathbf{A}\mathbf{v}_j = \lambda_j \mathbf{v}_j$. If the eigenvectors are linearly independent (which is not true for all matrices), the matrix inverse $\mathbf{V}^{-1}$ exists and one can therefore write $\mathbf{A}$ as

$$\mathbf{A} = \mathbf{V}\Lambda\mathbf{V}^{-1}, \tag{1.11}$$

which is called the eigenvalue decomposition of the matrix $\mathbf{A}$.

Symmetric matrices (such that $\mathbf{A} = \mathbf{A}^T$) have very nice properties regarding their eigenvalues and eigenvectors.

[1] The characteristic polynomial $Q_N(\lambda) = \det(\lambda\mathbf{1} - \mathbf{A})$ always has a coefficient 1 in front of its highest power ($Q_N(\lambda) = \lambda^N + O(\lambda^{N-1})$), such polynomials are called *monic*.

- They have exactly N eigenvalues when counted with their multiplicity.
- All their eigenvalues and eigenvectors are real.
- Their eigenvectors are orthogonal and can be chosen to be orthonormal (i.e. $\mathbf{v}_i^T \mathbf{v}_j = \delta_{ij}$). Here we assume that for degenerate eigenvalues we pick an orthogonal set of corresponding eigenvectors.

If we choose orthonormal eigenvectors, the matrix $\mathbf{V}$ has the property $\mathbf{V}^T\mathbf{V} = \mathbf{1}$ ($\Rightarrow \mathbf{V}^T = \mathbf{V}^{-1}$). Hence it is an *orthogonal matrix* $\mathbf{V} = \mathbf{O}$ and Eq. (1.11) reads

$$\mathbf{A} = \mathbf{O}\Lambda\mathbf{O}^T, \tag{1.12}$$

where Λ is a diagonal matrix containing the eigenvalues associated with the eigenvectors in the columns of $\mathbf{O}$. A symmetric matrix can be diagonalized by an orthogonal matrix. Remark that an $N \times N$ orthogonal matrix is fully parameterized by $N(N-1)/2$ "angles", whereas Λ contains N diagonal elements. So the total number of parameters of the diagonal decomposition is $N(N-1)/2 + N$, which is identical, as it should be, to the number of different elements of a symmetric $N \times N$ matrix.

Let us come back to our dynamical system example, Eq. (1.2). One basic question is to know whether the perturbation $\mathbf{y}$ will grow with time, or decay with time. The answer to this question is readily given by the eigenvalues of $\mathbf{A}$. For simplicity, we assume $\mathbf{F}$ to be a gradient such that $\mathbf{A}$ is symmetric. Since the eigenvectors of $\mathbf{A}$ are orthonormal, one can decompose $\mathbf{y}$ in term of the $\mathbf{v}$'s as

$$\mathbf{y}(t) = \sum_{i=1}^{N} c_i(t)\mathbf{v}_i. \tag{1.13}$$

Taking the dot product of Eq. (1.2) with $\mathbf{v}_i$ then shows that the dynamics of the coefficients $c_i(t)$ are decoupled and given by

$$\frac{\mathrm{d}c_i}{\mathrm{d}t} = \lambda_i c_i, \tag{1.14}$$

where λ_i is the eigenvalue associated with $\mathbf{v}_i$. Therefore, any component of the initial perturbation $\mathbf{y}(t=0)$ that is along an eigenvector with positive eigenvalue will grow exponentially with time, until the linearized approximation leading to Eq. (1.2) breaks down. Conversely, components along directions with negative eigenvalues decrease exponentially with time. An equilibrium $\mathbf{x}^*$ is called stable provided all eigenvalues are negative, and marginally stable if some eigenvalues are zero while all others are negative.

The important message carried by the example above is that diagonalizing a matrix amounts to finding a way to decouple the different degrees of freedom, and convert a matrix equation into a set of N scalar equations, as Eqs. (1.14). We will see later that the same idea holds for covariance matrices as well: their diagonalization allows one to find a set of uncorrelated vectors. This is usually called Principal Component Analysis (PCA).

Exercise 1.1.1 Instability of eigenvalues of non-symmetric matrices

Consider the $N \times N$ square band diagonal matrix $\mathbf{M}_0$ defined by $[\mathbf{M}_0]_{ij} = 2\delta_{i,j-1}$:

$$\mathbf{M}_0 = \begin{pmatrix} 0 & 2 & 0 & \cdots & 0 \\ 0 & 0 & 2 & \cdots & 0 \\ 0 & 0 & 0 & \ddots & 0 \\ 0 & 0 & 0 & \cdots & 2 \\ 0 & 0 & 0 & \cdots & 0 \end{pmatrix}. \tag{1.15}$$

(a) Show that $\mathbf{M}_0^N = 0$ and so all the eigenvalues of $\mathbf{M}_0$ must be zero. Use a numerical eigenvalue solver for non-symmetric matrices and confirm numerically that this is the case for $N = 100$.

(b) If $\mathbf{O}$ is an orthogonal matrix ($\mathbf{O}\mathbf{O}^T = \mathbf{1}$), $\mathbf{O}\mathbf{M}_0\mathbf{O}^T$ has the same eigenvalues as $\mathbf{M}_0$. Following Exercise 1.2.4, generate a random orthogonal matrix $\mathbf{O}$. Numerically find the eigenvalues of $\mathbf{O}\mathbf{M}_0\mathbf{O}^T$. Do you get the same answer as in (a)?

(c) Consider $\mathbf{M}_1$ whose elements are all equal to those of $\mathbf{M}_0$ except for one element in the lower left corner $[\mathbf{M}_1]_{N,1} = (1/2)^{N-1}$. Show that $\mathbf{M}_1^N = \mathbf{1}$; more precisely, show that the characteristic polynomial of $\mathbf{M}_1$ is given by $\det(\mathbf{M}_1 - \lambda\mathbf{1}) = \lambda^N - 1$, therefore $\mathbf{M}_1$ has N distinct eigenvalues equal to the N complex roots of unity $\lambda_k = e^{2\pi i k/N}$.

(d) For N greater than about 60, $\mathbf{O}\mathbf{M}_0\mathbf{O}^T$ and $\mathbf{O}\mathbf{M}_1\mathbf{O}^T$ are indistinguishable to machine precision. Compare numerically the eigenvalues of these two rotated matrices.

1.1.3 Singular Values

A non-symmetric, square matrix cannot in general be decomposed as $\mathbf{A} = \mathbf{O}\Lambda\mathbf{O}^T$, where Λ is a diagonal matrix and $\mathbf{O}$ an orthogonal matrix. One can however find a very useful alternative decomposition as

$$\mathbf{A} = \mathbf{V}\mathbf{S}\mathbf{U}^T, \tag{1.16}$$

where $\mathbf{S}$ is a non-negative diagonal matrix, whose elements are called the *singular values* of $\mathbf{A}$, and $\mathbf{U}, \mathbf{V}$ are two real, orthogonal matrices. Whenever $\mathbf{A}$ is symmetric positive semi-definite, one has $\mathbf{S} = \Lambda$ and $\mathbf{U} = \mathbf{V}$.

Equation (1.16) also holds for rectangular $N \times T$ matrices, where $\mathbf{V}$ is $N \times N$ orthogonal, $\mathbf{U}$ is $T \times T$ orthogonal and $\mathbf{S}$ is $N \times T$ diagonal as defined below. To construct the *singular value decomposition* (SVD) of $\mathbf{A}$, we first introduce two matrices $\mathbf{B}$ and $\widehat{\mathbf{B}}$, defined as $\mathbf{B} := \mathbf{A}\mathbf{A}^T$ and $\widehat{\mathbf{B}} = \mathbf{A}^T\mathbf{A}$. It is plain to see that these matrices are symmetric, since

$\mathbf{B}^T = (\mathbf{A}\mathbf{A}^T)^T = \mathbf{A}^{TT}\mathbf{A}^T = \mathbf{B}$ (and similarly for $\widehat{\mathbf{B}}$). They are also positive semi-definite as for any vector $\mathbf{x}$ we have $\mathbf{x}^T\mathbf{B}\mathbf{x} = ||\mathbf{A}^T\mathbf{x}||^2 \geq 0$.

We can show that $\mathbf{B}$ and $\widehat{\mathbf{B}}$ have the same non-zero eigenvalues. In fact, let $\lambda > 0$ be an eigenvalue of $\mathbf{B}$ and $\mathbf{v} \neq 0$ is the corresponding eigenvector. Then we have, by definition,

$$\mathbf{A}\mathbf{A}^T\mathbf{v} = \lambda\mathbf{v}. \tag{1.17}$$

Let $\mathbf{u} = \mathbf{A}^T\mathbf{v}$, then we can get from the above equation that

$$\mathbf{A}^T\mathbf{A}\mathbf{A}^T\mathbf{v} = \lambda\mathbf{A}^T\mathbf{v} \Rightarrow \widehat{\mathbf{B}}\mathbf{u} = \lambda\mathbf{u}. \tag{1.18}$$

Moreover,

$$\|\mathbf{u}\|^2 = \mathbf{v}^T\mathbf{A}\mathbf{A}^T\mathbf{v} = \mathbf{v}^T\mathbf{B}\mathbf{v} \neq 0 \Rightarrow \mathbf{u} \neq 0. \tag{1.19}$$

Hence λ is also an eigenvalue of $\widehat{\mathbf{B}}$. Note that for degenerate eigenvalues λ of $\mathbf{B}$, an orthogonal set of corresponding eigenvectors $\{\mathbf{v}_\ell\}$ gives rise to an orthogonal set $\{\mathbf{A}^T\mathbf{v}_\ell\}$ of eigenvectors of $\widehat{\mathbf{B}}$. Hence the multiplicity of λ in $\widehat{\mathbf{B}}$ is at least that of $\mathbf{B}$. Similarly, we can show that any non-zero eigenvalue of $\widehat{\mathbf{B}}$ is also an eigenvalue of $\mathbf{B}$. This finishes the proof of the claim.

Note that $\mathbf{B}$ has at most N non-zero eigenvalues and $\widehat{\mathbf{B}}$ has at most T non-zero eigenvalues. Thus by the above claim, if $T > N$, $\widehat{\mathbf{B}}$ has at least $T - N$ zero eigenvalues, and if $T < N$, $\mathbf{B}$ has at least $N - T$ zero eigenvalues. We denote the other $\min\{N,T\}$ eigenvalues of $\mathbf{B}$ and $\widehat{\mathbf{B}}$ by $\{\lambda_k\}_{1\leq k\leq\min\{N,T\}}$. Then the SVD of $\mathbf{A}$ is expressed as Eq. (1.16), where $\mathbf{V}$ is the $N \times N$ orthogonal matrix consisting of the N normalized eigenvectors of $\mathbf{B}$, $\mathbf{U}$ is the $T \times T$ orthogonal matrix consisting of the T normalized eigenvectors of $\widehat{\mathbf{B}}$, and $\mathbf{S}$ is an $N \times T$ rectangular diagonal matrix with $S_{kk} = \sqrt{\lambda_k} \geq 0$, $1 \leq k \leq \min\{N,T\}$ and all other entries equal to zero.

For instance, if $N < T$, we have

$$\mathbf{S} = \begin{pmatrix} \sqrt{\lambda_1} & 0 & 0 & 0 & \cdots & 0 \\ 0 & \sqrt{\lambda_2} & 0 & 0 & \cdots & 0 \\ 0 & 0 & \ddots & 0 & \cdots & 0 \\ 0 & 0 & 0 & \sqrt{\lambda_N} & \cdots & 0 \end{pmatrix}. \tag{1.20}$$

Although (non-degenerate) normalized eigenvectors are unique up to a sign, the choice of the positive sign for the square-root $\sqrt{\lambda_k}$ imposes a condition on the combined sign for the left and right singular vectors $\mathbf{v}_k$ and $\mathbf{u}_k$. In other words, simultaneously changing both $\mathbf{v}_k$ and $\mathbf{u}_k$ to $-\mathbf{v}_k$ and $-\mathbf{u}_k$ leaves the matrix $\mathbf{A}$ invariant, but for non-zero singular values one cannot individually change the sign of either $\mathbf{v}_k$ or $\mathbf{u}_k$.

The recipe to find the SVD, Eq. (1.16), is thus to diagonalize both $\mathbf{A}\mathbf{A}^T$ (to obtain $\mathbf{V}$ and $\mathbf{S}^2$) and $\mathbf{A}^T\mathbf{A}$ (to obtain $\mathbf{U}$ and again $\mathbf{S}^2$). It is insightful to again count the number of parameters involved in this decomposition. Consider a general $N \times T$ matrix with $T \geq N$ (the case $N \geq T$ follows similarly). The N eigenvectors of $\mathbf{A}\mathbf{A}^T$ are generically unique up to a sign, while for $T - N > 0$ the matrix $\mathbf{A}^T\mathbf{A}$ will have a degenerate eigenspace associated with the eigenvalue 0 of size $T - N$, hence its eigenvectors are only unique up to an arbitrary rotation in $T - N$ dimension. So generically the SVD decomposition amounts to writing the NT elements of $\mathbf{A}$ as

$$NT \equiv \frac{1}{2}N(N-1) + N + \frac{1}{2}T(T-1) - \frac{1}{2}(T-N)(T-N-1). \tag{1.21}$$

The interpretation of Eq. (1.16) for $N \times N$ matrices is that one can always find an orthonormal basis of vectors $\{\mathbf{u}\}$ such that the application of a matrix A amounts to a rotation (or an improper rotation) of $\{\mathbf{u}\}$ into another orthonormal set $\{\mathbf{v}\}$, followed by a dilation of each $\mathbf{v}_k$ by a positive factor $\sqrt{\lambda_k}$.

Normal matrices are such that $\mathbf{U} = \mathbf{V}$. In other words, $\mathbf{A}$ is normal whenever $\mathbf{A}$ commutes with its transpose: $\mathbf{A}\mathbf{A}^T = \mathbf{A}^T\mathbf{A}$. Symmetric, skew-symmetric and orthogonal matrices are normal, but other cases are possible. For example a 3×3 matrix such that each row and each column has exactly two elements equal to 1 and one element equal to 0 is normal.

1.2 Some Useful Theorems and Identities

In this section, we state without proof very useful theorems on eigenvalues and matrices.

1.2.1 Gershgorin Circle Theorem

Let $\mathbf{A}$ be a real matrix, with elements $\mathbf{A}_{ij}$. Define R_i as $R_i = \sum_{j \neq i} |\mathbf{A}_{ij}|$, and $\mathcal{D}_i$ a disk in the complex plane centered on $\mathbf{A}_{ii}$ and of radius R_i. Then every eigenvalue of $\mathbf{A}$ lies within at least one disk $\mathcal{D}_i$. For example, for the matrix

$$\mathbf{A} = \begin{pmatrix} 1 & -0.2 & 0.2 \\ -0.3 & 2 & -0.2 \\ 0 & 1.1 & 3 \end{pmatrix}, \tag{1.22}$$

the three circles are located on the real axis at $x = 1, 2$ and 3 with radii 0.4, 0.5 and 1.1 respectively (see Fig. 1.1).

In particular, eigenvalues corresponding to eigenvectors with a maximum amplitude on i lie within the disk $\mathcal{D}_i$.

1.2.2 The Perron–Frobenius Theorem

Let $\mathbf{A}$ be a real matrix, with all its elements positive $\mathbf{A}_{ij} > 0$. Then the top eigenvalue $\lambda_{\max}$ is unique and real (all other eigenvalues have a smaller real part). The corresponding top eigenvector $\mathbf{v}^*$ has all its elements positive:

$$\mathbf{A}\mathbf{v}^* = \lambda_{\max}\mathbf{v}^*; \qquad \mathbf{v}^*_k > 0, \forall k. \tag{1.23}$$

The top eigenvalue satisfies the following inequalities:

$$\min_i \sum_j \mathbf{A}_{ij} \leq \lambda_{\max} \leq \max_i \sum_j \mathbf{A}_{ij}. \tag{1.24}$$

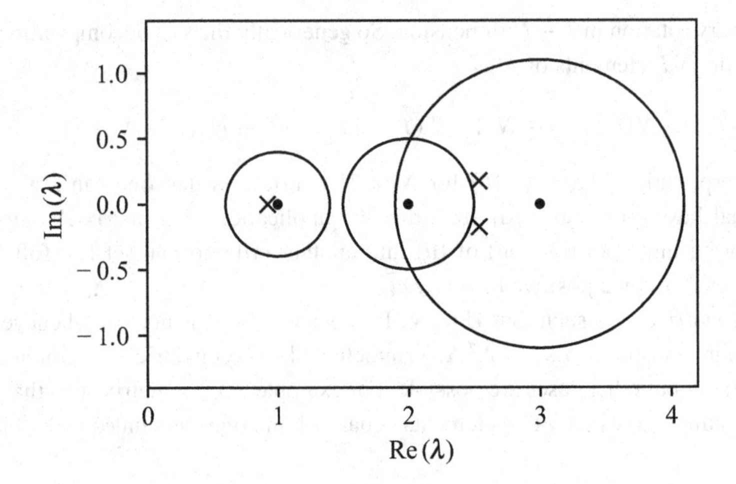

Figure 1.1 The three complex eigenvalues of the matrix (1.22) (crosses) and its three Gershgorin circles. The first eigenvalue $\lambda_1 \approx 0.92$ falls in the first circle while the other two $\lambda_{2,3} \approx 2.54 \pm 0.18i$ fall in the third one.

Application: Suppose $\mathbf{A}$ is a stochastic matrix, such that all its elements are positive and satisfy $\sum_i \mathbf{A}_{ij} = 1,\ \forall j$. Then clearly the vector $\vec{1}$ is an eigenvector of $\mathbf{A}^T$, with eigenvalue $\lambda = 1$. But since the Perron–Frobenius can be applied to $\mathbf{A}^T$, the inequalities (1.24) ensure that λ *is* the top eigenvalue of $\mathbf{A}^T$, and thus also of $\mathbf{A}$. All the elements of the corresponding eigenvector $\mathbf{v}^*$ are positive, and describe the stationary state of the associated Master equation, i.e.

$$P_i^* = \sum_j \mathbf{A}_{ij} P_j^* \longrightarrow P_i^* = \frac{\mathbf{v}_i^*}{\sum_k \mathbf{v}_k^*}. \tag{1.25}$$

Exercise 1.2.1 Gershgorin and Perron–Frobenius

Show that the upper bound in Eq. (1.24) is a simple consequence of the Gershgorin theorem.

1.2.3 The Eigenvalue Interlacing Theorem

Let $\mathbf{A}$ be an $N \times N$ symmetric matrix (or more generally Hermitian matrix) with eigenvalues $\lambda_1 \geq \lambda_2 \cdots \geq \lambda_N$. Consider the $N-1 \times N-1$ submatrix $\mathbf{A}_{\setminus i}$ obtained by removing the ith row and ith columns of $\mathbf{A}$. Its eigenvalues are $\mu_1^{(i)} \geq \mu_2^{(i)} \cdots \geq \mu_{N-1}^{(i)}$. Then the following *interlacing* inequalities hold:

$$\lambda_1 \geq \mu_1^{(i)} \geq \lambda_2 \cdots \geq \mu_{N-1}^{(i)} \geq \lambda_N. \tag{1.26}$$

Very recently, a formula relating eigenvectors to eigenvalues was (re-)discovered. Calling $\mathbf{v}_i$ the eigenvector of $\mathbf{A}$ associated with λ_i, one has[2]

$$\left|(\mathbf{v}_i)_j\right|^2 = \frac{\prod_{k=1}^{N-1} \lambda_i - \mu_k^{(j)}}{\prod_{\ell=1,\ell\neq i}^{N} \lambda_i - \lambda_\ell}. \tag{1.27}$$

1.2.4 Sherman–Morrison Formula

The Sherman–Morrison formula gives the inverse of a matrix $\mathbf{A}$ perturbed by a rank-1 perturbation:

$$(\mathbf{A} + \mathbf{u}\mathbf{v}^T)^{-1} = \mathbf{A}^{-1} - \frac{\mathbf{A}^{-1}\mathbf{u}\mathbf{v}^T\mathbf{A}^{-1}}{1 + \mathbf{v}^T\mathbf{A}^{-1}\mathbf{u}}, \tag{1.28}$$

valid for any invertible matrix $\mathbf{A}$ and vectors $\mathbf{u}$ and $\mathbf{v}$ such that the denominator does not vanish. This is a special case of the Woodbury identity, which reads

$$\left(\mathbf{A} + \mathbf{U}\mathbf{C}\mathbf{V}^T\right)^{-1} = \mathbf{A}^{-1} - \mathbf{A}^{-1}\mathbf{U}\left(\mathbf{C}^{-1} + \mathbf{V}^T\mathbf{A}^{-1}\mathbf{U}\right)^{-1}\mathbf{V}^T\mathbf{A}^{-1}, \tag{1.29}$$

where $\mathbf{U},\mathbf{V}$ are $N \times K$ matrices and $\mathbf{C}$ is a $K \times K$ matrix. Equation (1.28) corresponds to the case $K = 1$.

The associated Sherman–Morrison determinant lemma reads

$$\det(\mathbf{A} + \mathbf{v}\mathbf{u}^T) = \det\mathbf{A} \cdot \left(1 + \mathbf{u}^T\mathbf{A}^{-1}\mathbf{v}\right) \tag{1.30}$$

for invertible $\mathbf{A}$.

Exercise 1.2.2 Sherman–Morrison
Show that Eq. (1.28) is correct by multiplying both sides by $(\mathbf{A} + \mathbf{u}\mathbf{v}^T)$.

1.2.5 Schur Complement Formula

The Schur complement, also called inversion by partitioning, relates the blocks of the inverse of a matrix to the inverse of blocks of the original matrix. Let $\mathbf{M}$ be an invertible matrix which we divide in four blocks as

$$\mathbf{M} = \begin{pmatrix} \mathbf{M}_{11} & \mathbf{M}_{12} \\ \mathbf{M}_{21} & \mathbf{M}_{22} \end{pmatrix} \text{ and } \mathbf{M}^{-1} = \mathbf{Q} = \begin{pmatrix} \mathbf{Q}_{11} & \mathbf{Q}_{12} \\ \mathbf{Q}_{21} & \mathbf{Q}_{22} \end{pmatrix}, \tag{1.31}$$

where $[\mathbf{M}_{11}] = n \times n$, $[\mathbf{M}_{12}] = n \times (N-n)$, $[\mathbf{M}_{21}] = (N-n) \times n$, $[\mathbf{M}_{22}] = (N-n) \times (N-n)$, and $\mathbf{M}_{22}$ is invertible. The integer n can take any values from 1 to $N-1$.

[2] See: P. Denton, S. Parke, T. Tao, X. Zhang, *Eigenvalues from Eigenvectors: a survey of a basic identity in linear algebra, arXiv:1908.03795.*

Then the upper left $n \times n$ block of $\mathbf{Q}$ is given by

$$\mathbf{Q}_{11}^{-1} = \mathbf{M}_{11} - \mathbf{M}_{12}(\mathbf{M}_{22})^{-1}\mathbf{M}_{21}, \tag{1.32}$$

where the right hand side is called the Schur complement of the block $\mathbf{M}_{22}$ of the matrix $\mathbf{M}$.

Exercise 1.2.3 Combining Schur and Sherman–Morrison

In the notation of Eq. (1.31) for $n = 1$ and any $N > 1$, combine the Schur complement of the lower right block with the Sherman–Morrison formula to show that

$$\mathbf{Q}_{22} = (\mathbf{M}_{22})^{-1} + \frac{(\mathbf{M}_{22})^{-1}\mathbf{M}_{21}\mathbf{M}_{12}(\mathbf{M}_{22})^{-1}}{\mathbf{M}_{11} - \mathbf{M}_{12}(\mathbf{M}_{22})^{-1}\mathbf{M}_{21}}. \tag{1.33}$$

1.2.6 Function of a Matrix and Matrix Derivative

In our study of random matrices, we will need to extend real or complex scalar functions to take a symmetric matrix $\mathbf{M}$ as its argument. The simplest way to extend such a function is to apply it to each eigenvalue of the matrix $\mathbf{M} = \mathbf{O}\Lambda\mathbf{O}^T$:

$$F(\mathbf{M}) = \mathbf{O}F(\Lambda)\mathbf{O}^T, \tag{1.34}$$

where $F(\Lambda)$ is the diagonal matrix where we have applied the function F to each (diagonal) entry of Λ. The function $F(\mathbf{M})$ is now a matrix valued function of a matrix. Scalar polynomial functions can obviously be extended directly as

$$F(x) = \sum_{k=0}^{K} a_k x^k \Rightarrow F(\mathbf{M}) = \sum_{k=0}^{K} a_k \mathbf{M}^k, \tag{1.35}$$

but this is equivalent to applying the polynomial to the eigenvalues of $\mathbf{M}$. By extension, when the Taylor series of the function $F(x)$ converges for every eigenvalue of $\mathbf{M}$ the matrix Taylor series coincides with our definition.

Taking the trace of $F(\mathbf{M})$ will yield a matrix function that returns a scalar. This construction is rotationally invariant in the following sense:

$$\operatorname{Tr} F(\mathbf{U}\mathbf{M}\mathbf{U}^T) = \operatorname{Tr} F(\mathbf{M}) \text{ for any } \mathbf{U}\mathbf{U}^T = \mathbf{1}. \tag{1.36}$$

We can take the derivative of a scalar-valued function $\operatorname{Tr} F(\mathbf{M})$ with respect to each element of the matrix $\mathbf{M}$:

$$\frac{\mathrm{d}}{\mathrm{d}[\mathbf{M}]_{ij}} \operatorname{Tr}(F(\mathbf{M})) = [F'(\mathbf{M})]_{ij} \Rightarrow \frac{\mathrm{d}}{\mathrm{d}\mathbf{M}} \operatorname{Tr}(F(\mathbf{M})) = F'(\mathbf{M}). \tag{1.37}$$

Equation (1.37) is easy to derive when $F(x)$ is a monomial $a_k x^k$ and by linearity for polynomial or Taylor series $F(x)$.

1.2.7 Jacobian of Simple Matrix Transformations

Suppose one transforms an $N \times N$ matrix $\mathbf{A}$ into another $N \times N$ matrix $\mathbf{B}$ through some function of the matrix elements. The Jacobian of the transformation is defined as the determinant of the partial derivatives:

$$\mathbb{G}_{ij,k\ell} = \frac{\partial \mathbf{B}_{k\ell}}{\partial \mathbf{A}_{ij}}. \tag{1.38}$$

The simplest case is just multiplication by a scalar: $\mathbf{B} = \alpha\mathbf{A}$, leading to $\mathbb{G}_{ij,k\ell} = \alpha\delta_{ik}\delta_{j\ell}$. $\mathbb{G}$ is therefore the tensor product of $\alpha\mathbf{1}$ with $\mathbf{1}$, and its determinant is thus equal to α^N. Not much more difficult is the case of an orthogonal transformation $\mathbf{B} = \mathbf{O}\mathbf{A}\mathbf{O}^T$, for which $\mathbb{G}_{ij,k\ell} = \mathbf{O}_{ik}\mathbf{O}_{j\ell}$. $\mathbb{G}$ is now the tensor product $\mathbb{G} = \mathbf{O} \otimes \mathbf{O}$ and therefore its determinant is unity.

Slightly more complicated is the case where $\mathbf{B} = \mathbf{A}^{-1}$. Using simple algebra, one readily obtains, for symmetric matrices,

$$\mathbb{G}_{ij,k\ell} = \frac{1}{2}[\mathbf{A}^{-1}]_{ik}[\mathbf{A}^{-1}]_{j\ell} + \frac{1}{2}[\mathbf{A}^{-1}]_{i\ell}[\mathbf{A}^{-1}]_{jk}. \tag{1.39}$$

Let us now assume that $\mathbf{A}$ has eigenvalues λ_α and eigenvectors $\mathbf{v}_\alpha$. One can easily diagonalize $\mathbb{G}_{ij,k\ell}$ within the symmetric sector, since

$$\sum_{k\ell} \mathbb{G}_{ij,k\ell}\left[\mathbf{v}_{\alpha,k}\mathbf{v}_{\beta,\ell} + \mathbf{v}_{\alpha,\ell}\mathbf{v}_{\beta,k}\right] = \frac{1}{\lambda_\alpha\lambda_\beta}\left[\mathbf{v}_{\alpha,i}\mathbf{v}_{\beta,j} + \mathbf{v}_{\alpha,j}\mathbf{v}_{\beta,i}\right]. \tag{1.40}$$

So the determinant of $\mathbb{G}$ is simply $\prod_{\alpha,\beta\geq\alpha}(\lambda_\alpha\lambda_\beta)^{-1}$. Taking the logarithm of this product helps avoiding counting mistakes, and finally leads to the result

$$\det \mathbb{G} = (\det \mathbf{A})^{-N-1}. \tag{1.41}$$

Exercise 1.2.4 Random Matrices

We conclude this chapter on deterministic matrices with a numerical exercise on random matrices. Most of the results of this exercise will be explored theoretically in the following chapters.

- Let $\mathbf{M}$ be a random real symmetric orthogonal matrix, that is an $N \times N$ matrix satisfying $\mathbf{M} = \mathbf{M}^T = \mathbf{M}^{-1}$. Show that all the eigenvalues of $\mathbf{M}$ are ± 1.
- Let $\mathbf{X}$ be a Wigner matrix, i.e. an $N \times N$ real symmetric matrix whose diagonal and upper triangular entries are IID Gaussian random numbers with zero mean and variance σ^2/N. You can use $\mathbf{X} = \sigma(\mathbf{H} + \mathbf{H}^T)/\sqrt{2N}$ where $\mathbf{H}$ is a non-symmetric $N \times N$ matrix with IID standard Gaussians.
- The matrix $\mathbf{P}_+$ is defined as $\mathbf{P}_+ = \frac{1}{2}(\mathbf{M} + \mathbf{1}_N)$. Convince yourself that $\mathbf{P}_+$ is the projector onto the eigenspace of $\mathbf{M}$ with eigenvalue $+1$. Explain the effect of the matrix $\mathbf{P}_+$ on eigenvectors of $\mathbf{M}$.

- An easy way to generate a random matrix $\mathbf{M}$ is to generate a Wigner matrix (independent of $\mathbf{X}$), diagonalize it, replace every eigenvalue by its sign and reconstruct the matrix. The procedure does not depend on the σ used for the Wigner.
- We consider a matrix $\mathbf{E}$ of the form $\mathbf{E} = \mathbf{M}+\mathbf{X}$. To wit, $\mathbf{E}$ is a noisy version of $\mathbf{M}$. The goal of the following is to understand numerically how the matrix $\mathbf{E}$ is corrupted by the Wigner noise. Using the computer language of your choice, for a large value of N (as large as possible while keeping computing times below one minute), for three interesting values of σ of your choice, do the following numerical analysis.

(a) Plot a histogram of the eigenvalues of $\mathbf{E}$, for a single sample first, and then for many samples (say 100).

(b) From your numerical analysis, in the large N limit, for what values of σ do you expect a non-zero density of eigenvalues near zero.

(c) For every normalized eigenvector $\mathbf{v}_i$ of $\mathbf{E}$, compute the norm of the vector $\mathbf{P}_+\mathbf{v}_i$. For a single sample, do a scatter plot of $|\mathbf{P}_+\mathbf{v}_i|^2$ vs λ_i (its eigenvalue). Turn your scatter plot into an approximate conditional expectation value (using a histogram) including data from many samples.

(d) Build an estimator $\Xi(\mathbf{E})$ of $\mathbf{M}$ using only data from $\mathbf{E}$. We want to minimize the error $\mathcal{E} = \frac{1}{N}||(\Xi(\mathbf{E}) - \mathbf{M})||_F^2$ where $||A||_F^2 = \mathrm{Tr}\mathbf{A}\mathbf{A}^T$. Consider first $\Xi_1(\mathbf{E}) = \mathbf{E}$ and then $\Xi_0(\mathbf{E}) = 0$. What is the error $\mathcal{E}$ of these two estimators? Try to build an ad-hoc estimator $\Xi(\mathbf{E})$ that has a lower error $\mathcal{E}$ than these two.

(e) Show numerically that the eigenvalues of $\mathbf{E}$ are not IID. For each sample $\mathbf{E}$ rank its eigenvalues $\lambda_1 < \lambda_2 < \cdots < \lambda_N$. Consider the eigenvalue spacing $s_k = \lambda_k - \lambda_{k-1}$ for eigenvalues in the bulk ($.2N < k < .3N$ and $.7N < k < .8N$). Make a histogram of $\{s_k\}$ including data from 100 samples. Make 100 pseudo-IID samples: mix eigenvalues for 100 different samples and randomly choose N from the $100N$ possibilities, do not choose the same eigenvalue twice for a given pseudo-IID sample. For each pseudo-IID sample, compute s_k in the bulk and make a histogram of the values using data from all 100 pseudo-IID samples. (Bonus) Try to fit an exponential distribution to these two histograms. The IID case should be well fitted by the exponential but not the original data (not IID).

2

Wigner Ensemble and Semi-Circle Law

In many circumstances, the matrices that are encountered are large, and with no particular structure. Physicist Eugene Wigner postulated that one can often replace a large complex (but deterministic) matrix by a typical element of a certain ensemble of random matrices. This bold proposal was made in the context of the study of large complex atomic nuclei, where the "matrix" is the Hamiltonian of the system, which is a Hermitian matrix describing all the interactions between the neutrons and protons contained in the nucleus. At the time, these interactions were not well known; but even if they had been, the task of diagonalizing the Hamiltonian to find the energy levels of the nucleus was so formidable that Wigner looked for an alternative. He suggested that we should abandon the idea of finding precisely all energy levels, but rather rephrase the question as a *statistical question*: what is the probability to find an energy level within a certain interval, what is the probability that the distance between two successive levels is equal to a certain value, etc.? The idea of Wigner was that the answer to these questions could be, to some degree, universal, i.e. independent of the specific Hermitian matrix describing the system, provided it was *complex enough*. If this is the case, why not replace the Hamiltonian of the system by a purely random matrix with the correct symmetry properties? In the case of time-reversal invariant quantum systems, the Hamiltonian is a real symmetric matrix (of infinite size). In the presence of a magnetic field, the Hamiltonian is a complex, Hermitian matrix (see Section 3.1.1). In the presence of "spin–orbit coupling", the Hamiltonian is symplectic (see Section 3.1.2).

This idea has been incredibly fruitful and has led to the development of a subfield of mathematical physics called "random matrix theory". In this book we will study the properties of some ensembles of random matrices. We will mostly focus on symmetric matrices with real entries as those are the most commonly encountered in data analysis and statistical physics. For example, Wigner's idea has been transposed to glasses and spin-glasses, where the interaction between pairs of atoms or pairs of spins is often replaced by a real symmetric, random matrix (see Section 13.4). In other cases, the randomness stems from noisy observations. For example, when one wants to measure the covariance matrix of the returns of a large number of assets using a sample of finite length (for example the 500 stocks of the S&P500 using 4 years of daily data, i.e. $4 \times 250 = 1000$ data points per stock), there is inevitably some measurement noise that pollutes the determination of said covariance matrix. We will be confronted with this precise problem in Chapters 4 and 17.

In the present chapter and the following one, we will investigate the simplest of all ensembles of random matrices, which was proposed by Wigner himself in the context recalled above. These are matrices where all elements are Gaussian random variables, with the only constraint that the matrix is real symmetric (the Gaussian orthogonal ensemble, GOE), complex Hermitian (the Gaussian unitary ensemble, GUE) or symplectic (the Gaussian symplectic ensemble, GSE).

2.1 Normalized Trace and Sample Averages

We first generalize the notion of expectation value and moments from classical probabilities to large random matrices. We could simply consider the moments $\mathbb{E}[\mathbf{A}^k]$ but that object is very large ($N \times N$ dimensional). It is not clear how to interpret it as $N \to \infty$. It turns out that the correct analog of the expectation value is the normalized trace operator $\tau(.)$, defined as

$$\tau(\mathbf{A}) := \frac{1}{N}\mathbb{E}[\operatorname{Tr}\mathbf{A}]. \tag{2.1}$$

The normalization by $1/N$ is there to make the normalized trace operator finite as $N \to \infty$. For example for the identity matrix, $\tau(\mathbf{1}) = 1$ independently of the dimension and our definition therefore makes sense as $N \to \infty$. When using the notation $\tau(\mathbf{A})$ we will only consider the dominant term as $N \to \infty$, implicitly taking the large N limit.

For a polynomial function of a matrix $F(\mathbf{A})$ or by extension for a function that can be written as a power series, the trace of the function can be computed on the eigenvalues:

$$\frac{1}{N}\operatorname{Tr} F(\mathbf{A}) = \frac{1}{N}\sum_{k=1}^{N} F(\lambda_k). \tag{2.2}$$

In the following, we will denote as $\langle . \rangle$ the average over the eigenvalues of a single matrix $\mathbf{A}$ (sample), i.e.

$$\langle F(\lambda) \rangle := \frac{1}{N}\sum_{k=1}^{N} F(\lambda_k). \tag{2.3}$$

For large random matrices, many scalar quantities such as $\tau(F(\mathbf{A}))$ do not fluctuate from sample to sample, or more precisely such fluctuations go to zero in the large N limit. Physicists speak of this phenomenon as *self-averaging* and mathematicians speak of *concentration of measure*.

$$\tau(F(\mathbf{A})) = \frac{1}{N}\mathbb{E}[\operatorname{Tr} F(\mathbf{A})] \approx \langle F(\lambda) \rangle \text{ for a single } \mathbf{A}. \tag{2.4}$$

When the eigenvalues of a random matrix $\mathbf{A}$ converge to a well-defined density $\rho(\lambda)$, we can write

$$\tau(F(\mathbf{A})) = \int \rho(\lambda)F(\lambda)\mathrm{d}\lambda. \tag{2.5}$$

Using $F(\mathbf{A}) = \mathbf{A}^k$, we can define the kth moment of a random matrix by $m_k := \tau(\mathbf{A}^k)$. The first moment m_1 is simply the normalized trace of $\mathbf{A}$, while $m_2 = 1/N \sum_{ij} \mathbf{A}_{ij}^2$ is the normalized sum of the squares of all the elements. The square-root of m_2 satisfies the axioms of a norm and is called the *Frobenius norm* of $\mathbf{A}$:

$$||A||_\mathrm{F} := \sqrt{m_2}. \tag{2.6}$$

2.2 The Wigner Ensemble

2.2.1 Moments of Wigner Matrices

We will define a Wigner matrix $\mathbf{X}$ as a symmetric matrix ($\mathbf{X} = \mathbf{X}^T$) with Gaussian entries with zero mean. In a symmetric matrix there are really two types of elements: diagonal and off-diagonal, which can have different variances. Diagonal elements have variance σ_d^2 and off-diagonal elements have variance σ_od^2. Note that $\mathbf{X}_{ij} = \mathbf{X}_{ji}$ so they are not independent variables.

In fact, the elements in a Wigner matrix do not need to be Gaussian or even to be IID, as there are many weaker (more general) definitions of the Wigner matrix that yield the same final statistical results in the limit of large matrices $N \to \infty$. For the purpose of this introductory book we will stick to the strong Gaussian hypothesis.

The first few moments of our Wigner matrix $\mathbf{X}$ are given by

$$\tau(\mathbf{X}) = \frac{1}{N}\mathbb{E}[\mathrm{Tr}\,\mathbf{X}] = \frac{1}{N}\mathrm{Tr}\,\mathbb{E}[\mathbf{X}] = 0, \tag{2.7}$$

$$\tau(\mathbf{X}^2) = \frac{1}{N}\mathbb{E}[\mathrm{Tr}\,\mathbf{X}\mathbf{X}^T] = \frac{1}{N}\mathbb{E}\left[\sum_{ij=1}^{N} \mathbf{X}_{ij}^2\right] = \frac{1}{N}[N(N-1)\sigma_\mathrm{od}^2 + N\sigma_\mathrm{d}^2]. \tag{2.8}$$

The term containing σ_od^2 dominates when the two variances are of the same order of magnitude. So for a Wigner matrix we can pick any variance we want on the diagonal (as long as it is small with respect to $N\sigma_\mathrm{od}^2$). We want to normalize our Wigner matrix so that its second moment is independent of the size of the matrix (N). Let us pick

$$\sigma_\mathrm{od}^2 = \sigma^2/N. \tag{2.9}$$

For σ_d^2 the natural choice seems to be $\sigma_\mathrm{d}^2 = \sigma^2/N$. However, we will rather choose $\sigma_\mathrm{d}^2 = 2\sigma^2/N$, which is easy to generate numerically and more importantly respects rotational invariance for finite N, as we show in the next subsection. The ensemble described here (with the choice $\sigma_\mathrm{d}^2 = 2\sigma_\mathrm{od}^2$) is called the *Gaussian orthogonal ensemble* or GOE.[1]

[1] Some authors define a GOE matrix to have $\sigma^2 = 1$ others as $\sigma^2 = N$. For us a GOE matrix can have any variance and is thus synonymous with the Gaussian rotationally invariant Wigner matrix.

To generate a GOE matrix numerically, first generate a non-symmetric random square matrix $\mathbf{H}$ of size N with IID $\mathcal{N}(0,\sigma^2/(2N))$ coefficients. Then let the Wigner matrix $\mathbf{X}$ be $\mathbf{X} = \mathbf{H} + \mathbf{H}^T$. The matrix $\mathbf{X}$ will then be symmetric with diagonal variance twice the off-diagonal variance. The reason is that off-diagonal terms are sums of two independent Gaussian variables, so the variance is doubled. Diagonal elements, on the other hand, are equal to twice the original variables H_{ii} and so their variance is multiplied by 4.

With any choice of $\sigma_{\rm d}^2$ we have

$$\tau(\mathbf{X}^2) = \sigma^2 + O(1/N), \tag{2.10}$$

and hence we will call the parameter σ^2 the variance of the Wigner matrix.

The third moment $\tau(\mathbf{X}^3) = 0$ from the fact that the Gaussian distribution is even. Later we will show that

$$\tau(\mathbf{X}^4) = 2\sigma^4. \tag{2.11}$$

For standard Gaussian variables $\mathbb{E}[x^4] = 3\sigma^4$, this implies that the eigenvalue density of a Wigner is **not** Gaussian. What is this eigenvalue distribution? As we will show many times over in this book, it is given by the semi-circle law, originally derived by Wigner himself:

$$\rho(\lambda) = \frac{\sqrt{4\sigma^2 - \lambda^2}}{2\pi\sigma^2} \quad \text{for } -2\sigma < \lambda < 2\sigma. \tag{2.12}$$

2.2.2 Rotational Invariance

We remind the reader that to rotate a vector $\mathbf{v}$, one applies a rotation matrix $\mathbf{O}$: $\mathbf{w} = \mathbf{O}\mathbf{v}$ where $\mathbf{O}$ is an orthogonal matrix $\mathbf{O}^T = \mathbf{O}^{-1}$ (i.e. $\mathbf{O}\mathbf{O}^T = \mathbf{1}$). Note that in general $\mathbf{O}$ is not symmetric. To rotate the basis in which a matrix is written, one writes $\widetilde{\mathbf{X}} = \mathbf{O}\mathbf{X}\mathbf{O}^T$. The eigenvalues of $\widetilde{\mathbf{X}}$ are the same as those of $\mathbf{X}$. The eigenvectors are $\{\mathbf{O}\mathbf{v}\}$ where $\{\mathbf{v}\}$ are the eigenvectors of $\mathbf{X}$.

A rotationally invariant random matrix ensemble is such that the matrix $\mathbf{O}\mathbf{X}\mathbf{O}^T$ is as probable as the matrix $\mathbf{X}$ itself, i.e. $\mathbf{O}\mathbf{X}\mathbf{O}^T \overset{\text{in law}}{=} \mathbf{X}$.

Let us show that the construction $\mathbf{X} = \mathbf{H} + \mathbf{H}^T$ with a Gaussian IID matrix $\mathbf{H}$ leads to a rotationally invariant ensemble. First, note an important property of Gaussian variables, namely that a Gaussian IID vector $\mathbf{v}$ (a white multivariate Gaussian vector) is rotationally invariant. The reason is that $\mathbf{w} = \mathbf{O}\mathbf{v}$ is again a Gaussian vector (since sums of Gaussians are still Gaussian), with covariance given by

$$\mathbb{E}[w_i w_j] = \sum_{k\ell} O_{ik} O_{j\ell} \mathbb{E}[v_k v_\ell] = \sum_{k\ell} O_{ik} O_{j\ell} \delta_{k\ell} = [\mathbf{O}\mathbf{O}^T]_{ij} = \delta_{ij}. \tag{2.13}$$

Now, write

$$\mathbf{X} = \mathbf{H} + \mathbf{H}^T, \tag{2.14}$$

where $\mathbf{H}$ is a square matrix filled with IID Gaussian random numbers. Each column of $\mathbf{H}$ is rotationally invariant: $\mathbf{OH} \overset{\text{in law}}{=} \mathbf{H}$ and the matrix $\mathbf{OH}$ is row-wise rotationally invariant: $\mathbf{OHO}^T \overset{\text{in law}}{=} \mathbf{OH}$. So $\mathbf{H}$ is rotationally invariant as a matrix. Now

$$\mathbf{OXO}^T = \mathbf{O}(\mathbf{H}+\mathbf{H}^T)\mathbf{O}^T \overset{\text{in law}}{=} \mathbf{H}+\mathbf{H}^T = \mathbf{X}, \tag{2.15}$$

which shows that the Wigner ensemble with $\sigma_{\text{d}}^2 = 2\sigma_{\text{od}}^2$ is rotationally invariant for any matrix size N. More general definitions of the Wigner ensemble (including non-Gaussian ensembles) are only asymptotically rotationally invariant (i.e. when $N \to \infty$).

Another way to see the rotational invariance of the Wigner ensemble is to look at the joint law of matrix elements:

$$P(\{X_{ij}\}) = \left(\frac{1}{2\pi\sigma_{\text{d}}^2}\right)^{N/2} \left(\frac{1}{2\pi\sigma_{\text{od}}^2}\right)^{N(N-1)/4} \exp\left\{-\sum_{i=1}^{N}\frac{X_{ii}^2}{2\sigma_{\text{d}}^2} - \sum_{i<j}^{N}\frac{X_{ij}^2}{2\sigma_{\text{od}}^2}\right\}, \tag{2.16}$$

where only the diagonal and upper triangular elements are independent variables. With the choice $\sigma_{\text{od}}^2 = \sigma^2/N$ and $\sigma_{\text{d}}^2 = 2\sigma^2/N$ this becomes

$$P(\{X_{ij}\}) \propto \exp\left\{-\frac{N}{4\sigma^2}\operatorname{Tr}\mathbf{X}^2\right\}. \tag{2.17}$$

Under the change of variable $\mathbf{X} \to \widetilde{\mathbf{X}} = \mathbf{OXO}^T$ the argument of the exponential is invariant, because the trace of a matrix is independent of the basis, and because the Jacobian of the transformation is equal to 1 (see Section 1.2.7), therefore $\mathbf{X} \overset{\text{in law}}{=} \mathbf{OXO}^T$.

By the same argument any matrix whose joint probability density of its elements can be written as $P(\{M_{ij}\}) \propto \exp\{-N \operatorname{Tr} V(\mathbf{M})\}$, where $V(.)$ is an arbitrary function, will be rotationally invariant. We will study such matrix ensembles in Chapter 5.

2.3 Resolvent and Stieltjes Transform

2.3.1 Definition and Basic Properties

In this section we introduce the Stieltjes transform of a matrix. It will give us information about all the moments of the random matrix and also about the density of its eigenvalues in the large N limit. First we need to define the matrix resolvent.

Given an $N \times N$ real symmetric matrix $\mathbf{A}$, its *resolvent* is given by

$$\mathbf{G}_{\mathbf{A}}(z) = (z\mathbf{1} - \mathbf{A})^{-1}, \tag{2.18}$$

where z is a complex variable defined away from all the (real) eigenvalues of $\mathbf{A}$ and $\mathbf{1}$ denotes the identity matrix. Then the Stieltjes transform of $\mathbf{A}$ is given by[2]

$$g_N^{\mathbf{A}}(z) = \frac{1}{N}\operatorname{Tr}(\mathbf{G}_{\mathbf{A}}(z)) = \frac{1}{N}\sum_{k=1}^{N}\frac{1}{z-\lambda_k}, \tag{2.19}$$

[2] In mathematical literature, the Stieltjes transform is more commonly defined as $s_{\mathbf{A}}(z) = -(1/N)\operatorname{Tr}\mathbf{G}_{\mathbf{A}}(z)$, i.e. with an extra minus sign. Some authors prefer the name Cauchy transform.

where λ_k are the eigenvalues of $\mathbf{A}$. The subscript N indicates that this is the finite N Stieltjes transform of a single realization of $\mathbf{A}$. When it is clear from context which matrix we consider we will drop the superscript $\mathbf{A}$ and write $g_N(z)$.

Let us see why the Stieltjes transform gives useful information about the density of eigenvalues of $\mathbf{A}$. For a given random matrix $\mathbf{A}$, we can define the empirical spectral distribution (ESD) also called the sample eigenvalue density:

$$\rho_N(\lambda) = \frac{1}{N}\sum_{k=1}^{N}\delta(\lambda - \lambda_k), \tag{2.20}$$

where $\delta(x)$ is the Dirac delta function. Then the Stieltjes transform can be written as

$$g_N(z) = \int_{-\infty}^{+\infty} \frac{\rho_N(\lambda)}{z-\lambda}\mathrm{d}\lambda. \tag{2.21}$$

Note that $g_N(z)$ is well defined for any $z \notin \{\lambda_k : 1 \leq k \leq N\}$. In particular, it is well behaved at ∞:

$$g_N(z) = \sum_{k=0}^{\infty} \frac{1}{z^{k+1}}\frac{1}{N}\operatorname{Tr}(\mathbf{A}^k), \qquad \frac{1}{N}\operatorname{Tr}(\mathbf{A}^0) = 1. \tag{2.22}$$

We will consider random matrices $\mathbf{A}$ such that, for large N, the normalized traces of powers of $\mathbf{A}$ converge to their expectation values, which are deterministic numbers:

$$\lim_{N\to\infty} \frac{1}{N}\operatorname{Tr}(\mathbf{A}^k) = \tau(\mathbf{A}^k). \tag{2.23}$$

We then expect that, for large enough z, the function $g_{\mathbf{A}}(z)$ converges to a deterministic limit $\mathfrak{g}(z)$ defined as $\mathfrak{g}(z) = \lim_{N\to\infty}\mathbb{E}[g_N(z)]$, whose Taylor series is

$$\mathfrak{g}(z) = \sum_{k=0}^{\infty} \frac{1}{z^{k+1}}\tau(\mathbf{A}^k), \tag{2.24}$$

for z away from the real axis.

Thus $\mathfrak{g}(z)$ is a moment generating function of $\mathbf{A}$. In other words, the knowledge of $\mathfrak{g}(z)$ near infinity is equivalent to the knowledge of all the moments of $\mathbf{A}$. To the level of rigor of this book, the knowledge of all the moments of $\mathbf{A}$ is equivalent to the knowledge of the density of its eigenvalues. For any function $F(x)$ defined over the support of the eigenvalues $[\lambda_-, \lambda_+]$ of $\mathbf{A}$ we can compute its expectation:

$$\tau(F(\mathbf{A})) = \int_{\lambda_-}^{\lambda_+} \rho(\lambda)F(\lambda)\mathrm{d}\lambda; \qquad \rho(\lambda) := \mathbb{E}[\rho_{\mathbf{A}}(\lambda)]. \tag{2.25}$$

Alternatively we can approximate the function $F(x)$ arbitrarily well by a polynomial $Q(x) = a_0 + a_1x + \cdots + a_Kx^K$ and find

$$\tau(F(\mathbf{A})) \approx \tau(Q(\mathbf{A})) = \sum_{k=0}^{K} a_k\tau(\mathbf{A}^k). \tag{2.26}$$

To recap, we only need to know $\mathfrak{g}(z)$ in the neighborhood of $|z| \to \infty$ to know all the moments of $\mathbf{A}$ and these moments tell us everything about $\rho(\lambda)$. In computing the Stieltjes transform in concrete cases, we will often make use of that fact and only estimate it for very large values of z.

The Stieltjes transform also gives the negative moments when they exist. If the eigenvalues of $\mathbf{A}$ satisfy $\min_\ell \lambda_\ell > c$ for some constant $c > 0$, then the inverse moments of $\mathbf{A}$ exist and are given by the expansion of $\mathfrak{g}(z)$ around $z = 0$:

$$\mathfrak{g}(z) = -\sum_{k=0}^{\infty} z^k \tau(\mathbf{A}^{-k-1}). \tag{2.27}$$

In particular, we have

$$\mathfrak{g}(0) = -\tau(\mathbf{A}^{-1}). \tag{2.28}$$

Exercise 2.3.1 Stieltjes transform for shifted and scaled matrices

Let $\mathbf{A}$ be a random matrix drawn from a well-behaved ensemble with Stieltjes transform $\mathfrak{g}(z)$. What are the Stieltjes transforms of the random matrices $\alpha\mathbf{A}$ and $\mathbf{A} + \beta\mathbf{1}$ where α and β are non-zero real numbers and $\mathbf{1}$ the identity matrix?

2.3.2 Stieltjes Transform of the Wigner Ensemble

We are now ready to compute the Stieltjes transform of the Wigner ensemble. The first technique we will use is sometimes called the cavity method or the self-consistent equation. We will find a relation between the Stieltjes transform of a Wigner matrix of size N and one of size $N - 1$. In the large N limit, the two converge to the same limiting Stieltjes transform and give us a self-consistent equation that can be solved easily.

We would like to calculate $\mathfrak{g}_N^{\mathbf{X}}(z)$ when $\mathbf{X}$ is a Wigner matrix, with $\mathbf{X}_{ij} \sim \mathcal{N}(0, \sigma^2/N)$ and $\mathbf{X}_{ii} \sim \mathcal{N}(0, 2\sigma^2/N)$. In the large N limit, we expect that $\mathfrak{g}_N^{\mathbf{X}}(z)$ converges towards a well-defined limit $\mathfrak{g}(z)$.

We can use the Schur complement formula (1.32) to compute the $(1,1)$ element of the inverse of $\mathbf{M} = z\mathbf{1} - \mathbf{X}$. Then we have

$$\frac{1}{(\mathbf{G_X})_{11}} = \mathbf{M}_{11} - \sum_{k,l=2}^{N} \mathbf{M}_{1k} (\mathbf{M}_{22})^{-1}_{kl} \mathbf{M}_{l1}, \tag{2.29}$$

where the matrix $\mathbf{M}_{22}$ is the $(N-1) \times (N-1)$ submatrix of $\mathbf{M}$ with the first row and column removed. For large N, we argue that the right hand side is dominated by its expectation value with small ($\mathcal{O}(1/\sqrt{N})$) fluctuations. We will only compute its expectation value, but getting a more precise handle on its fluctuations would not be difficult. First, we note that $\mathbb{E}[\mathbf{M}_{11}] = z$. We then note that the entries of $\mathbf{M}_{22}$ are independent of the ones of $\mathbf{M}_{1i} = -\mathbf{X}_{1i}$. Thus we can first take the partial expectation over the $\{\mathbf{X}_{1i}\}$, and get

$$\mathbb{E}_{\{\mathbf{X}_{1i}\}}\left[\mathbf{M}_{1i}(\mathbf{M}_{22})^{-1}_{ij}\mathbf{M}_{1j}\right] = \frac{\sigma^2}{N}(\mathbf{M}_{22})^{-1}_{ii}\delta_{ij} \tag{2.30}$$

so we have

$$\mathbb{E}_{\{\mathbf{X}_{1i}\}}\left[\sum_{k,l=2}^{N}\mathbf{M}_{1k}(\mathbf{M}_{22})^{-1}_{kl}\mathbf{M}_{l1}\right] = \frac{\sigma^2}{N}\operatorname{Tr}\left((\mathbf{M}_{22})^{-1}\right). \tag{2.31}$$

Another observation is that $1/(N-1)\operatorname{Tr}\left((\mathbf{M}_{22})^{-1}\right)$ is the Stieltjes transform of a Wigner matrix of size $N-1$ and variance $\sigma^2(N-1)/N$. In the large N limit, the Stieltjes transform should be independent of the matrix size and the difference between N and $(N-1)$ is negligible. So we have

$$\mathbb{E}\left[\frac{1}{N}\operatorname{Tr}\left((\mathbf{M}_{22})^{-1}\right)\right] \to \mathfrak{g}(z). \tag{2.32}$$

We therefore have that $1/(\mathbf{G_X})_{11}$ equals a deterministic number with negligible fluctuations; hence in the large N limit we have

$$\mathbb{E}\left[\frac{1}{(\mathbf{G_X})_{11}}\right] = \frac{1}{\mathbb{E}[(\mathbf{G_X})_{11}]}. \tag{2.33}$$

From the rotational invariance of $\mathbf{X}$ and therefore of $\mathbf{G_X}$, all diagonal entries of $\mathbf{G_X}$ must have the same expectation value:

$$\mathbb{E}\left[(\mathbf{G_X})_{11}\right] = \frac{1}{N}\mathbb{E}[\operatorname{Tr}(\mathbf{G_X})] = \mathbb{E}[g_N] \to \mathfrak{g}. \tag{2.34}$$

Putting all the pieces together, we find that in the large N limit Eq. (2.29) becomes

$$\frac{1}{\mathfrak{g}(z)} = z - \sigma^2\mathfrak{g}(z). \tag{2.35}$$

Solving (2.35) we obtain that

$$\sigma^2\mathfrak{g}^2 - z\mathfrak{g} + 1 = 0 \Rightarrow \mathfrak{g} = \frac{z \pm \sqrt{z^2 - 4\sigma^2}}{2\sigma^2}. \tag{2.36}$$

We know that $\mathfrak{g}(z)$ should be analytic for large complex z but the square-root above can run into branch cuts. It is convenient to pull out a factor of z and express the square-root as a function of $1/z$ which becomes small for large z:

$$\mathfrak{g}(z) = \frac{z \pm z\sqrt{1 - 4\sigma^2/z^2}}{2\sigma^2}. \tag{2.37}$$

We can now choose the correct root: the $+$ sign gives an incorrect $\mathfrak{g}(z) \sim z/\sigma^2$ for large z while the $-$ sign gives $\mathfrak{g}(z) \sim 1/z$ for any large complex z as expected, so we have:

$$\mathfrak{g}(z) = \frac{z - z\sqrt{1 - 4\sigma^2/z^2}}{2\sigma^2}. \tag{2.38}$$

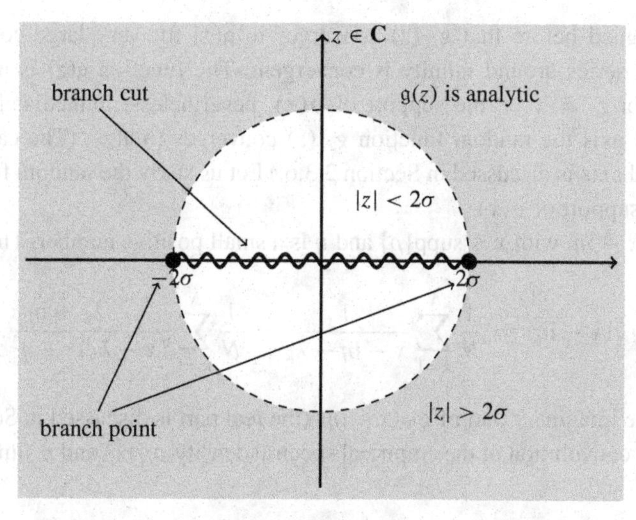

Figure 2.1 The branch cuts of the Wigner Stieltjes transform.

Note, for numerical applications, it is very important to pick the correct branch of the square-root. The function $\mathfrak{g}(z)$ is analytic for $|z| > 2\sigma$, the branch cuts of the square-root must therefore be confined to the interval $[-2\sigma, 2\sigma]$ (Fig. 2.1). We will come back to this problem of determining the correct branch of Stieltjes transform in Section 4.2.3.

It might seem strange that $\mathfrak{g}(z)$ given by Eq. (2.38) has no poles but only branch cuts. For finite N, the sample Stieltjes transform

$$g_N(z) := \frac{1}{N} \sum_{k=1}^{N} \frac{1}{z - \lambda_k} \tag{2.39}$$

has poles at the eigenvalues of $\mathbf{X}$. As $N \to \infty$, the poles fuse together and

$$\frac{1}{N} \sum_{k=1}^{N} \delta(x - \lambda_k) \sim \rho(x). \tag{2.40}$$

The density $\rho(x)$ can have extended support and/or isolated Dirac masses. Then as $N \to \infty$, we have

$$\mathfrak{g}(z) = \int_{\text{supp}\{\rho\}} \frac{\rho(x)\mathrm{d}x}{z - x}, \tag{2.41}$$

which is the Stieltjes transform of the limiting measure $\rho(x)$.

2.3.3 Convergence of Stieltjes near the Real Axis

It is natural to ask the following questions: how does $g_N(z)$ given by Eq. (2.39) converge to $\mathfrak{g}(z) = \int \frac{\rho(x)\mathrm{d}x}{z-x}$, and how do we recover $\rho(x)$ from $\mathfrak{g}(z)$?

We have argued before that $g_N(z)$ converges to $\mathfrak{g}(z)$ for very large complex z such that the Taylor series around infinity is convergent. The function $\mathfrak{g}(z)$ is not defined on the real axis for $z = x$ on the support of $\rho(x)$, nevertheless, immediately below (and above) the real axis the random function $g_N(z)$ converges to $\mathfrak{g}(z)$. (The case where z is right on the real axis is discussed in Section 2.3.6.) Let us study the random function $g_N(z)$ just below the support of $\rho(x)$.

We let $z = x - i\eta$, with $x \in \operatorname{supp}\{\rho\}$ and η is a small positive number. Then

$$g_N(x - i\eta) := \frac{1}{N}\sum_{k=1}^{N} \frac{1}{x - i\eta - \lambda_k} = \frac{1}{N}\sum_{k=1}^{N} \frac{x - \lambda_k + i\eta}{(x - \lambda_k)^2 + \eta^2}. \tag{2.42}$$

We focus on the imaginary part of $g_N(x - i\eta)$ (the real part is discussed in Section 19.5.2). Note that it is a convolution of the empirical spectral density $\rho_N(\lambda)$ and π times the Cauchy kernel:

$$\pi K_\eta(x) = \frac{\eta}{x^2 + \eta^2}. \tag{2.43}$$

The Cauchy kernel $K_\eta(x)$ is strongly peaked around zero with a window width of order η (Fig. 2.2). Since there are N eigenvalues lying inside the interval $[\lambda_-, \lambda_+]$, the typical eigenvalue spacing is of order $(\lambda_+ - \lambda_-)/N = O(N^{-1})$.

(1) Suppose $\eta \ll N^{-1}$. Then there are typically 0 or 1 eigenvalue within a window of size η around x. Then $\operatorname{Im} g_N$ will be affected by the fluctuations of single eigenvalues of $\mathbf{X}$, and hence it cannot converge to any deterministic function. (see Fig. 2.3).

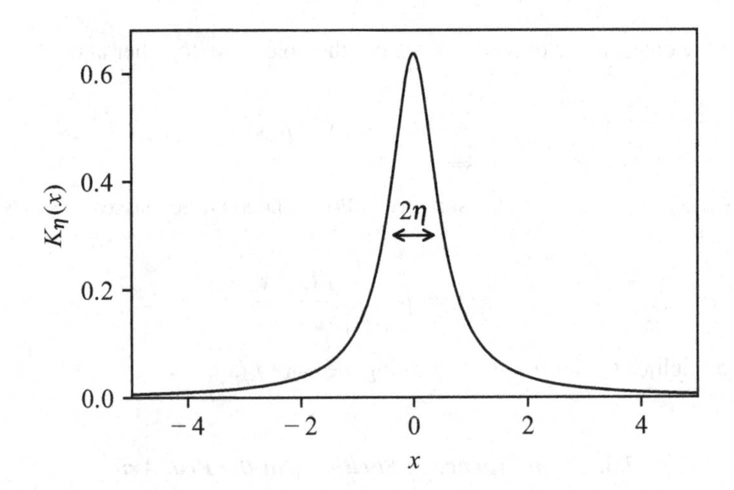

Figure 2.2 The Cauchy kernel for $\eta = 0.5$. It is strongly peaked around zero with a window width of order η. When $\eta \to 0$, the Cauchy kernel is a possible representation of Dirac's δ-function.

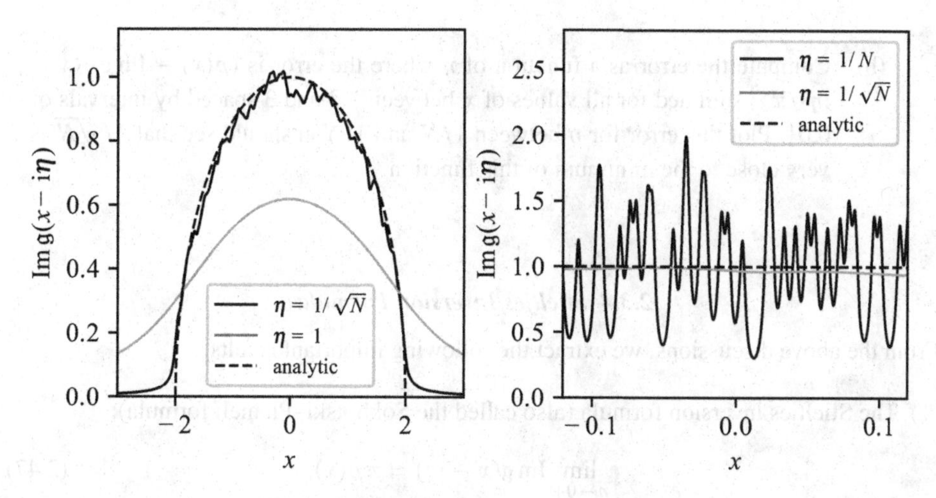

Figure 2.3 Imaginary part of $\mathfrak{g}(x - i\eta)$ for the Wigner ensemble. The analytic result for $\eta \to 0^+$ is compared with numerical simulations ($N = 400$). On the left for $\eta = 1/\sqrt{N}$ and $\eta = 1$. Note that for $\eta = 1$ the density is quite deformed. On the right (zoom near $x = 0$) for $\eta = 1/\sqrt{N}$ and $\eta = 1/N$. Note that for $\eta = 1/N$, the density fluctuates wildly as only a small number of (random) eigenvalues contribute to the Cauchy kernel.

(2) Suppose $N^{-1} \ll \eta \ll 1$ (e.g. $\eta = N^{-1/2}$). Then on a small scale $\eta \ll \Delta x \ll 1$, the density ρ is locally constant and there are a great number n of eigenvalues inside:

$$n \sim N\rho(x)\Delta x \gg N\eta \gg 1. \tag{2.44}$$

The law of large numbers allows us to replace the sum with an integral; we obtain that

$$\frac{1}{N} \sum_{k:\lambda_k \in [x-\Delta x, x+\Delta x]} \frac{i\eta}{(x-\lambda_k)^2 + \eta^2} \to i \int_{x-\Delta x}^{x+\Delta x} \frac{\rho(x)\eta dy}{(x-y)^2 + \eta^2} \to i\pi\rho(x), \tag{2.45}$$

where the last limit is obtained by writing $u = (y - x)/\eta$ and noting that as $\eta \to 0$ we have

$$\int_{-\infty}^{\infty} \frac{du}{u^2 + 1} = \pi. \tag{2.46}$$

Exercise 2.3.2 Finite N approximation and small imaginary part

Im $g_N(x - i\eta)/\pi$ is a good approximation to $\rho(x)$ for small positive η, where $g_N(z)$ is the sample Stieltjes transform ($g_N(z) = (1/N)\sum_k 1/(z - \lambda_k)$). Numerically generate a Wigner matrix of size N and $\sigma^2 = 1$.

(a) For three values of η, $\{1/N, 1/\sqrt{N}, 1\}$, plot Im $g_N(x - i\eta)/\pi$ and the theoretical $\rho(x)$ on the same plot for x between -3 and 3.

(b) Compute the error as a function of η where the error is $(\rho(x) - \operatorname{Im} g_N(x - i\eta)/\pi)^2$ summed for all values of x between -3 and 3 spaced by intervals of 0.01. Plot this error for η between $1/N$ and 1. You should see that $1/\sqrt{N}$ is very close to the minimum of this function.

2.3.4 Stieltjes Inversion Formula

From the above discussions, we extract the following important results:

(1) The Stieltjes inversion formula (also called the Sokhotski–Plemelj formula):

$$\lim_{\eta \to 0+} \operatorname{Im} \mathfrak{g}(x - i\eta) = \pi\rho(x). \tag{2.47}$$

(2) When applied to finite size Stieltjes transform $g_N(z)$, we should take $N^{-1} \ll \eta \ll 1$ for $g_N(x - i\eta)$ to converge to $\mathfrak{g}(x - i\eta)$ and for (2.47) to hold. Numerically, $\eta = N^{-1/2}$ works quite well.

We discuss briefly why $\eta = N^{-1/2}$ works best. First, we want η to be as small as possible such that the local density $\rho(x)$ is not blurred. If η is too large, one introduces a systematic error of order $\rho'(x)\eta$. On the other hand, we want $N\eta$ to be as large as possible such that we include the statistics of a sufficient number of eigenvalues so that we measure $\rho(x)$ accurately. In fact, the error between g_N and g is of order $\frac{1}{N\eta}$. Thus we want to minimize the total error $\mathcal{E}$ given by

$$\mathcal{E} = \rho'(x)\eta + \frac{1}{N\eta}, \qquad \rho'(x)\eta : \text{ systematic error,} \qquad \frac{1}{N\eta} : \text{ statistical error.} \tag{2.48}$$

Then it is easy to see that the total error is minimized when η is of order $1/\sqrt{N\rho'(x)}$.

2.3.5 Density of Eigenvalues of a Wigner Matrix

We go back to study the Stieltjes transform (2.38) of the Wigner matrix. Note that for $z = x - i\eta$ with $\eta \to 0$, $\mathfrak{g}(z)$ can only have an imaginary part if $\sqrt{x^2 - 4\sigma^2}$ is imaginary. Then, using (2.47), we get the Wigner semi-circle law:

$$\rho(x) = \frac{1}{\pi} \lim_{\eta \to 0+} \operatorname{Im} \mathfrak{g}(x - i\eta) = \frac{\sqrt{4\sigma^2 - x^2}}{2\pi\sigma^2}, \qquad -2\sigma \le x \le 2\sigma. \tag{2.49}$$

Note the following features of the semi-circle law (see Fig. 2.4): (1) asymptotically there is no eigenvalue for $x > 2\sigma$ and $x < -2\sigma$; (2) the eigenvalue density has square-root singularities near the edges: $\rho(x) \sim \sqrt{x + 2\sigma}$ near the left edge and $\rho(x) \sim \sqrt{2\sigma - x}$ near the right edge. For finite N, some eigenvalues are present in a small region of width $N^{-2/3}$ around the edges, see Section 14.1.

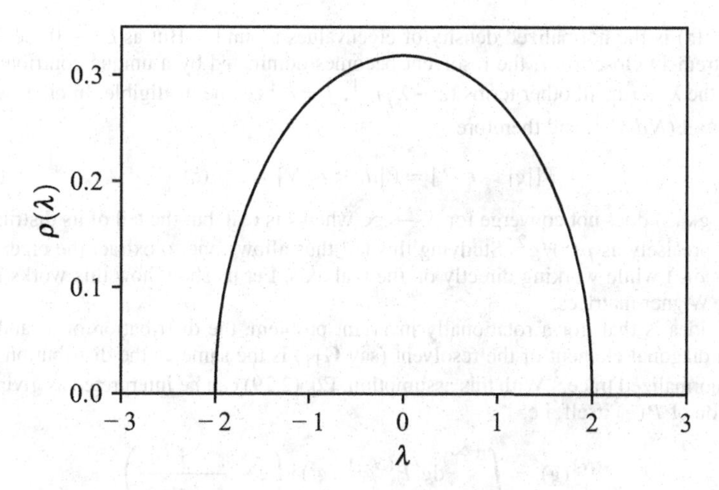

Figure 2.4 Density of eigenvalues of a Wigner matrix with $\sigma = 1$: the *semi-circle law.*

Exercise 2.3.3 From the moments to the density

A large random matrix has moments $\tau(\mathbf{A}^k) = 1/k$.

(a) Using Eq. (2.24) write the Taylor series of $\mathfrak{g}(z)$ around infinity.

(b) Sum the series to get a simple expression for $\mathfrak{g}(z)$. Hint: look up the Taylor series of $\log(1 + x)$.

(c) Where are the singularities of $\mathfrak{g}(z)$ on the real axis?

(d) Use Eq. (2.47) to find the density of eigenvalues $\rho(\lambda)$.

(e) Check your result by recomputing the moments and the Stieltjes transform from $\rho(\lambda)$.

(f) Redo all the above steps for a matrix whose odd moments are zero and even moments are $\tau(\mathbf{A}^{2k}) = 1$. Note that in this case the density $\rho(\lambda)$ has Dirac masses.

2.3.6 *Stieltjes Transform on the Real Axis*

What about computing the Stieltjes transform when z is real and inside the spectrum? This seems dangerous at first sight, since $g_N(z)$ diverges when z is equal to one of the eigenvalues of $\mathbf{X}$. As these eigenvalues become more and more numerous as N goes to infinity, z will always be very close to a pole of the resolvent $g_N(z)$. Interestingly, one can turn this predicament on its head and actually exploit these divergences. In a hand-waving manner, the probability that the difference $d_i = |z - \lambda_i|$ between z (now real) and a given eigenvalue λ_i is very small, is given by

$$\mathbb{P}[d_i < \epsilon/N] = 2\epsilon\rho(z), \tag{2.50}$$

where $\rho(z)$ is the normalized density of eigenvalues around z. But as $\epsilon \to 0$, i.e. when z is extremely close to λ_i, the resolvent becomes dominated by a unique contribution – that of the λ_i term, all other terms $(z - \lambda_j)^{-1}$, $j \neq i$, become negligible. In other words, $g_N(z) \approx \pm(Nd_i)^{-1}$, and therefore

$$\mathbb{P}[|g| > \epsilon^{-1}] = \mathbb{P}[d_i < \epsilon/N] = 2\epsilon\rho(z). \tag{2.51}$$

Hence, $g_N(z)$ does not converge for $N \to \infty$ when z is real, but the tail of its distribution decays precisely as $\rho(z)/g^2$. Studying this tail thus allows one to extract the eigenvalue density $\rho(z)$ while working directly on the real axis. Let us show how this works in the case of Wigner matrices.

The idea is that, for a rotationally invariant problem, the distribution of a randomly chosen diagonal element of the resolvent (say $\mathbf{G}_{11}$) is the same as the distribution $P(g)$ of the normalized trace.[3] With this assumption, Eq. (2.29) can be interpreted as giving the evolution of $P(g)$ itself, i.e.

$$P^{(N)}(g) = \int_{-\infty}^{+\infty} dg' P^{(N-1)}(g')\delta\left(g - \frac{1}{z - \sigma^2 g'}\right), \tag{2.52}$$

where we have used the fact that, for large N, $\sum_{k,\ell=2}^{N} \mathbf{M}_{1k}(\mathbf{M}_{22})^{-1}_{k\ell}\mathbf{M}_{\ell\ell} \to \sigma^2 g^{(N-1)}$. Now, this functional iteration admits the following Cauchy distribution as a fixed point:

$$P^{\infty}(g) = \frac{\rho(z)}{(g - \frac{z}{2\sigma^2})^2 + \pi^2\rho(z)}. \tag{2.53}$$

This simple result, that the resolvent of a Wigner matrix on the real axis is a Cauchy variable, calls for several comments. First, one finds that $P^{\infty}(g)$ indeed behaves as $\rho(z)/g^2$ for large g, as argued above. Second, it would have been entirely natural to find a Cauchy distribution for g had the eigenvalues been independent. Indeed, since g is then the sum of N random variables (i.e. the $1/d_i$'s) distributed with an inverse square power, the generalized CLT predicts that the resulting sum is Cauchy distributed. In the present case, however, the eigenvalues are strongly correlated – see Section 5.1.4. It was recently proven that the Cauchy distribution is in fact *super-universal* and holds for a wide class of point processes on the real axis, in particular for the eigenvalues of random matrices. It is in fact even true when these eigenvalues are strictly equidistant, with a random global shift.

Bibliographical Notes

- The overall content of this chapter is covered in many books and textbooks, see for example
  - M. L. Mehta. *Random Matrices*. Academic Press, San Diego, 3rd edition, 2004,
  - T. Tao. *Topics in Random Matrix Theory*. American Mathematical Society, Providence, Rhode Island, 2012,
  - G. W. Anderson, A. Guionnet, and O. Zeitouni. *An Introduction to Random Matrices*. Cambridge University Press, Cambridge, 2010,
  - G. Blower. *Random Matrices: High Dimensional Phenomena*. Cambridge University Press, Cambridge, 2009,

[3] This is not a trivial statement but it can be proven along the lines of Aizenman and Warzel [2015].

  - B. Eynard, T. Kimura, and S. Ribault. Random matrices. *preprint arXiv:1510.04430*, 2006,
  - G. Livan, M. Novaes, and P. Vivo. *Introduction to Random Matrices: Theory and Practice*. Springer, New York, 2018.
- For some historical papers, see
  - E. P. Wigner. On the statistical distribution of the widths and spacings of nuclear resonance levels. *Mathematical Proceedings of the Cambridge Philosophical Society*, 47(4):790–798, 1951,
  - F. J. Dyson. The threefold way: algebraic structure of symmetry groups and ensembles in quantum mechanics. *Journal of Mathematical Physics*, 3(6):1199–1215, 1962.
- About the Stieltjes transform on the real line and the super-universality of the Cauchy distribution, see
  - M. Aizenman and S. Warzel. On the ubiquity of the Cauchy distribution in spectral problems. *Probability Theory and Related Fields*, 163:61–87, 2015,
  - Y. V. Fyodorov and D. V. Savin. Statistics of impedance, local density of states, and reflection in quantum chaotic systems with absorption. *Journal of Experimental and Theoretical Physics Letters*, 80(12):725–729, 2004,
  - Y. V. Fyodorov and I. Williams. Replica symmetry breaking condition exposed by random matrix calculation of landscape complexity. *Journal of Statistical Physics*, 129(5-6):1081–1116, 2007,
  - J.-P. Bouchaud and M. Potters. Two short pieces around the Wigner problem. *Journal of Physics A: Mathematical and Theoretical*, 52(2):024001, 2018,

  and for a related discussion see also
  - M. Griniasty and V. Hakim. Correlations and dynamics in ensembles of maps: Simple models. *Physical Review E*, 49(4):2661, 1994.

3

More on Gaussian Matrices*

In the previous chapter, we dealt with the simplest of all Gaussian matrix ensembles, where entries are real, Gaussian random variables, and global symmetry is imposed. It was pointed out by Dyson that there exist precisely three division rings that contain the real numbers, namely, the real themselves, the complex numbers and the quaternions. He showed that this fact implies that there are only three acceptable ensembles of Gaussian random matrices with real eigenvalues: GOE, GUE and GSE. Each is associated with a Dyson index called β (1, 2 and 4, respectively) and except for this difference in β almost all of the results in this book (and many more) apply to the three ensembles. In particular their moments and eigenvalue density are the same as $N \to \infty$, while correlations and deviations from the asymptotic formulas follow families of laws with β as a parameter. In this chapter we will review the other two ensembles (GUE and GSE),[1] and also discuss the general moments of Gaussian random matrices, for which some interesting mathematical tools are available, that are useful beyond RMT.

3.1 Other Gaussian Ensembles

3.1.1 Complex Hermitian Matrices

For matrices with complex entries, the analog of a symmetric matrix is a (complex) Hermitian matrix. It satisfies $\mathbf{A}^\dagger = \mathbf{A}$ where the dagger operator is the combination of matrix transposition and complex conjugation. There are two important reasons to study complex Hermitian matrices. First they appear in many applications, especially in quantum mechanics. There, the energy and other observables are mapped into Hermitian operators, or Hermitian matrices for systems with a finite number of states. The first large N result of random matrix theory is the Wigner semi-circle law. As recalled in the introduction to Chapter 2, it was obtained by Wigner as he modeled the energy levels of complex heavy nuclei as a random Hermitian matrix.

[1] More recently, it was shown how ensembles with an arbitrary value of β can be constructed, see Dumitriu and Edelman [2002], Allez et al. [2012].

The other reason Hermitian matrices are important is mathematical. In the large N limit, the three ensembles (real, complex and quaternionic (see below)) behave the same way. But for finite N, computations and proofs are much simpler in the complex case. The main reason is that the Vandermonde determinant which we will introduce in Section 5.1.4 is easier to manipulate in the complex case. For this reason, most mathematicians discuss the complex Hermitian case first and treat the real and quaternionic cases as extensions. In this book we want to stay close to applications in data science and statistical physics, so we will discuss complex matrices only in the present chapter. In the rest of the book we will indicate in footnotes how to extend the result to complex Hermitian matrices.

A complex Hermitian matrix $\mathbf{A}$ has real eigenvalues and it can be diagonalized with a suitable unitary matrix $\mathbf{U}$. A unitary matrix satisfies $\mathbf{U}^\dagger\mathbf{U} = \mathbf{1}$. So $\mathbf{A}$ can be written as $\mathbf{A} = \mathbf{U}\Lambda\mathbf{U}^\dagger$, with Λ the diagonal matrix containing its N eigenvalues.

We want to build the complex Wigner matrix: a Hermitian matrix with IID Gaussian entries. We will choose a construction that has unitary invariance for every N. Let us study the unitary invariance of complex Gaussian vectors. First we need to define a complex Gaussian variable.

We say that the complex variable z is centered Gaussian with variance σ^2 if $z = x_\mathrm{r} + \mathrm{i}\, x_\mathrm{i}$ where x_r and x_i are centered Gaussian variables of variance $\sigma^2/2$. We have

$$\mathbb{E}[|z|^2] = \mathbb{E}[x_\mathrm{r}^2] + \mathbb{E}[x_\mathrm{i}^2] = \sigma^2. \tag{3.1}$$

A white complex Gaussian vector $\mathbf{x}$ is a vector whose components are IID complex centered Gaussians. Consider $\mathbf{y} = \mathbf{U}\mathbf{x}$ where $\mathbf{U}$ is a unitary matrix. Each of the components is a linear combination of Gaussian variables so $\mathbf{y}$ is Gaussian. It is relatively straightforward to show that each component has the same variance σ^2 and that there is no covariance between different components. Hence $\mathbf{y}$ is also a white Gaussian vector. The ensemble of a white complex Gaussian vector is invariant under unitary transformation.

To define the Hermitian Wigner matrix, we first define a (non-symmetric) square matrix $\mathbf{H}$ whose entries are centered complex Gaussian numbers and let $\mathbf{X}$ be the Hermitian matrix defined by

$$\mathbf{X} = \mathbf{H} + \mathbf{H}^\dagger. \tag{3.2}$$

If we repeat the arguments of Section 2.2.2, we can show that the ensemble of $\mathbf{X}$ is invariant under unitary transformation: $\mathbf{U}\mathbf{X}\mathbf{U}^\dagger \overset{\text{in law}}{=} \mathbf{X}$.

We did not specify the variance of the elements of $\mathbf{H}$. We would like $\mathbf{X}$ to be normalized as $\tau(\mathbf{X}^2) = \sigma^2 + O(1/N)$. Choosing the variance of the $\mathbf{H}$ as $\mathbb{E}[|\mathbf{H}_{ij}|^2] = 1/(2N)$ achieves precisely that.

The Hermitian matrix $\mathbf{X}$ has real diagonal elements with $\mathbb{E}[\mathbf{X}_{ii}^2] = 1/N$ and off-diagonal elements that are complex Gaussian with $\mathbb{E}[|\mathbf{X}_{ij}|^2] = 1/N$. In other words the real and imaginary parts of the off-diagonal elements of $\mathbf{X}$ have variance $1/(2N)$. We can put all this information together in the joint law of the matrix elements of the Hermitian matrix $\mathbf{H}$:

$$P(\{X_{ij}\}) \propto \exp\left\{-\frac{N}{2\sigma^2}\operatorname{Tr}\mathbf{X}^2\right\}. \tag{3.3}$$

This law is identical to the real symmetric case (Eq. 2.17) up to a factor of 2. We can then write both the symmetric and the Hermitian case as

$$P(\{X_{ij}\}) \propto \exp\left\{-\frac{\beta N}{4\sigma^2}\operatorname{Tr}\mathbf{X}^2\right\}, \tag{3.4}$$

where β is 1 or 2 respectively.

The complex Hermitian Wigner ensemble is called the *Gaussian unitary ensemble* or GUE.

The results of the previous chapter apply equally to the real symmetric and the complex Hermitian case. Both the self-consistent equation for the Stieltjes transform and the counting of non-crossing pair partitions (see next section) rely on the independence of the elements of the matrix and on the fact that $\mathbb{E}[|\mathbf{X}_{ij}|^2] = 1/N$, true in both cases. We then have that the Stieltjes transform of the two ensembles is the same and they have exactly the same semi-circle distribution of eigenvalues in the large N limit. The same will be true for the quaternionic case ($\beta = 4$) in the next section, and in fact for all values of β provided $N\beta \to \infty$ when $N \to \infty$, see Section 5.3.1:

$$\rho_\beta(\lambda) = \frac{\sqrt{4\sigma^2 - \lambda^2}}{2\pi\sigma^2}, \quad -2\sigma \le \lambda \le 2\sigma. \tag{3.5}$$

3.1.2 Quaternionic Hermitian Matrices

We will define here the quaternionic Hermitian matrices and the GSE. There are many fewer applications of quaternionic matrices than the more common real or complex matrices. We include this discussion here for completeness. In the literature the link between symplectic matrices and quaternions can be quite obscure for the novice reader. Except for the existence of an ensemble of matrices with $\beta = 4$ we will never refer to quaternionic matrices after this section, which can safely be skipped.

Quaternions are non-commutative extensions of the real and complex numbers. They are written as real linear combinations of the real number 1 and three abstract non-commuting objects (i, j, k) satisfying

$$\mathrm{i}^2 = \mathrm{j}^2 = \mathrm{k}^2 = \mathrm{ijk} = -1 \quad \Rightarrow \quad \mathrm{ij} = -\mathrm{ji} = \mathrm{k}, \quad \mathrm{jk} = -\mathrm{kj} = \mathrm{i}, \quad \mathrm{ki} = -\mathrm{ik} = \mathrm{j}. \tag{3.6}$$

So we can write a quaternion as $h = x_\mathrm{r} + \mathrm{i}\,x_\mathrm{i} + \mathrm{j}\,x_\mathrm{j} + \mathrm{k}\,x_\mathrm{k}$. If only x_r is non-zero we say that h is real. We define the quaternionic conjugation as $1^* = 1, \mathrm{i}^* = -\mathrm{i}, \mathrm{j}^* = -\mathrm{j}, \mathrm{k}^* = -\mathrm{k}$ so that the norm $|h|^2 := hh^* = x_\mathrm{r}^2 + x_\mathrm{i}^2 + x_\mathrm{j}^2 + x_\mathrm{k}^2$ is always real and non-negative. The abstract objects i, j and k can be represented as 2×2 complex matrices:

$$1 = \begin{pmatrix} 1 & 0 \\ 0 & 1 \end{pmatrix}, \quad \mathrm{i} = \begin{pmatrix} \mathrm{i} & 0 \\ 0 & -\mathrm{i} \end{pmatrix}, \quad \mathrm{j} = \begin{pmatrix} 0 & 1 \\ -1 & 0 \end{pmatrix}, \quad \mathrm{k} = \begin{pmatrix} 0 & \mathrm{i} \\ \mathrm{i} & 0 \end{pmatrix}, \tag{3.7}$$

where the i in the matrices is now the usual unit imaginary number.

Quaternions share all the algebraic properties of real and complex numbers except for commutativity (they form a *division ring*). Since matrices in general do not commute, matrices built out of quaternions behave like real or complex matrices.

A Hermitian quaternionic matrix is a square matrix $\mathbf{A}$ whose elements are quaternions and satisfy $\mathbf{A} = \mathbf{A}^\dagger$. Here the dagger operator is the combination of matrix transposition and quaternionic conjugation. They are diagonalizable and their eigenvalues are real. Matrices that diagonalize Hermitian quaternionic matrices are called *symplectic*. Written in terms of quaternions they satisfy $\mathbf{S}\mathbf{S}^\dagger = \mathbf{1}$.

Given representation of quaternions as 2×2 complex matrices, an $N \times N$ quaternionic Hermitian matrix $\mathbf{A}$ can be written as a $2N \times 2N$ complex matrix $\mathbf{Q}(\mathbf{A})$. We choose a representation where

$$\mathbf{Z} := \mathbf{Q}(1\mathrm{j}) = \begin{pmatrix} 0 & 1 \\ -1 & 0 \end{pmatrix}. \tag{3.8}$$

For a $2N \times 2N$ complex matrix $\mathbf{Q}$ to be the representation of a quaternionic Hermitian matrix it has to have two properties. First, quaternionic conjugation acts just like Hermitian conjugation so $\mathbf{Q}^\dagger = \mathbf{Q}$. Second, it has to be expressible as a real linear combination of unit quaternions. One can show that such matrices (and only them) satisfy

$$\mathbf{Q}^{\mathrm{R}} := \mathbf{Z}\mathbf{Q}^T\mathbf{Z}^{-1} = \mathbf{Q}^\dagger, \tag{3.9}$$

where $\mathbf{Q}^{\mathrm{R}}$ is called the dual of $\mathbf{Q}$. In other words an $N \times N$ Hermitian quaternionic matrix corresponds to a $2N \times 2N$ self-dual Hermitian matrix (i.e. $\mathbf{Q} = \mathbf{Q}^\dagger = \mathbf{Q}^{\mathrm{R}}$). In this $2N \times 2N$ representation symplectic matrices are complex matrices satisfying

$$\mathbf{S}\mathbf{S}^\dagger = \mathbf{S}\mathbf{S}^{\mathrm{R}} = \mathbf{1}. \tag{3.10}$$

To recap, a $2N \times 2N$ Hermitian self-dual matrix $\mathbf{Q}$ can be diagonalized by a symplectic matrix $\mathbf{S}$. Its $2N$ eigenvalues are real and they occur in pairs as they are the N eigenvalues of the equivalent Hermitian quaternionic $N \times N$ matrix.

We can now define the third Gaussian matrix ensemble, namely the *Gaussian symplectic ensemble* (GSE) consisting of Hermitian quaternionic matrices whose off-diagonal elements are quaternions with Gaussian distribution of zero mean and variance $\mathbb{E}[|X_{ij}|^2] = 1/N$. This means that each of the four components of each X_{ij} is a Gaussian number of zero mean and variance $1/(4N)$. The diagonal elements of $\mathbf{X}$ are real Gaussian numbers with zero mean and variance $1/(2N)$. As usual $\mathbf{X}_{ij} = \mathbf{X}_{ji}^*$ so only the upper (or lower) triangular elements are independent. The joint law for the elements of a GSE matrix with variance $\tau(\mathbf{X}^2) = \sigma^2$ is given by

$$P(\{X_{ij}\}) \propto \exp\left\{-\frac{N}{\sigma^2}\mathrm{Tr}\,\mathbf{X}^2\right\}, \tag{3.11}$$

which we identify with Eq. (3.4) with $\beta = 4$. This parameter $\beta = 4$ is a fundamental property of the symplectic group and will consistently appear in contrast with the orthogonal and unitary cases, $\beta = 1$ and $\beta = 2$ (see Section 5.1.4).

The parameter β can be interpreted as the randomness in the norm of the matrix elements. More precisely, we have

$$|X_{ij}|^2 = \begin{cases} x_{\mathrm{r}}^2 & \text{for real symmetric,} \\ x_{\mathrm{r}}^2 + x_{\mathrm{i}}^2 & \text{for complex Hermitian,} \\ x_{\mathrm{r}}^2 + x_{\mathrm{i}}^2 + x_{\mathrm{j}}^2 + x_{\mathrm{k}}^2 & \text{for quaternionic Hermitian,} \end{cases} \tag{3.12}$$

where $x_{\mathrm{r}}, x_{\mathrm{i}}, x_{\mathrm{j}}, x_{\mathrm{k}}$ are real Gaussian numbers such that $\mathbb{E}[|X_{ij}|^2] = 1$. We see that the fluctuations of $|X_{ij}|^2$ decrease with β (precisely $\mathbb{V}[|X_{ij}|^2] = 2/\beta$). By the law of large numbers (LLN), in the $\beta \to \infty$ limit (if such an ensemble existed) we would have $|X_{ij}|^2 = 1$ with no fluctuations.

Exercise 3.1.1 Quaternionic matrices of size one

The four matrices in Eq. (3.7) can be thought of as the 2×2 complex representations of the four unit quaternions.

(a) Define $\mathbf{Z} := \mathrm{j}$ and compute $\mathbf{Z}^{-1}$.

(b) Show that for all four matrices $\mathbf{Q}$, we have $\mathbf{Z}\mathbf{Q}^T\mathbf{Z}^{-1} = \mathbf{Q}^\dagger$ where the dagger here is the usual transpose plus complex conjugation.

(c) Convince yourself that, by linearity, any $\mathbf{Q}$ that is a real linear combination of the 2×2 matrices i, j, k and 1 must satisfy $\mathbf{Z}\mathbf{Q}^T\mathbf{Z}^{-1} = \mathbf{Q}^\dagger$.

(d) Give an example of a matrix $\mathbf{Q}$ that does not satisfy $\mathbf{Z}\mathbf{Q}^T\mathbf{Z}^{-1} = \mathbf{Q}^\dagger$.

3.1.3 The Ginibre Ensemble

The Gaussian orthogonal ensemble is such that all matrix elements of $\mathbf{X}$ are IID Gaussian, but with the strong constraint that $X_{ij} = X_{ji}$, which makes sure that all eigenvalues of $\mathbf{X}$ are real. What happens if we drop this constraint and consider a square matrix $\mathbf{H}$ with independent entries? In this case, one may choose two different routes, depending on the context.

- One route is simply to allow eigenvalues to be complex numbers. One can then study the eigenvalue distribution in the complex plane, so the distribution becomes a two-dimensional density. Some of the tools introduced in the previous chapter, such as the Sokhotski–Plemelj formula, can be generalized to complex eigenvalues. The final result is called the Girko circular law: the density of eigenvalues is constant within a disk centered at zero and of radius σ (see Fig. 3.1). In the general case where $\mathbb{E}[H_{ij}H_{ji}] = \rho\sigma^2$, the eigenvalues are confined within an ellipse of half-width $(1+\rho)\sigma$ along the real axis and $(1-\rho)\sigma$ in the imaginary direction, interpolating between a circle for $\rho = 0$ (independent entries) and a line segment on the real axis of length 4σ for $\rho = 1$ (symmetric matrices).
- The other route is to focus on the singular values of $\mathbf{H}$. One should thus study the real eigenvalues of $\mathbf{H}^T\mathbf{H}$ when $\mathbf{H}$ is a square random matrix made of independent Gaussian elements. This is precisely the Wishart problem that we will study in Chapter 4, for the special parameter value $q = 1$. Calling s the square-root of these real eigenvalues, the final result is a quarter-circle:

$$\rho(s) = \frac{\sqrt{4\sigma^2 - s^2}}{\pi\sigma^2}; \qquad s \in (0, 2\sigma). \tag{3.13}$$

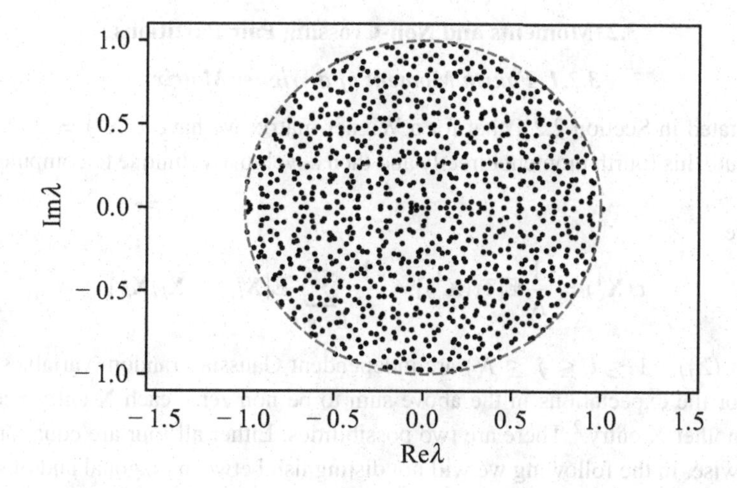

Figure 3.1 Complex eigenvalues of a random $N = 1000$ matrix taken from the Gaussian Ginibre ensemble, i.e. a non-symmetric matrix with IID Gaussian elements with variance $\sigma^2 = 1/N$. The dash line corresponds to the circle $|\lambda|^2 = 1$. As $N \to \infty$ the density becomes uniform in the complex unit disk. This distribution is called the circle law or sometimes, more accurately, the disk law.

Exercise 3.1.2 Three quarter-circle laws

Let $\mathbf{H}$ be a (non-symmetric) square matrix of size N whose entries are IID Gaussian random variable of variance σ^2/N. Then as a simple consequence of the above discussion the following three sets of numbers are distributed according to the quarter-circle law (3.13) in the large N limit. Define

$$w_i = |\lambda_i| \text{ where } \{\lambda_i\} \text{ are the eigenvalues of } \frac{\mathbf{H} + \mathbf{H}^T}{\sqrt{2}},$$

$$r_i = 2|\operatorname{Re}\lambda_i| \text{ where } \{\lambda_i\} \text{ are the eigenvalues of } \mathbf{H},$$

$$s_i = \sqrt{\lambda_i} \text{ where } \{\lambda_i\} \text{ are the eigenvalues of } \mathbf{H}\mathbf{H}^T.$$

(a) Generate a large matrix $\mathbf{H}$ with say $N = 1000$ and $\sigma^2 = 1$ and plot the histogram of the three above sets.

(b) Although these three sets of numbers converge to the same distribution there is no simple relation between them. In particular they are not equal. For a moderate N (10 or 20) examine the three sets and realize that they are all different.

3.2 Moments and Non-Crossing Pair Partitions

3.2.1 Fourth Moment of a Wigner Matrix

We have stated in Section 2.2.1 that for a Wigner matrix we have $\tau(\mathbf{X}^4) = 2\sigma^4$. We will now compute this fourth moment directly and then develop a technique to compute all other moments.

We have

$$\tau(\mathbf{X}^4) = \frac{1}{N}\mathbb{E}[\mathrm{Tr}(\mathbf{X}^4)] = \frac{1}{N}\sum_{i,j,k,l}\mathbb{E}[\mathbf{X}_{ij}\mathbf{X}_{jk}\mathbf{X}_{kl}\mathbf{X}_{li}]. \tag{3.14}$$

Recall that $(\mathbf{X}_{ij} : 1 \le i \le j \le N)$ are independent Gaussian random variables of mean zero. So for the expectations in the above sum to be non-zero, each $\mathbf{X}$ entry needs to be equal to another $\mathbf{X}$ entry.[2] There are two possibilities. Either all four are equal or they are equal pairwise. In the following we will not distinguish between diagonal and off-diagonal terms; as there are many more off-diagonal terms these terms always dominate.

(1) If $\mathbf{X}_{ij} = \mathbf{X}_{jk} = \mathbf{X}_{kl} = \mathbf{X}_{li}$, then

$$\mathbb{E}[\mathbf{X}_{ij}\mathbf{X}_{jk}\mathbf{X}_{kl}\mathbf{X}_{li}] = \frac{3\sigma^4}{N^2}, \tag{3.15}$$

and there are N^2 of them. Thus the total contribution from these terms is

$$\frac{1}{N}N^2\frac{3\sigma^4}{N^2} = \frac{3\sigma^4}{N} \to 0. \tag{3.16}$$

(2) Suppose there are two different pairs. Then there are three possibilities (see Fig. 3.2):

(i) $\mathbf{X}_{ij} = \mathbf{X}_{jk}$, $\mathbf{X}_{kl} = \mathbf{X}_{li}$, and $\mathbf{X}_{ij}$ is different than $\mathbf{X}_{li}$ (i.e. $j \neq l$). Then

$$(3.14) = \frac{1}{N}\sum_{i,j\neq l}\mathbb{E}[\mathbf{X}_{ij}^2\mathbf{X}_{il}^2] = \frac{1}{N}(N^3 - N^2)\left(\frac{\sigma^2}{N}\right)^2 \to \sigma^4 \tag{3.17}$$

as $N \to \infty$.

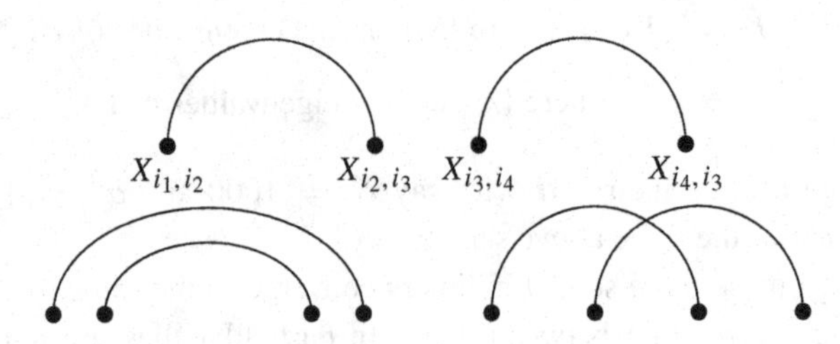

Figure 3.2 Graphical representation of the three terms contributing to $\tau(\mathbf{X}^4)$. The last one is a crossing partition and has a zero contribution.

[2] When we say that $\mathbf{X}_{ij} = \mathbf{X}_{kl}$, we mean that they are the same random variable; given that $\mathbf{X}$ is a symmetric matrix it means either ($i = k$ and $j = l$) or ($i = l$ and $j = k$).

(ii) $\mathbf{X}_{ij} = \mathbf{X}_{li}, \mathbf{X}_{jk} = \mathbf{X}_{kl}$, and $\mathbf{X}_{ij}$ is different than $\mathbf{X}_{jk}$ (i.e. $i \neq k$). Then

$$(3.14) = \frac{1}{N} \sum_{i \neq k, j} \mathbb{E}[\mathbf{X}_{ij}^2 \mathbf{X}_{jk}^2] = \frac{1}{N}(N^3 - N^2)\left(\frac{\sigma^2}{N}\right)^2 \to \sigma^4 \tag{3.18}$$

as $N \to \infty$.

(iii) $\mathbf{X}_{ij} = \mathbf{X}_{kl}, \mathbf{X}_{jk} = \mathbf{X}_{li}$, and $\mathbf{X}_{ij}$ is different than $\mathbf{X}_{jk}$ (i.e. $i \neq k$). Then we must have $i = l$ and $j = k$ from $\mathbf{X}_{ij} = \mathbf{X}_{kl}$, and $i = j$ and $k = l$ from $\mathbf{X}_{jk} = \mathbf{X}_{li}$. This gives a contradiction: there are no such terms.

In sum, we obtain that

$$\tau(\mathbf{X}^4) \to \sigma^4 + \sigma^4 = 2\sigma^4 \tag{3.19}$$

as $N \to \infty$, where the two terms come from the two non-crossing partitions, see Figure 3.2.

In the next technical section, we generalize this calculation to arbitrary moments of $\mathbf{X}$. Odd moments are zero by symmetry. Even moments $\tau(\mathbf{X}^{2k})$ can be written as sums over non-crossing diagrams (non-crossing pair partitions of $2k$ elements), where each such diagram contributes σ^{2k}. So

$$\tau(\mathbf{X}^{2k}) = C_k \sigma^{2k}, \tag{3.20}$$

where C_k are Catalan numbers, the number of such non-crossing diagrams. They satisfy

$$C_k = \sum_{j=1}^{k} C_{j-1} C_{k-j} = \sum_{j=0}^{k-1} C_j C_{k-j-1}, \tag{3.21}$$

with $C_0 = C_1 = 1$, and can be written explicitly as

$$C_k = \frac{1}{k+1}\binom{2k}{k}, \tag{3.22}$$

see Section 3.2.3.

3.2.2 Catalan Numbers: Counting Non-Crossing Pair Partitions

We would like to calculate all moments of $\mathbf{X}$. As written above, all the odd moments $\tau(\mathbf{X}^{2k+1})$ vanish (since the odd moments of a Gaussian random variable vanish). We only need to compute the even moments:

$$\tau(\mathbf{X}^{2k}) = \frac{1}{N}\mathbb{E}\left[\mathrm{Tr}(\mathbf{X}^{2k})\right] = \frac{1}{N} \sum_{i_1,\ldots,i_{2k}} \mathbb{E}\left(\mathbf{X}_{i_1 i_2} \mathbf{X}_{i_2 i_3} \ldots \mathbf{X}_{i_{2k} i_1}\right). \tag{3.23}$$

Since we assume that the elements of $\mathbf{X}$ are Gaussian, we can expand the above expectation value using Wick's theorem using the covariance of the $\{\mathbf{X}_{ij}\}$'s. The matrix $\mathbf{X}$ is symmetric, so we have to keep track of the fact that $\mathbf{X}_{ij}$ is the same variable as $\mathbf{X}_{ji}$. For this reason, using Wick's theorem proves quite tedious and we will not follow this route here.

From the Taylor series at infinity of the Stieltjes transform, we expect every even moment of $\mathbf{X}$ to converge to an $O(1)$ number as $N \to \infty$. We will therefore drop any $O(1/N)$ or smaller term as we proceed. In particular the difference of variance between diagonal and off-diagonal elements of $\mathbf{X}$ does not matter to first order in $1/N$.

In Eq. (3.23), each $\mathbf{X}$ entry must be equal to at least one another $\mathbf{X}$ entry, otherwise the expectation is zero. On the other hand, it is easy to show that for the partitions that contain at least one group with > 2 (actually ≥ 4) $\mathbf{X}$ entries that are equal to each other, their total contribution will be of order $O(1/N)$ or smaller (e.g. in case (1) of the previous section). Thus we only need to consider the cases where each $\mathbf{X}$ entry is paired to exactly one other $\mathbf{X}$ entry, which we also referred to as a pair partition.

We need to count the number of types of pairings of $2k$ elements that contribute to $\tau(\mathbf{X}^{2k})$ as $N \to \infty$. We associate to each pairing a diagram. For example, for $k = 3$, we have $5!! = 5 \cdot 3 \cdot 1 = 15$ possible pairings (see Fig. 3.3).

To compute the contribution of each of these pair partitions, we will compute the contribution of *non-crossing* pair partitions and argue that pair partitions with crossings do not contribute in the large N limit. First we need to define what is a non-crossing pair partition of $2k$ elements. A pair partition can be draw as a diagram where the $2k$ elements are points on a line and each point is joined with its pair partner by an arc drawn above that line. If at least two arcs cross each other the partition is called *crossing*, and *non-crossing* otherwise. In Figure 3.3 the five partitions on the left are non-crossing while the ten others are crossing.

In a non-crossing partition of size $2k$, there is always at least one pairing between consecutive points (the smallest arc). If we remove the first such pairing we get a non-crossing pair partition of $2k - 2$ elements. We can proceed in this way until we get to a paring of only two elements: the unique (non-crossing) pair partition contributing to (Fig. 3.4)

$$\tau(\mathbf{X}^2) = \sigma^2. \tag{3.24}$$

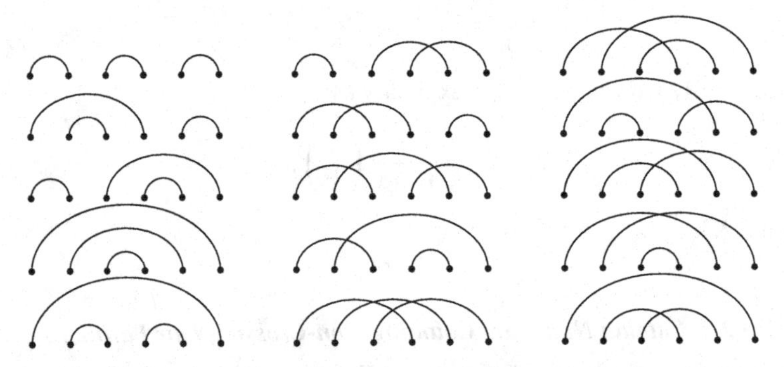

Figure 3.3 Graphical representation of the 15 terms contributing to $\tau(\mathbf{X}^6)$. Only the five on the left are non-crossing and have a non-zero contribution as $N \to \infty$.

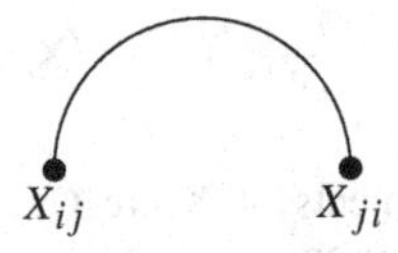

Figure 3.4 Graphical representation of the only term contributing to $\tau(\mathbf{X}^2)$. Note that the indices of two terms are already equal prior to pairing.

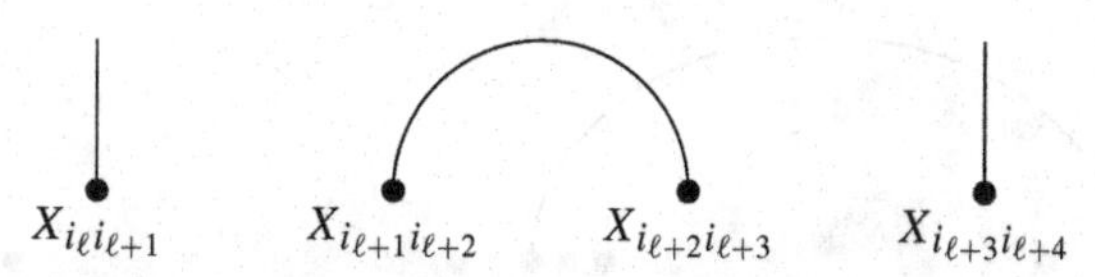

Figure 3.5 Zoom into the smallest arc of a non-crossing partition. The two middle matrices are paired while the other two could be paired together or to other matrices to the left and right respectively. After the pairing of $\mathbf{X}_{i_{\ell+1},i_{\ell+2}}$ and $\mathbf{X}_{i_{\ell+2},i_{\ell+3}}$, we have $i_{\ell+1} = i_{\ell+3}$ and the index $i_{\ell+2}$ is free.

We can use this argument to prove by induction that each non-crossing partition contributes a factor σ^{2k}. In Figure 3.5, consecutive elements $\mathbf{X}_{i_{\ell+1},i_{\ell+2}}$ and $\mathbf{X}_{i_{\ell+2},i_{\ell+3}}$ are paired; we want to evaluate that pair and remove it from the diagram. The variance contributes a factor σ^2/N. We can make two choices for index matching. First consider $i_{\ell+1} = i_{\ell+3}$ and $i_{\ell+2} = i_{\ell+2}$. In that case, the index $i_{\ell+2}$ is free and its summation contributes a factor of N. The identity $i_{\ell+1} = i_{\ell+3}$ means that the previous matrix $\mathbf{X}_{i_\ell,i_{\ell+1}}$ is now linked by matrix multiplication to the following matrix $\mathbf{X}_{i_{\ell+1},i_{\ell+4}}$. In other words we are left with σ^2 times a non-crossing partition of size $2k-2$, which contributes σ^{2k-2} by our induction hypothesis. The other choice of index matching, $i_{\ell+1} = i_{\ell+2} = i_{\ell+3}$, can be viewed as fixing a particular value for $i_{\ell+2}$ and is included in the sum over $i_{\ell+2}$ in the previous index matching. So by induction we do have that each non-crossing pair partition contributes σ^{2k}.

Before we discuss the contribution of crossing pair partitions, let's analyze in terms of powers of N the computation we just did for the non-crossing case. The computation of each term in $\tau(\mathbf{X}^{2k})$ involves $2k$ matrices that have in total $4k$ indices. The trace and the matrix multiplication forces $2k$ equalities among these indices. The normalization of the trace and the k variance terms gives a factor of σ^{2k}/N^{k+1}. To get a result of order 1 we need to be left with $k+1$ free indices whose summation gives a factor of N^{k+1}. Each k pairing imposes a matching between pairs of indices. For the first $k-1$ choice of pairing we managed to match one pair of indices that were already equal. At the last step we matched to pairs of indices that were already equal. Hence in total we added only $k+1$ equality constraints which left us with $k+1$ free indices as needed.

We can now argue that crossing pair partitions do not contribute in the large N limit. For crossing partition it is not possible to choose a matching at every step that matches a pair of indices that are already equal. If we use the previous algorithm of removing at each step the leftmost smallest arc, at some point, the smallest arc will have a crossing and we will be pairing to matrices that share no indices, adding two equality constraints at this step. The result will therefore be down by at least a factor of $1/N$ with respect to the non-crossing case. This argument is not really a proof but an intuition why this might be true.[3]

We can now complete our moments computation. Let

$$C_k := \#\text{ of non-crossing pairings of } 2k \text{ elements.} \tag{3.25}$$

Since every non-crossing pair partition contributes a factor σ^{2k}, summing over all non-crossing pairings we immediately get that

$$\tau(\mathbf{X}^{2k}) = C_k \sigma^{2k}. \tag{3.26}$$

[3] A more rigorous proof can be found in e.g. Anderson et al. [2010], Tao [2012] or Mingo and Speicher [2017]. In this last reference, the authors compute the moments of $\mathbf{X}$ exactly for every N (when $\sigma_{\text{d}}^2 = \sigma_{\text{od}}^2$).

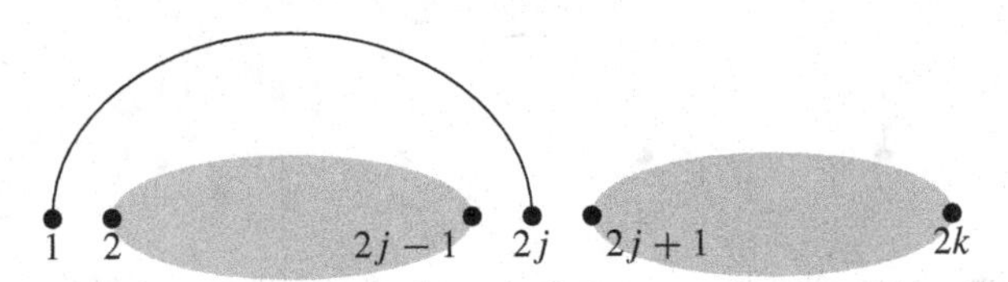

Figure 3.6 In a non-crossing pairing, the paring of site 1 with site $2j$ splits the graph into two disjoint non-crossing parings.

3.2.3 Recursion Relation for Catalan Numbers

In order to compute the Catalan numbers C_k, we will write a recursion relation for them. Take a non-crossing pairing, site 1 is linked to some even site $2j$ (it is easy to see that 1 cannot link to an odd site in order for the partition to be non-crossing). Then the diagram is split into two smaller non-crossing pairings of sizes $2(j-1)$ and $2(k-j)$, respectively (see Fig. 3.6). Thus we get the inductive relation[4]

$$C_k = \sum_{j=1}^{k} C_{j-1}C_{k-j} = \sum_{j=0}^{k-1} C_j C_{k-j-1}, \tag{3.27}$$

where we let $C_0 = C_1 = 1$. One can then prove by induction that C_k is given by the Catalan number:

$$C_k = \frac{1}{k+1}\binom{2k}{k}. \tag{3.28}$$

Using the Taylor series for the Stieltjes transform (2.22), we can use the Catalan number recursion relation to find an equation for the Stieltjes transform of the Wigner ensemble:

$$\mathfrak{g}(z) = \sum_{k=0}^{\infty} \frac{C_k}{z^{2k+1}}\sigma^{2k}. \tag{3.29}$$

Thus, using (3.27), we obtain that

$$\begin{aligned}
\mathfrak{g}(z) - \frac{1}{z} &= \sum_{k=1}^{\infty} \frac{\sigma^{2k}}{z^{2k+1}}\left(\sum_{j=0}^{k-1} C_j C_{k-j-1}\right) \\
&= \frac{\sigma^2}{z}\sum_{j=0}^{\infty}\frac{C_j}{z^{2j+1}}\sigma^{2j}\left(\sum_{k=j+1}^{\infty}\frac{C_{k-j-1}}{z^{2(k-j-1)+1}}\sigma^{2(k-j-1)}\right) \\
&= \frac{\sigma^2}{z}\left(\sum_{j=0}^{\infty}\frac{C_j}{z^{2j+1}}\sigma^{2j}\right)\left(\sum_{\ell=0}^{\infty}\frac{C_\ell}{z^{2\ell+1}}\sigma^{2\ell}\right) = \frac{\sigma^2}{z}\mathfrak{g}^2(z),
\end{aligned} \tag{3.30}$$

which gives the same self-consistent equation for $\mathfrak{g}(z)$ as in (2.35) and hence the same solution:

$$\mathfrak{g}(z) = \frac{z - z\sqrt{1-4\sigma^2/z^2}}{2\sigma^2}. \tag{3.31}$$

[4] Interestingly, this recursion relation is also found in the problem of RNA folding. For deep connections between the physics of RNA and RMT, see Orland and Zee [2002].

The same result could have been derived by substituting the explicit solution for the Catalan number Eq. (3.28) into (3.29), but this route requires knowledge of the Taylor series:

$$\sqrt{1-x} = 1 - \sum_{k=0}^{\infty} \frac{2}{k+1} \binom{2k}{k} \left(\frac{x}{4}\right)^{k+1}. \tag{3.32}$$

Exercise 3.2.1 Non-crossing pair partitions of eight elements

(a) Draw all the non-crossing pair partitions of eight elements. Hint: use the recursion expressed in Figure 3.6.

(b) If $\mathbf{X}$ is a unit Wigner matrix, what is $\tau(\mathbf{X}^8)$?

Bibliographical Notes

- Again, several books cover the content of this chapter, see for example
  - M. L. Mehta. *Random Matrices*. Academic Press, San Diego, 3rd edition, 2004,
  - G. Blower. *Random Matrices: High Dimensional Phenomena.* Cambridge University Press, Cambridge, 2009,
  - B. Eynard, T. Kimura, and S. Ribault. Random matrices. *preprint arXiv:1510.04430*, 2006,

  and in particular for a detailed discussion of the relation between non-crossing and large matrices, see
  - T. Tao. *Topics in Random Matrix Theory*. American Mathematical Society, Providence, Rhode Island, 2012,
  - G. W. Anderson, A. Guionnet, and O. Zeitouni. *An Introduction to Random Matrices*. Cambridge University Press, Cambridge, 2010,
  - J. A. Mingo and R. Speicher. *Free Probability and Random Matrices*. Springer, New York, 2017.
- Concerning the construction of generalized beta ensembles, see e.g.
  - I. Dumitriu and A. Edelman. Matrix models for beta ensembles. *Journal of Mathematical Physics*, 43(11):5830–5847, 2002,
  - R. Allez, J. P. Bouchaud, and A. Guionnet. Invariant beta ensembles and the Gauss-Wigner crossover. *Physical Review Letters*, 109(9):094102, 2012.
- On the Ginibre ensemble and the circular law, see
  - V. L. Girko. Circular law. *Theory of Probability and Its Applications*, 29(4):694–706, 1985,
  - H. J. Sommers, A. Crisanti, H. Sompolinsky, and Y. Stein. Spectrum of large random asymmetric matrices. *Physical Review Letters*, 60(19):1895–1899, 1988,
  - C. Bordenave and D. Chafaï. Around the circular law. *Probability Surveys*, 9, 2012.

- On the connection between non-crossing diagrams and RNA folding, see
  - P.-G. de Gennes. Statistics of branching and hairpin helices for the dAT copolymer. *Biopolymers*, 6(5):715–729, 1968,
  - H. Orland and A. Zee. RNA folding and large N matrix theory. *Nuclear Physics B*, 620(3):456–476, 2002.

4

Wishart Ensemble and Marčenko–Pastur Distribution

In this chapter we will study the statistical properties of large sample covariance matrices of some N-dimensional variables observed T times. More precisely, the empirical set consists of $N \times T$ data $\{x_i^t\}_{1\le i\le N, 1\le t\le T}$, where we have T observations and each observation contains N variables. Examples abound: we could consider the daily returns of N stocks, over a certain time period, or the number of spikes fired by N neurons during T consecutive time intervals of length Δt, etc. Throughout this book, we will use the notation q for the ratio N/T. When the number of observations is much larger than the number of variables, one has $q \ll 1$. If the number of observations is smaller than the number of variables (a case that can easily happen in practice), then $q > 1$.

In the case where $q \to 0$, one can faithfully reconstruct the "true" (or population) covariance matrix $\mathbf{C}$ of the N variables from empirical data. For $q = O(1)$, on the other hand, the empirical (or sample) covariance matrix is a strongly distorted version of $\mathbf{C}$, even in the limit of a large number of observations. This is not surprising since we are trying to estimate $O(N^2/2)$ matrix elements from $O(NT)$ observations. In this chapter, we will derive the well-known Marčenko–Pastur law for the eigenvalues of the sample covariance matrix for arbitrary values of q, in the "white" case where the population covariance matrix $\mathbf{C}$ is the identity matrix $\mathbf{C} = \mathbf{1}$.

4.1 Wishart Matrices

4.1.1 Sample Covariance Matrices

We assume that the observed variables x_i^t have zero mean. (Otherwise, we need to remove the sample mean $T^{-1}\sum_t x_i^t$ from x_i^t for each i. For simplicity, we will not consider this case.) Then the sample covariances of the data are given by

$$E_{ij} = \frac{1}{T}\sum_{t=1}^{T} x_i^t x_j^t. \tag{4.1}$$

Thus E_{ij} form an $N \times N$ matrix $\mathbf{E}$, called the *sample covariance matrix* (SCM), which we we write in a compact form as

$$\mathbf{E} = \frac{1}{T}\mathbf{H}\mathbf{H}^T, \tag{4.2}$$

where $\mathbf{H}$ is an $N \times T$ data matrix with entries $H_{it} = x_i^t$.

The matrix $\mathbf{E}$ is symmetric and positive semi-definite:

$$\mathbf{E} = \mathbf{E}^T, \quad \text{and} \quad \mathbf{v}^T\mathbf{E}\mathbf{v} = (1/T)\|\mathbf{H}^T\mathbf{v}\|^2 \geq 0, \tag{4.3}$$

for any $\mathbf{v} \in \mathbb{R}^N$. Thus $\mathbf{E}$ is diagonalizable and has all eigenvalues $\lambda_k^{\mathbf{E}} \geq 0$.

We can define another covariance matrix by transposing the data matrix $\mathbf{H}$:

$$\mathbf{F} = \frac{1}{N}\mathbf{H}^T\mathbf{H}. \tag{4.4}$$

The matrix $\mathbf{F}$ is a $T \times T$ matrix, it is also symmetric and positive semi-definite. If the index i $(1 < i < N)$ labels the variables and the index t $(1 < t < T)$ the observations, we can call the matrix $\mathbf{F}$ the covariance of the observations (as opposed to $\mathbf{E}$ the covariance of the variables). F_{ts} measures how similar the observations at t are to those at s – in the above example of neurons, it would measure how similar is the firing pattern at time t and at time s.

As we saw in Section 1.1.3, the matrices $T\mathbf{E}$ and $N\mathbf{F}$ have the same non-zero eigenvalues. Also the matrix $\mathbf{E}$ has at least $N - T$ zero eigenvalues if $N > T$ (and $\mathbf{F}$ has at least $T - N$ zero eigenvalues if $T > N$).

Assume for a moment that $N \leq T$ (i.e. $q \leq 1$), then we know that $\mathbf{F}$ has N (zero or non-zero) eigenvalues inherited from $\mathbf{E}$ and equal to $q^{-1}\lambda_k^{\mathbf{E}}$, and $T - N$ zero eigenvalues. This allows us to write an exact relation between the Stieltjes transforms of $\mathbf{E}$ and $\mathbf{F}$:

$$\begin{aligned} g_T^{\mathbf{F}}(z) &= \frac{1}{T}\sum_{k=1}^{T}\frac{1}{z - \lambda_k^{\mathbf{F}}} \\ &= \frac{1}{T}\left(\sum_{k=1}^{N}\frac{1}{z - q^{-1}\lambda_k^{\mathbf{E}}} + (T - N)\frac{1}{z - 0}\right) \\ &= q^2 g_N^{\mathbf{E}}(qz) + \frac{1-q}{z}. \end{aligned} \tag{4.5}$$

A similar argument with $T < N$ leads to the same Eq. (4.5) so it is actually valid for any value of q. The relationship should be true as well in the large N limit:

$$\mathfrak{g}_{\mathbf{F}}(z) = q^2\mathfrak{g}_{\mathbf{E}}(qz) + \frac{1-q}{z}. \tag{4.6}$$

4.1.2 First and Second Moments of a Wishart Matrix

We now study the SCM $\mathbf{E}$. Assume that the column vectors of $\mathbf{H}$ are drawn independently from a multivariate Gaussian distribution with mean zero and "true" (or "population") covariance matrix $\mathbf{C}$, i.e.

$$\mathbb{E}[H_{it}H_{js}] = C_{ij}\delta_{ts}, \tag{4.7}$$

with, again,

$$\mathbf{E} = \frac{1}{T}\mathbf{H}\mathbf{H}^T. \tag{4.8}$$

Sample covariance matrices of this type were first studied by the Scottish mathematician John Wishart (1898–1956) and are now called Wishart matrices.

Recall that if $(X_1, \ldots, X_{2n})$ is a zero-mean multivariate normal random vector, then by Wick's theorem,

$$\mathbb{E}[X_1 X_2 \cdots X_{2n}] = \sum_{\text{pairings}} \prod_{\text{pairs}} \mathbb{E}[X_i X_j] = \sum_{\text{pairings}} \prod_{\text{pairs}} \text{Cov}(X_i, X_j), \tag{4.9}$$

where $\sum_{\text{pairings}} \prod_{\text{pairs}}$ means that we sum over all distinct pairings of $\{X_1, \ldots, X_{2n}\}$ and each summand is the product of the n pairs.

First taking expectation, we obtain that

$$\mathbb{E}[E_{ij}] = \frac{1}{T}\mathbb{E}\left[\sum_{t=1}^{T} H_{it}H_{jt}\right] = \frac{1}{T}\sum_{t=1}^{T} C_{ij} = C_{ij}. \tag{4.10}$$

Thus, we have $\mathbb{E}[\mathbf{E}] = \mathbf{C}$: as it is well known, the SCM is an unbiased estimator of the true covariance matrix (at least when $\mathbb{E}[x_i^t] = 0$).

For the fluctuations, we need to study the higher order moments of $\mathbf{E}$. The second moment can be calculated as

$$\tau(\mathbf{E}^2) := \frac{1}{NT^2}\mathbb{E}\left[\text{Tr}(\mathbf{H}\mathbf{H}^T\mathbf{H}\mathbf{H}^T)\right] = \frac{1}{NT^2}\sum_{i,j,t,s}\mathbb{E}\left[H_{it}H_{jt}H_{js}H_{is}\right]. \tag{4.11}$$

Then by Wick's theorem, we have (see Fig. 4.1)

$$\begin{aligned}\tau(\mathbf{E}^2) &= \frac{1}{NT^2}\sum_{t,s}\sum_{i,j} C_{ij}^2 + \frac{1}{NT^2}\sum_{t,s}\sum_{i,j} C_{ii}C_{jj}\delta_{ts} + \frac{1}{NT^2}\sum_{t,s}\sum_{i,j} C_{ij}^2\delta_{ts} \\ &= \tau(\mathbf{C}^2) + \frac{N}{T}\tau(\mathbf{C})^2 + \frac{1}{T}\tau(\mathbf{C}^2).\end{aligned} \tag{4.12}$$

Suppose $N, T \to \infty$ with some fixed ratio $N/T = q$ for some constant $q > 0$. The last term on the right hand side then tends to zero and we get

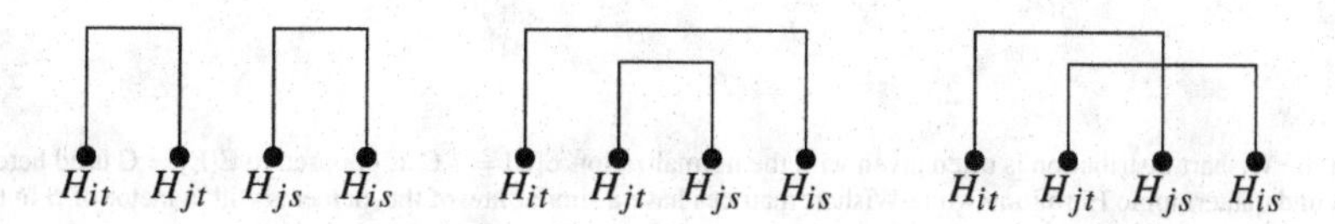

Figure 4.1 Graphical representation of the three Wick's contractions corresponding to the three terms in Eq. (4.12).

$$\tau(\mathbf{E}^2) \to \tau(\mathbf{C}^2) + q\tau(\mathbf{C})^2. \tag{4.13}$$

The variance of the SCM is greater than that of the true covariance by a term proportional to q. When $q \to 0$ we recover perfect estimation and the two matrices have the same variance. If $\mathbf{C} = \alpha\mathbf{1}$ (a multiple of the identity) then $\tau(\mathbf{C}^2) - \tau(\mathbf{C})^2 = 0$ but $\tau(\mathbf{E}^2) - \tau(\mathbf{E})^2 \to q\alpha^2$.

4.1.3 The Law of Wishart Matrices

Next, we give the joint distribution of elements of $\mathbf{E}$. For each fixed column of $\mathbf{H}$, the joint distribution of the elements is

$$P\left(\{H_{it}\}_{i=1}^N\right) = \frac{1}{\sqrt{(2\pi)^N \det \mathbf{C}}} \exp\left[-\frac{1}{2}\sum_{i,j} H_{it}(\mathbf{C})_{ij}^{-1} H_{jt}\right]. \tag{4.14}$$

Taking the product over $1 \le t \le T$ (since the columns are independent), we obtain

$$\begin{aligned} P(\mathbf{H}) &= \frac{1}{(2\pi)^{\frac{NT}{2}} \det \mathbf{C}^{T/2}} \exp\left[-\frac{1}{2}\operatorname{Tr}\left(\mathbf{H}^T\mathbf{C}^{-1}\mathbf{H}\right)\right] \\ &= \frac{1}{(2\pi)^{\frac{NT}{2}} \det \mathbf{C}^{T/2}} \exp\left[-\frac{T}{2}\operatorname{Tr}\left(\mathbf{E}\mathbf{C}^{-1}\right)\right]. \end{aligned} \tag{4.15}$$

Let us now make a change in variables $\mathbf{H} \to \mathbf{E}$. As shown in the technical paragraph 4.1.4, the Jacobian of the transformation is proportional to $(\det \mathbf{E})^{\frac{T-N-1}{2}}$. The following exact expression for the law of the matrix elements was obtained by Wishart:[1]

$$P(\mathbf{E}) = \frac{(T/2)^{NT/2}}{\Gamma_N(T/2)} \frac{(\det \mathbf{E})^{(T-N-1)/2}}{(\det \mathbf{C})^{T/2}} \exp\left[-\frac{T}{2}\operatorname{Tr}\left(\mathbf{E}\mathbf{C}^{-1}\right)\right], \tag{4.16}$$

where Γ_N is the multivariate gamma function. Note that the density is restricted to positive semi-definite matrices $\mathbf{E}$. The Wishart distribution can be thought of as the matrix generalization of the gamma distribution. Indeed for $N = 1$, $P(\mathbf{E})$ reduces to a such a distribution:

$$P_\gamma(x) = \frac{b^a}{\Gamma(a)} x^{a-1} e^{-bx}, \tag{4.17}$$

where $b = T/(2C)$ and $a = T/2$. Using the identity $\det \mathbf{E} = \exp(\operatorname{Tr}\log \mathbf{E})$, we can rewrite the above expression as[2]

[1] Note that the Wishart distribution is often given with the normalization $\mathbb{E}[\mathbf{E}] = T\mathbf{C}$ as opposed to $\mathbb{E}[\mathbf{E}] = \mathbf{C}$ used here.

[2] Complex and quaternionic Hermitian white Wishart matrices have a similar law of the elements with a factor of β in the exponential:

$$P(\mathbf{W}) \propto \exp\left[-\frac{\beta N}{2}\operatorname{Tr} V(\mathbf{W})\right] \text{ with } V(x) = \frac{N - T - 1 + 2/\beta}{N}\log x + \frac{T}{N}x, \tag{4.18}$$

with β equal to 1, 2 or 4 as usual. The large N limit $V(x)$ is the same in all three cases and is given by Eq. (4.21).

$$P(\mathbf{E}) = \frac{(T/2)^{NT/2}}{\Gamma_N(T/2)} \frac{1}{(\det \mathbf{C})^{T/2}} \exp\left[-\frac{T}{2}\operatorname{Tr}\left(\mathbf{E}\mathbf{C}^{-1}\right) + \frac{T-N-1}{2}\operatorname{Tr}\log\mathbf{E}\right]. \quad (4.19)$$

We will denote by $\mathbf{W}$ a SCM with $\mathbf{C} = \mathbf{1}$ and call such a matrix a *white Wishart* matrix. In this case, as $N, T \to \infty$ with $q := N/T$, we get that

$$P(\mathbf{W}) \propto \exp\left[-\frac{N}{2}\operatorname{Tr} V(\mathbf{W})\right], \quad (4.20)$$

where

$$V(\mathbf{W}) := (1 - q^{-1})\log\mathbf{W} + q^{-1}\mathbf{W}. \quad (4.21)$$

Note that the above $P(\mathbf{W})$ is rotationally invariant in the white case. In fact, if a vector $\mathbf{v}$ has Gaussian distribution $\mathcal{N}(0, \mathbf{1}_{N\times N})$, then $\mathbf{O}\mathbf{v}$ has the same distribution $\mathcal{N}(0, \mathbf{1}_{N\times N})$ for any orthogonal matrix $\mathbf{O}$. Hence $\mathbf{O}\mathbf{H}$ has the same distribution as $\mathbf{H}$, which shows that $\mathbf{O}\mathbf{E}\mathbf{O}^T$ has the same distribution as $\mathbf{E}$.

4.1.4 Jacobian of the Transformation $\mathbf{H} \to \mathbf{E}$

The aim here is to compute the volume $\Upsilon(\mathbf{E})$ corresponding to all $\mathbf{H}$'s such that $\mathbf{E} = T^{-1}\mathbf{H}\mathbf{H}^T$:

$$\Upsilon(\mathbf{E}) = \int \mathrm{d}\mathbf{H}\, \delta(\mathbf{E} - T^{-1}\mathbf{H}\mathbf{H}^T). \quad (4.22)$$

Note that this volume is the inverse of the Jacobian of the transformation $\mathbf{H} \to \mathbf{E}$. Next note that one can choose $\mathbf{E}$ to be diagonal, because one can always rotate the integral over $\mathbf{H}$ to an integral over $\mathbf{O}\mathbf{H}$, where $\mathbf{O}$ is the rotation matrix that makes $\mathbf{E}$ diagonal. Now, introducing the Fourier representation of the δ function for all $N(N+1)/2$ independent components of $\mathbf{E}$, one has

$$\Upsilon(\mathbf{E}) = \int \mathrm{d}\mathbf{H}\mathrm{d}\mathbf{A}\exp\left(\mathrm{i}\operatorname{Tr}(\mathbf{A}\mathbf{E} - T^{-1}\mathbf{A}\mathbf{H}\mathbf{H}^T)\right), \quad (4.23)$$

where $\mathbf{A}$ is the symmetric matrix of the corresponding Fourier variables, to which we add a small imaginary part proportional to $\mathbf{1}$ to make all the following integrals well defined. The Gaussian integral over $\mathbf{H}$ can now be performed explicitly for all $t = 1, \ldots, T$, leading to

$$\int \mathrm{d}\mathbf{H}\exp\left(-\mathrm{i}T^{-1}\operatorname{Tr}(\mathbf{A}\mathbf{H}\mathbf{H}^T)\right) \propto (\det\mathbf{A})^{-T/2}, \quad (4.24)$$

leaving us with

$$\Upsilon(\mathbf{E}) \propto \int \mathrm{d}\mathbf{A}\exp(\mathrm{i}\operatorname{Tr}(\mathbf{A}\mathbf{E}))(\det\mathbf{A})^{-T/2}. \quad (4.25)$$

We can change variables from $\mathbf{A}$ to $\mathbf{B} = \mathbf{E}^{\frac{1}{2}}\mathbf{A}\mathbf{E}^{\frac{1}{2}}$. The Jacobian of this transformation is

$$\begin{aligned}\prod_i \mathrm{d}\mathbf{A}_{ii}\prod_{j>i}\mathrm{d}\mathbf{A}_{ij} &= \prod_i \mathbf{E}_{ii}^{-1}\prod_{j>i}(\mathbf{E}_{ii}\mathbf{E}_{jj})^{-\frac{1}{2}}\prod_i \mathrm{d}\mathbf{B}_{ii}\prod_{j>i}\mathrm{d}\mathbf{B}_{ij}\\ &= (\det(\mathbf{E}))^{-\frac{N+1}{2}}\prod_i \mathrm{d}\mathbf{B}_{ii}\prod_{j>i}\mathrm{d}\mathbf{B}_{ij}. \end{aligned} \quad (4.26)$$

So finally,

$$\Upsilon(\mathbf{E}) \propto \left[\int \mathrm{d}\mathbf{B} \exp(\mathrm{i}\,\mathrm{Tr}(\mathbf{B}))\,(\det \mathbf{B})^{-T/2}\right] (\det(\mathbf{E}))^{\frac{T-N-1}{2}}, \tag{4.27}$$

as announced in the main text.

4.2 Marčenko–Pastur Using the Cavity Method

4.2.1 Self-Consistent Equation for the Resolvent

We first derive the asymptotic distribution of eigenvalues of the Wishart matrix with $\mathbf{C} = \mathbf{1}$, i.e. the Marčenko–Pastur distribution. We will use the same method as in the derivation of the Wigner semi-circle law in Section 2.3. In the case $\mathbf{C} = \mathbf{1}$, the $N \times T$ matrix $\mathbf{H}$ is filled with IID standard Gaussian random numbers and we have $\mathbf{W} = (1/T)\mathbf{H}\mathbf{H}^T$.

As in Section 2.3, we wish to derive a self-consistent equation satisfied by the Stieltjes transform:

$$\mathfrak{g}_{\mathbf{W}}(z) = \tau\left(\mathbf{G}_{\mathbf{W}}(z)\right), \quad \mathbf{G}_{\mathbf{W}}(z) := (z\mathbf{1} - \mathbf{W})^{-1}. \tag{4.28}$$

We fix a large N and first write an equation for the element 11 of $\mathbf{G}_{\mathbf{W}}(z)$. We will argue later that $\mathbf{G}_{11}(z)$ converges to $\mathfrak{g}(z)$ with negligible fluctuations. (We henceforth drop the subscript $\mathbf{W}$ as this entire section deals with the white Wishart case.)

Using again the Schur complement formula (1.32), we have that

$$\frac{1}{(\mathbf{G}(z))_{11}} = \mathbf{M}_{11} - \mathbf{M}_{12}(\mathbf{M}_{22})^{-1}\mathbf{M}_{21}, \tag{4.29}$$

where $\mathbf{M} := z\mathbf{1} - \mathbf{W}$, and the submatrices of size, respectively, $[\mathbf{M}_{11}] = 1 \times 1$, $[\mathbf{M}_{12}] = 1 \times (N-1)$, $[\mathbf{M}_{21}] = (N-1) \times 1$, $[\mathbf{M}_{22}] = (N-1) \times (N-1)$. We can expand the above expression and write

$$\frac{1}{(\mathbf{G}(z))_{11}} = z - \mathbf{W}_{11} - \frac{1}{T^2}\sum_{t,s=1}^{T}\sum_{j,k=2}^{N} \mathbf{H}_{1t}\mathbf{H}_{jt}(\mathbf{M}_{22})^{-1}_{jk}\mathbf{H}_{ks}\mathbf{H}_{1s}. \tag{4.30}$$

Note that the three matrices $\mathbf{M}_{22}$, $\mathbf{H}_{jt}$ $(j \geq 2)$ and $\mathbf{H}_{ks}$ $(k \geq 2)$ are independent of the entries H_{1t} for all t. We can write the last term on the right hand side as

$$\frac{1}{T}\sum_{t,s=1}^{N} \mathbf{H}_{1t}\Omega_{ts}\mathbf{H}_{1s} \quad \text{with} \quad \Omega_{ts} := \frac{1}{T}\sum_{j,k=2}^{N} \mathbf{H}_{jt}(\mathbf{M}_{22})^{-1}_{jk}\mathbf{H}_{ks}. \tag{4.31}$$

Provided $\gamma^2 := T^{-1}\operatorname{Tr}\Omega^2$ converges to a finite limit when $T \to \infty$,[3] one readily shows that the above sum converges to $T^{-1}\operatorname{Tr}\Omega$ with fluctuations of the order of $\gamma T^{-\frac{1}{2}}$. So we have, for large T,

$$\begin{aligned}\frac{1}{(\mathbf{G}(z))_{11}} &= z - \mathbf{W}_{11} - \frac{1}{T}\sum_{2\le j,k\le N}\frac{\sum_t H_{kt}H_{jt}}{T}(\mathbf{M}_{22})^{-1}_{jk} + O\left(T^{-\frac{1}{2}}\right)\\ &= z - \mathbf{W}_{11} - \frac{1}{T}\sum_{2\le j,k\le N}\mathbf{W}_{kj}(\mathbf{M}_{22})^{-1}_{jk} + O\left(T^{-\frac{1}{2}}\right)\\ &= z - 1 - \frac{1}{T}\operatorname{Tr}\mathbf{W}_2\mathbf{G}_2(z) + O\left(T^{-\frac{1}{2}}\right),\end{aligned} \tag{4.32}$$

where in the last step we have used the fact that $\mathbf{W}_{11} = 1 + O(T^{-\frac{1}{2}})$ and noted $\mathbf{W}_2$ and $\mathbf{G}_2(z)$ the SCM and resolvent of the $N-1$ variables excluding (1). We can rewrite the trace term:

$$\begin{aligned}\operatorname{Tr}(\mathbf{W}_2\mathbf{G}_2(z)) &= \operatorname{Tr}\left(\mathbf{W}_2(z\mathbf{1}-\mathbf{W}_2)^{-1}\right)\\ &= -\operatorname{Tr}\mathbf{1} + z\operatorname{Tr}\left((z\mathbf{1}-\mathbf{W}_2)^{-1}\right)\\ &= -\operatorname{Tr}\mathbf{1} + z\operatorname{Tr}\mathbf{G}_2(z).\end{aligned} \tag{4.33}$$

In the region where $\operatorname{Tr}\mathbf{G}(z)/N$ converges for large N to the deterministic $\mathfrak{g}(z)$, $\operatorname{Tr}\mathbf{G}_2(z)/N$ should also converge to the same limit as $\mathbf{G}_2(z)$ is just an $(N-1)\times(N-1)$ version of $\mathbf{G}(z)$. So in the region of convergence we have

$$\frac{1}{(\mathbf{G}(z))_{11}} = z - 1 + q - qz\mathfrak{g}(z) + O\left(N^{-\frac{1}{2}}\right), \tag{4.34}$$

where we have introduced $q = N/T = O(1)$, such that $N^{-\frac{1}{2}}$ and $T^{-\frac{1}{2}}$ are of the same order of magnitude. This last equation states that $1/\mathbf{G}_{11}(z)$ has negligible fluctuations and can safely be replaced by its expectation value, i.e.

$$\begin{aligned}\frac{1}{(\mathbf{G}(z))_{11}} &= \mathbb{E}\left[\frac{1}{(\mathbf{G}(z))_{11}}\right] + O\left(N^{-\frac{1}{2}}\right)\\ &= \frac{1}{\mathbb{E}\left[(\mathbf{G}(z))_{11}\right]} + O\left(N^{-\frac{1}{2}}\right).\end{aligned} \tag{4.35}$$

By rotational invariance of $\mathbf{W}$, we have

$$\mathbb{E}[\mathbf{G}(z)_{11}] = \frac{1}{N}\mathbb{E}[\operatorname{Tr}(\mathbf{G}(z))] \to \mathfrak{g}(z). \tag{4.36}$$

[3] It can be self-consistently checked from the solution below that $\lim_{T\to\infty}\gamma^2 = -q\mathfrak{g}'_{\mathbf{W}}(z)$.

In the large N limit we obtain the following self-consistent equation for $\mathfrak{g}(z)$:

$$\frac{1}{\mathfrak{g}(z)} = z - 1 + q - qz\mathfrak{g}(z). \tag{4.37}$$

4.2.2 Solution and Density of Eigenvalues

Solving (4.37) we obtain

$$\mathfrak{g}(z) = \frac{z + q - 1 \pm \sqrt{(z + q - 1)^2 - 4qz}}{2qz}. \tag{4.38}$$

The argument of the square-root is quadratic in z and its roots (the edge of spectrum) are given by

$$\lambda_\pm = (1 \pm \sqrt{q})^2. \tag{4.39}$$

Finding the correct branch is quite subtle, this will be the subject of Section 4.2.3. We will see that the form

$$\mathfrak{g}(z) = \frac{z - (1 - q) - \sqrt{z - \lambda_+}\sqrt{z - \lambda_-}}{2qz} \tag{4.40}$$

has all the correct analytical properties. Note that for $z = x - i\eta$ with $x \neq 0$ and $\eta \to 0$, $g(z)$ can only have an imaginary part if $\sqrt{(x - \lambda_+)(x - \lambda_-)}$ is imaginary. Then using (2.47), we get the famous Marčenko–Pastur distribution for the bulk:

$$\rho(x) = \frac{1}{\pi} \lim_{\eta \to 0+} \operatorname{Im} \mathfrak{g}(x - i\eta) = \frac{\sqrt{(\lambda_+ - x)(x - \lambda_-)}}{2\pi qx}, \quad \lambda_- < x < \lambda_+. \tag{4.41}$$

Moreover, by studying the behavior of Eq. (4.40) near $z = 0$ one sees that there is a pole at 0 when $q > 1$. This gives a delta mass as $z \to 0$:

$$\frac{q - 1}{q}\delta(x), \tag{4.42}$$

which corresponds to the $N - T$ trivial zero eigenvalues of E in the $N > T$ case. Combining the above discussions, the full Marčenko–Pastur law can be written as

$$\rho_{\mathrm{MP}}(x) = \frac{\sqrt{[(\lambda_+ - x)(x - \lambda_-)]_+}}{2\pi qx} + \frac{q - 1}{q}\delta(x)\Theta(q - 1), \tag{4.43}$$

where we denote $[a]_+ := \max\{a, 0\}$ for any $a \in \mathbb{R}$, and

$$\Theta(q - 1) := \begin{cases} 0, & \text{if } q \leq 1, \\ 1 & \text{if } q > 1. \end{cases} \tag{4.44}$$

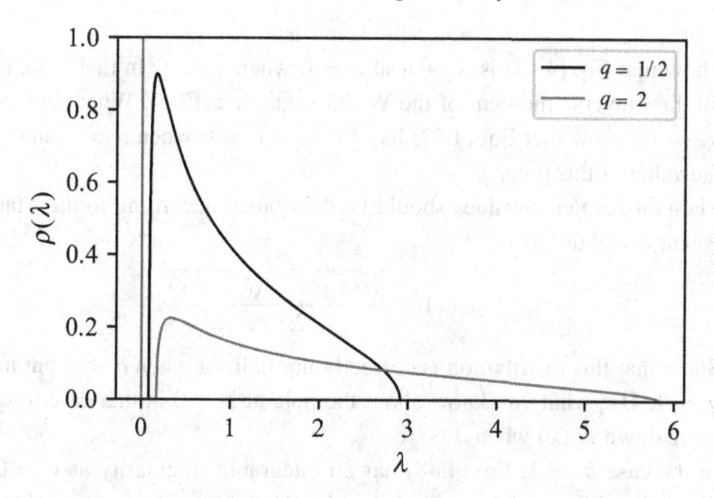

Figure 4.2 Marčenko–Pastur distribution: density of eigenvalues for a Wishart matrix for $q = 1/2$ and $q = 2$. Note that for $q = 2$ there is a Dirac mass at zero ($\frac{1}{2}\delta(\lambda)$). Also note that the two bulk densities are the same up to a rescaling and normalization $\rho_{1/q}(\lambda) = q^2 \rho_q(q\lambda)$.

Note that the Stieltjes transforms (Eq. (4.40)) for q and $1/q$ are related by Eq. (4.5). As a consequence the bulk densities for q and $1/q$ are the same when properly rescaled (see Fig. 4.2):

$$\rho_{1/q}(\lambda) = q^2 \rho_q(q\lambda). \tag{4.45}$$

Exercise 4.2.1 Properties of the Marčenko–Pastur solution

We saw that the Stieltjes transform of a large Wishart matrix (with $q = N/T$) should be given by

$$g(z) = \frac{z + q - 1 \pm \sqrt{(z + q - 1)^2 - 4qz}}{2qz}, \tag{4.46}$$

where the sign of the square-root should be chosen such that $g(z) \to 1/z$ when $z \to \pm\infty$.

(a) Show that the zeros of the argument of the square-root are given by $\lambda_\pm = (1 \pm \sqrt{q})^2$.

(b) The function

$$g(z) = \frac{z + q - 1 - \sqrt{(z - \lambda_-)}\sqrt{(z - \lambda_+)}}{2qz} \tag{4.47}$$

should have the right properties. Show that it behaves as $g(z) \to 1/z$ when $z \to \pm\infty$. By expanding in powers of $1/z$ up to $1/z^3$ compute the first and second moments of the Wishart distribution.

(c) Show that Eq. (4.47) is regular at $z = 0$ when $q < 1$. In that case, compute the first inverse moment of the Wishart matrix $\tau(\mathbf{E}^{-1})$. What happens when $q \to 1$? Show that Eq. (4.47) has a pole at $z = 0$ when $q > 1$ and compute the value of this pole.

(d) The non-zero eigenvalues should be distributed according to the Marčenko–Pastur distribution:

$$\rho_q(x) = \frac{\sqrt{(x - \lambda_-)(\lambda_+ - x)}}{2\pi q x}. \tag{4.48}$$

Show that this distribution is correctly normalized when $q < 1$ but not when $q > 1$. Use what you know about the pole at $z = 0$ in that case to correctly write down $\rho_q(x)$ when $q > 1$.

(e) In the case $q = 1$, Eq. (4.48) has an integrable singularity at $x = 0$. Write a simpler formula for $\rho_1(x)$. Let u be the square of an eigenvalue from a Wigner matrix of unit variance, i.e. $u = y^2$ where y is distributed according to the semi-circular law $\rho(y) = \sqrt{4 - y^2}/(2\pi)$. Show that u is distributed according to $\rho_1(x)$. This result is *a priori* not obvious as a Wigner matrix is symmetric while the square matrix $\mathbf{H}$ is generally not; nevertheless, moments of high-dimensional matrices of the form $\mathbf{H}\mathbf{H}^T$ are the same whether the matrix $\mathbf{H}$ is symmetric or not.

(f) Generate three matrices $\mathbf{E} = \mathbf{H}\mathbf{H}^T/T$ where the matrix $\mathbf{H}$ is an $N \times T$ matrix of IID Gaussian numbers of variance 1. Choose a large N and three values of T such that $q = N/T$ equals $\{1/2, 1, 2\}$. Plot a normalized histogram of the eigenvalues in the three cases vs the corresponding Marčenko–Pastur distribution; don't show the peak at zero. In the case $q = 2$, how many zero eigenvalues do you expect? How many do you get?

4.2.3 The Correct Root of the Stieltjes Transform

In our study of random matrices we will often encounter limiting Stieltjes transforms that are determined by quadratic or higher order polynomial equations, and the problem of choosing the correct solution (or branch) will come up repeatedly.

Let us go back to the unit Wigner matrix case where we found (see Section 2.3.2)

$$\mathfrak{g}(z) = \frac{z \pm \sqrt{z^2 - 4}}{2}. \tag{4.49}$$

On the one hand we want $\mathfrak{g}(z)$ that behaves like $1/z$ as $|z| \to \infty$ and we want the solution to be analytical everywhere but on the real axis in $[-2, 2]$. The square-root term must thus behave as $-z$ for real z when $z \to \pm\infty$. The standard definition of the square-root behaves as $\sqrt{z^2} \sim |z|$ and cannot be made to have the correct sign on both sides. Another issue with $\sqrt{z^2 - 4}$ is that it has a more extended branch cut than allowed. We expect the

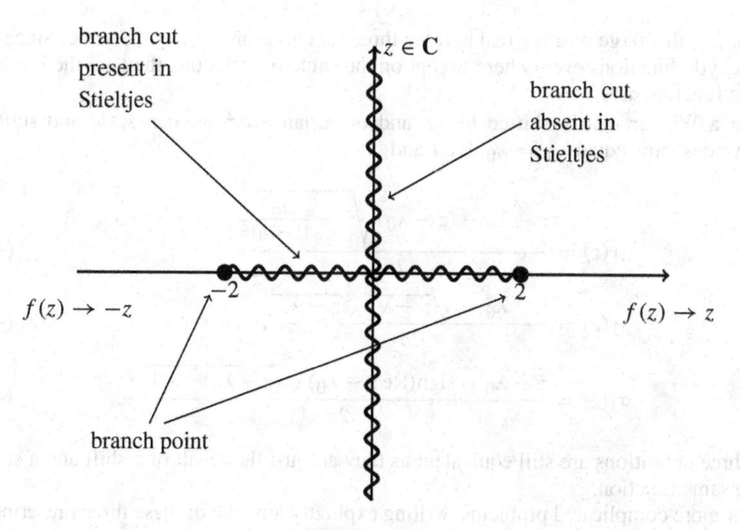

Figure 4.3 The branch cuts of $f(z) = \sqrt{z^2 - 4}$. The vertical branch cut (for z pure imaginary) should not be present in the Stieltjes transform of the Wigner ensemble. We have $f(\pm 0^+ + \mathrm{i}x) \approx \pm \mathrm{i}\sqrt{x^2 + 4}$; this branch cut can be eliminated by multiplying $f(z)$ by sign(Re z).

function $\mathfrak{g}(z)$ to be analytic everywhere except for real $z \in [-2, 2]$. The branch cut of a standard square-root is a set of points where its argument is real and negative. In the case of $\sqrt{z^2 - 4}$, this includes the interval $[-2, 2]$ as expected but also the pure imaginary line $z = \mathrm{i}x$ (Fig. 4.3). The finite N Stieltjes transform is perfectly regular on the imaginary axis so we expect its large N to be regular there as well.

For the unit Wigner matrix, there are at least three solutions to the branch problem:

$$\mathfrak{g}_1(z) = \frac{z - z\sqrt{1 - \frac{4}{z^2}}}{2}, \tag{4.50}$$

$$\mathfrak{g}_2(z) = \frac{z - \sqrt{z-2}\sqrt{z+2}}{2}, \tag{4.51}$$

$$\mathfrak{g}_3(z) = \frac{z - \mathrm{sign}(\mathrm{Re}\, z)\sqrt{z^2 - 4}}{2}. \tag{4.52}$$

All three definitions behave as $\mathfrak{g}(z) \sim 1/z$ at infinity. For the second one, we need to define the square-root of a negative real number. If we define it as $\mathrm{i}\sqrt{|z|}$, the two factors of i give a -1 for real $z < -2$. The three functions also have the correct branch cuts. For the first one, one can show that the argument of the square-root can be a negative real number only if $z \in (-2, 2)$, there are no branch cuts elsewhere in the complex plane. For the second one, there seems to be a branch cut for all real $z < 2$, but a closer inspection reveals that around real $z < -2$ the function has no discontinuity as one goes up and down the imaginary axis, as the two branch cuts exactly compensate each other. For the third one, the discontinuous sign function exactly cancels the branch cut on the pure imaginary axis (Fig. 4.3).

For z with a large positive real part the three functions are clearly the same. Since they are analytic functions everywhere except on the same branch cuts, they are the same and unique function $\mathfrak{g}(z)$.

For a Wigner matrix shifted by λ_0 and of variance σ^2 we can scale and shift the eigenvalues, now equal $\lambda_\pm = \lambda_0 \pm 2\sigma$ and find

$$\mathfrak{g}_1(z) = \frac{z - \lambda_0 - (z - \lambda_0)\sqrt{1 - \frac{4\sigma^2}{(z-\lambda_0)^2}}}{2\sigma}, \tag{4.53}$$

$$\mathfrak{g}_2(z) = \frac{z - \lambda_0 - \sqrt{z - \lambda_+}\sqrt{z - \lambda_-}}{2\sigma}, \tag{4.54}$$

$$\mathfrak{g}_3(z) = \frac{z - \lambda_0 - \text{sign}(\text{Re}\, z - \lambda_0)\sqrt{(z - \lambda_0)^2 - 4\sigma^2}}{2\sigma}. \tag{4.55}$$

The three definitions are still equivalent as they are just the result of a shift and a scaling of the same function.

For more complicated problems, writing explicitly any one of these three prescriptions can quickly become very cumbersome (except maybe in cases where $\lambda_+ + \lambda_- = 0$). We propose here a new notation. When finding the correct square-root of a second degree polynomial we will write

$$\begin{aligned}\sqrt[\oplus]{az^2 + bz + c} :&= \sqrt{a}\sqrt{z - \lambda_+}\sqrt{z - \lambda_-} \\ &= \sqrt{a}(z - \lambda_0)\sqrt{1 - \frac{\Delta}{(z - \lambda_0)^2}} \\ &= \text{sign}(\text{Re}\, z - \lambda_0)\sqrt{az^2 + bz + c},\end{aligned} \tag{4.56}$$

for $a > 0$ and where $\lambda_\pm = \lambda_0 \pm \sqrt{\Delta}$ are the roots of $az^2 + bz + c$ assumed to be real. While the notation is defined everywhere in the complex plane, it is easily evaluated for real arguments:

$$\sqrt[\oplus]{ax^2 + bx + c} = \begin{cases} -\sqrt{ax^2 + bx + c} & \text{for } x \le \lambda_-, \\ \sqrt{ax^2 + bx + c} & \text{for } x \ge \lambda_+. \end{cases} \tag{4.57}$$

The value on the branch cut is ill-defined but we have

$$\lim_{z \to x - \mathrm{i}0^+} \sqrt[\oplus]{az^2 + bz + c} = \mathrm{i}\sqrt{|ax^2 + bx + c|} \quad \text{for } \lambda_- < x < \lambda_+. \tag{4.58}$$

With our new notation, we can now safely write for the white Wishart:

$$\mathfrak{g}(z) = \frac{z + q - 1 - \sqrt[\oplus]{(z + q - 1)^2 - 4qz}}{2qz}, \tag{4.59}$$

or, more explicitly, using the second prescription:

$$\mathfrak{g}(z) = \frac{z + q - 1 - \sqrt{z - \lambda_+}\sqrt{z - \lambda_-}}{2qz}, \tag{4.60}$$

where $\lambda_\pm = (1 \pm \sqrt{q})^2$.

Exercise 4.2.2 Finding the correct root

(a) For the unit Wigner Stieltjes transform show that regardless of choice of sign in Eq. (4.49) the point $z = 2\mathrm{i}$ is located on a branch cut and the function is discontinuous at that point.

(b) Compute the value of Eqs. (4.50), (4.51) and (4.52) at $z = 2\mathrm{i}$. Hint: for $\mathfrak{g}_2(z)$ write $-2+2\mathrm{i} = \sqrt{8}\mathrm{e}^{3\mathrm{i}\pi/4}$ and similarly for $2+2\mathrm{i}$. The definition $\mathfrak{g}_3(z)$ is ambiguous for $z = 2\mathrm{i}$, compute the limiting value on both sides: $z = 0^+ + 2\mathrm{i}$ and $z = 0^- + 2\mathrm{i}$.

4.2.4 General (Non-White) Wishart Matrices

Recall our definition of a Wishart matrix from Section 4.1.2: a Wishart matrix is a matrix $\mathbf{E_C}$ defined as

$$\mathbf{E_C} = \frac{1}{T}\mathbf{H_C}\mathbf{H_C}^T, \tag{4.61}$$

where $\mathbf{H_C}$ is an $N \times T$ rectangular matrix with independent columns. Each column is a random Gaussian vector with covariance matrix $\mathbf{C}$; $\mathbf{E_C}$ corresponds to the sample (empirical) covariance matrix of variables characterized by a population (true) covariance matrix $\mathbf{C}$.

To understand the case where the true matrix $\mathbf{C}$ is different from the identity we first discuss how to generate a multivariate Gaussian vector with covariance matrix $\mathbf{C}$. We diagonalize $\mathbf{C}$ as

$$\mathbf{C} = \mathbf{O}\Lambda\mathbf{O}^T, \quad \Lambda = \begin{pmatrix} \sigma_1^2 & & \\ & \ddots & \\ & & \sigma_N^2 \end{pmatrix}. \tag{4.62}$$

The square-root of $\mathbf{C}$ can be defined as[4]

$$\mathbf{C}^{\frac{1}{2}} = \mathbf{O}\Lambda^{\frac{1}{2}}\mathbf{O}^T, \quad \Lambda^{\frac{1}{2}} = \begin{pmatrix} \sigma_1 & & \\ & \ddots & \\ & & \sigma_N \end{pmatrix}. \tag{4.63}$$

We now generate N IID unit Gaussian random variables x_i, $1 \le i \le N$, which form a random column vector $\mathbf{x}$ with entries x_i. Then we can generate the vector $\mathbf{y} = \mathbf{C}^{\frac{1}{2}}\mathbf{x}$. We claim that $\mathbf{y}$ is a multivariate Gaussian vector with covariance matrix $\mathbf{C}$. In fact, $\mathbf{y}$ is a linear combination of multivariate Gaussians, so it must itself be multivariate Gaussian. On the other hand, we have, using $\mathbb{E}[\mathbf{x}\mathbf{x}^T] = \mathbf{1}$,

[4] This is the canonical definition of the square-root of a matrix, but this definition is not unique – see the technical paragraph below.

$$\mathbb{E}(\mathbf{y}\mathbf{y}^T) = \mathbb{E}(\mathbf{C}^{\frac{1}{2}}\mathbf{x}\mathbf{x}^T\mathbf{C}^{\frac{1}{2}}) = \mathbf{C}. \tag{4.64}$$

By repeating the argument above for every column of $\mathbf{H}_\mathbf{C}$, $t = 1, \dots, T$, we see that this matrix can be written as $\mathbf{H}_\mathbf{C} = \mathbf{C}^{\frac{1}{2}}\mathbf{H}$, with $\mathbf{H}$ a rectangular matrix with IID unit Gaussian entries. The matrix $\mathbf{E}_\mathbf{C}$ is then equivalent to

$$\mathbf{E}_\mathbf{C} = \frac{1}{T}\mathbf{H}_\mathbf{C}\mathbf{H}_\mathbf{C}^T = \frac{1}{T}\mathbf{C}^{\frac{1}{2}}\mathbf{H}\mathbf{H}^T\mathbf{C}^{\frac{1}{2}} = \mathbf{C}^{\frac{1}{2}}\mathbf{W}_q\mathbf{C}^{\frac{1}{2}}, \tag{4.65}$$

where $\mathbf{W}_q = \frac{1}{T}\mathbf{H}\mathbf{H}^T$ is a white Wishart matrix with $q = N/T$.

We will see later that the above combination of matrices is called the *free product* of $\mathbf{C}$ and $\mathbf{W}$. Free probability will allow us to compute the resolvent and the spectrum in the case of a general $\mathbf{C}$ matrix.

The variables $\mathbf{x}$ defined above are called the "whitened" version of $\mathbf{y}$. If a zero mean random vector $\mathbf{y}$ has positive definite covariance matrix $\mathbf{C}$, we can define a *whitening* of $\mathbf{y}$ as a linear combination $\mathbf{x} = \mathbf{M}\mathbf{y}$ such that $\mathbb{E}[\mathbf{x}\mathbf{x}^T] = \mathbf{1}$. One can show that the matrix $\mathbf{M}$ satisfies $\mathbf{M}^T\mathbf{M} = \mathbf{C}^{-1}$ and has to be of the form $\mathbf{M} = \mathbf{O}\mathbf{C}^{-\frac{1}{2}}$, where $\mathbf{O}$ can be any orthogonal matrix and $\mathbf{C}^{\frac{1}{2}}$ the symmetric square-root of $\mathbf{C}$ defined above. Since $\mathbf{O}$ is arbitrary the procedure is not unique, which leads to three interesting choices for whitened varaibles:

- Perhaps the most natural one is the symmetric or *Mahalanobis* whitening where $\mathbf{M} = \mathbf{C}^{-\frac{1}{2}}$. In addition to being the only whitening scheme with a symmetric matrix $\mathbf{M}$, the white variables $\mathbf{x} = \mathbf{C}^{-\frac{1}{2}}\mathbf{y}$ are the closest to $\mathbf{y}$ in the following sense: the distance

$$||\mathbf{x} - \mathbf{y}||_{\mathbf{C}^\alpha} := \mathbb{E}\,\mathrm{Tr}\left[(\mathbf{x} - \mathbf{y})^T\mathbf{C}^\alpha(\mathbf{x} - \mathbf{y})\right] \tag{4.66}$$

  is minimal over all other choices of $\mathbf{O}$ for any α. The case $\alpha = -1$ is called the Mahalanobis norm.
- Triangular or Gram–Schmidt whitening where the vector $\mathbf{x}$ can be constructed using the Gram–Schmidt orthonormalization procedure. If one starts from the bottom with $x_N = y_N/\sqrt{C_{NN}}$, then the matrix $\mathbf{M}$ is upper triangular. The matrix $\mathbf{M}$ can be computed efficiently using the Cholesky decomposition of $\mathbf{C}^{-1}$. The Cholesky decomposition of a symmetric positive definite matrix $\mathbf{A}$ amounts to finding a lower triangular matrix $\mathbf{L}$ such that $\mathbf{L}\mathbf{L}^T = \mathbf{A}$. In the present case, $\mathbf{A} = \mathbf{C}^{-1}$ and the matrix $\mathbf{M}$ we are looking for is given by

$$\mathbf{M} = \mathbf{L}^T. \tag{4.67}$$

  This scheme has the advantage that the whitened variable x_k only depends on physical variables y_ℓ for $\ell \geq k$. In finance, for example, this allows one to construct whitened returns of a given stock using only the returns of itself and those of (say) more liquid stocks.
- Eigenvalue (or PCA) whitening where $\mathbf{O}$ corresponds to the eigenbasis of $\mathbf{C}$, i.e. such that $\mathbf{C} = \mathbf{O}\Lambda\mathbf{O}^T$ where Λ is diagonal. $\mathbf{M}$ is then computed as $\mathbf{M} = \Lambda^{-\frac{1}{2}}\mathbf{O}^T$. The whitened variables $\mathbf{x}$ are then called the normalized principal components of $\mathbf{y}$.

Bibliographical Notes

- For a historical perspective on Wishart matrices and the Marčenko–Pastur, see
  - J. Wishart. The generalised product moment distribution in samples from a normal multivariate population. *Biometrika*, 20A(1-2):32–52, 1928,
  - V. A. Marchenko and L. A. Pastur. Distribution of eigenvalues for some sets of random matrices. *Matematicheskii Sbornik*, 114(4):507–536, 1967.
- For more recent material on the content of this chapter, see e.g.
  - L. Pastur and M. Scherbina. *Eigenvalue Distribution of Large Random Matrices*. American Mathematical Society, Providence, Rhode Island, 2010,
  - Z. Bai and J. W. Silverstein. *Spectral Analysis of Large Dimensional Random Matrices*. Springer-Verlag, New York, 2010,
  - A. M. Tulino and S. Verdú. *Random Matrix Theory and Wireless Communications*. Now publishers, Hanover, Mass., 2004,
  - R. Couillet and M. Debbah. *Random Matrix Methods for Wireless Communications*. Cambridge University Press, Cambridge, 2011,
  - G. Livan, M. Novaes, and P. Vivo. *Introduction to Random Matrices: Theory and Practice*. Springer, New York, 2018,

  and also
  - J. W. Silverstein and S.-I. Choi. Analysis of the limiting spectral distribution of large dimensional random matrices. *Journal of Multivariate Analysis*, 54(2):295–309, 1995,
  - A. Sengupta and P. P. Mitra. Distributions of singular values for some random matrices. *Physical Review E*, 60(3):3389, 1999.

- A historical remark on the Cholesky decomposition: André-Louis Cholesky served in the French military as an artillery officer and was killed in battle a few months before the end of World War I; his discovery was published posthumously by his fellow officer Commandant Benoît in the *Bulletin Géodésique* (Wikipedia).

5
Joint Distribution of Eigenvalues

In the previous chapters, we have studied the moments, the Stieltjes transform and the eigenvalue density of two classical ensembles (Wigner and Wishart). These quantities in fact relate to *single eigenvalue* properties of these ensembles. By this we mean that the Stieltjes transform and the eigenvalue density are completely determined by the univariate law of eigenvalues but they do not tell us anything about the *correlations* between different eigenvalues.

In this chapter we will extend these results in two directions. First we will consider a larger class of rotationally invariant (or orthogonal) ensembles that contains Wigner and Wishart. Second we will study the joint law of all eigenvalues. In these models, the eigenvalues turn out to be strongly correlated and can be thought of as "particles" interacting through pairwise repulsion.

5.1 From Matrix Elements to Eigenvalues

5.1.1 Matrix Potential

Consider real symmetric random matrices $\mathbf{M}$ whose elements are distributed as the exponential of the trace of a certain matrix function $V(\mathbf{M})$, often called a potential by analogy with statistical physics:[1]

$$P(\mathbf{M}) = Z_N^{-1} \exp\left\{-\frac{N}{2} \operatorname{Tr} V(\mathbf{M})\right\}, \tag{5.1}$$

where Z_N is a normalization constant. These matrix ensembles are called orthogonal ensembles for they are rotationally invariant, i.e. invariant under orthogonal transformations.[2] For the Wigner ensemble, for example, we have (see Chapter 2)

[1] $V(\mathbf{M})$ is a matrix function best defined in the eigenbasis of $\mathbf{M}$ through a transformation of all its eigenvalues through a function of a scalar, $V(x)$, see Section 1.2.6.

[2] The results of this chapter extend to Hermitian ($\beta = 2$) or quarternion-Hermitian ($\beta = 4$) matrices with the simple introduction of a factor β in the probability distribution:

$$P(\mathbf{M}) \propto \exp\left\{-\frac{\beta N}{2} \operatorname{Tr} V(\mathbf{M})\right\}, \tag{5.2}$$

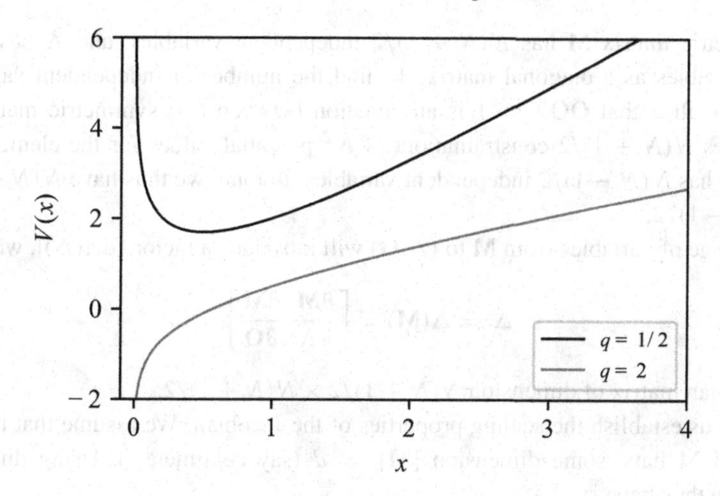

Figure 5.1 The Wishart matrix potential (Eq. (5.4)) for $q = 1/2$ and $q = 2$. The integration over positive semi-definite matrices imposes that the eigenvalues must be greater than or equal to zero. For $q < 1$ the potential naturally ensures that the eigenvalues are greater than zero and the constraint will not be explicitly needed in the computation. For $q \geq 1$, the constraint is needed to obtain a sensible result.

$$V(x) = \frac{x^2}{2\sigma^2}, \tag{5.3}$$

whereas the Wishart ensemble (at large N) is characterized by (see Chapter 4)

$$V(x) = \frac{x + (q-1)\log x}{q} \tag{5.4}$$

(see Fig. 5.1). We can also consider other matrix potentials, e.g.

$$V(x) = \frac{x^2}{2} + \frac{gx^4}{4}. \tag{5.5}$$

Note that $\operatorname{Tr} V(\mathbf{M})$ depends only on the eigenvalues of $\mathbf{M}$. We would like thus to write down the joint distribution of these eigenvalues alone. The key is to find the Jacobian of the change of variables from the entries of $\mathbf{M}$ to the eigenvalues $\{\lambda_1, \ldots, \lambda_N\}$.

5.1.2 Matrix Jacobian

Before computing the Jacobian of the transformation from matrix elements to eigenvalues and eigenvectors (or orthogonal matrices), let us count the number of variables in both parameterizations. Suppose $\mathbf{M}$ can be diagonalized as

$$\mathbf{M} = \mathbf{O}\Lambda\mathbf{O}^T. \tag{5.6}$$

this factor will match the factor of β from the Vandermonde determinant. These two other ensembles are called unitary ensembles and symplectic ensembles, respectively. Collectively they are called the beta ensembles.

The symmetric matrix $\mathbf{M}$ has $N(N+1)/2$ independent variables, and Λ has N independent variables as a diagonal matrix. To find the number of independent variables in $\mathbf{O}$ we first realize that $\mathbf{O}\mathbf{O}^T = \mathbf{1}$ is an equation between two symmetric matrices and thus imposes $N(N+1)/2$ constraints out of N^2 potential values for the elements of $\mathbf{O}$, therefore $\mathbf{O}$ has $N(N-1)/2$ independent variables. In total, we thus have $N(N+1)/2 = N + N(N-1)/2$.

The change of variables from $\mathbf{M}$ to $(\Lambda, \mathbf{O})$ will introduce a factor $|\det(\Delta)|$, where

$$\Delta := \Delta(\mathbf{M}) = \left[\frac{\partial \mathbf{M}}{\partial \Lambda}, \frac{\partial \mathbf{M}}{\partial \mathbf{O}}\right] \tag{5.7}$$

is the Jacobian matrix of dimension $N(N+1)/2 \times N(N+1)/2$.

First, let us establish the scaling properties of the Jacobian. We assume that the matrix elements of $\mathbf{M}$ have some dimension $[\mathbf{M}] = d$ (say centimeters). Using dimensional analysis, we thus have

$$[\mathcal{D}\mathbf{M}] = d^{N(N+1)/2}, \quad [\mathcal{D}\Lambda] = d^N, \quad [\mathcal{D}\mathbf{O}] = d^0, \tag{5.8}$$

since rotations are dimensionless. Hence we must have

$$[|\det(\Delta)|] \sim d^{N(N-1)/2}, \tag{5.9}$$

which has the dimension of an eigenvalue raised to the power $N(N-1)/2$, the number of distinct off-diagonal elements in $\mathbf{M}$.

We now compute this Jacobian exactly. First, notice that the Jacobian relates the volume "around" $(\Lambda, \mathbf{O})$ when Λ and $\mathbf{O}$ change by infinitesimal amounts, to the volume "around" $\mathbf{M}$ when its elements change by infinitesimal amounts. We note that volumes are invariant under rotations, so in order to compute the infinitesimal volume we can choose the rotation matrix $\mathbf{O}$ to be the identity matrix, which amounts to saying that we work in the basis where $\mathbf{M}$ is diagonal. Another way to see this is to note that the orthogonal transformation

$$\mathbf{M} \to \mathbf{U}^T\mathbf{M}\mathbf{U}; \qquad \mathbf{U}^T\mathbf{U} = \mathbf{1} \tag{5.10}$$

has a Jacobian equal to 1, see Section 1.2.7. One can always choose $\mathbf{U}$ such that $\mathbf{M}$ is diagonal.

5.1.3 Infinitesimal Rotations

For rotations $\mathbf{O}$ near the identity, we set

$$\mathbf{O} = \mathbf{1} + \epsilon\, \delta\mathbf{O}, \tag{5.11}$$

where ϵ is a small number and $\delta\mathbf{O}$ is some matrix. From the identity

$$\mathbf{1} = \mathbf{O}\mathbf{O}^T = \mathbf{1} + \epsilon(\delta\mathbf{O} + \delta\mathbf{O}^T) + \epsilon^2 \delta\mathbf{O}\delta\mathbf{O}^T, \tag{5.12}$$

we get $\delta\mathbf{O} = -\delta\mathbf{O}^T$ by comparing terms of the first order in ϵ, i.e. $\delta\mathbf{O}$ is skew-symmetric.[3] A convenient basis to write such infinitesimal rotations is

$$\epsilon\,\delta\mathbf{O} = \sum_{1\le k<l\le N} \theta_{kl}\mathbf{A}^{(kl)}, \tag{5.13}$$

where $\mathbf{A}^{(kl)}$ are the elementary skew-symmetric matrices such that $\mathbf{A}^{(kl)}$ has only two non-zero elements: $[\mathbf{A}^{(kl)}]_{kl} = 1$ and $[\mathbf{A}^{(kl)}]_{lk} = -1$:

$$\mathbf{A}^{(kl)} = \begin{pmatrix} 0 & \cdots & & & \cdots & 0 \\ \vdots & \ddots & & 1 & & \vdots \\ & & & & & \\ & & -1 & & & \\ \vdots & & & & \ddots & \vdots \\ 0 & \cdots & & & \cdots & 0 \end{pmatrix}. \tag{5.14}$$

An infinitesimal rotation is therefore fully described by $N(N-1)/2$ generalized "angles" θ_{kl}.

5.1.4 Vandermonde Determinant

Now, in the neighborhood of $(\Lambda, \mathbf{O} = \mathbf{1})$, the matrix $\mathbf{M} + \delta\mathbf{M}$ can be parameterized as

$$\mathbf{M} + \delta\mathbf{M} \approx \left(\mathbf{1} + \sum_{k,l} \theta_{kl}\mathbf{A}^{(kl)}\right)(\Lambda + \delta\Lambda)\left(\mathbf{1} - \sum_{k,l} \theta_{kl}\mathbf{A}^{(kl)}\right). \tag{5.15}$$

So to first order in $\delta\Lambda$ and θ_{kl},

$$\delta\mathbf{M} \approx \delta\Lambda + \sum_{k,l} \theta_{kl}\left[\mathbf{A}^{(kl)}\Lambda - \Lambda\mathbf{A}^{(kl)}\right]. \tag{5.16}$$

Using this local parameterization, we can compute the Jacobian matrix and find its determinant. For the diagonal contribution, we have

$$\frac{\partial M_{ij}}{\partial \Lambda_{nn}} = \delta_{in}\delta_{jn}, \tag{5.17}$$

i.e. perturbing a given eigenvalue only changes the corresponding diagonal element with slope 1.

For the rotation contribution, one has, for $k < l$ and $i < j$,

[3] The reader familiar with the analysis of compact Lie groups will recognize the statement that skew-symmetric matrices form the Lie algebra of $O(N)$.

$$\frac{\partial M_{ij}}{\partial \theta_{kl}} = \left(\mathbf{A}^{(kl)}\Lambda - \Lambda\mathbf{A}^{(kl)}\right)_{ij} = \begin{cases} \lambda_l - \lambda_k, & \text{if } i = k, j = l, \\ 0, & \text{otherwise.} \end{cases} \tag{5.18}$$

i.e. an infinitesimal rotation in the direction kl modifies only one distinct off-diagonal element ($\mathbf{M}_{kl} \equiv \mathbf{M}_{lk}$) with slope $\lambda_l - \lambda_k$. In particular, if two eigenvalues are the same ($\lambda_k = \lambda_l$) a rotation of the eigenvectors in that subspace has no effect on the matrix $\mathbf{M}$. This is expected since eigenvectors in a degenerate subspace are only defined up to a rotation within that subspace.

Finally, the $N(N+1)/2 \times N(N+1)/2$ determinant has its first N diagonal elements equal to unity, and the next $N(N-1)/2$ are equal to all possible pair differences $\lambda_i - \lambda_j$. Hence,

$$\Delta(\mathbf{M}) = \det \begin{pmatrix} 1 & & & & & & \\ & \ddots & & & & & \\ & & 1 & & & & \\ & & & \lambda_2 - \lambda_1 & & & \\ & & & & \lambda_3 - \lambda_1 & & \\ & & & & & \ddots & \\ & & & & & & \lambda_N - \lambda_{N-1} \end{pmatrix} = \prod_{k<\ell} (\lambda_\ell - \lambda_k). \tag{5.19}$$

The absolute value of Δ is then given by

$$|\Delta(\mathbf{M})| = \prod_{k<\ell} |\lambda_\ell - \lambda_k|. \tag{5.20}$$

We can check that this result has the expected dimension $d^{N(N-1)/2}$, since the product contains exactly $N(N-1)/2$ terms. The determinant $\Delta(\mathbf{M})$ is called the Vandermonde determinant as it is equal to the determinant of the following $N \times N$ Vandermonde matrix:

$$\begin{pmatrix} 1 & 1 & 1 & \dots & 1 \\ \lambda_1 & \lambda_2 & \lambda_3 & \dots & \lambda_N \\ \lambda_1^2 & \lambda_2^2 & \lambda_3^2 & \dots & \lambda_N^2 \\ \vdots & \vdots & \vdots & \ddots & \vdots \\ \lambda_1^{N-1} & \lambda_2^{N-1} & \lambda_3^{N-1} & \dots & \lambda_N^{N-1} \end{pmatrix}. \tag{5.21}$$

Since the above Jacobian has no dependence on the matrix $\mathbf{O}$, we can integrate out the rotation part of (5.1) to get the joint distribution of eigenvalues:

$$P(\{\lambda_i\}) \propto \prod_{k<l} |\lambda_k - \lambda_l| \exp\left\{-\frac{N}{2}\sum_{i=1}^{N} V(\lambda_i)\right\}. \tag{5.22}$$

A key feature of the above probability density is that the eigenvalues are not independent, since the term $\prod_{k<l} |\lambda_k - \lambda_l|$ indicates that the probability density vanishes when two eigenvalues tend towards one another. This can be interpreted as some effective "repulsion" between eigenvalues, as we will expand on now using an analogy with Coulomb gases.

Exercise 5.1.1 Vandermonde determinant for 2 × 2 matrices

In this exercise we will explicitly compute the Vandermonde determinant for 2×2 matrices. We define $\mathbf{O}$ and Λ as

$$\mathbf{O} = \begin{pmatrix} \cos(\theta) & \sin(\theta) \\ -\sin(\theta) & \cos(\theta) \end{pmatrix} \text{ and } \Lambda = \begin{pmatrix} \lambda_1 & 0 \\ 0 & \lambda_2 \end{pmatrix}. \tag{5.23}$$

Then any 2×2 symmetric matrix can be written as $\mathbf{M} = \mathbf{O}\Lambda\mathbf{O}^T$.

(a) Write explicitly $\mathbf{M}_{11}$, $\mathbf{M}_{12}$ and $\mathbf{M}_{22}$ as a function of λ_1, λ_2 and θ.

(b) Compute the 3×3 matrix Δ of partial derivatives of $\mathbf{M}_{11}$, $\mathbf{M}_{12}$ and $\mathbf{M}_{22}$ with respect to λ_1, λ_2 and θ.

(c) In the special cases where θ equals 0, $\pi/4$ and $\pi/2$ show that $|\det \Delta| = |\lambda_1 - \lambda_2|$. If you have the courage show that $|\det \Delta| = |\lambda_1 - \lambda_2|$ for all θ.

Exercise 5.1.2 Wigner surmise

Wigner was interested in the distribution of energy level spacings in heavy nuclei, which he modeled as the eigenvalues of a real symmetric random matrix (time reversal symmetry imposes that the Hamiltonian be real). Let $x = |\lambda_{k+1} - \lambda_k|$ for k in the bulk. In principle we can obtain the probability density of x by using Eq. (5.22) and integrating out all other variables. In practice it is very difficult to go much beyond $N = 2$. Since the $N = 2$ result (properly normalized) has the correct small x and large x behavior, Wigner surmised that it must be a good approximation at any N.

(a) For an $N = 2$ GOE matrix, i.e. $V(\lambda) = \lambda^2/2\sigma^2$, write the unnormalized law of λ_1 and λ_2, its two eigenvalues.

(b) Change variables to $\lambda_\pm = \lambda_2 \pm \lambda_1$, integrate out λ_+ and write the unnormalized law of $x = |\lambda_-|$.

(c) Normalize your law and choose σ such that $\mathbb{E}[x] = 1$; you should find

$$P(x) = \frac{\pi}{2} x \exp\left[-\frac{\pi}{4}x^2\right]. \tag{5.24}$$

(d) Using Eq. (5.26) redo the computation for GUE ($\beta = 2$). You should find

$$P(x) = \frac{32}{\pi^2} x^2 \exp\left[-\frac{4}{\pi}x^2\right]. \tag{5.25}$$

5.2 Coulomb Gas and Maximum Likelihood Configurations

5.2.1 A Coulomb Gas Analogy

The orthogonal ensemble defined in the previous section can be generalized to complex or quaternion Hermitian matrices. The corresponding joint distribution of eigenvalues is simply obtained by adding a factor of β (equal to 1, 2 or 4) to both the potential and the Vandermonde determinant:

$$P(\{\lambda_i\}) = Z_N^{-1}\left[\prod_{k<l}|\lambda_k-\lambda_l|^\beta\right]\exp\left\{-\frac{\beta}{2}\left[\sum_{i=1}^N NV(\lambda_i)\right]\right\}$$
$$= Z_N^{-1}\exp\left\{-\frac{\beta}{2}\left[\sum_{i=1}^N NV(\lambda_i)-\sum_{\substack{i,j=1\\ j\neq i}}^N \log|\lambda_i-\lambda_j|\right]\right\}. \tag{5.26}$$

This joint law is exactly the Gibbs–Boltzmann factor ($e^{-E/T}$) for a gas of N particles moving on a one-dimensional line, at temperature $T = 2/\beta$, whose potential energy is given by $NV(x)$ and that interact with each other via a pairwise repulsive force generated by the potential $V_R(x,y) = -\log(|x-y|)$. Formally, the repulsive term happens to be the Coulomb potential in two dimensions for particles that all have the same sign. In a truly one-dimensional problem, the Coulomb potential would read $V_{1d}(x,y) = -|x-y|$, but with a slight abuse of language one speaks about the eigenvalues of a random matrix as a *Coulomb gas* (in one dimension).

Even though we are interested in one particular value of β (namely $\beta = 1$), we can build an intuition by considering this system at various temperatures. At very low temperature (i.e. $\beta \to \infty$), the N particles all want to minimize their potential energy and sit at the minimum of $NV(x)$, but if they try to do so they will have to pay a high price in interaction energy as this energy increases as the particles get close to one another. The particles will have to spread themselves around the minimum of $NV(x)$ to minimize the sum of the potential and interaction energy and find the configuration corresponding to "mechanical equilibrium", i.e. such that the total force on each particle is zero. At non-zero temperature (finite β) the particles will fluctuate around this equilibrium solution. Since the repulsion energy diverges as any two eigenvalues get infinitely close, the particles will always avoid each other. Figure 5.2 shows a typical configuration of particles/eigenvalues for $N = 20$ at $\beta = 1$ in a quadratic potential (GOE matrix).

In the next section, we will study this equilibrium solution, which is exact at low temperature $\beta \to \infty$ or when $N \to \infty$, and is the maximum likelihood solution at finite β and finite N.

5.2.2 Maximum Likelihood Configuration and Stieltjes Transform

In the previous section, we saw that in the Coulomb gas analogy, $\beta \to \infty$ corresponds to the zero temperature limit, and that in this limit the eigenvalues *freeze* to the minimum of

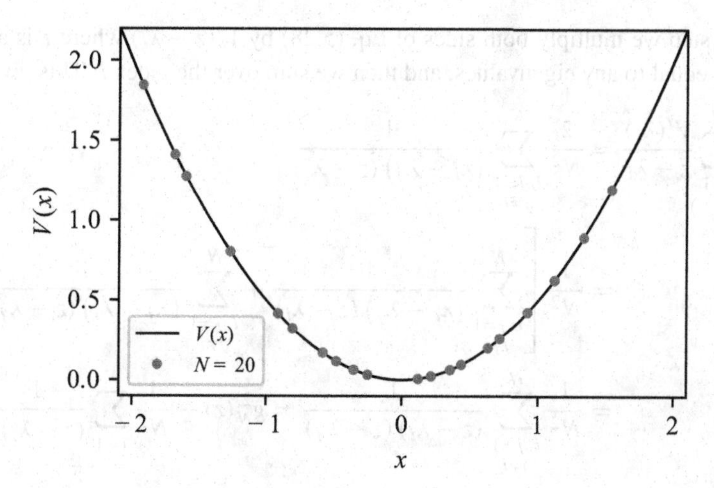

Figure 5.2 Representation of a typical $N = 20$ GOE matrix as a Coulomb gas. The full curve represents the potential $V(x) = x^2/2$ and the 20 dots, the positions of the eigenvalues of a typical configuration. In this analogy, the eigenvalues feel a potential $NV(x)$ and a repulsive pairwise interaction $V(x,y) = -\log(|x - y|)$. They fluctuate according to the Boltzmann weight $e^{-\beta E/2}$ with $\beta = 1$ in the present case.

the energy (potential plus interaction). We will argue that this freezing (or *concentration of the equilibrium measure*) also happens when $N \to \infty$ for fixed β.

Let us study the minimum energy configuration. We can rewrite Eq. (5.26) as

$$P(\{\lambda_i\}) \propto e^{\frac{1}{2}\beta N \mathcal{L}(\{\lambda_i\})}, \quad \mathcal{L}(\{\lambda_i\}) = -\sum_{i=1}^{N} V(\lambda_i) + \frac{1}{N} \sum_{\substack{i,j=1 \\ j \neq i}}^{N} \log|\lambda_i - \lambda_j|, \tag{5.27}$$

where $\mathcal{L}$ stands for "log-likelihood". For finite N and finite β, we can still consider the solution that maximizes $\mathcal{L}(\{\lambda_i\})$. This is the maximum likelihood solution, i.e. the configuration of $\{\lambda_i\}$ that has maximum probability. The maximum of $\mathcal{L}$ is determined by the equations

$$\frac{\partial \mathcal{L}}{\partial \lambda_i} = 0 \Rightarrow V'(\lambda_i) = \frac{2}{N} \sum_{\substack{j=1 \\ j \neq i}}^{N} \frac{1}{\lambda_i - \lambda_j}. \tag{5.28}$$

These are N coupled equations of N variables which can get very tedious to solve even for moderate values of N. In Exercise 5.2.1 we will find the solution for $N = 3$ in the Wigner case. The solution of these equations is the set of equilibrium positions of all the eigenvalues, i.e. the set of eigenvalues that maximizes the joint probability. To characterize this solution (which will allow us to obtain the density of eigenvalues), we will compute the Stieltjes transform of the $\{\lambda_i\}$ satisfying Eq. (5.28). The trick is to make algebraic manipulations to both sides of the equation to make the Stieltjes transform explicitly appear.

In a first step we multiply both sides of Eq. (5.28) by $1/(z-\lambda_i)$ where z is a complex variable not equal to any eigenvalues, and then we sum over the index i. This gives

$$\begin{aligned}\frac{1}{N}\sum_{i=1}^{N}\frac{V'(\lambda_i)}{z-\lambda_i} &= \frac{2}{N^2}\sum_{\substack{i,j=1\\ j\neq i}}^{N}\frac{1}{(\lambda_i-\lambda_j)(z-\lambda_i)}\\ &= \frac{1}{N^2}\left[\sum_{\substack{i,j=1\\ j\neq i}}^{N}\frac{1}{(\lambda_i-\lambda_j)(z-\lambda_i)}+\sum_{\substack{i,j=1\\ j\neq i}}^{N}\frac{1}{(\lambda_j-\lambda_i)(z-\lambda_j)}\right]\\ &= \frac{1}{N^2}\sum_{\substack{i,j=1\\ j\neq i}}^{N}\frac{1}{(z-\lambda_i)(z-\lambda_j)} = g_N^2(z)-\frac{1}{N^2}\sum_{i=1}^{N}\frac{1}{(z-\lambda_i)^2}\\ &= g_N^2(z)+\frac{g_N'(z)}{N},\end{aligned} \tag{5.29}$$

where $g_N(z)$ is the Stieltjes transform at finite N:

$$g_N(z) := \frac{1}{N}\sum_{i=1}^{N}\frac{1}{z-\lambda_i}. \tag{5.30}$$

We still need to handle the left hand side of the above equation. First we add and subtract $V'(z)$ on the numerator, yielding

$$\frac{1}{N}\sum_{i=1}^{N}\frac{V'(\lambda_i)}{z-\lambda_i} = V'(z)g_N(z)-\frac{1}{N}\sum_{i=1}^{N}\frac{V'(z)-V'(\lambda_i)}{z-\lambda_i} = V'(z)g_N(z)-\Pi_N(z), \tag{5.31}$$

where we have defined a new function $\Pi_N(z)$ as

$$\Pi_N(z) := \frac{1}{N}\sum_{i=1}^{N}\frac{V'(z)-V'(\lambda_i)}{z-\lambda_i}. \tag{5.32}$$

This does not look very useful as the equation for $g_N(z)$ will depend on some unknown function $\Pi_N(z)$ that depends on the eigenvalues whose statistics we are trying to determine. The key realization is that if $V'(z)$ is a polynomial of degree k then $\Pi_N(z)$ is also a polynomial and it has degree $k-1$. Indeed, for each i in the sum, $V'(z)-V'(\lambda_i)$ is a degree k polynomial having $z=\lambda_i$ as a zero, so $(V'(z)-V'(\lambda_i))/(z-\lambda_i)$ is a polynomial of degree $k-1$. $\Pi_N(z)$ is the sum of such polynomials so is itself a polynomial of degree $k-1$.

In fact, the argument is easy to generalize to the Laurent polynomials, i.e. such that $z^k V'(z)$ is a polynomial for some $k\in\mathbb{N}$. For example, in the Wishart case we have a Laurent polynomial

$$V'(z) = \frac{1}{q}\left(1+\frac{q-1}{z}\right). \tag{5.33}$$

Nevertheless, from now on we make the assumption that $V'(z)$ is a polynomial. We will later discuss how to relax this assumption.

Thus we get from Eq. (5.29) that

$$V'(z)g_N(z) - \Pi_N(z) = g_N^2(z) + \frac{g_N'(z)}{N} \tag{5.34}$$

for some polynomial Π_N of degree $\deg(V'(z)) - 1$, which needs to be determined self-consistently using Eq. (5.32). For a given $V'(z)$, the coefficients of Π_N are related to the moments of the $\{\lambda_i\}$, which themselves can be obtained from expanding $g_N(z)$ around infinity. In some cases, Eq. (5.34) can be solved exactly at finite N, for example in the case where $V(z) = z^2/2$, in which case the solution can be expressed in terms of Hermite polynomials – see Chapter 6. In the present chapter we will study this equation in the large N limit, which will allow us to derive a general formula for the limiting density of eigenvalues. Note that since Eq. (5.34) does not depend on the value of β, the corresponding eigenvalue density will also be independent of β.

Exercise 5.2.1 Maximum likelihood for 3 × 3 Wigner matrices

In this exercise we will write explicitly the three eigenvalues of the maximum likelihood configuration of a 3 × 3 GOE matrix. The potential for this ensemble is $V(x) = x^2/2$.

(a) Let $\lambda_1, \lambda_2, \lambda_3$ be the three maximum likelihood eigenvalues of the 3 × 3 GOE ensemble in decreasing order. By symmetry we expect $\lambda_3 = -\lambda_1$. What do you expect for λ_2?

(b) Consider Eq. (5.28). Assuming ($\lambda_3 = -\lambda_1$), check that your guess for λ_2 is indeed a solution. Now write the equation for λ_1 and solve it.

(c) Using your solution and the definition (5.30), show that the Stieltjes transform of the maximum likelihood configuration is given by

$$g_3(z) = \frac{z^2 - \frac{1}{3}}{z^3 - z}. \tag{5.35}$$

(d) In the simple case $V(x) = x^2/2$, the zero-degree polynomial $\Pi_N(z)$ is just a constant (independent of N) that can be evaluated from the definition (5.32). What is this constant?

(e) Verify that your $g_3(z)$ satisfies Eq. (5.34) with $N = 3$.

5.2.3 The Large N Limit

In the large N limit, $g_N(z)$ is self-averaging so computing $g_N(z)$ for the most likely configuration is the same as computing the average $\mathfrak{g}(z)$. As $N \to \infty$, Eq. (5.34) becomes

$$V'(z)\mathfrak{g}(z) - \Pi(z) = \mathfrak{g}^2(z). \tag{5.36}$$

Each value of N gives a different degree-$(k-1)$ polynomial $\Pi_N(z)$. From the definition (5.32), we can show that the coefficients of $\Pi_N(z)$ are related to the moments of the maximum likelihood configuration of size N. In the large N limit these moments converge so the sequence $\Pi_N(z)$ converges to a well-defined polynomial of degree $(k-1)$ which we call $\Pi(z)$.

Since Eq. (5.36) is quadratic in $\mathfrak{g}(z)$, its solution is given by

$$\mathfrak{g}(z) = \frac{1}{2}\left(V'(z) \pm \sqrt{V'(z)^2 - 4\Pi(z)}\right), \tag{5.37}$$

where we have to choose the branch where $\mathfrak{g}(z)$ goes to zero for large $|z|$.

The eigenvalues of $\mathbf{M}$ will be located where $\mathfrak{g}(z)$ has an imaginary part for z very close to the real axis. The first term $V'(z)$ is a real polynomial and is always real for real z. The expression $V'(z)^2 - 4\Pi(z)$ is also a real polynomial so $\mathfrak{g}(z)$ cannot be complex on the real axis unless $V'(z)^2 - 4\Pi(z) < 0$. In this case $\sqrt{V'(z)^2 - 4\Pi(z)}$ is purely imaginary. We conclude that, when x is such that $\rho(x) \neq 0$,

$$\mathrm{Re}(\mathfrak{g}(x)) := \fint \frac{\rho(\lambda)\mathrm{d}\lambda}{x-\lambda} = \frac{V'(x)}{2}, \tag{5.38}$$

where $\fint$ denotes the principal part of the integral. $\mathrm{Re}(\mathfrak{g}(x))/\pi$ is also called the Hilbert transform of $\rho(\lambda)$.[4]

We have thus shown that the Hilbert transform of the density of eigenvalues is (within its support) equal to $\pi/2$ times the derivative of the potential. We thus realize that the potential outside the support of the eigenvalue has no effect on the distribution of eigenvalues. This is natural in the Coulomb gas analogy. At equilibrium, the particles do not feel the potential away from where they are. One consequence is that we can consider potentials that are not confining at infinity as long as there is a confinement region and that all eigenvalues are within that region. For example, we will consider the quartic potential $V(x) = x^2/2 + \gamma x^4/4$. For small negative γ the region around $x = 0$ is convex. If all eigenvalues are found to be contained in that region, we can modify at will the potential away from it so that $V(x) \to +\infty$ for $|x| \to \infty$ and keep Eq. (5.1) normalizable, as a probability density should be.

Suppose now that we have a potential that is not a polynomial. In a finite region we can approximate it arbitrarily well by a polynomial of sufficiently high degree. If we choose the region of approximation such that for every successive approximation all eigenvalues lie in that region, we can take the limits of these approximations and find that Eq. (5.38) holds even if $V'(x)$ is not a polynomial.

We can also ask the reverse question. Given a density $\rho(x)$, does there exist a model from the orthogonal ensemble (or other β-ensemble) that has $\rho(\lambda)$ as its eigenvalue density? If the Hilbert transform of $\rho(x)$ is well defined, then the answer is yes and Eq. (5.38) gives the

[4] This is a slight abuse of terms, however, that can lead to paradoxes, if one extends Eq. (5.38) to the region where $\rho(x) = 0$. In this case, the right hand side of the equation is *not* equal to $V'(x)/2$. See also Appendix A.2.

corresponding potential. Note that the potential is only defined up to an additive constant (it can be absorbed in the normalization of Eq. (5.1)) so knowing its derivative is enough to compute $V(x)$. Note also that we only know the value of $V(x)$ on the support of $\rho(x)$; outside this support we can arbitrarily choose $V(x)$ provided it is convex and goes to infinity as $|x| \to \infty$.

Exercise 5.2.2 Matrix potential for the uniform density

In Exercise 2.3.3, we saw that the Stieltjes transform for a uniform density of eigenvalues between 0 and 1 is given by

$$\mathfrak{g}(z) = \log\left(\frac{z}{z-1}\right). \tag{5.39}$$

(a) By computing $\mathrm{Re}(\mathfrak{g}(x))$ for x between 0 and 1, find $V'(x)$ using Eq. (5.38).
(b) Compute the Hilbert transform of the uniform density to recover your answer in (a).
(c) From your answer in (a) and (b), show that the matrix potential is given by

$$V(x) = 2[(1-x)\log(1-x) + x\log(x)] + C \quad \text{for } 0 < x < 1, \tag{5.40}$$

where C is an arbitrary constant. Note that for $x < 0$ and $x > 1$ the potential should be completed by a convex function that goes to infinity as $|x| \to \infty$.

5.3 Applications: Wigner, Wishart and the One-Cut Assumption

5.3.1 Back to Wigner and Wishart

Now we apply the results of the previous section to the Gaussian orthogonal case where $V(z) = z^2/2\sigma^2$. In this simple case, $\Pi(z)$ can be computed from its definition without knowing the eigenvalues, since

$$V'(z) = \frac{z}{\sigma^2} \Rightarrow \Pi(z) = \frac{1}{\sigma^2}. \tag{5.41}$$

Then (5.37) gives

$$\mathfrak{g}(z) = \frac{z - \sqrt[\oplus]{z^2 - 4\sigma^2}}{2\sigma^2}, \tag{5.42}$$

which recovers, independently of the value of β, the Wigner result Eq. (2.38), albeit within a completely different framework (the notation $\sqrt[\oplus]{\cdot}$ was introduced in Section 4.2.3). In particular, the cavity method does not assume that the matrix ensemble is rotationally invariant, as we do here.

In the Wishart case, we only consider the case $q < 1$, otherwise ($q \geq 1$) the potential is not confining and we need to impose the positive semi-definiteness of the matrix to avoid eigenvalues running to minus infinity. We have

$$V'(z) = \frac{1}{q}\left(1 + \frac{q-1}{z}\right). \tag{5.43}$$

In this case $zV'(z)$ is of degree one, so $z\Pi(z)$ is a polynomial of degree zero:

$$\Pi(z) = \frac{c}{z} \tag{5.44}$$

for some constant c. Thus (5.37) then gives

$$\mathfrak{g}(z) = \frac{z + q - 1 - \sqrt[\oplus]{(z+q-1)^2 - 4cq^2 z}}{2qz}. \tag{5.45}$$

As $z \to +\infty$, this expression becomes

$$\mathfrak{g}(z) = \frac{cq}{z} + O(1/z^2). \tag{5.46}$$

Imposing $\mathfrak{g}(z) \sim z^{-1}$ gives $c = q^{-1}$. After some manipulations, we recover Eq. (4.40).

5.3.2 General Convex Potentials and the One-Cut Assumption

For more general polynomial potentials, finding an explicit solution for the limiting Stieltjes transform is a little bit more involved. We recall Eq. (5.37):

$$\mathfrak{g}(z) = \frac{1}{2}\left(V'(z) \pm \sqrt{V'(z)^2 - 4\Pi(z)}\right). \tag{5.47}$$

For a particular polynomial $V'(z)$, $\Pi(z)$ is a polynomial that depends on the moments of the matrix $\mathbf{M}$. The expansion of $\mathfrak{g}(z)$ near $z \to \infty$ will give a set of self-consistent equations for the coefficients of $\Pi(z)$.

The problem simplifies greatly if the support of density of eigenvalues is compact, i.e. if the density $\rho(\lambda)$ is non-zero for all λ's between two edges λ_- and λ_+. We expect this to be true if the potential $V(x)$ is convex. Indeed, by the Coulomb gas analogy we could place all eigenvalues near the minimum of $V(x)$ and let them find their equilibrium configuration by repelling each other. For a convex potential it is natural to assume that the equilibrium configuration would not have any gaps. This assumption is equivalent to assuming that the limiting Stieltjes transform has a single branch cut (from λ_- and λ_+), hence the name one-cut assumption.

So, for a convex polynomial potential $V(x)$, we expect that there exists a well-defined equilibrium density $\rho(\lambda)$ that is non-zero if and only if $\lambda_- < \lambda < \lambda_+$ and that $\mathfrak{g}(z)$ satisfies

$$\mathfrak{g}(z) = \int_{\lambda_-}^{\lambda_+} \frac{\rho(\lambda)}{z - \lambda} d\lambda. \tag{5.48}$$

From this equation we notice three important properties of $\mathfrak{g}(z)$:

- The function $\mathfrak{g}(z)$ is potentially singular at λ_- and λ_+.
- Near the real axis ($\mathrm{Im}\, z = 0^+$) $g(z)$ has an imaginary part if $z \in (\lambda_-, \lambda_+)$ and is real otherwise.
- The function $\mathfrak{g}(z)$ is analytic everywhere else.

If we go back to Eq. (5.37), we notice that any non-analytic behavior must come from the square-root. On the real axis, the only way $g(z)$ can have an imaginary part is if $D(z) := V'(z)^2 - 4\Pi(z) < 0$ for some values of z. So $D(z)$ (a polynomial of degree $2k$) must change sign at some values λ_- and λ_+, hence these must be zeros of the polynomial. On the real axis, the other possible zeros $D(z)$ can only be of even multiplicity (otherwise $D(z)$ would change sign). Elsewhere in the complex plane, zeros should also be of even multiplicity, otherwise $\sqrt{D(z)}$ would be singular at those zeros. In other words $D(z)$ must be of the form

$$D(z) = (z - \lambda_-)(z - \lambda_+)Q^2(z), \tag{5.49}$$

for some polynomial $Q(z)$ of degree $k - 1$ where k is the degree of $V'(z)$. We can therefore write $\mathfrak{g}(z)$ as

$$\mathfrak{g}(z) = \frac{V'(z) \pm Q(z)\sqrt{(z - \lambda_-)(z - \lambda_+)}}{2}, \tag{5.50}$$

where again $Q(z)$ is a polynomial with real coefficients of degree one less than $V'(z)$. The condition that

$$\mathfrak{g}(z) \to \frac{1}{z} \text{ when } |z| \to \infty \tag{5.51}$$

is now sufficient to compute $Q(z)$ and also $\lambda_\pm$ for a given potential $V(z)$. Indeed, expanding Eq. (5.50) near $z \to \infty$, the coefficients of $Q(z)$ and the values $\lambda_\pm$ must be such as to cancel the $k+1$ polynomial coefficients of $V'(z)$ and also ensure that the $1/z$ term has unit coefficient. This gives $k+2$ equations to determine the k coefficients of $Q(z)$ and the two edges $\lambda_\pm$, see next section for an illustration.

Once the polynomial $Q(x)$ is determined, we can read off the eigenvalue density:

$$\rho(\lambda) = \frac{Q(\lambda)\sqrt{(\lambda_+ - \lambda)(\lambda - \lambda_-)}}{2\pi} \quad \text{for} \quad \lambda_- \leq \lambda \leq \lambda_+. \tag{5.52}$$

We see that generically the eigenvalue density behaves as $\rho(\lambda_\pm \mp \delta) \propto \sqrt{\delta}$ near both edges of the spectrum. If by chance (or by construction) one of the edges is a zero of $Q(z)$, then the behavior changes to δ^θ near that edge, with $\theta = n + \frac{1}{2}$ and n the multiplicity of root of $Q(z)$. A potential with generic $\sqrt{\delta}$ behavior at the edge of the density is called non-critical. Other non-generic cases are called critical. In Section 14.1 we will see how the $\sqrt{\delta}$ edge singularity is smoothed over a region of width $N^{-2/3}$ for finite N. In the critical case, the smoothed region is of width $N^{-2/(3+2n)}$.

5.3.3 $M^2 + M^4$ Potential

One of the original motivations of Brézin, Itzykson, Parisi and Zuber to study the ensemble defined by Eq. (5.1) was to count the so-called planar diagrams in some field theories. To do so they considered the potential

$$V(x) = \frac{x^2}{2} + \frac{\gamma x^4}{4}. \tag{5.53}$$

We will not discuss how one can count planar diagrams from such a potential, but use this example to illustrate the general recipe given in the main text to compute the Stieltjes transform and the density of eigenvalues. Interestingly, for a certain value of γ, the edge singularity is $\delta^{3/2}$ instead of $\sqrt{\delta}$.

Since the potential is symmetric around zero we expect $\lambda_+ = -\lambda_- =: 2a$. We introduce this extra factor of 2, so that if $\gamma = 0$, we obtain the semi-circular law with $a = 1$. Since $V'(z) = z + \gamma z^3$ is a degree three polynomial, we write

$$Q(z) = a_0 + a_1 z + \gamma z^2, \tag{5.54}$$

where the coefficient of z^2 was chosen to cancel the γz^3 term at infinity. Expanding Eq. (5.50) near $z \to \infty$ and imposing $g(z) = 1/z + O(1/z^2)$ we get

$$\begin{aligned} a_1 &= 0, \\ 1 - a_0 + 2\gamma a^2 &= 0, \\ 2a^4\gamma + 2a^2 a_0 &= 2. \end{aligned} \tag{5.55}$$

Solving for a_0, we find

$$\mathfrak{g}(z) = \frac{z + \gamma z^3 - (1 + 2\gamma a^2 + \gamma z^2) \sqrt[\oplus]{z^2 - 4a^2}}{2}, \tag{5.56}$$

where a is a solution of

$$3\gamma a^4 + a^2 - 1 = 0 \quad \Rightarrow \quad a^2 = \frac{\sqrt{1 + 12\gamma} - 1}{6\gamma}. \tag{5.57}$$

The density of eigenvalues for the potential (5.53) reads

$$\rho(\lambda) = \frac{(1 + 2\gamma a^2 + \gamma\lambda^2)\sqrt{4a^2 - \lambda^2}}{2\pi} \quad \text{for} \quad \gamma > -\frac{1}{12}, \tag{5.58}$$

with a defined as above. For positive values of γ, the potential is confining (it is convex and grows faster than a logarithm for $z \to \pm\infty$). In that case the equation for a always has a solution, so the Stieltjes transform and the density of eigenvalues are well defined; see Figure 5.3. For small negative values of γ, the problem still makes sense. The potential is convex near zero and the eigenvalues will stay near zero as long as the repulsion does not push them too far in the non-convex region.

There is a critical value of γ at $\gamma_c = -1/12$, which corresponds to $a = \sqrt{2}$. At this critical point, $\mathfrak{g}_c(z)$ and the density are given by

$$\mathfrak{g}_c(z) = \frac{z^3}{24}\left[\left(1 - \frac{8}{z^2}\right)^{\frac{3}{2}} - 1 + \frac{12}{z^2}\right] \quad \text{and} \quad \rho_c(\lambda) = \frac{\left(8 - \lambda^2\right)^{3/2}}{24\pi}. \tag{5.59}$$

At this point the density of eigenvalues at the upper edge ($\lambda_+ = 2\sqrt{2}$) behaves as $\rho(\lambda) \sim (2\sqrt{2} - \lambda)_+^\theta$ with $\theta = 3/2$ and similarly at the lower edge ($\lambda_- = -2\sqrt{2}$). For values of

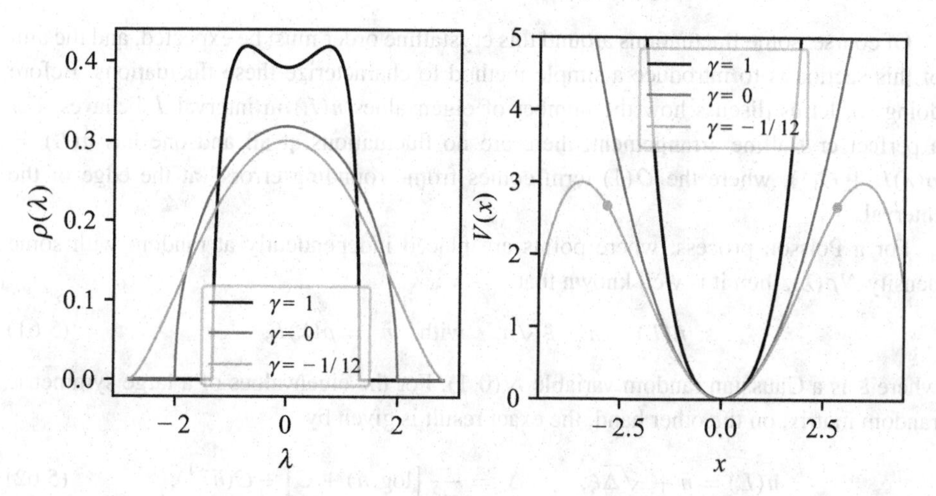

Figure 5.3 (left) Density of eigenvalues for the potential $V(x) = \frac{1}{2}x^2 + \frac{\gamma}{4}x^4$ for three values of γ. For $\gamma = 1$, even if the minimum of the potential is at $\lambda = 0$, the density develops a double hump due to the repulsion of the eigenvalues. $\gamma = 0$ corresponds to the Wigner case (semi-circle law). Finally $\gamma = -1/12$ is the critical value of γ. At this point the density is given by Eq. (5.59). For smaller values of γ the density does not exist. (right) Shape of the potential for the same three values of γ. The dots on the bottom curve indicate the edges of the critical spectrum.

γ more negative than γ_c, there are no real solutions for a and Eq. (5.56) ceases to make sense. In the Coulomb gas analogy, the eigenvalues push each other up to a critical point after which they run off to infinity. There is no simple argument that gives the location of the critical point (except for doing the above computation). It is given by a delicate balance between the repulsion of the eigenvalues and the confining potential. In particular it is not given by the point $V'(2a) = 0$ as one might naively expect. Note that at the critical point $V''(2a) = -1$, so we are already outside the convex region.

5.4 Fluctuations Around the Most Likely Configuration

5.4.1 Fluctuations of All Eigenvalues

The most likely positions of the N eigenvalues are determined by the equations (5.28) that we derived in Section 5.2.2. These equations balance a Coulomb (repulsive) term and a confining potential. On distances d of order $1/N$, the Coulomb force on each "charge" is of order $d^{-1} \sim N$, whereas the confining force V' is of order unity. Therefore, the Coulomb force is dominant at small scales and the most likely positions must be locally equidistant. This is expected to be true within small enough intervals

$$\mathcal{I} := [\lambda - L/2N, \lambda + L/2N], \tag{5.60}$$

where L is sufficiently small, such that the average density $\rho(\lambda)$ does not vary too much, i.e. $\rho'(\lambda)L/N \ll \rho(\lambda)$.

Of course, some fluctuations around this crystalline order must be expected, and the aim of this section is to introduce a simple method to characterize these fluctuations. Before doing so, let us discuss how the number of eigenvalues $n(L)$ in interval I behaves. For a perfect crystalline arrangement, there are no fluctuations at all and one has $n(L) = \rho(\lambda)L + O(1)$, where the $O(1)$ term comes from "rounding errors" at the edge of the interval.

For a Poisson process, where points are placed independently at random with some density $N\rho(\lambda)$, then it is well known that

$$n(L) = \bar{n} + \xi\sqrt{\bar{n}}, \quad \text{with} \quad \bar{n} := \rho(\lambda)L, \tag{5.61}$$

where ξ is a Gaussian random variable $\mathcal{N}(0,1)$. For the eigenvalues of a large symmetric random matrix, on the other hand, the exact result is given by

$$n(L) = \bar{n} + \sqrt{\Delta}\xi, \qquad \Delta := \frac{2}{\pi^2}\left[\log(\bar{n}) + C\right] + O(\bar{n}^{-1}), \tag{5.62}$$

where C is a numerical constant.[5] This result means that the typical fluctuations of the number of eigenvalues is of order $\sqrt{\log \bar{n}}$ for large L, much smaller than the Poisson result $\sqrt{\bar{n}}$. In fact, the growth of Δ with L is so slow that one can think of the arrangement of eigenvalues as "quasi-crystalline", even after factoring in fluctuations.

Let us see how this $\sqrt{\log \bar{n}}$ behavior of the fluctuations can be captured by computing the Hessian matrix $\mathbf{H}$ of the log-likelihood $\mathcal{L}$ defined in Eq. (5.27). One finds

$$\mathbf{H}_{ij} := -\frac{\partial^2 \mathcal{L}}{\partial\lambda_i \partial\lambda_j} = \begin{cases} V''(\lambda_i) + \frac{1}{N}\sum_{k\neq i} \frac{2}{(\lambda_i-\lambda_k)^2} & (i = j), \\ -\frac{2}{N}\frac{1}{(\lambda_i-\lambda_j)^2} & (i \neq j). \end{cases}$$

This Hessian matrix should be evaluated at the maximum of $\mathcal{L}$, i.e. for the most likely configurations of the λ_i. If we call ϵ_i/N the *deviation* of λ_i away from its most likely value, and assume all ϵ to be small enough, their joint distribution can be approximated by the following multivariate Gaussian distribution:

$$P(\{\epsilon_i\}) \propto \exp\left[-\frac{\beta}{4N}\sum_{i,j}\epsilon_i \mathbf{H}_{ij}\epsilon_j\right], \tag{5.63}$$

from which one can obtain the covariance of the deviations as

$$\mathbb{E}[\epsilon_i\epsilon_j] = \frac{2N}{\beta}[\mathbf{H}^{-1}]_{ij}. \tag{5.64}$$

Since the most likely positions of the λ_i are locally equidistant, one can approximate $\lambda_i - \lambda_j$ as $(i-j)/N\rho$ in the above expression. This is justified when the contribution of far-away eigenvalues to $\mathbf{H}^{-1}$ is negligible in the region of interest, which will indeed be the case because the off-diagonal elements of $\mathbf{H}$ decay fast enough (i.e. as $(i-j)^{-2}$).

[5] For $\beta = 1$, one finds $C = \log(2\pi) + \gamma + 1 + \frac{\pi^2}{8}$, where γ is Euler's constant, see Mehta [2004].

For simplicity, let us consider the case where $V(x) = x^2/2$, in which case $V''(x) = 1$. The matrix $\mathbf{H}$ is then a Toeplitz matrix, up to unimportant boundary terms. It can be diagonalized using plane waves (see Appendix A.3), with eigenvalues given by[6]

$$\mu_q = 1 + 4N\rho^2 \sum_{\ell=1}^{N-1} \frac{1 - \cos \frac{2\pi q \ell}{N}}{\ell^2}, \qquad q = 0, 1, \ldots, N-1, \tag{5.65}$$

where ρ is the local average eigenvalue density. In the large N limit, the (convergent) sum can be replaced by an integral:

$$\mu_q = 1 + 4N\rho^2 \times \frac{2\pi |q|}{N} \int_0^\infty du \frac{1 - \cos u}{u^2} = 1 + 4\pi^2\rho^2 |q|. \tag{5.66}$$

The eigenvalues of $\mathbf{H}^{-1}$ are then given by $1/\mu_q$ and the covariance of the deviations for $i - j = n$ is obtained from Eq. (5.64) as an inverse Fourier transform. After transforming $q \to uN/2\pi$ this reads

$$\mathbb{E}[\epsilon_i \epsilon_j] = \frac{2N}{\beta} \frac{1}{N} \int_{-\pi}^{\pi} \frac{du}{2\pi} \frac{e^{-iun}}{N^{-1} + 2\pi\rho^2 |u|}. \tag{5.67}$$

Now, the fluctuating distance d_{ij} between eigenvalues i and $j = i + n$ is, by definition of the ϵ,

$$d_{ij} = \frac{n}{N\rho} + \frac{\epsilon_i - \epsilon_j}{N}. \tag{5.68}$$

Its variance is obtained, for $N \to \infty$, from

$$\mathbb{E}[(\epsilon_i - \epsilon_j)^2] \approx \frac{2}{\beta\pi^2(\rho N)^2} \int_0^\pi du \frac{1 - \cos un}{u} \approx_{n \ll 1} \frac{2}{\beta\pi^2\rho^2} \log n. \tag{5.69}$$

The variables ϵ are thus long-ranged, log-correlated Gaussian variables. Interestingly, there has been a flurry of activity concerning this problem in recent years (see references at the end of this chapter).

Finally, the fluctuating local density of eigenvalues can be computed from the number of eigenvalues within a distance d_{ij}, i.e.

$$\frac{1}{N} \frac{n}{d_{ij}} \approx \rho + \rho^2 \frac{\epsilon_i - \epsilon_j}{n} \qquad (n \gg 1). \tag{5.70}$$

Its variance is thus given by

$$\mathbb{V}[\rho] = \frac{2\rho^2}{\beta\pi^2 n^2} \log n. \tag{5.71}$$

Hence, the fluctuation of the number of eigenvalues in a fixed interval of size L/N and containing on average $\bar{n} = \rho L$ eigenvalues is

$$\mathbb{V}[\rho L] = L^2 \mathbb{V}[\rho] = \frac{2}{\beta\pi^2} \log \bar{n}. \tag{5.72}$$

This argument recovers the leading term of the exact result for all values of β (compare with Eq. (5.62) for $\beta = 1$).

[6] In fact, the Hessian $\mathbf{H}$ can be diagonalized exactly in this case, without any approximations, see Agarwal et al. [2019] and references therein.

5.4.2 Large Deviations of the Top Eigenvalue

Another interesting question concerns the fluctuations of the top eigenvalue of a matrix drawn from a β-ensemble. Consider Eq. (5.26) with $\lambda_{\max}$ isolated:

$$P(\lambda_{\max};\{\lambda_i\}) = P_{N-1}(\{\lambda_i\}) \exp\left\{-\frac{N\beta}{2}\left[V(\lambda_{\max}) - \frac{2}{N}\sum_{i=1}^{N-1}\log(\lambda_{\max}-\lambda_i)\right]\right\}, \quad (5.73)$$

with

$$P_{N-1}(\{\lambda_i\}) := Z_N^{-1} \exp\left\{-\frac{\beta}{2}\left[\sum_{i=1}^{N-1} NV(\lambda_i) - \sum_{\substack{i,j=1 \\ j\neq i}}^{N-1} \log|\lambda_i - \lambda_j|\right]\right\}. \quad (5.74)$$

At large N, this probability is dominated by the most likely configuration, which is determined by Eq. (5.27) with $\lambda_{\max}$ removed. Clearly, the most likely positions $\{\lambda_i^*\}$ of the $N-1$ other eigenvalues are only changed by an amount $O(N^{-1})$, but since the log-likelihood is close to its maximum, the change of $\mathcal{L}$ (i.e. the quantity in the exponential) is only of order N^{-2} and we will neglect it. Then one has the following *large deviation* expression:

$$\frac{P(\lambda_{\max};\{\lambda_i^*\})}{P(\lambda_+;\{\lambda_i^*\})} := \exp\left[-\frac{N\beta}{2}\Phi(\lambda_{\max})\right], \quad (5.75)$$

with

$$\Phi(x) = V(x) - V(\lambda_+) - \frac{2}{N}\sum_{i=1}^{N-1}\log(x-\lambda_i^*) + \frac{2}{N}\sum_{i=1}^{N-1}\log(\lambda_+ - \lambda_i^*), \quad (5.76)$$

where λ_+ is the top edge of the spectrum. Note that $\Phi(\lambda_+) = 0$ by construction. To deal with the large N limit of this expression, we take the derivative of $\Phi(x)$ with respect to x to find

$$\Phi'(x) = V'(x) - \frac{2}{N}\sum_{i=1}^{N-1}\frac{1}{x-\lambda_i^*} \overset{N\to\infty}{\to} V'(x) - 2\mathfrak{g}(x). \quad (5.77)$$

Hence,

$$\Phi(x) = \int_{\lambda_+}^{x} \left(V'(s) - 2\mathfrak{g}(s)\right) \mathrm{d}s. \quad (5.78)$$

When the potential $V'(x)$ is a polynomial of degree $k \geq 1$, we can use Eq. (5.37), yielding

$$\Phi(x) = \int_{\lambda_+}^{x} \sqrt{V'(s)^2 - 4\Pi(s)}\, \mathrm{d}s, \quad (5.79)$$

where λ_+ is the largest zero of the polynomial $V'(s)^2 - 4\Pi(s)$. Since $\Pi(s)$ is a polynomial of order $k-1$, for large s one always has

$$V'(s)^2 \gg |\Pi(s)|, \quad (5.80)$$

and therefore

$$\Phi(x) \approx V(x). \tag{5.81}$$

As expected, the probability to observe a top eigenvalue very far from the bulk is dominated by the potential term, and the Coulomb interaction no longer plays any role. When $x - \lambda_+$ is small but still much larger than any inverse power of N, we have

$$\sqrt{V'(s)^2 - 4\Pi(s)} \approx C(s - \lambda_+)^\theta, \tag{5.82}$$

where C is a constant and θ depends on the type of root of $V'(s)^2 - 4\Pi(s)$ at $s = \lambda_+$. For a single root, which is the typical case, $\theta = 1/2$. Performing the integral we get

$$\Phi(\lambda_{\max}) = \frac{C}{\theta + 1}(\lambda_{\max} - \lambda_+)^{\theta+1}, \qquad \lambda_{\max} - \lambda_+ \ll 1. \tag{5.83}$$

Note that the constant C can be determined from the eigenvalue density near (but slightly below) λ_+, to wit, $\pi\rho(\lambda) \approx C(\lambda_+ - \lambda)^\theta$.

For a generic edge with $\theta = 1/2$, one finds that $\Phi(\lambda_{\max}) \propto (\lambda_{\max} - \lambda_+)^{3/2}$, and thus the probability to find $\lambda_{\max}$ just above λ_+ decays as

$$P(\lambda_{\max}) \sim \exp\left[-\frac{2\beta C}{3} u^{3/2}\right]; \qquad u = N^{2/3}(\lambda_{\max} - \lambda_+), \tag{5.84}$$

where we have introduced the rescaled variable u in order to show that this probability decays on scale $N^{-2/3}$. We will further discuss this result in Section 14.1, where we will see that the $u^{3/2}$ behavior coincides with the right tail of the Tracy–Widom distribution.

For a unit Wigner, we have (see Fig. 5.4)

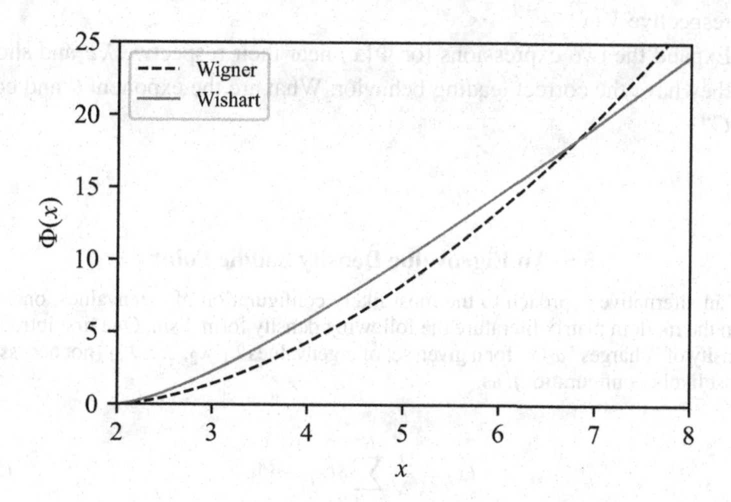

Figure 5.4 Large deviation function for the top eigenvalue $\Phi(x)$ for a unit Wigner and a Wishart matrix with $\lambda_+ = 2$ ($q = 3 - 2\sqrt{2}$). The Wishart curve was obtained by numerically integrating Eq. (5.78).

$$\Phi(x) = \frac{1}{2}x\sqrt{x^2-4} - 2\log\left(\frac{\sqrt{x^2-4}+x}{2}\right) \quad \text{for} \quad x > 2. \tag{5.85}$$

For a Wishart matrix with parameter $q = 1$,

$$\Phi(x) = \sqrt{(x-4)x} + 2\log\left(\frac{x-\sqrt{(x-4)x}-2}{2}\right) \quad \text{for} \quad x > 4. \tag{5.86}$$

Remember that the Stieltjes transform $\mathfrak{g}(z)$ and the density of eigenvalues $\rho(\lambda)$ are only sensitive to the potential function $V(x)$ for values in the support of ρ ($\lambda_- \le x \le \lambda_+$). The large deviation function $\Phi(x)$, on the other hand, depends on the value of the potential for $x > \lambda_+$. For Eq. (5.79) to hold, the same potential function must extend analytically outside the support of ρ. In Section 5.3.3, we considered a non-confining potential (when $\gamma < 0$). We argued that we could compute $\mathfrak{g}(z)$ and $\rho(\lambda)$ as if the potential were confining as long as we could replace the potential outside (λ_-, λ_+) by a convex function going to infinity. In that case, the function $\Phi(x)$ depends on the choice of regularization of the potential. Computing Eq. (5.78) with the non-confining potential would give nonsensical results.

Exercise 5.4.1 Large deviations for Wigner and Wishart

(a) Show that Eqs. (5.85) and (5.86) are indeed the large deviation function for the top eigenvalues of a unit Wigner and a Wishart $q = 1$. To do so, show that they satisfy Eq. (5.77) and that $\Phi(\lambda_+) = 0$ with $\lambda_+ = 2$ for Wigner and $\lambda_+ = 4$ for Wishart $q = 1$.

(b) Find the dominant contribution of both $\Phi(x)$ for large x. Compare to their respective $V(x)$.

(c) Expand the two expressions for $\Phi(x)$ near their respective λ_+ and show that they have the correct leading behavior. What are the exponent θ and constant C?

5.5 An Eigenvalue Density Saddle Point

As an alternative approach to the most likely configuration of eigenvalues, one often finds in the random matrix literature the following density formalism. One first introduces the density of "charges" $\omega(x)$ for a given set of eigenvalues $\lambda_1, \lambda_2, \ldots, \lambda_N$ (not necessarily the most likely configuration), as

$$\omega(x) = \frac{1}{N}\sum_{i=1}^{N} \delta(\lambda_i - x). \tag{5.87}$$

Expressed in terms of this density field, the joint distribution of eigenvalues, Eq. (5.26) can be expressed as

$$P(\{\omega\}) = Z^{-1} \exp\left(-\frac{\beta N^2}{2}\left[\int \mathrm{d}x\omega(x)V(x) - \fint \mathrm{d}x\mathrm{d}y\omega(x)\omega(y)\log|x-y|\right]\right.$$
$$\left. - N\int \mathrm{d}x\omega(x)\log\omega(x)\right), \tag{5.88}$$

where the last term is an entropy term, formally corresponding to the change of variables from the $\{\lambda_i\}$ to $\omega(x)$. Since this term is of order N, compared to the two first terms that are of order N^2, it is usually neglected.[7]

One then proceeds by looking for the density field that maximizes the term in the exponential, which is obtained by taking its functional derivative with respect to all $\omega(x)$:

$$\frac{\delta}{\delta\omega(x)}\left[\int \mathrm{d}y\omega(y)V(y) - \fint \mathrm{d}y\mathrm{d}y'\omega(y)\omega(y')\log|y-y'| - \zeta\int \mathrm{d}y\omega(y)\right]_{\omega^*} = 0, \tag{5.89}$$

where ζ is a Lagrange multiplier, used to impose the normalization condition $\int \mathrm{d}x\omega(x) = 1$. This leads to

$$V(x) = 2\fint \mathrm{d}y\omega^*(y)\log|x-y| + \zeta. \tag{5.90}$$

We can now take the derivative with respect to x to get

$$V'(x) = 2\fint \mathrm{d}y\frac{\omega^*(y)}{x-y}, \tag{5.91}$$

which is nothing but the continuum limit version of Eq. (5.28), and is identical to Eq. (5.38). Although this whole procedure looks somewhat ad hoc, it can be fully justified mathematically. In the mathematical literature, it is known as the large deviation formalism.

Equation (5.91) is a singular integral equation for $\omega(x)$ of the so-called Tricomi type. Such equations often have explicit solutions, see Appendix A.2. In the case where $V(x) = x^2/2$, one recovers, as expected, the semi-circle law:

$$\omega^*(x) = \frac{1}{2\pi}\sqrt{4-x^2}. \tag{5.92}$$

One interesting application of the density formalism is to investigate the case of Gaussian orthogonal random matrices conditioned to have all eigenvalues strictly positive. What is the probability for this to occur spontaneously? In such a case, what is the resulting distribution of eigenvalues?

The trick is to solve Eq. (5.91) with the constraint that $\omega(x<0) = 0$. This leads to

$$x = 2\fint_0^\infty \mathrm{d}y\frac{\omega^*(y)}{x-y}. \tag{5.93}$$

The solution to this truncated problem can also be found analytically using the general result of Appendix A.2. One finds

$$\omega^*(x) = \frac{1}{4\pi}\sqrt{\frac{\lambda_+ - x}{x}}(\lambda_+ + 2x), \quad \text{with} \quad \lambda_+ = \frac{4}{\sqrt{3}}. \tag{5.94}$$

The resulting density has a square-root divergence close to zero, which is a stigma of all the negative eigenvalues being pushed to the positive side. The right edge itself is pushed from 2 to $4/\sqrt{3}$, see Figure 5.5.

[7] Note however that when $\beta = c/N$, as considered in Allez et al. [2012], the entropy term must be retained.

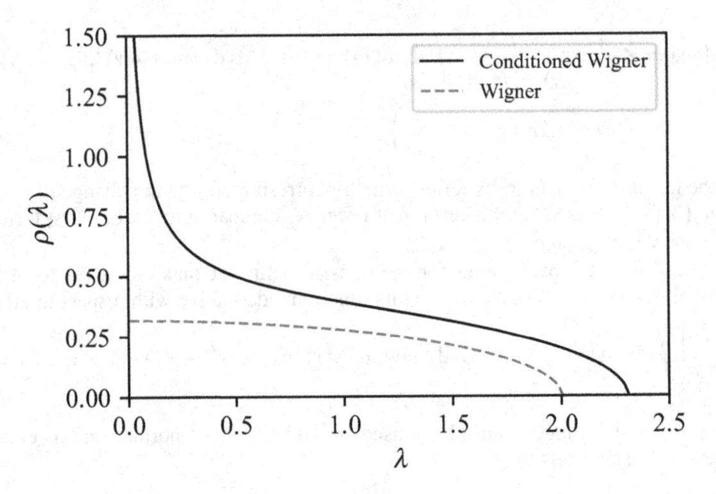

Figure 5.5 Density of eigenvalues for a Wigner matrix conditioned to be positive semi-definite. The positive part of the density of a standard Wigner is shown for comparison.

Injecting this solution back into Eq. (5.88) and comparing with the result corresponding to the standard semi-circle allows one to compute the probability for such an exceptional configuration to occur. After a few manipulations, one finds that this probability is given by $e^{-\beta C N^2}$ with $C = \log 3/4 \approx 0.2746....$ The probability that a Gaussian random matrix has by chance all its eigenvalues positive therefore decreases extremely fast with N.

Other constraints are possible as well, for example if one chooses to confine all eigenvalues in a certain interval $[\ell_-, \ell_+]$ with $\ell_- \geq \lambda_-$ or $\ell_+ \leq \lambda_+$. (In the cases where $\ell_- < \lambda_-$ and $\ell_+ > \lambda_+$ the confinement plays no role.) Let us study this problem in the case where there is no external potential at all, i.e. when $V(x) \equiv C$, where C is an arbitrary constant, but only confining walls. In this case, the general solution of Appendix A.2 immediately leads to

$$\omega^*(x) = \frac{1}{\pi \sqrt{(x - \ell_-)(\ell_+ - x)}}, \tag{5.95}$$

which has a square-root divergence at both edges. This law is called the arcsine law, which appears in different contexts, see Sections 7.2 and 15.3.1.

Note that the minimization of the quadratic form $\fint \mathrm{d}x\mathrm{d}y\omega(x)\omega(y)G(|x-y|)$ for a general power-law interaction $G(u) = u^{-\gamma}$, subject to the constraint $\int \mathrm{d}x\omega(x) = 1$, has been solved in the very different, financial engineering context of optimal execution with quadratic costs. The solution in that case reads

$$\omega^*(x) = A(\gamma)\ell^{-\gamma} \sqrt{(x - \ell_-)^{\gamma-1}(\ell_+ - x)^{\gamma-1}}, \tag{5.96}$$

with $\ell := \ell_- + \ell_+$ and $A(\gamma) := \gamma\Gamma[\gamma]/\Gamma^2[(1+\gamma)/2]$. The case $G(u) = -\log(u)$ formally corresponds to $\gamma = 0$, in which case one recovers Eq. (5.95).

Bibliographical Notes

- For a general introduction to the subject, see
  - M. L. Mehta. *Random Matrices*. Academic Press, San Diego, 3rd edition, 2004,
  - P. J. Forrester. *Log Gases and Random Matrices*. Princeton University Press, Princeton, NJ, 2010,

  and for more technical aspects,
  - B. Eynard, T. Kimura, and S. Ribault. Random matrices. *preprint arXiv:1510.04430*, 2006.
- One-cut solution and the $\mathbf{M}^2 + \mathbf{M}^4$ model:
  - E. Brézin, C. Itzykson, G. Parisi, and J.-B. Zuber. Planar diagrams. *Communications in Mathematical Physics*, 59(1):35–51, 1978.
- On the relation between log gases and the Calogero model, see
  - S. Agarwal, M. Kulkarni, and A. Dhar. Some connections between the classical Calogero-Moser model and the log gas. *Journal of Statistical Physics*, 176(6), 1463–1479, 2019.
- About spectral rigidity and log-fluctuations of eigenvalues, see
  - M. L. Mehta. *Random Matrices*. Academic Press, San Diego, 3rd edition, 2004,
  - M. V. Berry. Semiclassical theory of spectral rigidity. *Proceedings of the Royal Society of London, Series A*, 400(1819):229–251, 1985,

  for classical results, and
  - Y. V. Fyodorov and J. P. Keating. Freezing transitions and extreme values: Random matrix theory, and disordered landscapes. *Philosophical Transactions of the Royal Society A: Mathematical, Physical and Engineering Sciences*, 372(2007):2012053, 2014,
  - Y. V. Fyodorov, B. A. Khoruzhenko, and N. J. Simm. Fractional Brownian motion with Hurst index $H = 0$ and the Gaussian Unitary Ensemble. *The Annals of Probability*, 44(4):2980–3031, 2016,
  - R. Chhaibi and J. Najnudel. On the circle, $gmc^\gamma = c\beta e_\infty$ for $\gamma = \sqrt{2/\beta}$, $(\gamma \leq 1)$. *preprint arXiv:1904.00578*, 2019,

  for the recent spree of activity in the field of log-correlated variables.
- On the probability of a very large eigenvalue, see Chapter 14 and
  - S. N. Majumdar and M. Vergassola. Large deviations of the maximum eigenvalue for Wishart and Gaussian random matrices. *Physical Review Letters*, 102:060601, 2009.
- On the continuous density formalism, see
  - D. S. Dean and S. N. Majumdar. Extreme value statistics of eigenvalues of Gaussian random matrices. *Physical Review E*, 77:041108, 2008

  for a physicists' introduction, and

  - G. Ben Arous and A. Guionnet. Large deviations for Wigner's law and Voiculescu's non-commutative entropy. *Probability Theory and Related Fields*, 108(4):517–542, 1997,
  - G. W. Anderson, A. Guionnet, and O. Zeitouni. *An Introduction to Random Matrices*. Cambridge University Press, Cambridge, 2010

for more mathematical discussions in the context of large deviation theory. A case where the entropy term must be retained is discussed in

  - R. Allez, J. P. Bouchaud, and A. Guionnet. Invariant beta ensembles and the Gauss-Wigner crossover. *Physical Review Letters*, 109(9):094102, 2012.

- On the relation with optimal execution problems, see
  - J. Gatheral, A. Schied, and A. Slynko. Transient linear price impact and Fredholm integral equations. *Mathematical Finance*, 22(3):445–474, 2012,
  - J.-P. Bouchaud, J. Bonart, J. Donier, and M. Gould. *Trades, Quotes and Prices*. Cambridge University Press, Cambridge, 2nd edition, 2018.

6
Eigenvalues and Orthogonal Polynomials*

In this chapter, we investigate yet another route to shed light on the eigenvalue density of the Wigner and Wishart ensembles. We show (a) that the most probable positions of the Coulomb gas problem coincide with the zeros of Hermite polynomials in the Wigner case, and of Laguerre polynomials in the Wishart case; and (b) that the average (over randomness) of the characteristic polynomials (defined as $\det(z\mathbf{1} - \mathbf{X}_N)$) of Wigner or Wishart random matrices of size N obey simple recursion relations that allow one to express them as, respectively, Hermite and Laguerre polynomials. The fact that the two methods lead to the same result (at least for large N) reflects the fact that eigenvalues fluctuate very little around their most probable positions. Finally we show that for unitary ensembles $\beta = 2$, the expected characteristic polynomial is always an orthogonal polynomial with respect to some weight function related to the matrix potential.

6.1 Wigner Matrices and Hermite Polynomials

6.1.1 Most Likely Eigenvalues and Zeros of Hermite Polynomials

In the previous chapter, we established a general equation for the Stieltjes transform of the most likely positions of the eigenvalues of random matrices belonging to a general orthogonal ensemble, see Eq. (5.34). In the special case of a quadratic potential $V(x) = x^2/2$, this equation reads

$$z g_N(z) - 1 = g_N^2(z) + \frac{g_N'(z)}{N}. \tag{6.1}$$

This ordinary differential equation is of the *Ricatti type*,[1] and can be solved by setting $g_N(z) := \psi'(z)/N\psi(z)$. This yields, upon substitution,

$$\psi''(z) - Nz\psi'(z) + N^2\psi(z) = 0, \tag{6.2}$$

or, with $\psi(z) = \Psi(x = \sqrt{N}z)$,

$$\Psi''(x) - x\Psi'(x) + N\Psi(x) = 0. \tag{6.3}$$

[1] A Ricatti equation is a first order differential equation that is quadratic in the unknown function, in our case $g_N(z)$.

The solution of this last equation with the correct behavior for large x is the Hermite polynomial of order N. General Hermite polynomials $H_n(x)$ are defined as the nth order polynomial that starts as x^n and is orthogonal to all previous ones under the unit Gaussian measure:[2]

$$\int \frac{\mathrm{d}x}{\sqrt{2\pi}} H_n(x) H_m(x) \mathrm{e}^{-\frac{x^2}{2}} = 0 \text{ when } n \neq m. \tag{6.4}$$

The first few are given by

$$\begin{aligned} H_0(x) &= 1, \\ H_1(x) &= x, \\ H_2(x) &= x^2 - 1, \\ H_3(x) &= x^3 - 3x, \\ H_4(x) &= x^4 - 6x^2 + 3. \end{aligned} \tag{6.5}$$

In addition to the above ODE (6.3), they satisfy

$$\frac{\mathrm{d}}{\mathrm{d}x} H_n(x) = n H_{n-1}(x), \tag{6.6}$$

and the recursion

$$H_n(x) = x H_{n-1}(x) - (n-1) H_{n-2}(x), \tag{6.7}$$

which combined together recovers Eq. (6.3). Hermite polynomials can be written explicitly as

$$H_n(x) = \exp\left[-\frac{1}{2}\left(\frac{\mathrm{d}}{\mathrm{d}x}\right)^2\right] x^n = \sum_{m=0}^{\lfloor n/2 \rfloor} \frac{(-1)^m}{2^m} \frac{n!}{m!\,(n-2m)!} x^{n-2m}, \tag{6.8}$$

where $\lfloor n/2 \rfloor$ is the integer part of $n/2$.

Coming back to Eq. (6.1), we thus conclude that the exact solution for the Stieltjes transform $g_N(z)$ at finite N is

$$g_N(z) = \frac{H_N'(\sqrt{N}z)}{\sqrt{N} H_N(\sqrt{N}z)} \tag{6.9}$$

or, writing $H_N(x) = \prod_{i=1}^{N}(x - \sqrt{N} h_i^{(N)})$, where $\sqrt{N} h_i^{(N)}$ are the N (real) zeros of $H_N(x)$,

$$g_N(z) = \frac{1}{N} \sum_{i=1}^{N} \frac{1}{z - h_i^{(N)}}. \tag{6.10}$$

[2] Hermite polynomials can be defined using two different conventions for the unit Gaussian measure. We use here the "probabilists' Hermite polynomials", while the "physicists' convention" uses a Gaussian weight proportional to e^{-x^2}.

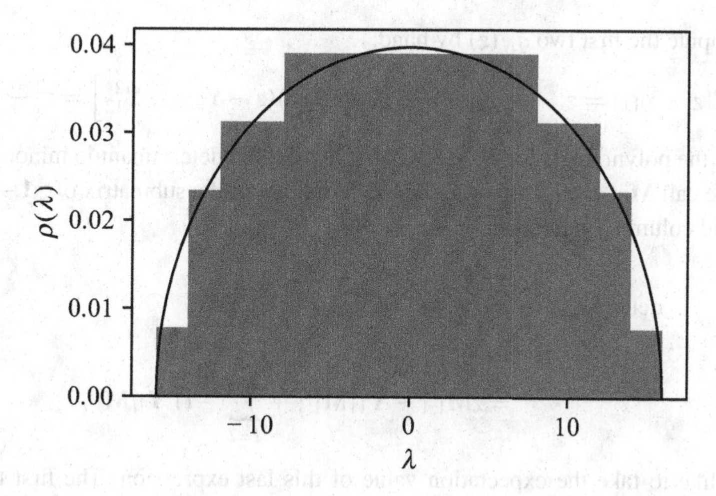

Figure 6.1 Histogram of the 64 zeros of $H_{64}(x)$. The full line is the asymptotic prediction from the semi-circle law.

Comparing with the definition of $g_N(z)$, Eq. (5.30), one concludes that the most likely positions of the Coulomb particles are exactly given by the zeros of Hermite polynomials (scaled by $\sqrt{N}$). This is a rather remarkable result, which holds for more general confining potentials, to which are associated different kinds of orthogonal polynomials. Explicit examples will be given later in this section for the Wishart ensemble, where we will encounter Laguerre polynomials (see also Chapter 7 for Jacobi polynomials).

Since we know that $g_N(z)$ converges, for large N, towards the Stieltjes transform of the semi-circle law, we can conclude that the rescaled zeros of Hermite polynomials are themselves distributed, for large N, according to the same semi-circle law. This classical property of Hermite polynomials is illustrated in Figure 6.1.

6.1.2 Expected Characteristic Polynomial of Wigner Matrices

In this section, we will show that the expected characteristic polynomial of a Wigner matrix $\mathbf{X}_N$, defined as $Q_N(z) := \mathbb{E}[\det(z\mathbf{1} - \mathbf{X}_N)]$, is given by the same Hermite polynomial as above. The idea is to write a recursion relation for $Q_N(z)$ by expanding the determinant in minors. Since we will be comparing Wigner matrices of different size, it is more convenient to work with unscaled matrices $\mathbf{Y}_N = \sqrt{N}\mathbf{X}_N$, i.e. symmetric matrices of size N with elements of zero mean and variance 1 (it will turn out that the variance of diagonal elements is actually irrelevant, and so can also be chosen to be 1). We define

$$q_N(z) := \mathbb{E}[\det(z\mathbf{1} - \mathbf{Y}_N)]. \tag{6.11}$$

Using $\det(\alpha\mathbf{A}) = \alpha^N \det(\mathbf{A})$, we then have

$$Q_N(z) = N^{-N/2} q_N(\sqrt{N}z). \tag{6.12}$$

We can compute the first two $q_N(z)$ by hand:

$$q_1(z) = \mathbb{E}[z - \mathbf{Y}_{11}] = z; \qquad q_2(z) = \mathbb{E}\left[(z - \mathbf{Y}_{11})(z - \mathbf{Y}_{22}) - \mathbf{Y}_{12}^2\right] = z^2 - 1. \quad (6.13)$$

To compute the polynomials for $N \geq 3$ we first expand the determinant in minors from the first line. We call $\mathbf{M}_{i,j}$ the ij-minor, i.e. the determinant of the submatrix of $z\mathbf{1} - \mathbf{Y}_N$ with the line i and column j removed:

$$\begin{aligned} \det(z\mathbf{1} - Y_N) &= \sum_{i=1}^{N} (-1)^{i+1}(z\delta_{i1} - \mathbf{Y}_{1i})\mathbf{M}_{1,i} \\ &= z\mathbf{M}_{1,1} - \mathbf{Y}_{11}\mathbf{M}_{1,1} + \sum_{i=2}^{N} (-1)^i \mathbf{Y}_{1i}\mathbf{M}_{1,i}. \end{aligned} \quad (6.14)$$

We would like to take the expectation value of this last expression. The first two terms are easy: the minor $\mathbf{M}_{1,1}$ is the same determinant with a Wigner matrix of size $N-1$, so $\mathbb{E}[\mathbf{M}_{1,1}] \equiv q_{N-1}(z)$; the diagonal element $\mathbf{Y}_{11}$ is independent from the rest of the matrix and its expectation is zero.

For the other terms in the sum, the minor $\mathbf{M}_{i,1}$ is not independent of $\mathbf{Y}_{i1}$. Indeed, because $\mathbf{X}_N$ is symmetric, the corresponding submatrix contains another copy of $\mathbf{Y}_{1i}$. Let us then expand $\mathbf{M}_{1,i}$ itself on the ith row, to make the other term $\mathbf{Y}_{i1}$ appear explicitly. For $i \neq 1$, we have

$$\begin{aligned} \mathbf{M}_{i,1} &= \sum_{j=1,\, j\neq i}^{N} (-1)^{i-j}\mathbf{Y}_{ij}\mathbf{M}_{1i,ij} \\ &= (-1)^{i-1}\mathbf{Y}_{i1}\mathbf{M}_{1i,i1} + \sum_{j=2,\, j\neq i}^{N} (-1)^{i-j}\mathbf{Y}_{ij}\mathbf{M}_{1i,ij}, \end{aligned} \quad (6.15)$$

where $\mathbf{M}_{ij,kl}$ is the "sub-minor", with rows i,j and columns k,l removed.

We can now take the expectation value of Eq. (6.14) by noting that $\mathbf{Y}_{1i}$ is independent of all the terms in Eq. (6.15) except the first one. We also realize that $\mathbf{M}_{1i,i1}$ is the same determinant with a Wigner matrix of size $N-2$ that is now independent of $\mathbf{Y}_{1i}$, so we have $\mathbb{E}[\mathbf{M}_{1i,i1}] = q_{N-2}(z)$. Putting everything together we get

$$q_N(z) := \mathbb{E}[\det(z\mathbf{1} - \mathbf{Y}_N)] = zq_{N-1}(z) - (N-1)q_{N-2}(z). \quad (6.16)$$

We recognize here precisely the recursion relation (6.7) that defines Hermite polynomials.

How should this result be interpreted? Suppose for one moment that the positions of the eigenvalues λ_i of $\mathbf{Y}_N$ were not fluctuating from sample to sample, and fixed to their most likely values λ_i^*. In this case, the expectation operator would not be needed and one would have

$$g_N(z) = \frac{\mathrm{d}}{\mathrm{d}z}\log Q_N(z) = \frac{H_N'(\sqrt{N}z)}{\sqrt{N}H_N(\sqrt{N}z)}, \quad (6.17)$$

recovering the result of the previous section. What is somewhat surprising is that

$$\mathbb{E}\left[\prod_{i=1}^{N}(z-\lambda_i)\right] \equiv \prod_{i=1}^{n}(z-\lambda_i^*) \tag{6.18}$$

even when fluctuations are accounted for. In particular, in the limit $N \to \infty$, the average Stieltjes transform should be computed from the average of the logarithm of the characteristic polynomial:

$$\mathfrak{g}(z) = \lim_{N\to\infty} \frac{1}{N}\mathbb{E}\left[\frac{\mathrm{d}}{\mathrm{d}z}\log\det(z\mathbf{1}-\mathbf{X}_N)\right]; \tag{6.19}$$

but the above calculation shows that one can compute the logarithm of the average characteristic polynomial instead. The deep underlying mechanism is the eigenvalue spectrum of random matrices is rigid – fluctuations around most probable positions are small.

Exercise 6.1.1 Hermite polynomials and moments of the Wigner

Show that (for $n \geq 4$)

$$Q_N(x) = x^n - \frac{n-1}{2}x^{n-2} + \frac{(n-1)(n-2)(n-3)}{8n}x^{n-4} + O\left(x^{n-6}\right), \tag{6.20}$$

therefore

$$g_N(z) = \frac{1}{z} - \frac{n-1}{N}\frac{1}{z^3} - \frac{(n-1)(2n-3)}{N}\frac{1}{z^5} + O\left(\frac{1}{z^7}\right), \tag{6.21}$$

so in the large N limit we recover the first few terms of the Wigner Stieltjes transform

$$\mathfrak{g}(z) = \frac{1}{z} - \frac{1}{z^3} - \frac{2}{z^5} + O\left(\frac{1}{z^7}\right). \tag{6.22}$$

6.2 Laguerre Polynomials

6.2.1 Most Likely Characteristic Polynomial of Wishart Matrices

Similarly to the case of Wigner matrices, the Stieltjes transform of the most likely positions of the Coulomb charges in the Wishart ensemble can be written as $g_N(z) := \psi'(z)/N\psi(z)$, where $\psi(z)$ is a monic polynomial of degree N satisfying

$$\psi''(x) - NV'(x)\psi'(x) + N^2\Pi_N(x)\psi(x) = 0, \tag{6.23}$$

with

$$NV'(x) = \frac{N-T-1+2\beta^{-1}}{x} + T, \tag{6.24}$$

and, using Eq. (5.32),

$$\Pi_N(x) = \frac{1}{N}\sum_{k=1}^{N} \frac{V'(x) - V'(\lambda_k^*)}{x - \lambda_k^*} = \frac{c_N}{x}, \tag{6.25}$$

where

$$c_N = -\frac{N - T - 1 + 2\beta^{-1}}{N^2}\sum_{k=1}^{N}\frac{1}{\lambda_k^*}. \tag{6.26}$$

Writing now $\psi(x) = \Psi(Tx)$ and $u = Tx$, Eq. (6.23) becomes

$$u\Psi''(u) - (N - T - 1 + 2\beta^{-1} + u)\Psi'(u) + \frac{N^2}{T}c_N\Psi(u) = 0. \tag{6.27}$$

This is the differential equation for the so-called associated Laguerre polynomials $L^{(\alpha)}$ with $\alpha = T - N - 2\beta^{-1}$. It has polynomial solutions of degree N if and only if the coefficient of the $\Psi(u)$ term is an integer equal to N (i.e. if $c_N = T/N$). The solution is then given by

$$\Psi(u) \propto L_N^{(T-N-2\beta^{-1})}(u), \tag{6.28}$$

where

$$L_n^{(\alpha)}(x) = x^{-\alpha}\frac{(\frac{\mathrm{d}}{\mathrm{d}x} - 1)^n}{n!}x^{\alpha+n}. \tag{6.29}$$

Note that associated Laguerre polynomials are orthogonal with respect to the measure $x^\alpha \mathrm{e}^{-x}$, i.e.

$$\int_0^\infty \mathrm{d}x x^\alpha \mathrm{e}^{-x} L_n^{(\alpha)}(x)L_m^{(\alpha)}(x) = \delta_{nm}\frac{(n+\alpha)!}{n!}. \tag{6.30}$$

Given that the standard associated Laguerre polynomials have as a leading term $(-1)^N/N!\,z^N$ and that $\psi(x)$ is monic, we finally find

$$\psi(x) = \begin{cases} (-1)^N N!\,T^{-N}L^{(T-N-2)}(Tx) & \text{real symmetric} \quad (\beta = 1), \\ (-1)^N N!\,T^{-N}L^{(T-N-1)}(Tx) & \text{complex Hermitian} \quad (\beta = 2). \end{cases} \tag{6.31}$$

Hence, the most likely positions of the Coulomb–Wishart charges are given by the zeros of associated Laguerre polynomials, exactly as the most likely positions of the Coulomb–Wigner charges are given by the zeros of Hermite polynomials.

We should nevertheless check that $c_N = T/N$ is compatible with Eq. (6.26), i.e. that the following equality holds:

$$\frac{T}{N} = \frac{\alpha+1}{N^2}\sum_{k=1}^{N}\frac{1}{\lambda_k^*}, \tag{6.32}$$

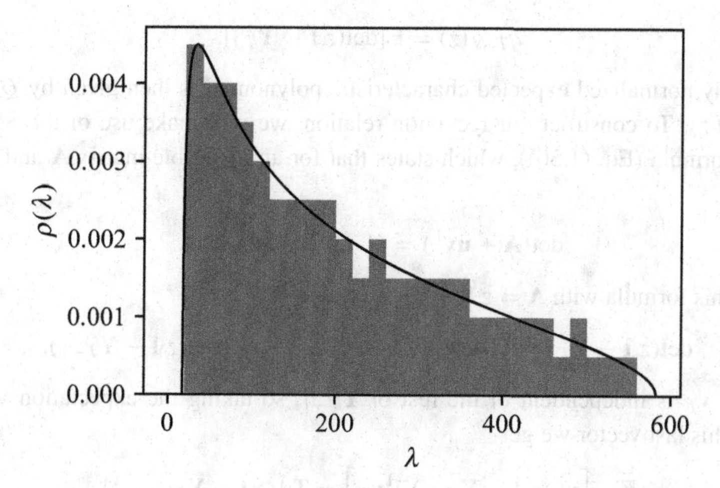

Figure 6.2 Histogram of the 100 zeros of $L_{100}^{(100)}(x)$. The full line is the Marčenko–Pastur distribution (4.43) with $q = \frac{1}{2}$ scaled by a factor $T = 200$.

where the λ_k^* are the zeros of $T^{-1}L(Tx)$, i.e. $\lambda_k^* = \ell_k^{(\alpha)}/T$, where $\ell_k^{(\alpha)}$ are the zeros of the associate Laguerre polynomials $L_N^{(\alpha)}$, which indeed obey the following relation:[3]

$$\frac{1}{N}\sum_{k=0}^{N}\frac{1}{\ell_k^{(\alpha)}} = \frac{1}{\alpha+1}. \tag{6.33}$$

From the results of Section 5.3.1, we thus conclude that the zeros of the Laguerre polynomials $L^{(T-N-2)}(Tx)$ converge to a Wishart distribution with $q = N/T$. Figure 6.2 shows the histogram of zeros of $L_{100}^{(100)}(x)$ with the asymptotic prediction for large N and T. Note that $\alpha \approx N(q^{-1} - 1)$ in that limit.

6.2.2 Average Characteristic Polynomial

As in the Wigner case we would like to get a recursion relation for $Q_{q,N}(z) := \mathbb{E}[\det(z\mathbf{1} - \mathbf{W}_q^{(N)})]$, where $\mathbf{W}_q^{(N)}$ is a white Wishart matrix of size N and parameter $q = T/N$. This time the recursion will be over T at N fixed. So we keep N fixed (we will drop the (N) index to keep the notation light) and consider an unnormalized white Wishart matrix:

$$\mathbf{Y}_T = \sum_{t=1}^{T}\mathbf{v}_t\mathbf{v}_t^T, \tag{6.34}$$

where $\mathbf{v}_t$ are T N-dimensional independent random vectors uniformly distributed on the sphere. We want to compute

[3] See e.g. Alıcı and Taeli [2015], where other inverse moments of the $\ell_k^{(\alpha)}$ are also derived.

$$q_{T,N}(z) = \mathbb{E}[\det(z\mathbf{1} - \mathbf{Y}_T)]. \tag{6.35}$$

The properly normalized expected characteristic polynomial is then given by $Q_{q,N}(z) = T^{-n} q_{T,N}(Tz)$. To construct our recursion relation, we will make use of the Shermann–Morrison formula (Eq. (1.30)), which states that for an invertible matrix $\mathbf{A}$ and vectors $\mathbf{u}$ and $\mathbf{v}$,

$$\det(\mathbf{A} + \mathbf{u}\mathbf{v}^T) = (1 + \mathbf{v}^T \mathbf{A}^{-1}\mathbf{u}) \det \mathbf{A}. \tag{6.36}$$

Applying this formula with $\mathbf{A} = z\mathbf{1} - \mathbf{Y}_{T-1}$, we get

$$\det(z\mathbf{1} - \mathbf{Y}_T) = (1 - \mathbf{v}_T^T (z\mathbf{1} - \mathbf{Y}_{T-1})^{-1}\mathbf{v}_T) \det(z\mathbf{1} - \mathbf{Y}_{T-1}). \tag{6.37}$$

The vector $\mathbf{v}_T$ is independent of the rest of $\mathbf{Y}_{T-1}$, so taking the expectation value with respect to this last vector we get

$$\mathbb{E}_{\mathbf{v}_T}\left[\mathbf{v}_T^T (z\mathbf{1} - \mathbf{Y}_{T-1})^{-1}\mathbf{v}_T\right] = \mathrm{Tr}[(z\mathbf{1} - \mathbf{Y}_{T-1})^{-1}]. \tag{6.38}$$

Now, using once again the general relation (easily derived in the basis where $\mathbf{A}$ is diagonal),

$$\mathrm{Tr}[(z\mathbf{1} - \mathbf{A})^{-1}] \det(z\mathbf{1} - \mathbf{A}) = \frac{\mathrm{d}}{\mathrm{d}z} \det(z\mathbf{1} - \mathbf{A}), \tag{6.39}$$

with $\mathbf{A} = \mathbf{Y}_{T-1}$, we can take the expectation value of Eq. (6.37). We obtain

$$q_{T,N}(z) = \left(1 - \frac{\mathrm{d}}{\mathrm{d}z}\right) q_{T-1,N}(z). \tag{6.40}$$

To start the recursion relation, we note that $\mathbf{Y}_0$ is the N-dimensional zero matrix for which $q_0(z) = z^N$. Hence,[4]

$$q_{T,N}(z) = \left(1 - \frac{\mathrm{d}}{\mathrm{d}z}\right)^T z^N. \tag{6.41}$$

If we apply an extra $\left(1 - \frac{\mathrm{d}}{\mathrm{d}z}\right)$ to Eq. (6.41), we get the following recursion relation:

$$q_{T+1,N}(z) = q_{T,N}(z) - N q_{T,N-1}(z), \tag{6.42}$$

which is similar to the classic "three-point rule" for Laguerre polynomials:

$$L_N^{(\alpha+1)}(x) = L_N^{(\alpha)}(x) + L_{N-1}^{(\alpha+1)}(x). \tag{6.43}$$

This allows us to make the identification

$$q_{T,N}(z) = (-1)^N N!\, L_N^{(T-N)}(z). \tag{6.44}$$

The correctly normalized average characteristic polynomial finally reads:

$$Q_{T,N}(z) = (-1)^N T^{-N} N!\, L_N^{(T-N)}(Tz). \tag{6.45}$$

[4] This relation will be further discussed in the context of finite free convolutions, see Chapter 12.

Hence, the average characteristic polynomial of real Wishart matrices is a Laguerre polynomial, albeit with a slightly different value of α compared to the one obtained in Eq. (6.31) above ($\alpha = T - N$ instead of $\alpha = T - N - 2$). The difference however becomes small when $N, T \to \infty$.

6.3 Unitary Ensembles

In this section we will discuss the average characteristic polynomial for unitary ensembles, ensembles of complex Hermitian matrices that are invariant under unitary transformations. Although this book mainly deals with real symmetric matrices, more is known about complex Hermitian matrices, so we want to give a few general results about these matrices that do not have a known equivalent in the real case. The main reason unitary ensembles are easier to deal with than orthogonal ones has to do with the Vandermonde determinant which is needed to change variables from matrix elements to eigenvalues. Recall that

$$|\det(\Delta(\mathbf{M}))| = |\det \mathbf{V}|^{\beta} \text{ with } \mathbf{V} = \begin{pmatrix} 1 & 1 & 1 & \dots & 1 \\ \lambda_1 & \lambda_2 & \lambda_3 & \dots & \lambda_N \\ \lambda_1^2 & \lambda_2^2 & \lambda_3^2 & \dots & \lambda_N^2 \\ \vdots & \vdots & \vdots & \ddots & \vdots \\ \lambda_1^{N-1} & \lambda_2^{N-1} & \lambda_3^{N-1} & \dots & \lambda_N^{N-1} \end{pmatrix} \tag{6.46}$$

in the orthogonal case $\beta = 1$, the absolute value sign is needed to get the correct result. In the case $\beta = 2$, $(\det \mathbf{V})^2$ is automatically positive and no absolute value is needed. The absolute value for $\beta = 1$ is very hard to deal with analytically, while for $\beta = 2$ the Vandermonde determinant is a polynomial in the eigenvalues.

6.3.1 Complex Wigner

In Section 6.1.2 we have shown that the expected characteristic polynomial of a unit variance real Wigner matrix is given by the Nth Hermite polynomial properly rescaled. The argument relied on two facts: (i) the expectation value of any element of a Wigner matrix is zero and (ii) all matrix elements are independent, save for

$$\mathbb{E}[\mathbf{W}_{ij}\mathbf{W}_{ji}] = 1/N \text{ for } i \neq j. \tag{6.47}$$

These two properties are shared by complex Wigner matrices. Therefore, the expected characteristic polynomial of a complex Wigner matrix is the same as for a real Wigner of the same size. We have shown that for a real or complex Wigner of size N,

$$Q_N(z) := \mathbb{E}[\det(z\mathbf{1} - \mathbf{W})] = N^{-N/2} H_N(\sqrt{N}z), \tag{6.48}$$

where $H_N(x)$ is the Nth Hermite polynomial, i.e. the Nth monic polynomial orthogonal in the following sense:

$$\int_{-\infty}^{\infty} H_i(x)H_j(x)e^{-x^2/2} = 0 \text{ when } i \neq j. \tag{6.49}$$

We can actually absorb the factors of N in $Q_N(z)$ in the measure $\exp(-x^2/2)$ and realize that the polynomial $Q_N(z)$ is the Nth monic polynomial orthogonal with respect to the measure

$$w_N(x) = \exp(-Nx^2/2). \tag{6.50}$$

There are two important remarks to be made about the orthogonality of $Q_N(x)$ with respect to the measure $w_N(x)$. First, $Q_N(x)$ is the Nth in a series of orthogonal polynomials with respect to an N-dependent measure. In particular $Q_M(x)$ for $M \neq N$ is an orthogonal polynomial coming from a different measure. Second, the measure $w_N(x)$ is exactly the weight coming from the matrix potential $\exp(-\beta NV(\lambda)/2)$ for $\beta = 2$ and $V(x) = x^2/2$. In Section 6.3.3, we will see that these two statements are true for a general potential $V(x)$ when $\beta = 2$.

6.3.2 Complex Wishart

A complex white Wishart matrix can be written as a normalized sum of rank-1 complex Hermitian projectors:

$$\mathbf{W} = \frac{1}{T}\sum_{t=1}^{T} \mathbf{v}_t\mathbf{v}_t^\dagger, \tag{6.51}$$

where the vectors $\mathbf{v}_t$ are vectors of IID complex Gaussian numbers with zero mean and normalized as

$$\mathbb{E}[\mathbf{v}_t\mathbf{v}_t^\dagger] = \mathbf{1}. \tag{6.52}$$

The derivation of the average characteristic polynomial in the Wishart case in Section 6.2.2 only used the independence of the vectors $\mathbf{v}_t$ and the expectation value $\mathbb{E}[\mathbf{v}_t\mathbf{v}_t^T] = \mathbf{1}$. So, by replacing the matrix transposition by the Hermitian conjugation in the derivation we can show that the expected characteristic polynomial of a complex white Wishart of size N is also given by a Laguerre polynomial, as in Eq. (6.45). The Laguerre polynomials $L_k^{(T-N)}(x)$ are orthogonal in the sense of Eq. (6.30), with $\alpha = T - N$. As in the Wigner case, we can include the extra factor of T in the orthogonality weight and realize that the expected characteristic polynomial of a real or complex Wishart matrix is the Nth monic polynomial orthogonal with respect to the weight:

$$w_N(x) = x^{T-N}e^{-Tx} \text{ for } 0 \leq x < \infty. \tag{6.53}$$

This weight function is precisely the single eigenvalue weight, without the Vandermonde term, of a complex Hermitian white Wishart of size N (see the footnote on page 46).

Note that the normalization of the weight $w_N(x)$ is irrelevant: the condition that the polynomial is monic uniquely determines its normalization. Note as well that the real case is given by the same polynomials, i.e. polynomials that are orthogonal with respect to the *complex* weight, which is different from the real weight.

6.3.3 General Potential $V(x)$

The average characteristic polynomial for a matrix of size N in a unitary ensemble with potential $V(\mathbf{M})$ is given by

$$Q_N(x) := \mathbb{E}[\det(z\,\mathbf{1} - \mathbf{M})], \tag{6.54}$$

which we can express via the joint law of the eigenvalues of $\mathbf{M}$:

$$Q_N(z) \propto \int \mathrm{d}^N\mathbf{x} \prod_{k=1}^{N}(z - x_k)\, \Delta^2(\mathbf{x}) \mathrm{e}^{-N\sum_{k=1}^{N} V(x_k)}, \tag{6.55}$$

where $\Delta(\mathbf{x})$ is the Vandermonde determinant:

$$\Delta(\mathbf{x}) = \prod_{k<\ell}(x_\ell - x_k). \tag{6.56}$$

We do not need to worry about the normalization of the above expectation value as we know that $Q_N(z)$ is a monic polynomial of degree N. In other words, the condition $Q_N(z) = z^N + O(z^{N-1})$ is sufficient to properly normalize $Q_N(z)$. The first step of the computation is to combine one of the two Vandermonde determinants with the product of $(z - x_k)$:

$$\Delta^2(\mathbf{x}) \prod_{k=1}^{N}(z - x_k) = \Delta(\mathbf{x}) \prod_{k<\ell}(x_\ell - x_k) \prod_{k=1}^{N}(z - x_k) \equiv \Delta(\mathbf{x})\Delta(\mathbf{x}; z), \tag{6.57}$$

where $\Delta(\mathbf{x}; z)$ is a Vandermonde determinant of $N + 1$ variables, namely the N variables x_k and the extra variable z.

The second step is to write the determinants in the Vandermonde form:

$$\Delta(\mathbf{x}) = \det \begin{pmatrix} 1 & 1 & 1 & \dots & 1 \\ x_1 & x_2 & x_3 & \dots & x_N \\ x_1^2 & x_2^2 & x_3^2 & \dots & x_N^2 \\ \vdots & \vdots & \vdots & \ddots & \vdots \\ x_1^{N-1} & x_2^{N-1} & x_3^{N-1} & \dots & x_N^{N-1} \end{pmatrix}. \tag{6.58}$$

We can add or subtract to any line a multiple of any other line and not change the above determinant. By doing so we can transform all monomials x_ℓ^k into a monic polynomial of degree k of our choice, so we have

$$\Delta(\mathbf{x}) = \det \begin{pmatrix} 1 & 1 & 1 & \dots & 1 \\ p_1(x_1) & p_1(x_2) & p_1(x_3) & \dots & p_1(x_N) \\ p_2(x_1) & p_2(x_2) & p_2(x_3) & \dots & p_2(x_N) \\ \vdots & \vdots & \vdots & \ddots & \vdots \\ p_{N-1}(x_1) & p_{N-1}(x_2) & p_{N-1}(x_3) & \dots & p_{N-1}(x_N) \end{pmatrix}. \tag{6.59}$$

We will choose the polynomials $p_n(x)$ to be the monic polynomials orthogonal with respect to the measure $w(x) = \mathrm{e}^{-NV(x)}$, this will turn out to be extremely useful. We can now perform the integral of the vector $\mathbf{x}$ in the following expression:

$$Q_N(z) \propto \int \mathrm{d}^N\mathbf{x}\, \Delta(\mathbf{x})\Delta(\mathbf{x};z)\mathrm{e}^{-N\sum_{k=1}^{N} V(x_k)}. \tag{6.60}$$

If we expand the two determinants $\Delta(\mathbf{x})$ and $\Delta(\mathbf{x};z)$ as signed sums over all permutations and take their product, we realize that in each term each variable x_k will appear exactly twice in two polynomials, say $p_n(x_k)$ and $p_m(x_k)$, but by orthogonality we have

$$\int \mathrm{d}x_k\, p_n(x_k)p_m(x_k)\mathrm{e}^{-NV(x_k)} = Z_n\delta_{mn}, \tag{6.61}$$

where Z_n is a normalization constant that will not matter in the end. The only terms that will survive are those for which every x_k appears in the same polynomial in both determinants. For this to happen, the variable z must appear as $p_N(z)$, the only polynomial not in the first determinant. So this trick allows us to conclude with very little effort that

$$Q_N(z) \propto p_N(z). \tag{6.62}$$

But since both $Q_N(z)$ and $p_N(z)$ are monic, they must be equal. We have just shown that for a Hermitian matrix $\mathbf{M}$ of size N drawn from a unitary ensemble with potential $V(x)$,

$$\mathbb{E}[\det(z\,\mathbf{1} - \mathbf{M})] = p_N(z), \tag{6.63}$$

where $p_N(x)$ is the Nth monic orthogonal polynomial with respect to the measure $\mathrm{e}^{-NV(x)}$.

It is possible to generalize this result to expectation of products of characteristic polynomials evaluated at K different points z_k, which allows one to study the joint distribution of K eigenvalues. We give here the result without proof.[5] We first define the expectation value of a product of K characteristic polynomials:

$$F_K(z_1, z_2, \dots, z_K) := \mathbb{E}[\det(z_1\mathbf{1} - \mathbf{M})\det(z_2\mathbf{1} - \mathbf{M})\dots\det(z_K\mathbf{1} - \mathbf{M})]. \tag{6.64}$$

The multivariate function F_K can be expressed as a determinant of orthogonal polynomials:

$$F_K(z_1, z_2, \dots, z_K) = \frac{1}{\Delta}\det \begin{pmatrix} p_N(z_1) & p_N(z_2) & \dots & p_N(z_K) \\ p_{N+1}(z_1) & p_{N+1}(z_2) & \dots & p_{N+1}(z_K) \\ \vdots & \vdots & \ddots & \vdots \\ p_{N+K-1}(z_1) & p_{N+K-1}(z_2) & \dots & p_{N+K-1}(z_K) \end{pmatrix}, \tag{6.65}$$

[5] See Brézin and Hikami [2011] for a derivation. Note that their formula equivalent to Eq. (6.67) is missing the K-dependent constant factor.

where $\Delta := \Delta(z_1, z_2, \ldots, z_K)$ is the usual Vandermonde determinant and the $p_\ell(x)$ are the monic orthogonal polynomials orthogonal with respect to $e^{-NV(x)}$. When $K = 1$, $\Delta = 1$ by definition and we recover our previous result $F_1(z) = p_N(z)$. When the arguments of F_K are not all different, Eq. (6.65) gives an undetermined result (0/0) but the limit is well defined. A useful case is when all arguments are equal:

$$F_K(z) := F_K(z, z, \ldots, z) = \mathbb{E}[\det(z\mathbf{1} - \mathbf{M})^K]. \tag{6.66}$$

Taking the limit of Eq. (6.65) we find the rather simple result

$$F_K(z) = \frac{1}{\prod_{\ell=0}^{K-1} \ell!} \det \begin{pmatrix} p_N(z) & p'_N(z) & \cdots & p_N^{(K-1)}(z) \\ p_{N+1}(z) & p'_{N+1}(z) & \cdots & p_{N+1}^{(K-1)}(z) \\ \vdots & \vdots & \ddots & \vdots \\ p_{N+K-1}(z) & p'_{N+K-1}(z) & \cdots & p_{N+K-1}^{(K-1)}(z) \end{pmatrix}, \tag{6.67}$$

where $p_\ell^{(k)}(x)$ is the kth derivative of the ℓth polynomial. In particular the average-square characteristic polynomial is given by

$$F_2(z) = p_N(z)p'_{N+1}(z) - p'_N(z)p_{N+1}(z). \tag{6.68}$$

Exercise 6.3.1 Variance of the Characteristic Polynomial of a 2×2 Hermitian Wigner Matrix

(a) Show that the characteristic polynomial of a 2×2 Hermitian Wigner matrix is given by

$$Q_2^{\mathbf{W}}(z) = (z - w_{11})(z - w_{22}) - (w_{12}^{\mathrm{R}})^2 - (w_{12}^{\mathrm{I}})^2, \tag{6.69}$$

where $w_{11}, w_{22}, w_{12}^{\mathrm{R}}$ and w_{12}^{I} are four real independent Gaussian random numbers with variance 1 for the first two and 1/2 for the other two.

(b) Compute directly the mean and the variance of $Q_2^{\mathbf{W}}(z)$.

(c) Use Eqs. (6.63) and (6.68) and the first few Hermite polynomials given in Section 6.1.1 to obtain the same result, namely $\mathbb{V}[Q_2^{\mathbf{W}}(z)] = 2z^2 + 2$.

Bibliographical Notes

- On orthogonal polynomials, definitions and recursion relations:

  - G. Szegő. *Orthogonal Polynomials*. AMS Colloquium Publications, volume 23. American Mathematical Society, 1975,
  - D. Zwillinger, V. Moll, I. Gradshteyn, and I. Ryzhik, editors. *Table of Integrals, Series, and Products (Eighth Edition)*. Academic Press, New York, 2014,
  - R. Beals and R. Wong. *Special Functions and Orthogonal Polynomials*. Cambridge Studies in Advanced Mathematics. Cambridge University Press, Cambridge, 2016.

- On the relation between Hermite polynomials, log gases and the Calogero model:
  - S. Agarwal, M. Kulkarni, and A. Dhar. Some connections between the classical Calogero-Moser model and the log gas. *Journal of Statistical Physics*, 176(6), 1463–1479, 2019.
- Orthogonal polynomials and random matrix theory:
  - M. L. Mehta. *Random Matrices*. Academic Press, San Diego, 3rd edition, 2004,
  - E. Brézin and S. Hikami. Characteristic polynomials. In *The Oxford Handbook of Random Matrix Theory*. Oxford University Press, Oxford, 2011,
  - P. Deift. *Orthogonal Polynomials and Random Matrices: A Riemann-Hilbert Approach*. Courant Institute, New York, 1999.
- For identities obeyed by the zeros of Laguerre polynomials:
  - H. Alıcı and H. Taeli. Unification of Stieltjes-Calogero type relations for the zeros of classical orthogonal polynomials. *Mathematical Methods in the Applied Sciences*, 38(14):3118–3129, 2015.

7
The Jacobi Ensemble*

So far we have encountered two classical random matrix ensembles, namely Wigner and Wishart. They are, respectively, the matrix equivalents of the Gaussian and the gamma distribution. For example a 1×1 Wigner matrix is a single Gaussian random number and a 1×1 Wishart is a gamma distributed number (see Eq. (4.16) with $N = 1$). We also saw in Chapter 6 that these ensembles are intimately related to classical orthogonal polynomials, respectively Hermite and Laguerre. The Gaussian distribution and its associated Hermite polynomials appear very naturally in contexts where the underlying variable is unbounded above and below. Gamma distributions and Laguerre polynomials appear in problems where the variable is bounded from below (e.g. positive variables). Variables that are bounded both from above and from below have their own natural distribution and associated classical orthogonal polynomials, namely the beta distribution and Jacobi polynomials.

In this chapter, we introduce a third classical random matrix ensemble: the Jacobi ensemble. It is the random matrix equivalent of the beta distribution (and hence often called matrix variate beta distribution). It will turn out to be strongly linked to Jacobi orthogonal polynomials.

Jacobi matrices appear in multivariate analysis of variance and hence the Jacobi ensemble is sometimes called the MANOVA ensemble. An important special case of the Jacobi ensemble is the arcsine law which we already encountered in Section 5.5, and will again encounter in Section 15.3.1. It is the law governing Coulomb repelling eigenvalues with no external forces save for two hard walls. It also shows up in simple problems of matrix addition and multiplications for matrices with only two eigenvalues.

7.1 Properties of Jacobi Matrices

7.1.1 Construction of a Jacobi Matrix

A beta-distributed random variable $x \in (0, 1)$ has the following law:

$$P_{c_1,c_2}(x) = \frac{\Gamma(c_1)\Gamma(c_2)}{\Gamma(c_1 + c_2)} x^{c_1-1}(1 - x)^{c_2-1}, \tag{7.1}$$

where $c_1 > 0$ and $c_2 > 0$ are two parameters characterizing the law.

To generalize (7.1) to matrices, we could define $\mathbf{J}$ as a matrix generated from a beta ensemble with a matrix potential that tends to $V(x) = -\log(P_{c_1,c_2}(x))$ in the large N limit. Although this is indeed the result we will get in the end, we would rather use a more constructive approach that will give us a sensible definition of the matrix $\mathbf{J}$ at finite N and for the three standard values of β.

A beta(c_1, c_2) random number can alternatively be generated from two gamma-distributed variables:

$$x = \frac{w_1}{w_1 + w_2}, \qquad w_{1,2} \sim \text{Gamma}(c_{1,2}, 1). \tag{7.2}$$

The same relation can be rewritten as

$$x = \frac{1}{1 + w_1^{-1} w_2}. \tag{7.3}$$

This is the formula we need for our matrix generalization. An unnormalized white Wishart with $T = cN$ is the matrix generalization of a Gamma$(c, 1)$ random variable. Combining two such matrices as above will give us our Jacobi random matrix $\mathbf{J}$. One last point before we proceed, we need to symmetrize the combination $w_1^{-1} w_2$ to yield a symmetric matrix. We choose $\sqrt{w_2} w_1^{-1} \sqrt{w_2}$ which makes sense as Wishart matrices (like gamma-distributed numbers) are positive definite.

We can now define the Jacobi matrix. Let $\mathbf{E}$ be the symmetrized product of a white Wishart and the inverse of another independent white Wishart, both without the usual $1/T$ normalization:

$$\mathbf{E} = \widetilde{\mathbf{W}}_2^{1/2} \widetilde{\mathbf{W}}_1^{-1} \widetilde{\mathbf{W}}_2^{1/2}, \quad \text{where} \quad \widetilde{\mathbf{W}}_{1,2} := \mathbf{H}_{1,2} \mathbf{H}_{1,2}^T. \tag{7.4}$$

The two matrices $\mathbf{H}_{1,2}$ are rectangular matrices of standard Gaussian random numbers with aspect ratio $c_1 = T_1/N$ and $c_2 = T_2/N$ (note that the usual aspect ratio is $q = c^{-1}$). The standard Jacobi matrix is defined as

$$\mathbf{J} = (\mathbf{1} + \mathbf{E})^{-1}. \tag{7.5}$$

A Jacobi matrix has all its eigenvalues between 0 and 1. For the matrices $\widetilde{\mathbf{W}}_{1,2}$ to make sense we need $c_{1,2} > 0$. In addition, to ensure that $\widetilde{\mathbf{W}}_1$ is invertible we need to impose $c_1 > 1$. It turns out that we can relax that assumption later and the ensemble still makes sense for any $c_{1,2} > 0$.

7.1.2 Joint Law of the Elements

The joint law of the elements of a Jacobi matrix for $\beta = 1, 2$ or 4 is given by

$$P_\beta(\mathbf{J}) = c_J^{\beta, T_1, T_2} [\det(\mathbf{J})]^{\beta(T_1 - N + 1)/2 - 1} [\det(\mathbf{1} - \mathbf{J})]^{\beta(T_2 - N + 1)/2 - 1}, \tag{7.6}$$

$$c_J^{\beta, T_1, T_2} = \prod_{j=1}^{N} \frac{\Gamma(1 + \beta/2)\Gamma(\beta(T_1 + T_2 - N + j)/2)}{\Gamma(1 + \beta j/2)\Gamma(\beta(T_1 - N + j)/2)\Gamma(\beta(T_2 - N + j)/2)}, \tag{7.7}$$

over the space of matrices of the proper symmetry such that both $\mathbf{J}$ and $\mathbf{1}-\mathbf{J}$ are positive definite.

To obtain this result, one needs to know the law of Wishart matrices (Chapter 4) and the law of a matrix given the law of its inverse (Chapter 1).

Here is the derivation in the real symmetric case. We first write the law of the matrix $\mathbf{E}$ by realizing that for a fixed matrix $\widetilde{\mathbf{W}}_1$, the matrix $\mathbf{E}/T_2$ is a Wishart matrix with $T = T_2$ and true covariance $\widetilde{\mathbf{W}}_1^{-1}$. From Eq. (4.16), we thus have

$$P(\mathbf{E}|\mathbf{W}_1) = \frac{(\det \mathbf{E})^{(T_2-N-1)/2}(\det \widetilde{\mathbf{W}}_1)^{T_2/2}}{2^{NT_2/2}\Gamma_N(T_2/2)} \exp\left[-\frac{1}{2}\operatorname{Tr}\left(\mathbf{E}\widetilde{\mathbf{W}}_1\right)\right]. \tag{7.8}$$

The matrix $\widetilde{\mathbf{W}}_1/T_1$ is itself a Wishart with probability

$$P\left(\widetilde{\mathbf{W}}_1\right) = \frac{(\det \widetilde{\mathbf{W}}_1)^{(T_1-N-1)/2}}{2^{NT_1/2}\Gamma_N(T_1/2)} \exp\left[-\frac{1}{2}\operatorname{Tr}\left(\widetilde{\mathbf{W}}_1\right)\right]. \tag{7.9}$$

Averaging Eq. (7.8) with respect to $\widetilde{\mathbf{W}}_1$ we find

$$\begin{aligned} P(\mathbf{E}) &= \frac{(\det \mathbf{E})^{(T_2-N-1)/2}}{2^{N(T_1+T_2)/2}\Gamma_N(T_1/2)\Gamma_N(T_2/2)} \\ &\quad \times \int d\mathbf{W} (\det \mathbf{W})^{(T_1+T_2-N-1)/2} \exp\left[-\frac{1}{2}\operatorname{Tr}\left((\mathbf{1}+\mathbf{E})\mathbf{W}\right)\right]. \end{aligned} \tag{7.10}$$

We can perform the integral over $\mathbf{W}$ by realizing that $\mathbf{W}/T$ is a Wishart matrix with $T = T_1 + T_2$ and true covariance $\mathbf{C} = (\mathbf{1}+\mathbf{E})^{-1}$, see Eq. (4.16). We just need to introduce the correct power of $\det \mathbf{C}$ and numerical factors to make the integral equal to 1. Thus,

$$P(\mathbf{E}) = \frac{\Gamma_N((T_1+T_2)/2)}{\Gamma_N(T_1/2)\Gamma_N(T_2/2)} (\det \mathbf{E})^{(T_2-N-1)/2} (\det(\mathbf{1}+\mathbf{E}))^{-(T_1+T_2)/2}. \tag{7.11}$$

The miracle that for any N one can integrate exactly the product of a Wishart matrix and the inverse of another Wishart matrix will appear again in the Bayesian theory of SCM (see Section 18.3).

Writing $\mathbf{E}_+ = \mathbf{E} + \mathbf{1}$ we find

$$P(\mathbf{E}_+) = \frac{\Gamma_N((T_1+T_2)/2)}{\Gamma_N(T_1/2)\Gamma_N(T_2/2)} (\det(\mathbf{E}_+ - \mathbf{1}))^{(T_2-N-1)/2} (\det \mathbf{E}_+)^{-(T_1+T_2)/2}. \tag{7.12}$$

Finally we want $\mathbf{J} := \mathbf{E}_+^{-1}$. The law of the inverse of a symmetric matrix $\mathbf{A} = \mathbf{M}^{-1}$ of size N is given by (see Section 1.2.7)

$$P_{\mathbf{A}}(\mathbf{A}) = P_{\mathbf{M}}(\mathbf{A}^{-1}) \det(\mathbf{A})^{-N-1}. \tag{7.13}$$

Hence,

$$P(\mathbf{J}) = \frac{\Gamma_N((T_1+T_2)/2)}{\Gamma_N(T_1/2)\Gamma_N(T_2/2)} \left(\det(\mathbf{J}^{-1} - \mathbf{1})\right)^{(T_2-N-1)/2} (\det \mathbf{J})^{(T_1+T_2)/2-N-1}. \tag{7.14}$$

Using $\det(\mathbf{J}^{-1} - \mathbf{1})\det(\mathbf{J}) = \det(\mathbf{1} - \mathbf{J})$ we can reorganize the powers of the determinants and get

$$P(\mathbf{J}) = \frac{\Gamma_N((T_1+T_2)/2)}{\Gamma_N(T_1/2)\Gamma_N(T_2/2)} (\det(\mathbf{1} - \mathbf{J}))^{(T_2-N-1)/2} (\det \mathbf{J})^{(T_1-N-1)/2}, \tag{7.15}$$

which is equivalent to Eq. (7.6) for $\beta = 1$.

7.1.3 Potential and Stieltjes Transform

The Jacobi ensemble is a beta ensemble satisfying Eq. (5.2) with matrix potential

$$V(x) = -\frac{T_1 - N + 1 - 2/\beta}{N}\log(x) - \frac{T_2 - N + 1 - 2/\beta}{N}\log(1-x). \tag{7.16}$$

The derivative of the potential, in the large N limit, is given by

$$V'(x) = \frac{c_1 - 1 - (c_1 + c_2 - 2)x}{x(x-1)}. \tag{7.17}$$

The function $V'(x)$ is not a polynomial or a Laurent polynomial but the function $x(x-1)V'(x)$ is a degree one polynomial. With a slight modification of the argument of Section 5.2.2, we can show that $x(x-1)\Pi(x) = r + sx$ is a degree one polynomial. In the large N limit we then have

$$\mathfrak{g}(z) = \frac{c_1 - 1 - (c_+ - 2)z + \sqrt[\oplus]{c_+^2 z^2 - 2(c_1 c_+ + c_-)z + (c_1 - 1)^2}}{2z(z-1)}, \tag{7.18}$$

where we have used the fact that we need $s = 0$ and $r = 1 + c_1 + c_2$ to get a $1/z$ behavior at infinity; we used the shorthand $c_\pm = c_2 \pm c_1$.

From the large z limit of (7.18) one can read off the normalized trace (or the average eigenvalue):

$$\tau(\mathbf{J}) = \frac{c_1}{c_1 + c_2}, \tag{7.19}$$

which is equal to one-half when $c_1 = c_2$. For $c_1 > 1$ and $c_2 > 1$, there are no poles, and eigenvalues exist only when the argument of the square-root is negative. The density of eigenvalues is therefore given by (see Fig. 7.1)

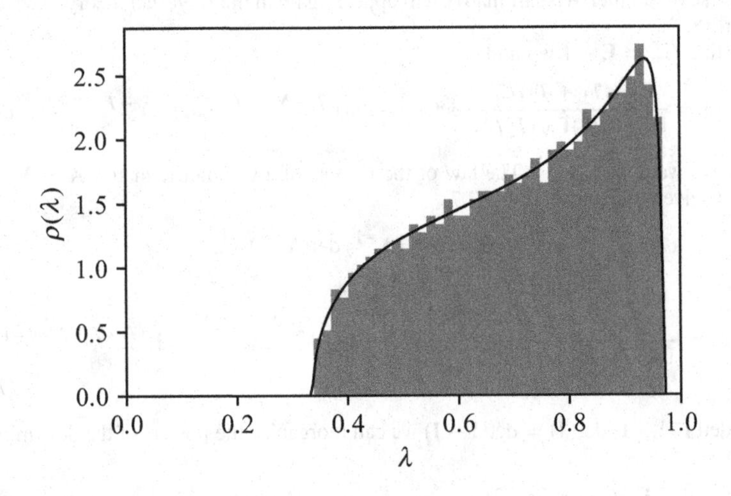

Figure 7.1 Density of eigenvalues for the Jacobi ensemble with $c_1 = 5$ and $c_2 = 2$. The histogram is a simulation of a single $N = 1000$ matrix with the same parameters.

$$\rho(\lambda) = c_+ \frac{\sqrt{(\lambda_+ - \lambda)(\lambda - \lambda_-)}}{2\pi\lambda(1-\lambda)}, \tag{7.20}$$

where the edges of the spectrum are given by

$$\lambda_\pm = \frac{c_1 c_+ + c_- \pm 2\sqrt{c_1 c_2 (c_+ - 1)}}{c_+^2}. \tag{7.21}$$

For $0 < c_1 < 1$ or $0 < c_2 < 1$, Eq. (7.18) will have Dirac deltas at $z = 0$ or $z = 1$, depending on cases (see Exercise 7.1.1).

In the symmetric case $c_1 = c_2 = c$, we have explicitly

$$\mathfrak{g}(z) = \frac{(c-1)(1-2z) + \sqrt[\oplus]{c^2(2z-1)^2 - c(c-2)}}{2z(z-1)}. \tag{7.22}$$

The density for $c \geq 1$ has no Dirac mass and is given by

$$\rho(\lambda) = c\frac{\sqrt{(\lambda_+ - \lambda)(\lambda - \lambda_-)}}{\pi\lambda(1-\lambda)} \tag{7.23}$$

with the edges given by

$$\lambda_\pm = \frac{1}{2} \pm \frac{\sqrt{2c-1}}{2c}. \tag{7.24}$$

Note that the distribution is symmetric around $\lambda = 1/2$ (see Fig. 7.2).

As $c \to 1$, the edges tend to 0 and 1 and we recover the arcsine law:

$$\rho(\lambda) = \frac{1}{\pi\sqrt{\lambda(1-\lambda)}}. \tag{7.25}$$

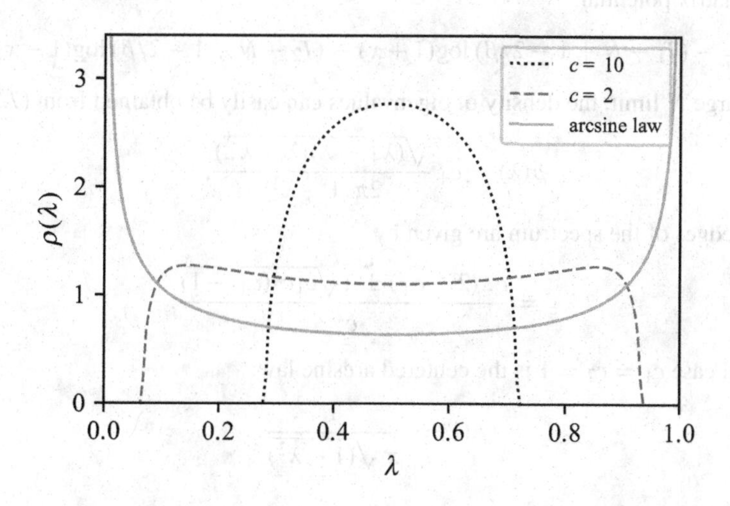

Figure 7.2 Density of eigenvalues for a Jacobi matrix in the symmetric case ($c_1 = c_2 = c$) for $c = 20, 2$ and 1. The case $c = 1$ is the arcsine law.

Exercise 7.1.1 Dirac masses in the Jacobi density

(a) Assuming that $c_1 > 1$ and $c_2 > 1$ show that there are no poles in Eq. (7.18) at $z = 0$ or $z = 1$ by showing that the numerator vanishes for these two values of z.

(b) The parameters $c_{1,2}$ can be smaller than 1 (as long as they are positive). Show that in that case $\mathfrak{g}(z)$ can have poles at $z = 0$ and/or $z = 1$ and find the residue at these poles.

7.2 Jacobi Matrices and Jacobi Polynomials

7.2.1 Centered-Range Jacobi Ensemble

The standard Jacobi matrix defined above has all its eigenvalues between 0 and 1. We would like to use another definition of the Jacobi matrix with eigenvalues between -1 and 1. This will make easier the link with orthogonal polynomials. We define the centered-range Jacobi matrix[1]

$$\mathbf{J}_{\mathrm{c}} = 2\mathbf{J} - 1. \tag{7.26}$$

This definition is equivalent to

$$\mathbf{J}_{\mathrm{c}} = \widetilde{\mathbf{W}}_+^{-1/2}(\widetilde{\mathbf{W}}_-)\widetilde{\mathbf{W}}_+^{-1/2}, \text{ where } \widetilde{\mathbf{W}}_\pm = \mathbf{H}_1\mathbf{H}_1^T \pm \mathbf{H}_2\mathbf{H}_2^T, \tag{7.27}$$

with $\mathbf{H}_{1,2}$ as above.

The matrix $\mathbf{J}_{\mathrm{c}}$ is still a member of a beta ensemble satisfying Eq. (5.2) with a slightly modified matrix potential:

$$NV(x) = -(T_1 - N + 1 - 2/\beta)\log(1+x) - (T_2 - N + 1 - 2/\beta)\log(1-x). \tag{7.28}$$

In the large N limit, the density of eigenvalues can easily be obtained from (7.20):

$$\rho(\lambda) = c_+ \frac{\sqrt{(\lambda_+ - \lambda)(\lambda - \lambda_-)}}{2\pi(1-\lambda^2)}, \tag{7.29}$$

where the edges of the spectrum are given by

$$\lambda_\pm = \frac{c_-(2 - c_+) \pm 4\sqrt{c_1 c_2 (c_+ - 1)}}{c_+^2}. \tag{7.30}$$

The special case $c_1 = c_2 = 1$ is the centered arcsine law:

$$\rho(\lambda) = \frac{1}{\pi\sqrt{(1-\lambda^2)}}. \tag{7.31}$$

[1] Note that the matrix $\mathbf{J}_{\mathrm{c}}$ is not necessarily centered in the sense $\tau(\mathbf{J}_{\mathrm{c}}) = 0$ but the potential range of its eigenvalues is $[-1, 1]$ centered around zero.

7.2.2 Average Expected Characteristic Polynomial

In Section 6.3.3, we saw that the average characteristic polynomial when $\beta = 2$ is the Nth monic polynomial orthogonal to the weight $w(x) \propto \exp(-NV(x))$. For the centered-range Jacobi matrix we have

$$w(x) = (1+x)^{T_1-N}(1-x)^{T_2-N}. \tag{7.32}$$

The Jacobi polynomials $P_n^{(a,b)}(x)$ are precisely orthogonal to such weight functions with $a = T_2 - N$ and $b = T_1 - N$. (Note that unfortunately the standard order of the parameters is inverted with respect to Jacobi matrices.) Jacobi polynomials satisfy the following differential equation:

$$(1-x^2)y'' + (b-a-(a+b+2)x)y' + n(n+a+b+1)y = 0. \tag{7.33}$$

This equation has polynomial solutions if and only if n is an integer. The solution is then $y \propto P_n^{(a,b)}(x)$.

The first three Jacobi polynomials are

$$\begin{aligned} P_0^{(a,b)}(x) &= 1, \\ P_1^{(a,b)}(x) &= \frac{(a+b+2)}{2}x + \frac{a-b}{2}, \\ P_2^{(a,b)}(x) &= \frac{(a+b+3)(a+b+4)}{8}(x-1)^2 + \frac{(a+2)(a+b+3)}{2}(x-1) \\ &\quad + \frac{(a+1)(a+2)}{2}. \end{aligned} \tag{7.34}$$

The normalization of Jacobi polynomials is arbitrary but in the standard normalization they are not monic. The coefficient of x^n in $P_n^{(a,b)}(x)$ is

$$a_n = \frac{\Gamma[a+b+2n+1]}{2^n n!\,\Gamma[a+b+n-1]}. \tag{7.35}$$

In summary we have (for $\beta = 2$)

$$\mathbb{E}\left[\det[z\mathbf{1} - \mathbf{J}_c]\right] = \frac{2^N N!\,\Gamma[T_1+T_2+1-N]}{\Gamma[T_1+T_2+1]} P_N^{(T_2-N,T_1-N)}(z). \tag{7.36}$$

Note that we must have $T_1 \geq N$ and $T_2 \geq N$.

When $T_1 = T_2 = N$ (i.e. $c_1 = c_2 = 1$, corresponding to the arcsine law), the polynomials $P_N^{(0,0)}(z)$ are called Legendre polynomials $P_N(z)$:[2]

$$\mathbb{E}\left[\det[z\mathbf{1} - \mathbf{Y}_c]\right] = \frac{2^N (N!)^2}{(2N)!} P_N(z). \tag{7.38}$$

[2] Legendre polynomials are defined as the polynomial solution of

$$\frac{d}{dx}\left[(1-x^2)\frac{dP_n(x)}{dx}\right] + n(n+1)P_n(x) = 0, \qquad P_n(1) = 1. \tag{7.37}$$

7.2.3 Maximum Likelihood Configuration at Finite N

In Chapter 6, we studied the most likely configuration of eigenvalues for the beta ensemble at finite N. We saw that the finite-N Stieltjes transform of this solution $g_N(z)$ is related to a monic polynomial $\psi(x)$ via

$$g_N(z) = \frac{1}{N}\sum_{i=1}^{N}\frac{1}{z-\lambda_i} = \frac{\psi'(z)}{N\psi(z)}, \quad \text{where} \quad \psi(x) = \prod_{i=1}^{N}(x-\lambda_i). \tag{7.39}$$

The polynomial $\psi(x)$ satisfies Eq. (6.23) which we recall here:

$$\psi''(x) - NV'(x)\psi'(x) + N^2\Pi_N(x)\psi(x) = 0, \tag{7.40}$$

where the function $\Pi_N(x)$ is defined by Eq. (5.32). For the case of the centered-range Jacobi ensemble we have

$$NV'(x) = \frac{a-b+(a+b+2)x}{1-x^2}, \tag{7.41}$$

where we have anticipated the result by introducing the notation $a = T_2 - N - 2/\beta$ and $b = T_1 - N - 2/\beta$. The function $\Pi_N(x)$ is given by

$$\Pi_N(x) = \frac{r_N + s_N x}{1-x^2}. \tag{7.42}$$

We will see below that the coefficient s_N is zero because of the symmetry $\{c_1, c_2, \lambda_k\} \to \{c_2, c_1, -\lambda_k\}$ of the most likely solution.

The equation for $\psi(x)$ becomes

$$(1-x^2)\psi''(x) + (b-a-(a+b+2)x)\psi'(x) + r_N N^2\psi(x) = 0. \tag{7.43}$$

We recognize the differential equation satisfied by the Jacobi polynomials (7.33). Its solutions are polynomials only if $r_N N = N + a + b + 1$, which implies that $r_N = c_1 + c_2 - 1 + (1-4\beta)/N$. This is consistent with the large N limit $r = c_1 + c_2 - 1$ in Section 7.1.3. The solutions are given by

$$\psi(x) \propto P_N^{(a,b)}(x), \tag{7.44}$$

with the proportionality constant chosen such that $\psi(x)$ is monic.

In the special case $T_1 = T_2 = N + 2/\beta$, i.e. $T = N + 2$ for real symmetric matrices and $T = N + 1$ for complex Hermitian matrices, we have $a = b = 0$ and the polynomials reduce to Legendre polynomials.

Another special case corresponds to Chebyshev polynomials:

$$T_n(x) = P_n^{\left(-\frac{1}{2},-\frac{1}{2}\right)}(x) \quad \text{and} \quad U_n(x) = P_n^{\left(\frac{1}{2},\frac{1}{2}\right)}(x), \tag{7.45}$$

where $T_n(x)$ and $U_n(x)$ are the Chebyshev polynomials of first and second kind respectively. Since $a = T_2 - N - 2/\beta$ and $b = T_1 - N - 2/\beta$, they appear as solutions for $T_1 = T_2 = N + 2/\beta - 1/2$ (first kind) or $T_1 = T_2 = N + 2/\beta + 1/2$ (second kind). These values of $T_1 = T_2$ are not integers but we can still consider the matrix potential

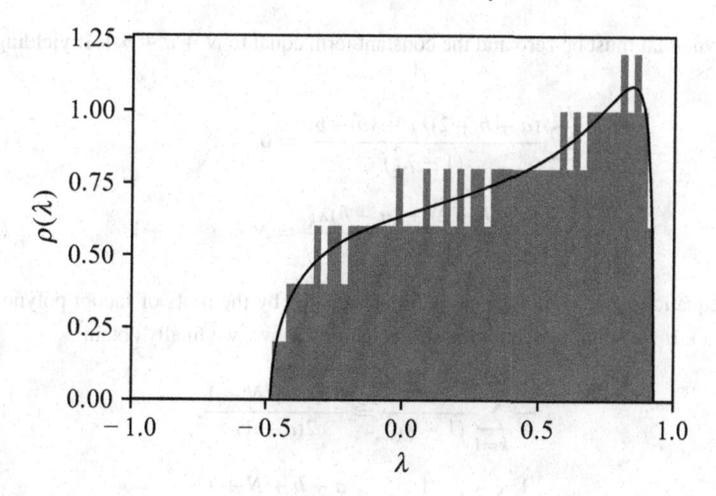

Figure 7.3 Histogram of the 200 zeros of $P_{200}^{(200,600)}(x)$. The full line is the Jacobi eigenvalue density, Eq. (7.29), with $c_1 = 4$ and $c_2 = 2$.

given by Eq. (7.28) without the explicit Wishart matrix construction. In the large N limit, the density of the zeros of the Jacobi polynomials $P_N^{(T_2-N,T_1-N)}(x)$ is given by Eq. (7.29) with $c_{1,2} = T_{1,2}/N$. Figure 7.3 shows a histogram of the zeros of $P_{200}^{(200,600)}(x)$.

When $c_1 \to 1$ and $c_2 \to 1$, the density becomes the centered arcsine law. As a consequence, we have shown that the zeros of Chebyshev (both kinds) and Legendre polynomials (for which $T_1 = T_2 = N + O(1)$) are distributed according to the centered arcsine law in the large N limit.

We have seen that in order for Eq. (7.40) to have polynomial solutions we must have

$$N(1-x^2)\Pi_N(x) = N + a + b + 1, \tag{7.46}$$

where the function $\Pi_N(x)$ is defined from the most likely configuration or, equivalently, the roots of the Jacobi polynomial $P_N^{(a,b)}(z)$ by

$$\Pi_N(x) = \frac{1}{N}\sum_{k=0}^{N} \frac{V'(x) - V'(\lambda_k)}{x - \lambda_k}. \tag{7.47}$$

From these expressions we can find a relationship that roots of Jacobi polynomials must satisfy. Indeed, injecting Eq. (7.41), we find

$$N(1-x^2)\Pi_N(x) = \sum_{k=1}^{N} \frac{(a-b)(x^2-\lambda_k^2) + (a+b+2)(x-\lambda_k)(1+x\lambda_k)}{(1-\lambda_k^2)(x-\lambda_k)}. \tag{7.48}$$

For each k the numerator is a second degree polynomial in x that is zero at $x = \lambda_k$, canceling the $x - \lambda_k$ factor in the denominator, so the whole expression is a first degree polynomial. Equating this expression to Eq. (7.47), we find that the term linear in x of this polynomial must be zero and the constant term equal to $N+a+b+1$, yielding two equations:

$$\frac{1}{N}\sum_{k=1}^{N}\frac{(a+b+2)\lambda_k+(a-b)}{(1-\lambda_k^2)}=0, \tag{7.49}$$

$$\frac{1}{N}\sum_{k=1}^{N}\frac{(a+b+2)+(a-b)\lambda_k}{(1-\lambda_k^2)}=N+a+b+1. \tag{7.50}$$

These equations give us non-trival relations satisfied by the roots of Jacobi polynomials $P_N^{(a,b)}(x)$. If we sum or subtract the two equations above, we finally obtain[3]

$$\frac{1}{N}\sum_{k=1}^{N}\frac{1}{(1-\lambda_k)}=\frac{a+b+N+1}{2(a+1)}, \tag{7.51}$$

$$\frac{1}{N}\sum_{k=1}^{N}\frac{1}{(1+\lambda_k)}=\frac{a+b+N+1}{2(b+1)}. \tag{7.52}$$

7.2.4 Discrete Laplacian in One Dimension and Chebyshev Polynomials

Chebyshev polynomials and the arcsine law are also related via a simple deterministic matrix: the discrete Laplacian in one dimension, defined as

$$\Delta=\frac{1}{2}\begin{pmatrix} 2 & -1 & 0 & \cdots & 0 & 0\\ -1 & 2 & -1 & \cdots & 0 & 0\\ 0 & -1 & 2 & \cdots & 0 & 0\\ 0 & 0 & -1 & \cdots & 0 & 0\\ 0 & 0 & 0 & \ddots & 0 & -1\\ 0 & 0 & 0 & \cdots & -1 & 2 \end{pmatrix}. \tag{7.53}$$

We will see that the spectrum of $\Delta-\mathbf{1}$ at large N is again given by the arcsine law. One way to obtain this result is to modify the top-right and bottom-left corner elements by adding -1. The modified matrix is then a circulant matrix which can be diagonalized exactly (see Exercise 7.2.1 and Appendix A.3).

We will use a different route which will also uncover the link to Chebyshev polynomials. We will compute the characteristic polynomial $Q_N(z)$ of $\Delta-\mathbf{1}$ for all values of N by induction. The first two are

$$Q_1(z)=z, \tag{7.54}$$

$$Q_2(z)=z^2-\frac{1}{4}. \tag{7.55}$$

[3] These relations were recently obtained in Alıcı and Taeli [2015].

For $N \geq 3$ we can write a recursion relation by expanding in minors the first line of the determinant of $z\mathbf{1} - \Delta + \mathbf{1}$. The (11)-minor is just $Q_{N-1}(z)$. The first column of the (12)-minor only has one element equal to $1/2$; if we expand this column its only minor is $Q_{N-2}(z)$. We find

$$Q_N(z) = zQ_{N-1}(z) - \frac{1}{4}Q_{N-2}(z). \tag{7.56}$$

This simple recursion relation is similar to that of Chebyshev polynomials $U_N(x)$:

$$U_N(x) = 2xU_{N-1} - U_{N-2}(x). \tag{7.57}$$

The standard Chebyshev polynomials are not monic but have leading term $U_N(x) \sim 2^N x^N$. Monic Chebyshev ($\widetilde{U}_N(x) = 2^{-N}U_N(x)$) in fact precisely satisfy Eq. (7.56). Given our first two polynomials are the monic Chebyshev of the second kind, we conclude that $Q_N(z) = 2^{-N}U_N(z)$ for all N. The eigenvalues of $\Delta - \mathbf{1}$ at size N are therefore given by the zeros of the Nth Chebyshev polynomial of the second kind. In the large N limit those are distributed according to the centered arcsine law. QED.

We will see in Section 15.3.1 that the sum of two random symmetric orthogonal matrices also has eigenvalues distributed according to the arcsine law.

Exercise 7.2.1 Diagonalizing the Discrete Laplacian

Consider $\mathbf{M} = \Delta - \mathbf{1}$ with $-1/2$ added to the top-right and bottom-left corners.

(a) Show that the vectors $[\mathbf{v}_k]_j = e^{i2\pi kj}$ are eigenvectors of $\mathbf{M}$ with eigenvalues $\lambda_k = \cos(2\pi k)$.

(b) Show that the eigenvalue density of $\mathbf{M}$ in the large N limit is given by the centered arcsine law (7.31).

Bibliographical Notes

- For general references on orthogonal polynomials, see
  - R. Beals and R. Wong. *Special Functions and Orthogonal Polynomials*. Cambridge Studies in Advanced Mathematics. Cambridge University Press, Cambridge, 2016,
  - D. Zwillinger, V. Moll, I. Gradshteyn, and I. Ryzhik, editors. *Table of Integrals, Series, and Products (Eighth Edition)*. Academic Press, New York, 2014.
- On relations obeyed by zeros of Jacobi polynomials, see
  - H. Alıcı and H. Taeli. Unification of Stieltjes-Calogero type relations for the zeros of classical orthogonal polynomials. *Mathematical Methods in the Applied Sciences*, 38(14):3118–3129, 2015.
- On the MANOVA method, see e.g.

  - R. T. Warne. A primer on multivariate analysis of variance (MANOVA) for behavioral scientists. *Practical Assessment, Research & Evaluation*, 19(17):1–10, 2014.
- On the spectrum of Jacobi matrices using Coulomb gas methods, see
  - H. M. Ramli, E. Katzav, and I. P. Castillo. Spectral properties of the Jacobi ensembles via the Coulomb gas approach. *Journal of Physics A: Mathematical and Theoretical*, 45(46):465005, 2012.

and references therein.

Part II

Sums and Products of Random Matrices

8

Addition of Random Variables and Brownian Motion

In the following chapters we will be interested in the properties of sums (and products) of random matrices. Before embarking on this relatively new field, the present chapter will quickly review some classical results concerning sums of random *scalars*, and the corresponding continuous time limit that leads to the Brownian motion and stochastic calculus.

8.1 Sums of Random Variables

Let us thus consider $X = X_1+X_2$ where X_1 and X_2 are two random variables, independent, and distributed according to, respectively, $P_1(x_1)$ and $P_2(x_2)$. The probability that X is equal to x (to within $\mathrm{d}x$) is given by the sum over all combinations of x_1 and x_2 such that $x_1 + x_2 = x$, weighted by their respective probabilities. The variables X_1 and X_2 being independent, the joint probability that $X_1 = x_1$ and $X_2 = x - x_1$ is equal to $P_1(x_1)P_2(x - x_1)$, from which one obtains

$$P^{(2)}(x) = \int P_1(x')P_2(x - x')\,\mathrm{d}x'. \tag{8.1}$$

This equation defines the *convolution* between P_1 and P_2, which we will write $P^{(2)} = P_1 \star P_2$. The generalization to the sum of N independent random variables is immediate. If $X = X_1 + X_2 + \cdots + X_N$ with X_i distributed according to $P_i(x_i)$, the distribution of X is obtained as

$$P^{(N)}(x) = \int \prod_{i=1}^{N} \mathrm{d}x_i\, P_1(x_1)P_2(x_2)\ldots P_N(x_N)\delta\left(x - \sum_{i=1}^{N} x_i\right), \tag{8.2}$$

where $\delta(.)$ is the Dirac delta function. The analytical or numerical manipulations of Eqs. (8.1) and (8.2) are much eased by the use of Fourier transforms, for which convolutions become simple products. The equation $P^{(2)}(x) = [P_1 \star P_2](x)$, reads, in Fourier space,

$$\varphi^{(2)}(k) = \int \mathrm{e}^{ik(x-x'+x')} \int P_1(x')P_2(x - x')\,\mathrm{d}x'\,\mathrm{d}x \equiv \varphi_1(k)\varphi_2(k), \tag{8.3}$$

where $\varphi(k)$ denotes the Fourier transform of the corresponding probability density $P(x)$. It is often called its characteristic (or generating) function. Since the characteristic functions multiply, their logarithms add, i.e. the function $H(k)$ defined below is additive:

$$H(k) := \log \varphi(k) = \log \mathbb{E}[e^{ikX}]. \tag{8.4}$$

It allows one to recovers its so called *cumulants* c_n (provided they are finite) through

$$c_n := (-i)^n \left. \frac{d^n}{dz^n} H(k) \right|_{k=0}. \tag{8.5}$$

The cumulants c_n are polynomial combinations of the moments m_p with $p \leq n$. For example $c_1 = m_1$ is the mean of the distribution and $c_2 = m_2 - m_1^2 = \sigma^2$ its variance. It is clear that the mean of the sum of two random variables (independent or not) is equal to the sum of the individual means. The mean is thus additive under convolution. The same is true for the variance, but only for independent variables.

More generally, from the additive property of $H(k)$ all the cumulants of two independent distributions simply add. The additivity of cumulants is a consequence of the linearity of derivation. The cumulants of a given law convoluted N times with itself thus follow the simple rule $c_{n,N} = N c_{n,1}$, where the $\{c_{n,1}\}$ are the cumulants of the elementary distribution P_1.

An important case is when P_1 is a Gaussian distribution,

$$P_1(x) = \frac{1}{\sqrt{2\pi\sigma^2}} e^{-\frac{(x-m)^2}{2\sigma^2}}, \tag{8.6}$$

such that $\log \varphi_1(k) = imk - \sigma^2 k^2/2$. The Gaussian distribution is such that all cumulants of order ≥ 3 are zero. This property is clearly preserved under convolution: the sum of Gaussian random variables remains Gaussian. Conversely, one can always write a Gaussian variable as a sum of an arbitrary number of Gaussian variables: Gaussian variables are infinitely divisible.

In the following, we will consider infinitesimal Gaussian variables, noted dB, such that $\mathbb{E}[dB] = 0$ and $\mathbb{E}[dB^2] = dt$, where $dt \to 0$ is an infinitesimal quantity, which we will interpret as an infinitesimal time increment. In other words, dB is a mean zero Gaussian random variable which has fluctuations of order $\sqrt{dt}$.

8.2 Stochastic Calculus

8.2.1 Brownian Motion

The starting point of stochastic calculus is the Brownian motion (also called Wiener process) X_t, which is a Gaussian random variable of mean μt and variance $\sigma^2 t$. From the infinite divisibility property of Gaussian variables, one can always write

$$X_{t_k} = \sum_{\ell=0}^{k-1} \mu \delta t + \sum_{\ell=0}^{k-1} \sigma \delta B_\ell, \tag{8.7}$$

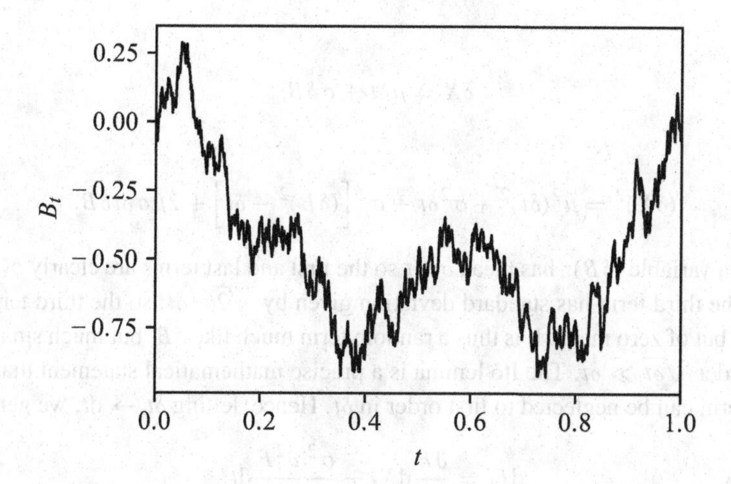

Figure 8.1 An example of Brownian motion.

where $t_k = kt/N$, $0 \le k \le N$, $\delta t = T/N$ and $\delta B_\ell \sim \mathcal{N}(0, \delta t)$ for each ℓ. By construction $X(t_N) = X_t$. In the limit $N \to \infty$, we have $\delta t \to \mathrm{d}t$, $\delta B_k \to \mathrm{d}B$ and X_{t_k} becomes a continuous time process with

$$\mathrm{d}X_t = \mu \mathrm{d}t + \sigma \mathrm{d}B_t, \qquad X_0 = 0, \tag{8.8}$$

where $\mathrm{d}B_t$ are independent, infinitesimal Gaussian variables as defined above (see Fig. 8.1). The process X_t is continuous but nowhere differentiable. Note that X_t and $X_{t'}$ are not independent but their increments are, i.e. $X_{t'}$ and $X_t - X_{t'}$ are independent whenever $t' < t$.

Note the convention that X_{t_k} is built from *past increments* δB_ℓ for $\ell < k$ but does not include δB_k. This convention is called the Itô prescription.[1] Its main advantage is that X_t is independent of the equal-time $\mathrm{d}B_t$, but this comes at a price: the usual chain rule for differentiation has to be corrected by the so-called Itô term, which we now discuss.

8.2.2 Itô's Lemma

We now study the behavior of functions $F(X_t)$ of a Wiener process X_t. Because $\mathrm{d}B^2$ is of order $\mathrm{d}t$, and not $\mathrm{d}t^2$, one has to be careful when evaluating derivatives of functions of X_t.

Given a twice differentiable function $F(.)$, we consider the process $F(X_t)$. Reverting for a moment to a discretized version of the process, one has

$$F(X(t + \delta t)) = F(X_t) + \delta X\, F'(X_t) + \frac{(\delta X)^2}{2} F''(X_t) + o(\delta t), \tag{8.9}$$

[1] In the Stratonovich prescription, half of δB_k contributes to X_{t_k}. In this prescription, the Itô lemma is not needed, i.e. the chain rule applies without any correction term, but the price to pay is a correlation between X_t and $\mathrm{d}B(t)$. We will not use the Stratonovich prescription in this book.

where

$$\delta X = \mu \delta t + \sigma \delta B, \tag{8.10}$$

and

$$(\delta X)^2 = \mu^2 (\delta t)^2 + \sigma^2 \delta t + \sigma^2 \Big[(\delta B)^2 - \delta t \Big] + 2\mu\sigma \delta t \delta B. \tag{8.11}$$

The random variable $(\delta B)^2$ has mean $\sigma^2 \delta t$ so the first and last terms are clearly $o(\delta t)$ when $\delta t \to 0$. The third term has standard deviation given by $\sqrt{2}\sigma^2 \delta t$; so the third term is also of order δt but of zero mean. It is thus a random term much like δB, but much smaller since δB is of order $\sqrt{\delta t} \gg \delta t$. The Itô lemma is a precise mathematical statement that justifies why this term can be neglected to first order in δt. Hence, letting $\delta t \to \mathrm{d}t$, we get

$$\mathrm{d}F_t = \frac{\partial F}{\partial X} \mathrm{d}X_t + \frac{\sigma^2}{2} \frac{\partial^2 F}{\partial^2 X} \mathrm{d}t. \tag{8.12}$$

When compared to ordinary calculus, there is a correction term – the Itô term – that depends on the second order derivative of F.

More generally, we can consider a general Itô process where μ and σ themselves depend on X_t and t, i.e.

$$\mathrm{d}X_t = \mu(X_t, t)\mathrm{d}t + \sigma(X_t, t)\mathrm{d}B_t. \tag{8.13}$$

Then, for functions $F(X, t)$ that may have an explicit time dependence, one has

$$\mathrm{d}F_t = \frac{\partial F}{\partial X} \mathrm{d}X_t + \left[\frac{\partial F}{\partial t} + \frac{\sigma^2(X_t, t)}{2} \frac{\partial^2 F}{\partial^2 X} \right] \mathrm{d}t. \tag{8.14}$$

Itô's lemma can be extended to functions of several stochastic variables. Consider a collection of N independent stochastic variables $\{X_{i,t}\}$ (written vectorially as $\mathbf{X}_t$) and such that

$$\mathrm{d}X_{i,t} = \mu_i(\mathbf{X}_t, t)\mathrm{d}t + \mathrm{d}W_{i,t}, \tag{8.15}$$

where $\mathrm{d}W_{i,t}$ are Wiener noises such that

$$\mathbb{E}\left[\mathrm{d}W_{i,t} \mathrm{d}W_{j,t}\right] := C_{ij}(\mathbf{X}_t, t)\mathrm{d}t. \tag{8.16}$$

The vectorial form of Itô's lemma states that the time evolution of a function $F(\mathbf{X}_t, t)$ is given by the sum of three contributions:

$$\mathrm{d}F_t = \sum_{i=1}^{N} \frac{\partial F}{\partial X_i} \mathrm{d}X_{i,t} + \left[\frac{\partial F}{\partial t} + \sum_{i,j=1}^{N} \frac{C_{ij}(\mathbf{X}_t, t)}{2} \frac{\partial^2 F}{\partial X_i \partial X_j} \right] \mathrm{d}t. \tag{8.17}$$

The formula simplifies when all the Wiener noises are independent, in which case the Itô term only contains the second derivatives $\partial^2 F / \partial^2 X_i$.

8.2.3 Variance as a Function of Time

As an illustration of how to use Itô's formula let us recompute the time dependent variance of X_t. Assume $\mu = 0$ and choose $F(X) = X^2$. Applying Eq. (8.12), we get that

$$\mathrm{d}F_t = 2X_t\mathrm{d}X_t + \sigma^2\mathrm{d}t \Rightarrow F(X_t) = 2\int_0^t \sigma X_s \mathrm{d}B_s + \sigma^2 t. \tag{8.18}$$

In order to take the expectation value of this equation, we scrutinize the term $\mathbb{E}[X_s\mathrm{d}B_s]$. As alluded to above, the random infinitesimal element $\mathrm{d}B_s$ does not contribute to X_s, which only depends on $\mathrm{d}B_{s'<s}$. Therefore $\mathbb{E}[X_s\mathrm{d}B_s] = 0$, and, as expected,

$$\mathbb{E}[X_t^2] = \mathbb{E}[F(X_t)] = \sigma^2 t. \tag{8.19}$$

The Brownian motion has a variance from the origin that grows linearly with time. The same result can of course be derived directly from the integrated form $X_t = \sigma B_t$, where B_t is a Gaussian random number of variance equal to t.

8.2.4 Gaussian Addition

Itô's lemma can be used to compute a special case of the law of addition of independent random variables, namely when one of the variables is Gaussian. Consider the random variable $Z = Y + X$, where Y is some random variable, and X is an independent Gaussian $(X \sim \mathcal{N}(\mu, \sigma^2))$. The law of Z is uniquely determined by its characteristic function:

$$\varphi(k) := \mathbb{E}[\mathrm{e}^{ikZ}]. \tag{8.20}$$

We now let $Z \to Z_t$ be a Brownian motion with $Z_0 = Y$:

$$\mathrm{d}Z_t = \mu\mathrm{d}t + \sigma\mathrm{d}B_t, \quad Z_0 = Y. \tag{8.21}$$

Note that $Z_{t=1}$ has the same law as Z. The idea is now to study the function $F(Z_t) := \mathrm{e}^{ikZ_t}$ using Itô's lemma, Eq. (8.14). Hence,

$$\mathrm{d}F_t = ik\mathrm{e}^{ikZ_t}\mathrm{d}Z_t - \frac{k^2\sigma^2}{2}\mathrm{e}^{ikZ_t}\mathrm{d}t = \left(ik\mu F - \frac{k^2\sigma^2}{2}F\right)\mathrm{d}t + ikF\mathrm{d}B_t. \tag{8.22}$$

Taking the expectation value, writing $\varphi_t(k) = \mathbb{E}[F(t)]$, and noting that the differential d is a linear operator and therefore commutes with the expectation value, we obtain

$$\mathrm{d}\varphi_t(k) = \left(ik\mu - \frac{k^2\sigma^2}{2}\right)\varphi_t(k)\mathrm{d}t, \tag{8.23}$$

or

$$\frac{1}{\varphi_t(k)}\frac{\mathrm{d}}{\mathrm{d}t}\varphi_t(k) = \frac{\mathrm{d}}{\mathrm{d}t}\log(\varphi_t(k)) = \left(ik\mu - \frac{k^2\sigma^2}{2}\right). \tag{8.24}$$

From its solution at $t = 1$, we get

$$\log(\varphi_1(k)) = \log(\varphi_0(k)) + ik\mu - \frac{k^2\sigma^2}{2}. \tag{8.25}$$

Recognizing the last two terms in the right hand side as the characteristic function of a Gaussian variable, we recover the fact that the log-characteristic function is additive under the addition of independent random variables. Although the result is true in general, the calculation above using stochastic calculus is only valid if one of the random variable is Gaussian.

8.2.5 The Langevin Equation

We would like to construct a stochastic process for a variable X_t such that in the steady-state regime the values of X_t are drawn from a given probability distribution $P(x)$. To build our stochastic process, let us first consider the simple Brownian motion with unit variance per unit time:

$$\mathrm{d}X_t = \mathrm{d}B_t. \tag{8.26}$$

As revealed by Eq. (8.19) the variance of X_t grows linearly with time and the process never reaches a stationary state. To make it stationary we need a mechanism to limit the variance of X_t. We cannot 'subtract' variance but we can reduce X_t by scaling. If at every infinitesimal time step we replace $X_{t+\mathrm{d}t}$ by $X_{t+\mathrm{d}t}/\sqrt{1+\mathrm{d}t}$, the variance of X_t will remain equal to unity. We also know that the distribution of X_t is Gaussian (if the initial condition is Gaussian or constant). With this extra rescaling, X_t is still Gaussian at every step, so clearly this will describe the stationary state of our rescaled process. As a stochastic differential equation, we have, neglecting terms of order $(\mathrm{d}t)^{3/2}$

$$\mathrm{d}X_t = \mathrm{d}B_t + \frac{X_t}{\sqrt{1+\mathrm{d}t}} - X_t = \mathrm{d}B_t - \frac{1}{2}X_t\mathrm{d}t. \tag{8.27}$$

This stationary version of the random walk is the Ornstein–Uhlenbeck process (see Fig. 8.2). A physical interpretation of this equation is that of a particle located at X_t moving in a viscous medium subjected to random forces $\mathrm{d}B_t/\mathrm{d}t$ and a deterministic harmonic force ("spring") $-X_t/2$. The viscous medium is such that velocity (and not acceleration) is proportional to force.

We would like to generalize the above formalism to generate any distribution $P(x)$ for the distribution of X in the stationary state. One way to do so is to change the linear force $-X_t/2$ to a general non-linear force $F(X_t) := -V'(X_t)/2$, where we have written the force as the derivative of a potential V and introduced a factor of 2 which will prove to be convenient. If the potential is convex, the force will drive the particle towards the minimum of the potential while the noise $\mathrm{d}B_t$ will drive the particle away. We expect that this system will reach a steady state. Our stochastic equation is now

$$\mathrm{d}X_t = \mathrm{d}B_t + F(X_t)\mathrm{d}t. \tag{8.28}$$

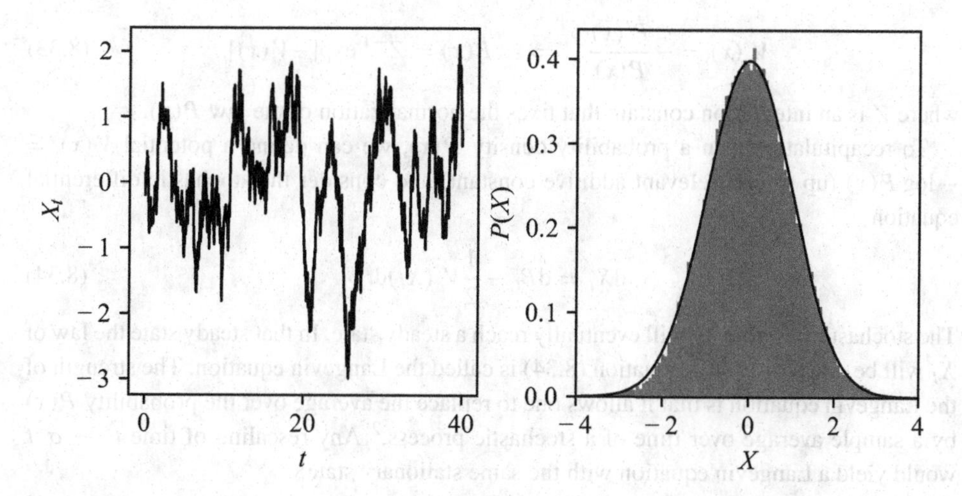

Figure 8.2 (left) A simulation of the Langevin equation for the Ornstein–Uhlenbeck process (8.27) with 50 steps per unit time. Note that the correlation time is $\tau_c = 2$ here, so excursions away from zero typically take 2 time units to mean-revert back to zero. Farther excursions take longer to come back. (right) Histogram of the values of X_t for the same process simulated up to $t = 2000$ and comparison with the normal distribution. The agreement varies from sample to sample as a rare far excursion can affect the sample distribution even for $t = 2000$.

What is the distribution of X_t in the steady state? To find out, let us consider an arbitrary test function $f(X_t)$ and see how it behaves in the steady state. Using Itô's lemma, Eq. (8.12), we have

$$df(X_t) = f'(X_t)\left[dB_t - \frac{1}{2}V'(X_t)dt\right] + \frac{1}{2}f''(X_t)dt. \tag{8.29}$$

Taking the expectation value on both sides and demanding that $d\mathbb{E}[f_t]/dt = 0$ in the steady state we find

$$\mathbb{E}\left[f'(X_t)V'(X_t)\right] = \mathbb{E}\left[f''(X_t)\right]. \tag{8.30}$$

This must be true for any function f. In order to infer the corresponding stationary distribution $P(x)$, let us write $h(x) = f'(x)$ and write these expectation values as

$$\int h(x)V'(x)P(x)dx = \int h'(x)P(x)dx. \tag{8.31}$$

Since we want to relate an integral of h to one of h' we should use integration by parts:

$$\int h'(x)P(x)dx = -\int h(x)P'(x)dx = -\int h(x)\frac{P'(x)}{P(x)}P(x)dx. \tag{8.32}$$

Since Eq. (8.31) is true for any function $h(x)$ we must have

$$V'(x) = -\frac{P'(x)}{P(x)} \quad \Rightarrow \quad P(x) = Z^{-1}\exp[-V(x)], \tag{8.33}$$

where Z is an integration constant that fixes the normalization of the law $P(x)$.

To recapitulate, given a probability density $P(x)$, we can define a potential $V(x) = -\log P(x)$ (up to an irrelevant additive constant) and consider the stochastic differential equation

$$\mathrm{d}X_t = \mathrm{d}B_t - \frac{1}{2}V'(X_t)\mathrm{d}t. \tag{8.34}$$

The stochastic variable X_t will eventually reach a steady state. In that steady state the law of X_t will be given by $P(x)$. Equation (8.34) is called the Langevin equation. The strength of the Langevin equation is that it allows one to replace the average over the probability $P(x)$ by a sample average over time of a stochastic process.[2] Any rescaling of time $t \to \sigma^2 t$ would yield a Langevin equation with the same stationary state:

$$\mathrm{d}X_t = \sigma\mathrm{d}B_t - \frac{\sigma^2}{2}V'(X_t)\mathrm{d}t. \tag{8.35}$$

We have learned another useful fact from Eq. (8.30): the random variable $V'(X)$ acts as a derivative with respect to X under the expectation value. In that sense $V'(X)$ can be considered the conjugate variable to X.

It is very straightforward to generalize our one-dimensional Langevin equation to a set of N variables $\{X_i\}$ that are drawn from the joint law $P(\mathbf{x}) = Z^{-1}\exp[-V(\mathbf{x})]$. We get

$$\mathrm{d}X_i = \sigma\mathrm{d}B_i + \frac{\sigma^2}{2}\frac{\partial}{\partial X_i}\log P(\mathbf{x})\,\mathrm{d}t, \tag{8.36}$$

where we have dropped the subscript t for clarity.

Exercise 8.2.1 Langevin equation for Student's t-distributions

The family of Student's t-distributions, parameterized by the tail exponent μ, is given by the probability density

$$P_\mu(x) = Z_\mu^{-1}\left(1 + \frac{x^2}{\mu}\right)^{-\frac{\mu+1}{2}} \quad \text{with } Z_\mu^{-1} = \frac{\Gamma\left(\frac{\mu+1}{2}\right)}{\sqrt{\mu\pi}\,\Gamma\left(\frac{\mu}{2}\right)}. \tag{8.37}$$

(a) What is the potential $V(x)$ and its derivative $V'(x)$ for these laws?

(b) Using Eq. (8.30), show that for a t-distributed variable x we have

$$\mathbb{E}\left[\frac{x^2}{x^2+\mu}\right] = \frac{1}{1+\mu}. \tag{8.38}$$

[2] A process for which the time evolution samples the entire set of possible values according to the stationary probability is called ergodic. A discussion of the condition for ergodicity is beyond the scope of this book.

(c) Write the Langevin equation for a Student's t-distribution. What is the $\mu \to \infty$ limit of this equation?
(d) Simulate your Langevin equation for $\mu = 3$, 20 time steps per unit time and run the simulation for 20 000 units of time. Make a normalized histogram of the sampled values of X_t and compare with the law for $\mu = 3$ given above.
(e) Compared to the Gaussian process (Ornstein–Uhlenbeck), the Student t-process has many more short excursions but the long excursions are much longer than the Gaussian ones. Explain this behavior by comparing the function $V'(x)$ in the two cases. Describe their relative small $|x|$ and large $|x|$ behavior.

8.2.6 The Fokker–Planck Equation

It is interesting to derive, from the Langevin equation Eq. (8.28), the so-called Fokker–Planck equation that describes the dynamical evolution of the time dependent probability density $P(x,t)$ of the random variable X_t. The trick is to use Eq. (8.29) again, with $f(x)$ an arbitrary test function. Taking expectations, one finds

$$\mathrm{d}\mathbb{E}[f(X_t)] = \mathbb{E}\left[f'(X_t)F(X_t)\mathrm{d}t\right] + \frac{1}{2}\mathbb{E}[f''(X_t)]\mathrm{d}t, \tag{8.39}$$

where we have used the fact that, in the Itô convention, $\mathbb{E}[f'(X_t)\mathrm{d}B_t] = 0$. But by definition of $P(x,t)$, one also has

$$\mathbb{E}[f(X_t)] := \int f(x)P(x,t)\mathrm{d}x. \tag{8.40}$$

Hence,

$$\int f(x)\frac{\partial P(x,t)}{\partial t}\mathrm{d}x = \int f'(x)F(x)P(x,t)\mathrm{d}x + \frac{1}{2}\int f''(x)P(x,t)\mathrm{d}x. \tag{8.41}$$

Integrating by parts the right-hand side leads to

$$\int f(x)\frac{\partial P(x,t)}{\partial t}\mathrm{d}x = -\int f(x)\frac{\partial F(x)P(x,t)}{\partial x}\mathrm{d}x + \frac{1}{2}\int f(x)\frac{\partial^2 P(x,t)}{\partial x^2}\mathrm{d}x. \tag{8.42}$$

Since this equation holds for an arbitrary test function $f(x)$, it must be that

$$\frac{\partial P(x,t)}{\partial t} = -\frac{\partial F(x)P(x,t)}{\partial x} + \frac{\sigma^2}{2}\frac{\partial^2 P(x,t)}{\partial x^2}, \tag{8.43}$$

which is called the Fokker–Planck equation. We have reintroduced an arbitrary value of σ here, to make the equation more general. One can easily check that when $F(x) = -V'(x)/2$, the stationary state of this equation, such that the left hand side is zero, is

$$P(x) = Z^{-1}\exp[-V(x)/\sigma^2], \tag{8.44}$$

as expected from the previous section.

Bibliographical Notes

- A general introduction to probability theory:
  - W. Feller. *An Introduction to Probability Theory and Its Applications*. Wiley, 1968.
- For introductory textbooks on the Langevin and Fokker–Planck equation, see
  - C. W. Gardiner. *Handbook of Stochastic Methods for Physics, Chemistry and the Natural Sciences*, volume 13 of Springer Series in Synergetics. Springer-Verlag, Berlin, 2004,
  - N. V. Kampen. *Stochastic Processes in Physics and Chemistry*. North Holland, Amsterdam, 2007.
- The central limit theorem:
  - P. Lévy. *Théorie de l'addition des variables aléatoires*. Gauthier Villars, Paris, 1937–1954,
  - B. V. Gnedenko and A. N. Kolmogorov. *Limit Distributions for Sums of Independent Random Variables*. Addison-Wesley Publishing Co., Reading, Mass., London, Don Mills., Ont., 1968.
- For a physicist discussion of the central limit theorem, see
  - J.-P. Bouchaud and M. Potters. *Theory of Financial Risk and Derivative Pricing: From Statistical Physics to Risk Management*. Cambridge University Press, Cambridge, 2nd edition, 2003.

9

Dyson Brownian Motion

In this chapter we would like to start our investigation of the addition of random matrices, which will lead to the theory of so-called "free matrices". This topic has attracted substantial interest in recent years and will be covered in the next chapters.

We will start by studying how a fixed large matrix (random or not) is modified when one adds a Wigner matrix. The elements of a Wigner matrix are Gaussian random numbers and, as we saw in the previous chapter, each of them can be written as a sum of Gaussian numbers with smaller variance. By pushing this reasoning to the limit we can write the addition of a Wigner matrix as a continuous process of addition of infinitesimal Wigner matrices. Such a matrix Brownian motion process, viewed through the lens of eigenvalues and eigenvectors, is called a Dyson Brownian motion (DBM) after the physicist Freeman Dyson who first introduced it in 1962.

9.1 Dyson Brownian Motion I: Perturbation Theory

9.1.1 Perturbation Theory: A Short Primer

We begin by recalling how perturbation theory works for eigenvalues and eigenvectors. This is a standard topic in elementary quantum mechanics, but is not necessarily well known in other circles. Let us consider a matrix $\mathbf{H} = \mathbf{H}_0 + \epsilon \mathbf{H}_1$, where $\mathbf{H}_0$ is a real symmetric matrix whose eigenvalues and eigenvectors are assumed to be known, and $\mathbf{H}_1$ is a real symmetric matrix that gives the perturbation, with ϵ a small parameter. (In quantum mechanics, $\mathbf{H}_0$ and $\mathbf{H}_1$ are complex Hermitian, but the final equations below are the same in both cases.)

Suppose $\lambda_{i,0}$, $1 \leq i \leq N$, are the eigenvalues of $\mathbf{H}_0$ and $\mathbf{v}_{i,0}$, $1 \leq i \leq N$, are the corresponding eigenvectors. We assume that the perturbed eigenvalues and eigenvectors are given by the series expansion (in ϵ):

$$\lambda_i = \lambda_{i,0} + \sum_{k=1}^{\infty} \epsilon^k \lambda_{i,k}, \quad \mathbf{v}_i = \mathbf{v}_{i,0} + \sum_{k=1}^{\infty} \epsilon^k \mathbf{v}_{i,k}, \tag{9.1}$$

with the constraint that

$$\|\mathbf{v}_i\| = \|\mathbf{v}_{i,0}\| = 1, \quad 1 \leq i \leq N. \tag{9.2}$$

Since the quantity $\|\mathbf{v}_i\|$ is constant, its first order variation with respect to ϵ must be zero. This constraint gives that $\mathbf{v}_{i,1} \perp \mathbf{v}_{i,0}$.

We assume that the $\lambda_{i,0}$ are all different, i.e. we consider non-degenerate perturbation theory. Then, plugging (9.1) into

$$\mathbf{H}\mathbf{v}_i = \lambda_i \mathbf{v}_i \tag{9.3}$$

and matching the left and right hand side term by term in powers of ϵ, one obtains

$$\lambda_i = \lambda_{i,0} + \epsilon(\mathbf{H}_1)_{ii} + \epsilon^2 \sum_{\substack{j=1 \\ j \neq i}}^{N} \frac{|(\mathbf{H}_1)_{ij}|^2}{\lambda_{i,0} - \lambda_{j,0}} + O(\epsilon^3), \tag{9.4}$$

where $(\mathbf{H}_1)_{ij} := \mathbf{v}_{j,0}^T \mathbf{H}_1 \mathbf{v}_{i,0}$, and

$$\mathbf{v}_i = \mathbf{v}_{i,0} + \epsilon \sum_{\substack{j=1 \\ j \neq i}}^{N} \frac{(\mathbf{H}_1)_{ij}}{\lambda_{i,0} - \lambda_{j,0}} \mathbf{v}_{j,0} + O(\epsilon^2). \tag{9.5}$$

Notice that the first order correction to $\mathbf{v}_i$ is indeed perpendicular to $\mathbf{v}_{i,0}$ as it does not have a component in that direction.

9.1.2 A Stochastic Differential Equation for Eigenvalues

Next we use the above formulas to derive the so-called Dyson Brownian motion (DBM), which gives the evolution of eigenvalues of a random matrix plus a Wigner ensemble whose variance grows linearly with time. Let $\mathbf{M}_0$ be the initial matrix (random or not), and $\mathbf{X}_1$ be a unit Wigner matrix that is independent of $\mathbf{M}_0$. Then we study, using (9.4), the eigenvalues of

$$\mathbf{M} = \mathbf{M}_0 + \sqrt{\mathrm{d}t}\, \mathbf{X}_1, \tag{9.6}$$

where $\mathrm{d}t$ is a small quantity which will be interpreted as a differential time step.

The derivation of the SDE for eigenvalues is much simpler if we use the rotational invariance of the Wigner ensemble. The matrix $\mathbf{X}_1$ has the same law in any basis, we therefore choose to express it in the diagonal basis of $\mathbf{M}_0$. In order to do so we must work with the exact rotationally invariant Wigner ensemble where the diagonal variance is twice the off-diagonal variance.

First, for the first order term (in terms of ϵ), we have

$$(\mathbf{X}_1)_{ii} := \mathbf{v}_{i,0}^T \mathbf{X}_1 \mathbf{v}_{i,0} \sim \mathcal{N}\left(0, \frac{2}{N}\right). \tag{9.7}$$

Note that the $(\mathbf{X}_1)_{ii}$ are independent for different i's.

Then we study the second order term. We have

$$(\mathbf{X}_1)_{ji} := \mathbf{v}_{j,0}^T \mathbf{X}_1 \mathbf{v}_{i,0} \sim \mathcal{N}\left(0, \frac{1}{N}\right). \tag{9.8}$$

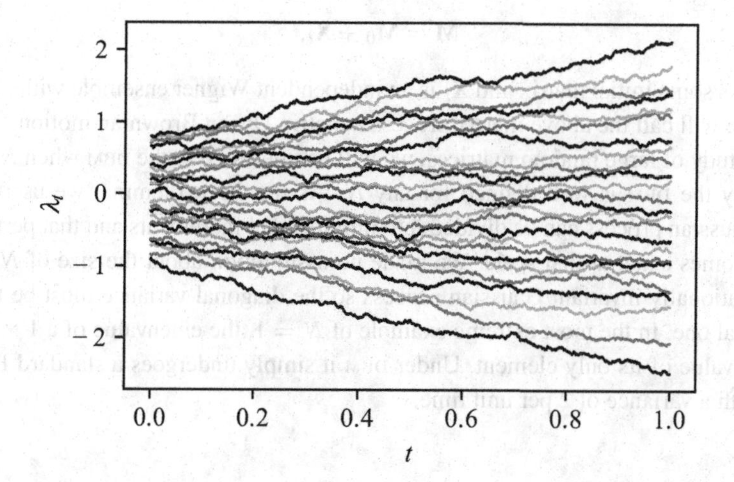

Figure 9.1 A simulation of DBM for an $N = 25$ matrix starting for a Wigner matrix with $\sigma^2 = 1/4$ and evolving for one unit of time.

So $|(\mathbf{X}_1)_{ji}|^2$ is a random variable with mean $1/N$ and with fluctuations around the mean also of order $1/N$. As in Section 8.2.2, one can argue that these fluctuations are negligible when integrated over time in the limit $\mathrm{d}t \to 0$. In other words, $|(\mathbf{X}_1)_{ji}|^2$ can be treated as deterministic.

Now using (9.4), we get that

$$\mathrm{d}\lambda_i = \sqrt{\frac{2}{\beta N}}\mathrm{d}B_i + \frac{1}{N}\sum_{\substack{j=1\\ j\neq i}}^{N} \frac{\mathrm{d}t}{\lambda_i - \lambda_j}, \tag{9.9}$$

where $\mathrm{d}B_i$ denotes a Brownian increment that comes from the $(\mathbf{X}_1)_{ii}$ term, and we have added a factor β for completeness (equal to 1 for real symmetric matrices). This is the fundamental evolution equation for eigenvalues in a fictitious time that describes how much of a Wigner matrix one progressively adds to the "initial" matrix $\mathbf{M}_0$ (see Fig. 9.1). The astute reader will probably have recognized in the second term a Coulomb force deriving from the logarithmic Coulomb potential $\log|\lambda_i - \lambda_j|$ encountered in Chapter 5.

One can also derive a similar process for the eigenvectors that we give here for $\beta = 1$:

$$\mathrm{d}\mathbf{v}_i = \frac{1}{\sqrt{N}}\sum_{\substack{j=1\\ j\neq i}}^{N} \frac{\mathrm{d}B_{ij}}{\lambda_i - \lambda_j}\mathbf{v}_j - \frac{1}{2N}\sum_{\substack{j=1\\ j\neq i}}^{N} \frac{\mathrm{d}t}{(\lambda_i - \lambda_j)^2}\mathbf{v}_i, \tag{9.10}$$

where $\mathrm{d}B_{ij} = \mathrm{d}B_{ji}$ $(i \neq j)$ is a symmetric collection of Brownian motions, independent of each other and of the $\{\mathrm{d}B_i\}$ above.

The formulas (9.9) and (9.10) give the Dyson Brownian motion for the stochastic evolution of the eigenvalues and eigenvectors of matrices of the form

$$\mathbf{M} = \mathbf{M}_0 + \mathbf{X}_t, \tag{9.11}$$

where $\mathbf{M}_0$ is some initial matrix and $\mathbf{X}_t$ is an independent Wigner ensemble with parameter $\sigma^2 = t$. We will call the above matrix process a matrix Dyson Brownian motion.

In our study of large random matrices, we will be interested in the DBM when N is large, but actually the DBM is well defined for any N. As for the Itô lemma, we used the fact that the Gaussian process can be divided into infinitesimal increments and that perturbation theory becomes exact at that scale. We made no assumption about the size of N. We did need a rotationally invariant Gaussian process so the diagonal variance must be twice the off-diagonal one. In the most extreme example of $N = 1$, the eigenvalue of a 1×1 matrix is just the value of its only element. Under DBM it simply undergoes a standard Brownian motion with a variance of 2 per unit time.

Exercise 9.1.1 Variance as a function of time under DBM

Consider the Dyson Brownian motion for a finite N matrix:

$$\mathrm{d}\lambda_i = \sqrt{\frac{2}{N}}\mathrm{d}B_i + \frac{1}{N}\sum_{\substack{j=1 \\ j \neq i}}^{N} \frac{\mathrm{d}t}{\lambda_i - \lambda_j} \tag{9.12}$$

and the function $F(\{\lambda_i\})$ that computes the second moment:

$$F(\{\lambda_i\}) = \frac{1}{N}\sum_{i=1}^{N} \lambda_i^2. \tag{9.13}$$

(a) Write down the stochastic process for $F(\{\lambda_i\})$ using the Itô vectorial formula (8.17). In the case at hand F does not depend explicitly on time and $\sigma_i^2 = 2/N$. You will need to use the following identity:

$$2\sum_{\substack{i,j=1 \\ j \neq i}}^{N} \frac{\lambda_i}{\lambda_i - \lambda_j} = \sum_{\substack{i,j=1 \\ j \neq i}}^{N} \frac{\lambda_i - \lambda_j}{\lambda_i - \lambda_j} = N(N-1). \tag{9.14}$$

(b) Take the expectation value of your equation and show that $F(t) := \mathbb{E}[F(\{\lambda_i(t)\})]$ follows

$$F(t) = F(0) + \frac{N+1}{N}t. \tag{9.15}$$

Do not assume that N is large.

9.2 Dyson Brownian Motion II: Itô Calculus

Another way to derive the Dyson Brownian motion for the eigenvalues is to consider the matrix Brownian motion (9.11) as a Brownian motion of all the elements of the matrix $\mathbf{X}$. We have to treat the diagonal and off-diagonal elements separately because we want to use the rotationally invariant Wigner matrix (GOE) with diagonal variance equal to twice the off-diagonal variance. Also, only half of the off-diagonal elements are independent (the matrix $\mathbf{X}$ is symmetric). We have

$$\mathrm{d}X_{kk} = \sqrt{\frac{2}{N}}\mathrm{d}B_{kk} \quad \text{and} \quad \mathrm{d}X_{k\ell} = \sqrt{\frac{1}{N}}\mathrm{d}B_{k\ell} \quad \text{for} \quad k < \ell, \tag{9.16}$$

where $\mathrm{d}B_{kk}$ and $\mathrm{d}B_{k\ell}$ are N and $N(N-1)/2$ independent unit Brownian motions.

Each eigenvalue λ_i is a (complicated) function of all the matrix elements of $\mathbf{X}$. We can use the vectorial form of Itô's lemma (8.17) to write a stochastic differential equation (SDE) for $\lambda_i(\mathbf{X})$:

$$\mathrm{d}\lambda_i = \sum_{k=1}^{N} \frac{\partial \lambda_i}{\partial X_{kk}} \sqrt{\frac{2}{N}}\mathrm{d}B_{kk} + \sum_{\substack{k=1\\ \ell=k+1}}^{N} \frac{\partial \lambda_i}{\partial X_{k\ell}} \sqrt{\frac{1}{N}}\mathrm{d}B_{k\ell} + \sum_{k=1}^{N} \frac{\partial^2 \lambda_i}{\partial X_{kk}^2}\frac{\mathrm{d}t}{N} + \sum_{\substack{k=1\\ \ell=k+1}}^{N} \frac{\partial^2 \lambda_i}{\partial X_{k\ell}^2}\frac{\mathrm{d}t}{2N}. \tag{9.17}$$

The key is to be able to compute the following partial derivatives:

$$\frac{\partial \lambda_i}{\partial X_{kk}}; \qquad \frac{\partial \lambda_i}{\partial X_{k\ell}}; \qquad \frac{\partial^2 \lambda_i}{\partial X_{kk}^2}; \qquad \frac{\partial^2 \lambda_i}{\partial X_{k\ell}^2}, \tag{9.18}$$

where $k < \ell$.

Since $\mathbf{X}_t$ is rotational invariant, we can rotate the basis such that $\mathbf{X}_0$ is diagonal:

$$\mathbf{X}_0 = \mathrm{diag}(\lambda_1(0), \ldots, \lambda_N(0)). \tag{9.19}$$

In order to compute the partial derivatives above, we can consider adding one small element to the matrix $\mathbf{X}_0$, and compute the corresponding change in eigenvalues. We first perturb diagonal elements and later we will deal with off-diagonal elements.

A shift of the kth diagonal entry of $\mathbf{X}_0$ by δX_{kk} affects λ_i with $i = k$ in a linear fashion but leaves all other eigenvalues unaffected:

$$\lambda_i \to \lambda_i + \delta X_{kk}\delta_{ki}. \tag{9.20}$$

Thus we have

$$\frac{\partial \lambda_i}{\partial X_{kk}} = \delta_{ik}; \qquad \frac{\partial^2 \lambda_i}{\partial X_{kk}^2} = 0. \tag{9.21}$$

Next we discuss how a perturbation in an off-diagonal entry of $\mathbf{X}_0$ can affect the eigenvalues. A perturbation of the $(k\ell)$ entry by $\delta X_{k\ell} = \delta X_{\ell k}$ entry leads to the following matrix:

$$\mathbf{X}_0 + \delta\mathbf{X} = \begin{pmatrix} \lambda_1 & & & & & \\ & \ddots & & & & \\ & & \lambda_k & & \delta X_{k\ell} & \\ & & & \ddots & & \\ & & \delta X_{k\ell} & & \lambda_\ell & \\ & & & & & \ddots & \\ & & & & & & \lambda_N \end{pmatrix}. \tag{9.22}$$

Since this matrix is block diagonal (after a simple permutation), the eigenvalues in the $N-2$ other 1×1 blocks are not affected by the perturbation, so trivially

$$\frac{\partial \lambda_i}{\partial X_{k\ell}} = 0; \qquad \frac{\partial^2 \lambda_i}{\partial X_{k\ell}^2} = 0, \qquad \forall i \neq k, \ell. \tag{9.23}$$

On the other hand, the eigenvalues of the block

$$\begin{pmatrix} \lambda_k & \delta X_{k\ell} \\ \delta X_{k\ell} & \lambda_\ell \end{pmatrix} \tag{9.24}$$

are modified and and exact diagonalization gives

$$\lambda_\pm = \frac{\lambda_k + \lambda_\ell}{2} \pm \frac{\lambda_k - \lambda_\ell}{2} \sqrt{1 + \frac{4(\delta X_{k\ell})^2}{(\lambda_k - \lambda_\ell)^2}}. \tag{9.25}$$

We can expand this result to second order in $\delta X_{k\ell}$ to find

$$\lambda_k \to \lambda_k + \frac{(\delta X_{k\ell})^2}{\lambda_k - \lambda_\ell} \quad \text{and} \quad \lambda_\ell \to \lambda_\ell + \frac{(\delta X_{k\ell})^2}{\lambda_\ell - \lambda_k}. \tag{9.26}$$

We thus readily see that the first partial derivative of λ_i with respect to any off-diagonal element is zero:

$$\frac{\partial \lambda_i}{\partial X_{k\ell}} = 0 \text{ for } k < \ell. \tag{9.27}$$

For the second derivative, on the other hand, we obtain

$$\frac{\partial^2 \lambda_i}{\partial X_{k\ell}^2} = \frac{2\delta_{ik}}{\lambda_i - \lambda_\ell} + \frac{2\delta_{i\ell}}{\lambda_i - \lambda_k} \text{ for } k < \ell. \tag{9.28}$$

Of the two terms on the right hand side, the first term exists only if $\ell > i$ while the second term is only present when $k < i$. So, for a given i, only $N-1$ terms of the form $2/(\lambda_i - \lambda_j)$ are present (note that the problematic term $i = j$ is absent). Putting everything back into Eq. (9.17), we find, with $\beta = 1$ here,

$$\mathrm{d}\lambda_i = \sqrt{\frac{2}{\beta N}} \mathrm{d}B_i + \frac{1}{N} \sum_{\substack{j=1 \\ j \neq i}}^{N} \frac{\mathrm{d}t}{\lambda_i - \lambda_j}, \tag{9.29}$$

where $\mathrm{d}B_i$ are independent Brownian motions (the old $\mathrm{d}B_{kk}$ for $k = i$). We have thus precisely recovered Equation (9.9) using Itô's calculus.

9.3 The Dyson Brownian Motion for the Resolvent

9.3.1 A Burgers' Equation for the Stieltjes Transform

Consider a matrix $\mathbf{M}_t$ that undergoes a DBM starting from a matrix $\mathbf{M}_0$. At each time t, $\mathbf{M}_t$ can be viewed as the sum of the matrix $\mathbf{M}_0$ and a Wigner matrix $\mathbf{X}_t$ of variance t:

$$\mathbf{M}_t = \mathbf{M}_0 + \mathbf{X}_t. \tag{9.30}$$

In order to characterize the spectrum of the matrix $\mathbf{M}_t$, one should compute its Stieltjes transform $\mathfrak{g}(z)$, which is the expectation value of the large N limit of function $g_N(z)$, defined by

$$g_N(z, \{\lambda_i\}) := \frac{1}{N}\sum_{i=1}^{N}\frac{1}{z-\lambda_i}. \tag{9.31}$$

g_N is thus a function of all eigenvalues $\{\lambda_i\}$ that undergo DBM while z is just a constant parameter. Since the eigenvalues evolve with time g_N is really a function of both z and t. We can use Itô's lemma to write a SDE for $g_N(z, \{\lambda_i\})$. The ingredients are, as usual now, the following partial derivatives:

$$\frac{\partial g_N}{\partial \lambda_i} = \frac{1}{N}\frac{1}{(z-\lambda_i)^2}; \quad \frac{\partial^2 g_N}{\partial \lambda_i^2} = \frac{2}{N}\frac{1}{(z-\lambda_i)^3}. \tag{9.32}$$

We can now apply Itô's lemma (8.17) using the dynamical equation for eigenvalues (9.9) to find

$$\mathrm{d}g_N = \frac{1}{N}\sqrt{\frac{2}{N}}\sum_{i=1}^{N}\frac{\mathrm{d}B_i}{(z-\lambda_i)^2} + \frac{1}{N^2}\sum_{\substack{i,j=1\\ j\neq i}}^{N}\frac{\mathrm{d}t}{(z-\lambda_i)^2(\lambda_i-\lambda_j)} + \frac{2}{N^2}\sum_{i=1}^{N}\frac{\mathrm{d}t}{(z-\lambda_i)^3}. \tag{9.33}$$

We now massage the second term to arrive at a symmetric form in i and j. In order to do so, note that i and j are dummy indices that are summed over, so we can rename $i \to j$ and vice versa and get the same expression. Adding the two versions and dividing by 2, we get that this term is equal to

$$\begin{aligned}
\frac{1}{N^2}\sum_{\substack{i,j=1\\ j\neq i}}^{N}\frac{\mathrm{d}t}{(z-\lambda_i)^2(\lambda_i-\lambda_j)} &= \frac{1}{2N^2}\sum_{\substack{i,j=1\\ j\neq i}}^{N}\left[\frac{\mathrm{d}t}{(z-\lambda_i)^2(\lambda_i-\lambda_j)} + \frac{\mathrm{d}t}{(z-\lambda_j)^2(\lambda_j-\lambda_i)}\right] \\
&= \frac{1}{2N^2}\sum_{\substack{i,j=1\\ j\neq i}}^{N}\frac{(2z-\lambda_i-\lambda_j)\mathrm{d}t}{(z-\lambda_i)^2(z-\lambda_j)^2} = \frac{1}{N^2}\sum_{\substack{i,j=1\\ j\neq i}}^{N}\frac{\mathrm{d}t}{(z-\lambda_i)(z-\lambda_j)^2} \\
&= \frac{1}{N^2}\sum_{i,j=1}^{N}\frac{\mathrm{d}t}{(z-\lambda_i)(z-\lambda_j)^2} - \frac{1}{N^2}\sum_{i=1}^{N}\frac{\mathrm{d}t}{(z-\lambda_i)^3} = -g_N\frac{\partial g_N}{\partial z}\mathrm{d}t - \frac{1}{N^2}\sum_{i=1}^{N}\frac{\mathrm{d}t}{(z-\lambda_i)^3}.
\end{aligned} \tag{9.34}$$

Note that very similar manipulations have been used in Section 5.2.2. Thus we have

$$\begin{aligned}
\mathrm{d}g_N &= \frac{1}{N}\sqrt{\frac{2}{N}}\sum_{i=1}^{N}\frac{\mathrm{d}B_i}{(z-\lambda_i)^2} - g_N\frac{\partial g_N}{\partial z}\mathrm{d}t + \frac{1}{N^2}\sum_{i=1}^{N}\frac{\mathrm{d}t}{(z-\lambda_i)^3} \\
&= \frac{1}{N}\sqrt{\frac{2}{N}}\sum_{i=1}^{N}\frac{\mathrm{d}B_i}{(z-\lambda_i)^2} - g_N\frac{\partial g_N}{\partial z}\mathrm{d}t + \frac{1}{2N}\frac{\partial^2 g_N}{\partial z^2}\mathrm{d}t.
\end{aligned} \tag{9.35}$$

Taking now the expectation of this equation (such that the $\mathrm{d}B_i$ term vanishes), we get

$$\mathbb{E}[\mathrm{d}g_N(z)] = -\mathbb{E}\left[g_N(z)\frac{\partial g_N(z)}{\partial z}\right]\mathrm{d}t + \frac{1}{2N}\mathbb{E}\left[\frac{\partial^2 g_N(z)}{\partial z^2}\right]\mathrm{d}t. \tag{9.36}$$

This equation is exact for any N. We can now take the $N \to \infty$ limit, where $g_N(z) \to \mathfrak{g}_t(z)$. Using the fact that the Stieltjes transform is self-averaging, we get a PDE for the time dependent Stieltjes transform $\mathfrak{g}_t(z)$:

$$\frac{\partial \mathfrak{g}_t(z)}{\partial t} = -\mathfrak{g}_t(z)\frac{\partial \mathfrak{g}_t(z)}{\partial z}. \tag{9.37}$$

Equation (9.37) is called the inviscid Burgers' equation. Such an equation can in fact develop singularities, so it often needs to be regularized by a viscosity term, which is in fact present for finite N: it is precisely the last term of Eq. (9.36). Equation (9.37) can be exactly solved using the methods of characteristics. This will be the topic of Section 10.1 in the next chapter.

9.3.2 The Evolution of the Resolvent

Let us now consider the full matrix resolvent of $\mathbf{M}_t$, defined as $\mathbf{G}_t(z) = (z\mathbf{1} - \mathbf{M}_t)^{-1}$. Clearly, the quantity $\mathfrak{g}_t(z)$ studied above is simply the trace of $\mathbf{G}_t(z)$, but $\mathbf{G}_t(z)$ also contains information on the evolution of eigenvectors. Since each element of $\mathbf{G}_t$ depends on all the elements of $\mathbf{M}_t$, one can again use Itô's calculus to derive an evolution equation for $\mathbf{G}_t(z)$. The calculation is more involved because one needs to carefully keep track of all indices. In this technical section, we sketch the derivation of Dyson Brownian motion for the resolvent and briefly discuss the result, which will be used further in Section 10.1.

Since $\mathbf{M}_t = \mathbf{M}_0 + \mathbf{X}_t$, the Itô lemma gives

$$\mathrm{d}G_{ij}(z) = \sum_{k,\ell=1}^{N} \frac{\partial G_{ij}}{\partial M_{k\ell}}\mathrm{d}X_{k\ell} + \frac{1}{2}\sum_{k,\ell,m,n=1}^{N} \frac{\partial^2 G_{ij}}{\partial M_{k\ell}\partial M_{mn}}\mathrm{d}[X_{k\ell}X_{mn}], \tag{9.38}$$

where the last term denotes the covariation of $X_{k\ell}$ and X_{mn}, and we have considered M_{kl} and M_{lk} to be independent variables following 100% correlated Brownian motions. Next, we compute the derivatives

$$\frac{\partial G_{ij}}{\partial M_{k\ell}} = \frac{1}{2}\left[G_{ik}G_{j\ell} + G_{jk}G_{i\ell}\right], \tag{9.39}$$

from which we deduce the second derivatives

$$\frac{\partial^2 G_{ij}}{\partial M_{k\ell}\partial M_{mn}} = \frac{1}{4}\left[(G_{im}G_{kn} + G_{im}G_{kn})G_{j\ell} + \cdots\right], \tag{9.40}$$

where we have not written the other GGG products obtained by applying Eq. (9.39) twice. Now, using the properties of the Brownian noise, the quadratic covariation reads

$$\mathrm{d}[X_{k\ell}X_{mn}] = \frac{\mathrm{d}t}{N}\left(\delta_{km}\delta_{\ell n} + \delta_{kn}\delta_{\ell m}\right), \tag{9.41}$$

so that we get from (9.38) and taking into account symmetries:

$$\mathrm{d}G_{ij}(z,t) = \sum_{k,\ell=1}^{N} G_{ik}G_{j\ell}\mathrm{d}X_{k\ell} + \frac{1}{N}\sum_{k,\ell=1}^{N}\left(G_{ik}G_{\ell k}G_{\ell j} + G_{ik}G_{kj}G_{\ell\ell}\right)\mathrm{d}t. \tag{9.42}$$

If we now take the average over the Brownian motions $dX_{k\ell}$, we find the following evolution for the average resolvent:

$$\partial_t \mathbb{E}[\mathbf{G}_t(z)] = \operatorname{Tr} \mathbf{G}_t(z)\, \mathbb{E}[\mathbf{G}_t^2(z)] + \frac{1}{N}\mathbb{E}[\mathbf{G}_t^3(z)]. \tag{9.43}$$

Now, one can notice that

$$\mathbf{G}^2(z,t) = -\partial_z \mathbf{G}(z,t); \qquad \mathbf{G}^3(z,t) = \frac{1}{2}\partial^2_{zz}\mathbf{G}(z,t), \tag{9.44}$$

which hold even before averaging. By sending $N \to \infty$, we then obtain the following matrix PDE for the resolvent:

$$\partial_t \mathbb{E}[\mathbf{G}_t(z)] = -\mathfrak{g}_t(z)\, \partial_z \mathbb{E}[\mathbf{G}_t(z)], \quad \text{with} \quad \mathbb{E}[\mathbf{G}_0(z)] = \mathbf{G}_{\mathbf{M}_0}(z). \tag{9.45}$$

Note that this equation is *linear* in $\mathbf{G}_t(z)$ once the Stieltjes transform $\mathfrak{g}_t(z)$ is known. Taking the trace of Eq. (9.45) immediately leads back to the Burgers' equation (9.37) for $\mathfrak{g}_t(z)$ itself, as expected.

9.4 The Dyson Brownian Motion with a Potential

9.4.1 A Modified Langevin Equation for Eigenvalues

In this section, we modify Dyson Brownian motion by adding a potential such that the stationary state of these interacting random walks coincides with the eigenvalue measure of β-ensembles, namely

$$P(\{\lambda_i\}) = Z_N^{-1} \exp\left\{-\frac{\beta}{2}\left[\sum_{i=1}^{N} N V(\lambda_i) - \sum_{\substack{i,j=1\\ j\neq i}}^{N} \log|\lambda_i - \lambda_j|\right]\right\}. \tag{9.46}$$

The general vectorial Langevin equation (8.36) leading to such an equilibrium with $\sigma^2 = 2/N$ immediately gives us the following DBM in a potential $V(\lambda)$:

$$d\lambda_k = \sqrt{\frac{2}{N}}\, dB_k + \left(\frac{1}{N}\sum_{\substack{\ell=1\\ \ell\neq k}}^{N} \frac{\beta}{\lambda_k - \lambda_\ell} - \frac{\beta}{2}V'(\lambda_k)\right) dt, \tag{9.47}$$

which recovers Eq. (9.9) in the absence of a potential. See Figure 9.2 for an illustration.

Dyson Brownian motion in a potential has many applications. Numerically it can be used to generate matrices for an arbitrary potential and an arbitrary value of β, a task not obvious *a priori* from the definition (9.46). Figure 9.2 shows a simulation of the matrix potential studied in Section 5.3.3. Note that DBM generates the correct density of eigenvalues; it also generates the proper statistics for the joint distribution of eigenvalues.

It is interesting to see how Burgers' equation for the Stieltjes transform, Eq. (9.37), is changed in the case where $V(\lambda) = \lambda^2/2$, i.e. in the standard GOE case $\beta = 1$. Redoing the steps leading to Eq. (9.36) with the extra V' term in the right hand side of Eq. (9.47) modifies the Burgers' equation into

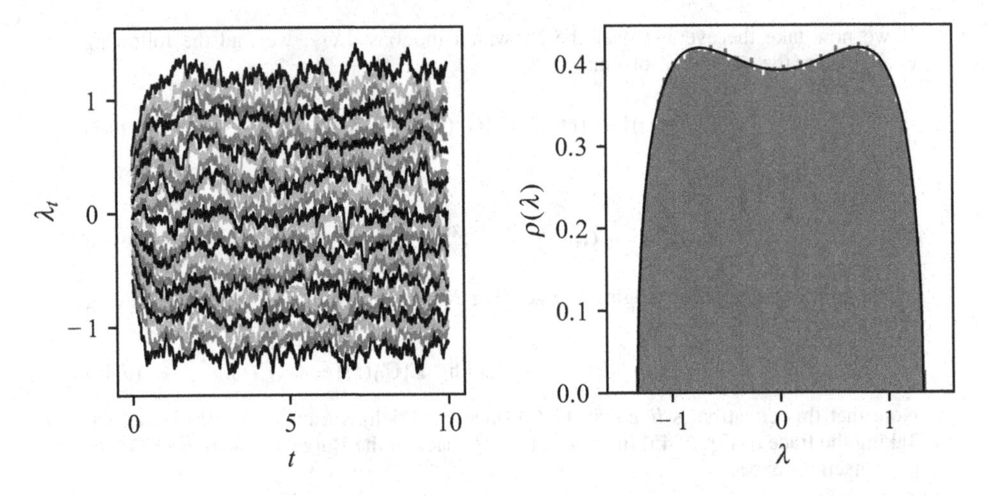

Figure 9.2 (left) A simulation of DBM with a potential for an $N = 25$ matrix starting from a Wigner matrix with $\sigma^2 = 1/10$ and evolving within the potential $V(x) = \frac{x^2}{2} + \frac{x^4}{4}$ for 10 units of time. Note that the steady state is quickly reached (within one or two units of time). (right) Histogram of the eigenvalues for the same process for $N = 400$ and 200 discrete steps per unit time. The histogram is over all matrices from time 3 to 10 (560 000 points). The agreement with the theoretical density (Eq. (5.58)) is very good.

$$\frac{\partial \mathfrak{g}_t(z)}{\partial t} = -\mathfrak{g}_t(z)\frac{\partial \mathfrak{g}_t(z)}{\partial z} + \frac{1}{2}\frac{\partial(z\mathfrak{g}_t(z))}{\partial z}. \tag{9.48}$$

The solution to this equation will be discussed in the next chapter.

More theoretically, DBM can be used in proofs of *local universality*, which is one of the most important results in random matrix theory. Local universality is the concept that many properties of the joint law of eigenvalues do not depend on the specifics of the random matrix in question, provided one looks at them on a scale ξ comparable to the average distance between eigenvalues, i.e. on scales $N^{-1} \lesssim \xi \ll 1$. Many such properties arise from the logarithmic eigenvalue repulsion and indeed depend only on the symmetry class (β) of the model.

Another useful property of DBM is its speed of convergence to the steady state. With time normalized as in Eq. (9.47), global properties (such as the density of eigenvalues) converge in a time of order 1, as we discuss in the next subsection. Local properties on the other hand (e.g. eigenvalue spacing) converge much faster, in a time of order $1/N$, i.e. as soon as the eigenvalues have "collided" a few times with one another.[1]

[1] The time needed for two Brownian motions a distance d apart to meet for the first time is of order d^2/σ^2. In our case $d = 1/N$ (the typical distance between two eigenvalues) and $\sigma^2 = 2/N$. The typical collision time is therefore $(2N)^{-1}$. Note however that eigenvalues actually never cross under Eq. (9.47), but the corresponding eigenvectors strongly mix when such quasi-collisions occur.

Exercise 9.4.1 Moments under DBM with a potential

Consider the moments of the eigenvalues as a function of time:

$$F_k(t) = \frac{1}{N}\sum_{i=1}^{N} \lambda_i^k(t), \tag{9.49}$$

for eigenvalues undergoing DBM under a potential $V(x)$, Eq. (9.47), in the orthogonal case $\beta = 1$. In this exercise you will need to show (and to use) the following identity:

$$2\sum_{\substack{i,j=1\\ j\neq i}}^{N} \frac{\lambda_i^k}{\lambda_i - \lambda_j} = \sum_{\substack{i,j=1\\ j\neq i}}^{N} \frac{\lambda_i^k - \lambda_j^k}{\lambda_i - \lambda_j} = \sum_{\substack{i,j=1\\ j\neq i}}^{N} \sum_{\ell=0}^{k} \lambda_i^{\ell}\lambda_j^{k-\ell}. \tag{9.50}$$

(a) Using Itô calculus, write a SDE for $F_2(t)$.

(b) By taking the expectation value of your equation, show that

$$\frac{\mathrm{d}}{\mathrm{d}t}\mathbb{E}[F_2(t)] = 1 - \mathbb{E}\left[\frac{1}{N}\sum_{i=1}^{N} \lambda_i(t)V'(\lambda_i(t))\right] + \frac{1}{N}. \tag{9.51}$$

(c) In the Wigner case, $V'(x) = x$, find the steady-state value of $\mathbb{E}[F_2(t)]$ for any finite N.

(d) For a random matrix $\mathbf{X}$ drawn from a generic potential $V(x)$, show that in the large N limit, we have

$$\tau\left[V'(\mathbf{X})\mathbf{X}\right] = 1, \tag{9.52}$$

where τ is the expectation value of the normalized trace defined by (2.1).

(e) Show that this equation is consistent with $\tau(\mathbf{W}) = 1$ for a Wishart matrix whose potential is given by Eq. (5.4).

(f) In the large N limit, find a general expression for $\tau[V'(\mathbf{X})\mathbf{X}^k]$ by writing the steady-state equation for $\mathbb{E}[F_{k+1}(t)]$; you can neglect the Itô term. The first two should be given by

$$\tau[V'(\mathbf{X})\mathbf{X}^2] = 2\tau[\mathbf{X}] \quad \text{and} \quad \tau[V'(\mathbf{X})\mathbf{X}^3] = 2\tau[\mathbf{X}^2] + \tau[\mathbf{X}]^2. \tag{9.53}$$

(g) In the unit Wigner case $V'(x) = x$, show that your relation in (f) is equivalent to the Catalan number inductive relation (3.27), with $\tau(\mathbf{X}^{2m}) = C_m$ and $\tau(\mathbf{X}^{2m+1}) = 0$.

9.4.2 The Fokker–Planck Equation for DBM

From the Langevin equation, Eq. (9.47), one can derive the Fokker–Planck equation describing the time evolution of the joint distribution of eigenvalues, $P(\{\lambda_i\}, t)$. It reads

$$\frac{\partial P}{\partial t} = \frac{1}{N}\sum_{i=1}^{N}\frac{\partial}{\partial \lambda_i}\left[\frac{\partial P}{\partial \lambda_i} - \mathcal{F}_i P\right], \tag{9.54}$$

where we use P as an abbreviation for $P(\{\lambda_i\}, t)$, and, for a quadratic confining potential $V(\lambda) = \lambda^2/2$, a generalized force given by

$$\mathcal{F}_i := \beta \sum_{\substack{j=1 \\ j\neq i}}^{N} \frac{1}{\lambda_i - \lambda_j} - \frac{N\beta\lambda_i}{2}. \tag{9.55}$$

The trick now is to introduce an auxiliary function $W(\{\lambda_i\}, t)$, defined as

$$P(\{\lambda_i\}, t) := \exp\left[\frac{\beta}{4}\sum_{\substack{i,j=1 \\ j\neq i}}^{N} \log|\lambda_i - \lambda_j| - \frac{\beta N}{8}\sum_{i=1}^{N}\lambda_i^2\right] W(\{\lambda_i\}, t). \tag{9.56}$$

Then after a little work, one finds the following evolution equation for W:[2]

$$\frac{\partial W}{\partial t} = \frac{1}{N}\sum_{i=1}^{N}\left[\frac{\partial^2 W}{\partial \lambda_i^2} - \mathcal{V}_i W\right], \tag{9.58}$$

with

$$\mathcal{V}_i(\{\lambda_j\}) := \frac{\beta^2 N^2}{16}\lambda_i^2 - \frac{\beta(2-\beta)}{4}\sum_{\substack{j=1 \\ j\neq i}}^{N}\frac{1}{(\lambda_i - \lambda_j)^2} - \frac{N\beta}{4}\left(1 + \frac{\beta(N-1)}{2}\right). \tag{9.59}$$

Looking for W_Γ such that

$$\frac{\partial W_\Gamma}{\partial t} = -\Gamma W_\Gamma, \tag{9.60}$$

one finds that W_Γ is the solution of the following eigenvalue problem:

$$\frac{2}{N}\sum_{i=1}^{N}\left[-\frac{1}{2}\frac{\partial^2 W_\Gamma}{\partial \lambda_i^2} + \frac{1}{2}\mathcal{V}_i W_\Gamma\right] = \Gamma W_\Gamma. \tag{9.61}$$

One notices that Eq. (9.61) is a (real) Schrodinger equation for N interacting "particles" in a quadratic potential, with an interacting potential that depends on the inverse of the square distance between particles. This is called the Calogero model, which happens to be exactly soluble in one dimension, both classically and quantum mechanically. In particular, the whole spectrum of this Hamiltonian is known, and given by[3]

$$\Gamma(n_1, n_2, \ldots, n_N) = \frac{\beta}{2}\left(\sum_{i=1}^{N} n_i - \frac{N(N-1)}{2}\right); \qquad 0 \le n_1 < n_2 \cdots < n_N. \tag{9.62}$$

[2] One has to use, along the line, the following identity:

$$\sum_{i\neq j\neq k}\frac{1}{\lambda_i - \lambda_j}\frac{1}{\lambda_i - \lambda_k} \equiv 0. \tag{9.57}$$

[3] Because $W_\Gamma(\{\lambda_i\})$ must vanish as two λ's coincide, one must choose the so-called fermionic branch of the spectrum.

The smallest eigenvalue corresponds to $n_1 = 0, n_2 = 1, \ldots, n_N = N - 1$ and is such that $\Gamma \equiv 0$. This corresponds to the equilibrium state of the Fokker–Planck equation:

$$W_0(\{\lambda_i\}) = \exp\left[\frac{\beta}{4}\sum_{\substack{i,j=1\\ j\neq i}}^{N}\log|\lambda_i - \lambda_j| - \frac{\beta N}{8}\sum_{i=1}^{N}\lambda_i^2\right]. \tag{9.63}$$

All other "excited states" have positive Γ's, corresponding to exponentially decaying modes (in time) of the Fokker–Planck equation. The smallest, non-zero value of Γ is such that $n_N = N$, all other values of n_i being unchanged. Hence Γ_1 is equal to $\beta/2$.

In conclusion, we have explicitly shown that the equilibration time of the DBM in a quadratic potential is equal to $2/\beta$. As announced at the end of the previous section, the density of eigenvalues indeed converges in a time of order unity.

The case $\beta = 2$ is particularly simple, since the interaction term in $\mathcal{V}_i$ disappears completely. We will use this result to show that, in the absence of a quadratic potential, the time dependent joint distribution of eigenvalues, $P(\{\lambda_i\}, t)$, can be expressed, for $\beta = 2$, as a simple determinant: this is the so-called *Karlin–McGregor* representation, see next section.

9.5 Non-Intersecting Brownian Motions and the Karlin–McGregor Formula

Let $p(y,t|x)$ be the probability density that a Brownian motion starting at x at $t = 0$ is at y at time t

$$p(y,t|x) = \frac{1}{\sqrt{2\pi t}}\exp\left(-\frac{(x-y)^2}{2t}\right), \tag{9.64}$$

where we set $\sigma^2 = 1$. Note that $p(y,t|x)$ obeys the diffusion equation

$$\frac{\partial p(y,t|x)}{\partial t} = \frac{1}{2}\frac{\partial^2 p(y,t|x)}{\partial y^2}. \tag{9.65}$$

We now consider N independent Brownian motions starting at $\mathbf{x} = (x_1, x_2, \ldots x_N)$ at $t = 0$, with $x_1 > x_2 > \ldots > x_N$. The Karlin–McGregor formula states that the probability $P_{KM}(\mathbf{y}, t|\mathbf{x})$ that these Brownian motions have reached positions $\mathbf{y} = (y_1 > y_2 > \ldots > y_N)$ at time t without ever intersecting between 0 and t is given by the following determinant:

$$P_{\text{KM}}(\mathbf{y}, t|\mathbf{x}) = \begin{vmatrix} p(y_1,t|x_1) & p(y_1,t|x_2) & \ldots & p(y_1,t|x_N) \\ p(y_2,t|x_1) & p(y_2,t|x_2) & \ldots & p(y_2,t|x_N) \\ \vdots & \vdots & \ddots & \vdots \\ p(y_N,t|x_1) & p(y_N,t|x_2) & \ldots & p(y_N,t|x_N) \end{vmatrix}. \tag{9.66}$$

One can easily prove this by noting that the determinant involves sums of products of N terms $p(y_i,t|x_j)$, each product Π involving one and only one y_i. Each product Π therefore obeys the multidimensional diffusion equation:

$$\frac{\partial \Pi}{\partial t} = \frac{1}{2}\sum_{i=1}^{N}\frac{\partial^2 \Pi}{\partial y_i^2}. \tag{9.67}$$

Being the sum of such products, $P_{\rm KM}(\mathbf{y},t|\mathbf{x})$ also obeys the same diffusion equation, as it should since we consider N independent Brownian motions:

$$\frac{\partial P_{\rm KM}(\mathbf{y},t|\mathbf{x})}{\partial t} = \frac{1}{2}\sum_{i=1}^{N} \frac{\partial^2 P_{\rm KM}(\mathbf{y},t|\mathbf{x})}{\partial y_i^2}, \tag{9.68}$$

The determinant structure also ensures that $P_{\rm KM}(\mathbf{y},t|\mathbf{x}) = 0$ as soon as any two y's are equal. Finally, because the x's and the y's are ordered, $P_{\rm KM}(\mathbf{y},t=0|\mathbf{x})$ obeys the correct initial condition.

Note that the total survival probability $\mathcal{P}(t;\mathbf{x}) := \int d\mathbf{y} P_{\rm KM}(\mathbf{y},t|\mathbf{x})$ decreases with time, since as soon as two Brownian motions meet, the corresponding trajectory is killed. In fact, one can show that $\mathcal{P}(t;\mathbf{x}) \sim t^{-N(N-1)/4}$ at large times.

Now, the probability $P(\mathbf{y},t|\mathbf{x})$ that these N independent Brownian motions end at $\mathbf{y}$ at time t conditional on the fact that the paths never ever intersect, i.e. for any time between $t=0$ and $t=\infty$, turns out to be given by a very similar formula:

$$P(\mathbf{y},t|\mathbf{x}) = \frac{\Delta(\mathbf{y})}{\Delta(\mathbf{x})} P_{\rm KM}(\mathbf{y},t|\mathbf{x}), \tag{9.69}$$

where $\Delta(\mathbf{x})$ is the Vandermonde determinant $\prod_{i<j}(x_j - x_i)$ (and similarly for $\Delta(\mathbf{y})$).

What we want to show is that $P(\mathbf{y},t|\mathbf{x})$ is the solution of the Fokker–Planck equation for the Dyson Brownian motion, Eq. (9.54), with $\beta = 2$ and in the absence of any confining potential. Indeed, as shown above, $P_{\rm KM}(\mathbf{y},t|\mathbf{x})$ obeys the diffusion equation for N independent Brownian motions with the annihilation boundary condition $P_{\rm KM}(\mathbf{y},t|\mathbf{x}) = 0$ when $y_i = y_j$ for any given pair i,j.

Now compare with the definition Eq. (9.56) of W for the Dyson Brownian motions with $\beta = 2$ and without any confining potential:

$$P(\{\lambda_i\},t) := \exp\left[\frac{1}{2}\sum_{\substack{j=1\\ j\neq i}}^{N} \log|\lambda_i - \lambda_j|\right] W(\{\lambda_i\},t) \equiv \Delta(\{\lambda_i\}) W(\{\lambda_i\},t). \tag{9.70}$$

From Eq. (9.58) we see that in the present case $W(\{\lambda_i\},t)$ also obeys the diffusion equation for N independent Brownian motions. Since $P(\{\lambda_i\},t) \sim |\lambda_i - \lambda_j|^2$ when two eigenvalues are close, we also see that $W(\{\lambda_i\},t)$ vanishes linearly whenever two eigenvalues meet, and therefore obeys the same boundary conditions as $P_{\rm KM}(\mathbf{y},t|\mathbf{x})$ with $y_i = \lambda_i$.

The conclusion is therefore that the Dyson Brownian motion without external forces is, for $\beta = 2$, equivalent to N Brownian motions constrained to never ever intersect.

Bibliographical Notes

- The Dyson Brownian motion was introduced in
  - F. J. Dyson. A Brownian-motion model for the eigenvalues of a random matrix. *Journal of Mathematical Physics*, 3:1191–1198, 1962.
- It is not often discussed in books on random matrix theory. The subject is treated in
  - J. Baik, P. Deift, and T. Suidan. *Combinatorics and Random Matrix Theory*. American Mathematical Society, Providence, Rhode Island, 2016,
  - L. Erdős and H.-T. Yau. *A Dynamical Approach to Random Matrix Theory*. American Mathematical Society, Providence, Rhode Island, 2017,

the latter discusses DBM with a potential beyond the Ornstein–Uhlenbeck model.

- On the Burgers' equation for the Stieltjes transform, see
  - L. C. G. Rodgers and Z. Shi. Interacting Brownian particles and the Wigner law. *Probability Theory and Related Fields*, 95:555–570, 1993.
- On the evolution of the matrix resolvent, see
  - R. Allez, J. Bun, and J.-P. Bouchaud. The eigenvectors of Gaussian matrices with an external source. *preprint arXiv:1412.7108*, 2014,
  - J. Bun, J.-P. Bouchaud, and M. Potters. Cleaning large correlation matrices: Tools from random matrix theory. *Physics Reports*, 666:1–109, 2017.
- On the universality of the local properties of eigenvalues, see e.g.
  - E. Brézin and A. Zee. Universality of the correlations between eigenvalues of large random matrices. *Nuclear Physics B*, 402(3):613–627, 1993,
  - L. Erdős. Universality of Wigner random matrices: A survey of recent results. *Russian Mathematical Surveys*, 66(3):507, 2011.
- On the Calogero model, see
  - F. Calogero. Solution of the one-dimensional n-body problem with quadratic and/or inversely quadratic pair potentials. *Journal of Mathematical Physics*, 12:419–436, 1971,
  - P. J. Forrester. *Log Gases and Random Matrices*. Princeton University Press, Princeton, NJ, 2010,
  - A. P. Polychronakos. The physics and mathematics of Calogero particles. *Journal of Physics A: Mathematical General*, 39(41):12793–12845, 2006,

  and http://www.scholarpedia.org/article/Calogero-Moser_system
- On the Karlin–McGregor formula, see
  - S. Karlin and J. McGregor. Coincidence probabilities. *Pacific Journal of Mathematics*, 9(4):1141–1164, 1959,
  - P.-G. de Gennes. Soluble model for fibrous structures with steric constraints. *The Journal of Chemical Physics*, 48(5):2257–2259, 1968,

  and for recent applications
  - T. Gautié, P. Le Doussal, S. N. Majumdar, and G. Schehr. Non-crossing Brownian paths and Dyson Brownian motion under a moving boundary. *Journal of Statistical Physics*, 177(5):752–805, 2019,

  and references therein.

10
Addition of Large Random Matrices

In this chapter we seek to understand how the eigenvalue density of the sum of two large random matrices **A** and **B** can be obtained from their individual densities. In the case where, say, **A** is a Wigner matrix **X**, the Dyson Brownian motion formalism of the previous chapter allows us to swiftly answer that question. We will see that a particular transform of the density of **B**, called the R-transform, appears naturally. We then show that the R-transform appears in the more general context where the eigenbases of **A** and **B** are related by a random rotation matrix **O**. In this case, one can construct a Fourier transform for matrices, which allows us to define the analog of the generating function for random variables. As in the case of IID random variables, the logarithm of this matrix generating function is additive when one adds two randomly rotated, large matrices. The derivative of this object turns out to be the R-transform, leading to the central result of the present chapter (and of the more abstract theory of free variables, see Chapter 11): the R-transform of the sum of two randomly rotated, large matrices is equal to the sum of R-transforms of each individual matrix.

10.1 Adding a Large Wigner Matrix to an Arbitrary Matrix

Let $\mathbf{M}_t = \mathbf{M}_0 + \mathbf{X}_t$ be the sum of a large matrix $\mathbf{M}_0$ and a large Wigner matrix $\mathbf{X}_t$, such that the variance of each element grows as t. This defines a Dyson Brownian motion as described in the previous chapter, see Eq. (9.11). We have shown in Section 9.3.1 that in this case the Stieltjes transform $\mathfrak{g}_t(z)$ of $\mathbf{M}_t$ satisfies the Burgers' equation:

$$\frac{\partial \mathfrak{g}_t(z)}{\partial t} = -\mathfrak{g}_t(z)\frac{\partial \mathfrak{g}_t(z)}{\partial z}, \tag{10.1}$$

with initial condition $\mathfrak{g}_0(z) := \mathfrak{g}_{M_0}(z)$. We now proceed to show that the solution of this Burgers' equation can be simply expressed using an $\mathbf{M}_0$ dependent function: its R-transform.

Using the so-called method of characteristics, one can show that

$$\mathfrak{g}_t(z) = \mathfrak{g}_0(z - t\mathfrak{g}_t(z)). \tag{10.2}$$

If the method of characteristics is unknown to the reader, one can verify that (10.2) indeed satisfies Eq. (10.1) for any function $\mathfrak{g}_0(z)$. Indeed, let us compute $\partial_t \mathfrak{g}_t(z)$ and $\partial_z \mathfrak{g}_t(z)$ using Eq. (10.2):

$$\partial_t \mathfrak{g}_t(z) = \mathfrak{g}_0'(z - t\mathfrak{g}_t(z))\left[-\mathfrak{g}_t(z) - t\partial_t \mathfrak{g}_t(z)\right] \Rightarrow \partial_t \mathfrak{g}_t(z) = -\frac{\mathfrak{g}_t(z)\mathfrak{g}_0'(z - t\mathfrak{g}_t(z))}{1 + t\mathfrak{g}_0'(z - t\mathfrak{g}_t(z))}, \quad (10.3)$$

and

$$\partial_z \mathfrak{g}_t(z) = \mathfrak{g}_0'(z - t\mathfrak{g}_t(z))\left[1 - t\partial_z \mathfrak{g}_t(z)\right] \Rightarrow \partial_z \mathfrak{g}_t(z) = \frac{\mathfrak{g}_0'(z - t\mathfrak{g}_t(z))}{1 + t\mathfrak{g}_0'(z - t\mathfrak{g}_t(z))}, \quad (10.4)$$

such that Eq. (10.1) is indeed satisfied.

Example: Suppose $\mathbf{M}_0 = 0$. Then we have $\mathfrak{g}_0(z) = z^{-1}$. Plugging into (10.2), we obtain that

$$\mathfrak{g}_t(z) = \frac{1}{z - t\mathfrak{g}_t(z)}, \quad (10.5)$$

which is the self-consistent Eq. (2.35) in the Wigner case with $\sigma^2 = t$. Indeed, if we start with the zero matrix, then $\mathbf{M}_t = \mathbf{X}_t$ is just a Wigner with parameter $\sigma^2 = t$.

Back to the general case, we denote as $\mathfrak{z}_t(g)$ the inverse function[1] of $\mathfrak{g}_t(z)$. Now fix $g = \mathfrak{g}_t(z) = \mathfrak{g}_0(z - tg)$ and $z = \mathfrak{z}_t(g)$, we apply the function $\mathfrak{z}_0$ to g and get

$$\begin{aligned} \mathfrak{z}_0(g) &= z - tg = \mathfrak{z}_t(g) - tg, \\ \mathfrak{z}_t(g) &= \mathfrak{z}_0(g) + tg. \end{aligned} \quad (10.6)$$

The inverse of the Stieltjes transform of $\mathbf{M}_t$ is given by the inverse of that of $\mathbf{M}_0$ plus a simple shift tg. If we know $\mathfrak{g}_0(z)$ we can compute its inverse $\mathfrak{z}_0(g)$ and thus easily obtain $\mathfrak{z}_t(g)$, which after inversion hopefully recovers $\mathfrak{g}_t(z)$.

Example: Suppose $\mathbf{M}_0$ is a Wigner matrix with variance σ^2. We first want to compute the inverse of $\mathfrak{g}_0(z)$; to do so we use the fact that $\mathfrak{g}_0(z)$ satisfies Eq. (2.35), and we get that

$$\mathfrak{z}_0(g) = \sigma^2 g + \frac{1}{g}. \quad (10.7)$$

Then, by (10.6), we get that

$$\mathfrak{z}_t(g) = \mathfrak{z}_0(g) + tg = \left(\sigma^2 + t\right) g + \frac{1}{g}, \quad (10.8)$$

which is the inverse Stieltjes transform for Wigner matrices with variance $\sigma^2 + t$. In other words $\mathfrak{g}_t(z)$ satisfies the Wigner equation (2.35) with σ^2 replaced by $\sigma^2 + t$. This result is not surprising, each element of the sum of two Wigner matrices is just the sum of Gaussian random variables. So $\mathbf{M}_t$ is itself a Wigner matrix with the sum of the variances as its variance.

[1] We will discuss in Section 10.4 the invertibility of the function $\mathfrak{g}(z)$.

We can now tackle the more general case when the initial matrix is not necessarily Wigner. Call $\mathbf{B} = \mathbf{M}_t$ and $\mathbf{A} = \mathbf{M}_0$. Then by (10.6), we get

$$\mathfrak{z}_{\mathbf{B}}(g) = \mathfrak{z}_{\mathbf{A}}(g) + tg = \mathfrak{z}_{\mathbf{A}}(g) + \mathfrak{z}_{\mathbf{X}_t}(g) - \frac{1}{g}. \tag{10.9}$$

We now define the R-transform as

$$R(g) := \mathfrak{z}(g) - \frac{1}{g}. \tag{10.10}$$

Note that the R-transform of a Wigner matrix of variance t is simply given by

$$R_{\mathbf{X}}(g) = tg. \tag{10.11}$$

This definition allows us to rewrite Eq. (10.9) above as a nice additive relation between R-transforms:

$$R_{\mathbf{B}}(g) = R_{\mathbf{A}}(g) + R_{\mathbf{X}_t}(g). \tag{10.12}$$

In the next section we will generalize this law of addition to (large) matrices $\mathbf{X}$ that are not necessarily Wigner. The R-transform will prove to be a very powerful tool to study large random matrices. Some of its properties are left to be derived in Exercises 10.1.1 and 10.1.2 and will be further discussed in Chapter 15. We finish this section by computing the R-transform of a white Wishart matrix. Remember that its Stieltjes transform satisfies Eq. (4.37), i.e.

$$qz\mathfrak{g}^2 - (z - 1 + q)\mathfrak{g} + 1 = 0, \tag{10.13}$$

which can be written in terms of the inverse function $\mathfrak{z}(g)$:

$$\mathfrak{z}(g) = \frac{1}{1 - qg} + \frac{1}{g}. \tag{10.14}$$

From which we can read off the R-transform:

$$R_{\mathbf{W}}(g) = \frac{1}{1 - qg}. \tag{10.15}$$

Exercise 10.1.1 Taylor series for the R-transform

Let $\mathfrak{g}(z)$ be the Stieltjes transform of a random matrix $\mathbf{M}$:

$$\mathfrak{g}(z) = \tau\left((z\mathbf{1} - \mathbf{M})^{-1}\right) = \int_{\mathrm{supp}\{\rho\}} \frac{\rho(\lambda)\mathrm{d}\lambda}{z - \lambda}. \tag{10.16}$$

We saw that the power series of $g(z)$ around $z = \infty$ is given by the moments of $\mathbf{M}$ ($m_n := \tau(\mathbf{M}^n)$):

$$g(z) = \sum_{n=0}^{\infty} \frac{m_n}{z^{n+1}} \text{ with } m_0 \equiv 1. \tag{10.17}$$

Call $\mathfrak{z}(g)$ the functional inverse of $\mathfrak{g}(z)$ which is well defined in a neighborhood of $g = 0$. And define $R(g)$ as

$$R(g) = \mathfrak{z}(g) - 1/g. \tag{10.18}$$

(a) By writing the power series of $R(g)$ near zero, show that $R(g)$ is regular at zero and that $R(0) = m_1$. Therefore the power series of $R(g)$ starts at g^0:

$$R(g) = \sum_{n=1}^{\infty} \kappa_n g^{n-1}. \tag{10.19}$$

(b) Now assume $m_1 = \kappa_1 = 0$ and compute κ_2, κ_3 and κ_4 as a function of m_2, m_3 and m_4 in that case.

Exercise 10.1.2 Scaling of the R-transform

Using your answer from Exercise 2.3.1: If $\mathbf{A}$ is a random matrix drawn from a well-behaved ensemble with Stieltjes transform $\mathfrak{g}_\mathbf{A}(z)$ and R-transform $R_\mathbf{A}(g)$, what is the R-transform of the random matrices $\alpha\mathbf{A}$ and $\mathbf{A} + b\mathbf{1}$ where α and b are non-zero real numbers?

Exercise 10.1.3 Sum of symmetric orthogonal and Wigner matrices

Consider as in Exercise 1.2.4 a random symmetric orthogonal matrix $\mathbf{M}$ and a Wigner matrix $\mathbf{X}$ of variance σ^2. We are interested in the spectrum of their sum $\mathbf{E} = \mathbf{M} + \mathbf{X}$.

(a) Given that the eigenvalues of $\mathbf{M}$ are ± 1 and that in the large N limit each eigenvalue appears with weight $\frac{1}{2}$, write the limiting Stieltjes transform $\mathfrak{g}_\mathbf{M}(z)$.

(b) $\mathbf{E}$ can be thought of as undergoing Dyson Brownian motion starting at $\mathbf{E}(0) = \mathbf{M}$ and reaching the desired $\mathbf{E}$ at $t = \sigma^2$. Use Eq. (10.2) to write an equation for $\mathfrak{g}_\mathbf{E}(z)$. This will be a cubic equation in $\mathfrak{g}$.

(c) You can obtain the same equation using the inverse function $\mathfrak{z}_\mathbf{M}(g)$ of your answer in (a). Show that

$$\mathfrak{z}_\mathbf{M}(g) = \frac{1 + \sqrt{1 - 4g^2}}{2g}, \tag{10.20}$$

where one had to pick the root that makes $\mathfrak{z}(g) \sim 1/g$ near $g = 0$.

(d) Using Eq. (10.6), write $z_\mathbf{E}(g)$ and invert this relation to obtain an equation for $\mathfrak{g}_\mathbf{E}(z)$. You should recover the same equation as in (b).

(e) Eigenvalues of $\mathbf{E}$ will be located where your equation admits non-real solutions for real z. First look at $z = 0$; the equation becomes quadratic after factorizing a trivial root. Find a criterion for σ^2 such that the equation admits non-real solutions. Compare with your answer in Exercise 1.2.4 (b).

(f) At $\sigma^2 = 1$, the equation is still cubic but is somewhat simpler. A real cubic equation of the form $ax^3 + bx^2 + cx + d = 0$ will have non-real solutions iff $\Delta < 0$ where $\Delta = 18abcd - 4b^3d + b^2c^2 - 4ac^3 - 27a^2d^2$. Using this criterion show that for $\sigma^2 = 1$ the edges of the eigenvalue spectrum are given by $\lambda = \pm 3\sqrt{3}/2 \approx \pm 2.60$.

(g) Again at $\sigma^2 = 1$, the solution near $g(0) = 0$ can be expanded in fractional powers of z. Show that we have

$$g(z) = z^{1/3} + O(z), \text{ which implies } \rho(x) = \frac{\sqrt{3}}{2}\sqrt[3]{|x|}, \tag{10.21}$$

for x near zero.

(h) For $\sigma^2 = 1/2, 1$ and 2, solve numerically the cubic equation for $\mathfrak{g}_{\mathbf{E}}(z)$ for $z = x$ real and plot the density of eigenvalues $\rho(x) = |\,\mathrm{Im}(\mathfrak{g}_{\mathbf{E}}(x))|/\pi$ for one of the complex roots if present.

10.2 Generalization to Non-Wigner Matrices

10.2.1 Set-Up

In the previous section, we derived a formula for the Stieltjes transform of the sum of a Wigner matrix and an arbitrary matrix. We would like to find a generalization of this result to a larger class of matrices.

Take two $N \times N$ matrices: $\mathbf{A}$, with eigenvalues $\{\lambda_i\}_{1\le i\le N}$ and eigenvectors $\{\mathbf{v}_i\}_{1\le i\le N}$, and $\mathbf{B}$, with eigenvalues $\{\mu_i\}_{1\le i\le N}$ and eigenvectors $\{\mathbf{u}_i\}_{1\le i\le N}$. Then the eigenvalues of $\mathbf{C} = \mathbf{B} + \mathbf{A}$ will in general depend in a complicated way on the overlaps between the eigenvectors of $\mathbf{B}$ and the eigenvectors of $\mathbf{A}$. In the trivial case where $\mathbf{v}_i = \mathbf{u}_i$ for all i, we have that the eigenvalues of $\mathbf{B} + \mathbf{A}$ are simply given by $\nu_i = \lambda_i + \mu_i$. However, this is neither generic nor very interesting.

One important property of Wigner matrices is that their eigenvectors are Haar distributed, that is, the matrix of eigenvectors is distributed uniformly in the group $O(N)$ and each eigenvector is uniformly distributed on the unit sphere S^{N-1}. Thus, when N is large, it is very unlikely that any one of them will have a significant overlap with the eigenvectors of $\mathbf{B}$. This is the property that we want to keep in our generalization. We will study what happens for general matrices $\mathbf{B}$ and $\mathbf{A}$ when their eigenvectors are random with respect to one another. We will define this relative randomness notion (called "freeness") more precisely in the next chapter. Here, to ensure the randomness of the eigenvectors, we will apply a random rotation to the matrix $\mathbf{A}$ and define the free addition as

$$\mathbf{C} = \mathbf{B} + \mathbf{O}\mathbf{A}\mathbf{O}^T, \tag{10.22}$$

where $\mathbf{O}$ is a Haar distributed random orthogonal matrix. Then it is easy to see that $\mathbf{O}\mathbf{A}\mathbf{O}^T$ is rotational invariant since $\mathbf{O}'\mathbf{O}$ is also Haar distributed for any fixed $\mathbf{O}' \in O(N)$.

10.2.2 Matrix Fourier Transform

We saw in Section 8.1 that the function $H_X(t) = \log \mathbb{E} \exp(itX)$ is additive when one adds independent scalar variables. When X is a matrix, it is plausible that t should also be a matrix $\mathbf{T}$, but in the end we need to take the exponential of a scalar, so a possible candidate would be

$$I(\mathbf{X},\mathbf{T}) := \left\langle \exp\left(\frac{N}{2} \operatorname{Tr} \mathbf{TOXO}^T\right)\right\rangle_{\mathbf{O}}. \tag{10.23}$$

The notation $\langle\cdot\rangle_{\mathbf{O}}$ means that we average over all orthogonal matrices $\mathbf{O}$ (with a flat weight) normalized such that $\langle 1\rangle_{\mathbf{O}} = 1$. This defines the Haar measure on the group of orthogonal matrices. Equation (10.23) defines the so-called Harish-Chandra–Itzykson–Zuber (HCIZ) integral.[2] Note that by definition, $I(\mathbf{O}_1\mathbf{X}\mathbf{O}_1^T,\mathbf{T}) = I(\mathbf{X},\mathbf{O}_1\mathbf{T}\mathbf{O}_1^T) = I(\mathbf{X},\mathbf{T})$ for an arbitrary rotation matrix $\mathbf{O}_1$. This means that $I(\mathbf{X},\mathbf{T})$ only depends on the eigenvalues of $\mathbf{X}$ and $\mathbf{T}$.

Now consider $\mathbf{C} = \mathbf{B} + \mathbf{O}_1\mathbf{A}\mathbf{O}_1^T$ with a random $\mathbf{O}_1$. For large matrix sizes, the eigenvalue spectrum of $\mathbf{C}$ will turn out not to depend on the specific choice of $\mathbf{O}_1$, provided it is chosen according to the Haar measure. Therefore, one can average $I(\mathbf{C},\mathbf{T})$ over $\mathbf{O}_1$ and obtain

$$I(\mathbf{C},\mathbf{T}) = \left\langle \exp\left(\frac{N}{2} \operatorname{Tr} \mathbf{TO}(\mathbf{B}+\mathbf{O}_1\mathbf{A}\mathbf{O}_1^T)\mathbf{O}^T\right)\right\rangle_{\mathbf{O},\mathbf{O}_1} = I(\mathbf{B},\mathbf{T})I(\mathbf{A},\mathbf{T}), \tag{10.25}$$

where we have used that $\mathbf{OO}_1 = \mathbf{O}'$ is a random rotation independent from $\mathbf{O}$ itself. Hence we conclude that $\log I$ is additive in this case, as is the logarithm of the characteristic function in the scalar case.

For a general matrix $\mathbf{T}$, the HCIZ integral is quite complicated, as will be further discussed in Section 10.5. Fortunately, for our purpose we can choose the "Fourier" matrix $\mathbf{T}$ to be rank-1 and in this case the integral can be computed. A symmetric rank-1 matrix can be written as

$$\mathbf{T} = t\,\mathbf{v}\mathbf{v}^T, \tag{10.26}$$

where t is the eigenvalue and $\mathbf{v}$ is a unit vector. We will show that the large N behavior of $I(\mathbf{T},\mathbf{B})$ is given, in this case, by

$$I(\mathbf{T},\mathbf{B}) \approx \exp\left(\frac{N}{2} H_{\mathbf{B}}(t)\right), \tag{10.27}$$

for some function $H_{\mathbf{B}}(t)$ that depends on the particular matrix $\mathbf{B}$.

[2] The HCIZ can be defined with an integral over orthogonal, unitary or symplectic matrices. In the general case it is defined as

$$I_\beta(\mathbf{X},\mathbf{T}) := \left\langle \exp\left(\frac{N\beta}{2} \operatorname{Tr} \mathbf{XOTO}^\dagger\right)\right\rangle_{\mathbf{O}}, \tag{10.24}$$

with beta equal to 1, 2 or 4 and $\mathbf{O}$ is averaged over the corresponding group. The unitary $\beta = 2$ case is the most often studied, for which some explicit results are available.

More formally we define

$$H_{\mathbf{B}}(t) = \lim_{N\to\infty} \frac{2}{N} \log \left\langle \exp\left(\frac{tN}{2} \operatorname{Tr} \mathbf{v}\mathbf{v}^T \mathbf{O}\mathbf{B}\mathbf{O}^T\right)\right\rangle_{\mathbf{O}}. \tag{10.28}$$

If $\mathbf{C} = \mathbf{B} + \mathbf{A}$ where $\mathbf{A}$ is randomly rotated with respect to $\mathbf{B}$, the precise statement is that

$$H_{\mathbf{C}}(t) = H_{\mathbf{B}}(t) + H_{\mathbf{A}}(t), \tag{10.29}$$

i.e. the function H is additive. We now need to relate this function to the R-transform encountered in the previous section.

10.3 The Rank-1 HCIZ Integral

To get a useful theory, we need to have a concrete expression for this function $H_{\mathbf{B}}$. Without loss of generality, we can assume $\mathbf{B}$ is diagonal (in fact, we can diagonalize B and absorb the eigenmatrix into the orthogonal matrix $\mathbf{O}$ we integrate over). Moreover, for simplicity we assume that $t > 0$. Then $\mathbf{O}^T\mathbf{T}\mathbf{O}$ can be regarded as proportional to a random projector:

$$\mathbf{O}^T\mathbf{T}\mathbf{O} = \boldsymbol{\psi}\boldsymbol{\psi}^T, \tag{10.30}$$

with $\|\boldsymbol{\psi}\|^2 = t$ and $\boldsymbol{\psi}/\|\boldsymbol{\psi}\|$ uniformly distributed on the unit sphere. Then we make a change of variable $\boldsymbol{\psi} \to \boldsymbol{\psi}/\sqrt{N}$, and calculate

$$Z_t(\mathbf{B}) = \int \frac{\mathrm{d}^N\boldsymbol{\psi}}{(2\pi)^{N/2}} \delta\left(\|\boldsymbol{\psi}\|^2 - Nt\right) \exp\left(\frac{1}{2}\boldsymbol{\psi}^T\mathbf{B}\boldsymbol{\psi}\right), \tag{10.31}$$

where we have added a factor of $(2\pi)^{-N/2}$ for later convenience. Because $Z_t(\mathbf{B})$ is not properly normalized (i.e. $Z_t(0) \neq 1$), we will need to normalize it to compute $I(\mathbf{T}, \mathbf{B})$:

$$\left\langle \exp\left(\frac{N}{2} \operatorname{Tr} \mathbf{T}\mathbf{O}\mathbf{B}\mathbf{O}^T\right)\right\rangle_{\mathbf{O}} = \frac{Z_t(\mathbf{B})}{Z_t(0)}. \tag{10.32}$$

10.3.1 A Saddle Point Calculation

We can now express the Dirac delta as an integral over the imaginary axis:

$$\delta(x) = \int_{-\infty}^{\infty} \frac{\mathrm{e}^{-\mathrm{i}zx}}{2\pi} \mathrm{d}z = \int_{-\mathrm{i}\infty}^{\mathrm{i}\infty} \frac{\mathrm{e}^{-zx/2}}{4\mathrm{i}\pi} \mathrm{d}z.$$

Now let Λ be a parameter larger than the maximum eigenvalue of $\mathbf{B}$: $\Lambda > \lambda_{\max}(\mathbf{B})$. We introduce the factor

$$1 = \exp\left(-\frac{\Lambda\left(\|\boldsymbol{\psi}\|^2 - Nt\right)}{2}\right),$$

since $\|\boldsymbol{\psi}\|^2 = Nt$. Then, absorbing Λ into z, we get that

$$Z_t(\mathbf{B}) = \int_{\Lambda-i\infty}^{\Lambda+i\infty} \frac{dz}{4\pi} \int \frac{d^N \boldsymbol{\psi}}{(2\pi)^{N/2}} \exp\left(-\frac{1}{2}\boldsymbol{\psi}^T (z - \mathbf{B})\boldsymbol{\psi} + \frac{Nzt}{2}\right). \tag{10.33}$$

We can now perform the Gaussian integral over the vector $\boldsymbol{\psi}$:

$$\begin{aligned} Z_t(\mathbf{B}) &= \int_{\Lambda-i\infty}^{\Lambda+i\infty} \frac{dz}{4\pi} \det(z - \mathbf{B})^{-1/2} \exp\left(\frac{Nzt}{2}\right) \\ &= \int_{\Lambda-i\infty}^{\Lambda+i\infty} \frac{dz}{4\pi} \exp\left[\frac{N}{2}\left(zt - \frac{1}{N}\sum_k \log(z - \lambda_k(\mathbf{B}))\right)\right], \end{aligned} \tag{10.34}$$

where $\lambda_k(\mathbf{B})$, $1 \le k \le N$, are the eigenvalues of $\mathbf{B}$. Then we denote

$$F_t(z, \mathbf{B}) := zt - \frac{1}{N}\sum_k \log(z - \lambda_k(\mathbf{B})). \tag{10.35}$$

The integral in (10.34) is oscillatory, and by the stationary phase approximation (see Appendix A.1), it is dominated by the point where

$$\partial_z F_t(z, \mathbf{B}) = 0 \Rightarrow t - \frac{1}{N}\sum_k \frac{1}{z - \lambda_k(\mathbf{B})} = t - g_N^{\mathbf{B}}(z) = 0. \tag{10.36}$$

If $g_N^{\mathbf{B}}(z)$ can be inverted then we can express z as $\mathfrak{z}(t)$. For $x > \lambda_{\max}$, $g_N^{\mathbf{B}}(x)$ is monotonically decreasing and thus invertible. So for $t < g_N^{\mathbf{B}}(\lambda_{\max})$, a unique $\mathfrak{z}(t)$ exists and $\mathfrak{z}(t) > \lambda_{\max}$ (see Section 10.4). Since $F_t(z, \mathbf{B})$ is analytic to the right of $z = \lambda_{\max}$, we can deform the contour to reach this point (see Fig. 10.1). Using the saddle point formula (Eq. (A.3)), we have

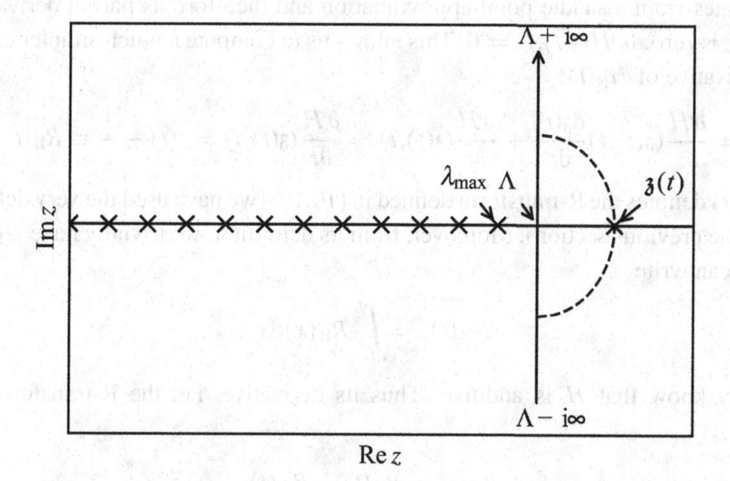

Figure 10.1 Graphical representation of the integral Eq. (10.34) in the complex plane. The crosses represent the eigenvalues of $\mathbf{B}$ and are singular points of the integrand. The integration is from $\Lambda - i\infty$ to $\Lambda + i\infty$ where $\Lambda > \lambda_{\max}$. The saddle point is at $z = \mathfrak{z}(t) > \lambda_{\max}$. Since the integrand is analytic right of $\lambda_{\max}$, the integration path can be deformed to go through $\mathfrak{z}(t)$.

$$Z_t(\mathbf{B}) \sim \frac{\sqrt{4\pi}/(4\pi)}{|N\partial_z^2 F(\mathfrak{z}(t),\mathbf{B})|^{1/2}} \exp\left[\frac{N}{2}\left(\mathfrak{z}(t)t - \frac{1}{N}\sum_k \log(\mathfrak{z}(t) - \lambda_k(\mathbf{B}))\right)\right]$$

$$\sim \frac{1}{2\sqrt{N\pi|g'_{\mathbf{B}}(\mathfrak{z}(t))|}} \exp\left[\frac{N}{2}\left(\mathfrak{z}(t)t - \frac{1}{N}\sum_k \log(\mathfrak{z}(t) - \lambda_k(\mathbf{B}))\right)\right]. \quad (10.37)$$

For the case $\mathbf{B} = 0$, we have $g_{\mathbf{B}}(z) = z^{-1} \Rightarrow \mathfrak{z}(t) = t^{-1}$, so we get

$$Z_t(0) \sim \frac{1}{2t\sqrt{N\pi}} \exp\left[\frac{N}{2}(1 + \log t)\right]. \quad (10.38)$$

In the large N limit, the prefactor in front of the exponential does not contribute to $H_{\mathbf{B}}(t)$ and we finally get

$$\lim_{N\to\infty} \frac{2}{N} \log\left\langle \exp\left(\frac{N}{2}\operatorname{Tr}\mathbf{TOBO}^T\right)\right\rangle_{\mathbf{O}} = \mathfrak{z}(t)t - 1 - \log t - \frac{1}{N}\sum_k \log(\mathfrak{z}(t) - \lambda_k(\mathbf{B})). \quad (10.39)$$

By the definition (10.28), we then get that

$$H_{\mathbf{B}}(t) = \mathcal{H}(\mathfrak{z}(t),t), \quad \mathcal{H}(z,t) := zt - 1 - \log t - \frac{1}{N}\sum_k \log(z - \lambda_k(\mathbf{B})). \quad (10.40)$$

10.3.2 Recovering R-Transforms

We found an expression for $H_{\mathbf{B}}(t)$ but in a form that is not easy to work with. But note $\mathcal{H}(z,t)$ comes from a saddle point approximation and therefore its partial derivative with respect to z is zero: $\partial_z\mathcal{H}(\mathfrak{z}(t),t) = 0$. This allows us to compute a much simpler expression for the derivative of $H_{\mathbf{B}}(t)$:

$$\frac{\mathrm{d}H_{\mathbf{B}}(t)}{\mathrm{d}t} = \frac{\partial\mathcal{H}}{\partial z}(\mathfrak{z}(t),t)\frac{\mathrm{d}\mathfrak{z}(t)}{\mathrm{d}t} + \frac{\partial\mathcal{H}}{\partial t}(\mathfrak{z}(t),t) = \frac{\partial\mathcal{H}}{\partial t}(\mathfrak{z}(t),t) = \mathfrak{z}(t) - \frac{1}{t} \equiv R_{\mathbf{B}}(t), \quad (10.41)$$

where $R_{\mathbf{B}}(t)$ denotes the R-transform defined in (10.10) (we have used the very definition of $\mathfrak{z}(t)$ from the previous section). Moreover, from its definition, we trivially have $H_{\mathbf{B}}(0) = 0$. Hence we can write

$$H_{\mathbf{B}}(t) := \int_0^t R_{\mathbf{B}}(x)\mathrm{d}x. \quad (10.42)$$

We already know that H is additive. Thus its derivative, i.e. the R-transform, is also additive:

$$R_{\mathbf{C}}(t) = R_{\mathbf{B}}(t) + R_{\mathbf{A}}(t), \quad (10.43)$$

as is the case when $\mathbf{A}$ is a Wigner matrix. This property is therefore valid as soon as $\mathbf{A}$ is "free" with respect to $\mathbf{B}$, i.e. when the basis that diagonalizes $\mathbf{A}$ is a random rotation of the basis that diagonalizes $\mathbf{B}$.

The discussion leading to Eq. (10.42) can be extended to the HCIZ integral (Eq. (10.23)), when the rank of the matrix $\mathbf{T}$ is very small compared to N. In this case we get[3]

$$I(\mathbf{T},\mathbf{B}) \approx \exp\left(\frac{N}{2}\sum_{i=1}^{n} H_{\mathbf{B}}(t_i)\right) = \exp\left(\frac{N}{2}\operatorname{Tr} H_{\mathbf{B}}(\mathbf{T})\right), \tag{10.45}$$

where t_i are the n non-zero eigenvalues of $\mathbf{T}$ and with the same $H_{\mathbf{B}}(t)$ as above. When $\mathbf{T}$ has rank-1 we recover that $\operatorname{Tr} H_{\mathbf{B}}(\mathbf{T}) = H_{\mathbf{B}}(t)$, where t is the sole non-zero eigenvalue of $\mathbf{T}$.

The above formalism is based on the assumption that $\mathfrak{g}(z)$ is invertible, which is generally only true when $t = \mathfrak{g}(z)$ is small enough. This corresponds to the case where z is sufficiently large. Recall that the expansion of $g(z)$ at large z has coefficients given by the moments of the random matrix by (2.22). On the other hand, the expansion of $H(t)$ around $t = 0$ will give coefficients called the free cumulants of the random matrix, which are important objects in the study of free probability, as we will show in the next chapter.

10.4 Invertibility of the Stieltjes Transform

The question of the invertibility of the Stieltjes transform arises often enough that it is worth spending some time discussing it. In Section 10.1, we used the inverse of the limiting Stieltjes transform $\mathfrak{g}(z)$ to solve Burgers' equation, which led to the introduction of the R-transform $R(g) = \mathfrak{z}(g) - 1/g$. In Section 10.3.1 we invoked the invertibility of the discrete Stieltjes transform $g_N(z)$ to compute the rank-1 HCIZ integral.

10.4.1 Discrete Stieltjes Transform

Recall the discrete Stieltjes transform of a matrix $\mathbf{A}$ with N eigenvalues $\{\lambda_k\}$:

$$g_N^{\mathbf{A}}(z) = \frac{1}{N}\sum_{k=1}^{N}\frac{1}{z-\lambda_k}. \tag{10.46}$$

This function is well defined for any z on the real axis except on the finite set $\{\lambda_k\}$. For $z > \lambda_{\max}$, each of the terms in the sum is positive and monotonically decreasing with z so $g_N^{\mathbf{A}}(z)$ is a positive monotonically decreasing function of z. As $z \to \infty$, $g_N^{\mathbf{A}}(z) \to 0$. By the same argument, for $z < \lambda_{\min}$, $g_N^{\mathbf{A}}(z)$ is a negative monotonically decreasing function of z tending to zero as z goes to minus infinity. Actually, the normalization of $g_N^{\mathbf{A}}(z)$ is such that we have

[3] The same computation can be done for any value of beta, yielding

$$I_\beta(\mathbf{T},\mathbf{B}) \approx \exp\left(\frac{N\beta}{2}\sum_{i=1}^{n} H_{\mathbf{B}}(t_i)\right) = \exp\left(\frac{N\beta}{2}\operatorname{Tr} H_{\mathbf{B}}(\mathbf{T})\right), \tag{10.44}$$

where $I_\beta(\mathbf{T},\mathbf{B})$ is defined in the footnote on page 141 and $\mathbf{T}$ has low rank.

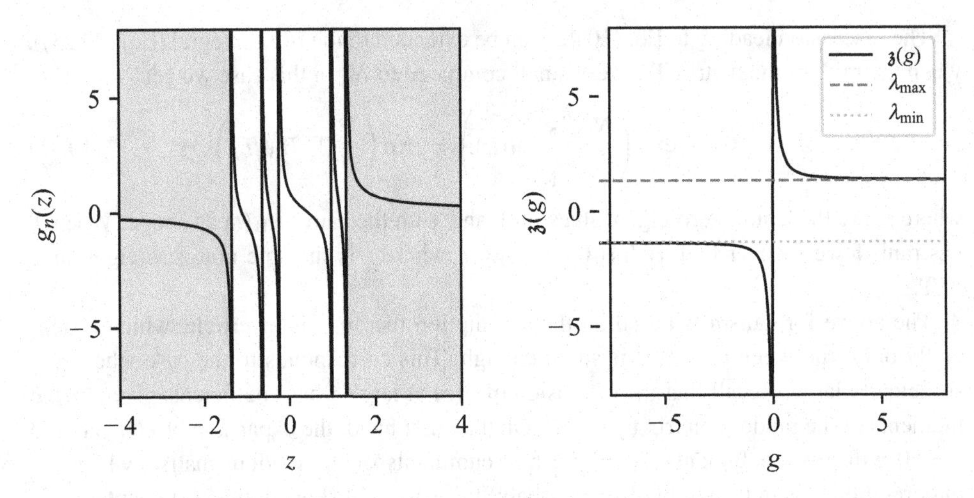

Figure 10.2 (left) A particular $g_N^{\mathbf{A}}(z)$ for $\mathbf{A}$ a Wigner matrix of size $N = 5$ shown for real values of z. The gray areas left of $\lambda_{\min}$ and right of $\lambda_{\max}$ show the values of z for which it is invertible. (right) The inverse function $\mathfrak{z}(g)$. Note that $g(z)$ behaves as $1/z$ near zero and tends to $\lambda_{\min}$ and $\lambda_{\max}$ as g goes to plus or minus infinity respectively.

$$g_N^{\mathbf{A}}(z) = \frac{1}{z} + O\left(\frac{1}{z^2}\right) \quad \text{when} \quad |z| \to \infty. \tag{10.47}$$

For large $|z|$, $g_N^{\mathbf{A}}(z)$ is thus invertible and its inverse behaves as

$$\mathfrak{z}(g) = \frac{1}{g} + \text{regular terms} \quad \text{when} \quad |g| \to 0. \tag{10.48}$$

If we consider values of $g_N^{\mathbf{A}}(z)$ for $z > \lambda_{\max}$, we realize that the function takes all possible positive values once and only once, from the extremely large (near $z = \lambda_{\max}$) to almost zero (when $z \to \infty$). Similarly, all possible negative values are attained when $z \in (-\infty, \lambda_{\min})$ (see Fig. 10.2 left). We conclude that the inverse function $\mathfrak{z}(g)$ exists for all non-zero values of g. The behavior of $g_N^{\mathbf{A}}(z)$ near $\lambda_{\min}$ and $\lambda_{\max}$ gives us the asymptotes

$$\lim_{g \to -\infty} \mathfrak{z}(g) = \lambda_{\min} \quad \text{and} \quad \lim_{g \to \infty} \mathfrak{z}(g) = \lambda_{\max}. \tag{10.49}$$

10.4.2 Limiting Stieltjes Transform

Let us now discuss the inverse function of the limiting Stieltjes transform $\mathfrak{g}(z)$. The limiting Stieltjes transform satisfies Eq. (2.41), which we recall here:

$$\mathfrak{g}(z) = \int_{\text{supp}\{\rho\}} \frac{\rho(x)\mathrm{d}x}{z - x}, \tag{10.50}$$

where $\rho(\lambda)$ is the limiting spectral distribution and may contain Dirac deltas. We denote $\lambda_\pm$ the edges of the support of ρ. We have that for $z > \lambda_+$, $\mathfrak{g}(z)$ is a positive, monotonically decreasing function of z. Similarly for $z < \lambda_-$, $\mathfrak{g}(z)$ is a negative, monotonically decreasing function of z. From the normalization of $\rho(\lambda)$, we again find that

$$\mathfrak{g}(z) = \frac{1}{z} + O\left(\frac{1}{z^2}\right) \quad \text{when} \quad |z| \to \infty. \tag{10.51}$$

Using the same arguments as for the discrete Stieltjes transform, we have that the inverse function $\mathfrak{z}(g)$ exists for small arguments and behaves as

$$\mathfrak{z}(g) = \frac{1}{g} + \text{regular terms} \quad \text{when} \quad |g| \to 0. \tag{10.52}$$

The behavior of $\mathfrak{g}(z)$ at $\lambda_\pm$ can be different from that of $g_N(z)$ at its extreme eigenvalues. The points $\lambda_\pm$ are singular points of $\mathfrak{g}(z)$. If the density near λ_+ goes to zero as $\rho(\lambda) \sim (\lambda_+ - \lambda)^\theta$ for some $\theta > 0$ (typically $\theta = 1/2$) then the integral (10.50) converges at $z = \lambda_+$ and $g_+ := \mathfrak{g}(\lambda_+)$ is a finite number. For $z < \lambda_+$ the function $\mathfrak{g}(z)$ has a branch cut and is ill defined for z on the real axis. The point $z = \lambda_+$ is an essential singularity of $\mathfrak{g}(z)$. The function is clearly no longer invertible for $z < \lambda_+$. Similarly, if $\rho(\lambda)$ grows as a positive power near λ_-, then $\mathfrak{g}(z)$ is invertible up to the point $g_- := \mathfrak{g}(\lambda_-)$.

If the density $\rho(\lambda)$ does not go to zero at one of its edges (or if it has a Dirac delta), the function $\mathfrak{g}(z)$ diverges at that edge. We may still define $g_\pm = \lim_{z\to\lambda_\pm} \mathfrak{g}(z)$ if we allow $g_\pm$ to be plus or minus infinity.

In all cases, the inverse function $\mathfrak{z}(g)$ exists in the range $g_- \leq g \leq g_+$, with the property

$$\mathfrak{z}(g_\pm) = \lambda_\pm. \tag{10.53}$$

In the unit Wigner case, we have $\lambda_\pm = \pm 2$ and $g_\pm = \pm 1$ and the inverse function $\mathfrak{z}(g)$ only exists between -1 and 1 (see Fig. 10.3).

10.4.3 The Inverse Stieltjes Transform for Larger Arguments

In some computations, as in the HCIZ integral, one needs the value of $\mathfrak{z}(g)$ beyond $g_\pm$. What can we say then? First of all, one should not be fooled by spurious solutions of the inversion problem. For example in the Wigner case we know that $\mathfrak{g}(z)$ satisfies

$$g + \frac{1}{g} - z = 0, \tag{10.54}$$

so we would be tempted to write

$$\mathfrak{z}(g) = g + \frac{1}{g} \tag{10.55}$$

for all g. But this is wrong as $g + 1/g$ is not the inverse of $\mathfrak{g}(z)$ for $|g| > 1$ (Fig. 10.3).

The correct way to extend $\mathfrak{z}(g)$ beyond $g_\pm$ is to realize that in most computation, we use $\mathfrak{g}(z)$ as an approximation for $g_N(z)$ for very large N. For $z > \lambda_+$ the function $g_N(z)$

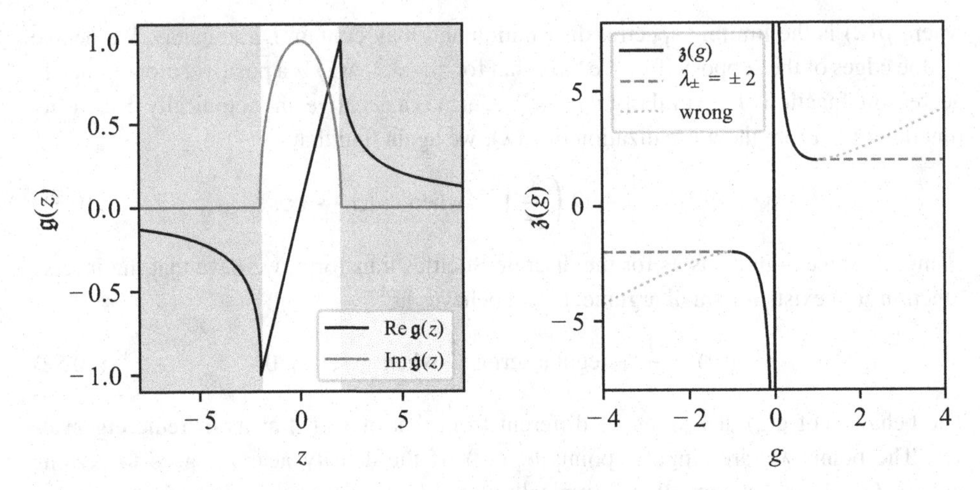

Figure 10.3 (left) The limiting function $\mathfrak{g}(z)$ for a Wigner matrix, a typical density that vanishes at its edges. The function is plotted against a real argument. In the white part of the graph, the function is ill defined and it is shown here for a small negative imaginary part of its argument. In the gray part ($z < \lambda_-$ and $z > \lambda_+$) the function is well defined, real and monotonic. It is therefore invertible. (right) The inverse function $\mathfrak{z}(g)$ only exists for $g_- \leq g \leq g_+$ and has a $1/g$ singularity at zero. The dashed lines show the extension of $\mathfrak{z}(g)$ to all values of g that are natural when we think of $\mathfrak{g}(z)$ as the limit of $g_N(z)$ with maximal and minimal eigenvalues $\lambda_\pm$. The dotted lines indicate the wrong branch of the solution of $\mathfrak{z}(g) = g + 1/g$.

converges to $\mathfrak{g}(z)$ and this approximation can be made arbitrarily good for large enough N. On the other hand we know that on the support of ρ, $g_N(z)$ does not converge to $\mathfrak{g}(z)$. The former has a series of simple poles at random locations, while the later has typically a branch cut.

At large but finite N, there will be a maximum eigenvalue $\lambda_{\max}$. This eigenvalue is random but to dominant order in N it converges to λ_+, the edge of the spectrum. For z above but very close to λ_+ we should think of $g_N(z)$ as

$$g_N(z) \approx \mathfrak{g}(z) + \frac{1}{N}\frac{1}{z - \lambda_{\max}} \approx \mathfrak{g}(z) + \frac{1}{N}\frac{1}{z - \lambda_+}. \tag{10.56}$$

Because $1/N$ goes to zero, the correction above does not change the limiting value of $\mathfrak{g}(z)$ at any finite distance from λ_+. On the other hand, this correction does change the behavior of the inverse function $\mathfrak{z}(g)$. We now have

$$\lim_{z \to \lambda_+} g_N(z) \to \infty \quad \text{and} \quad \mathfrak{z}(g) = \lambda_+ \text{ for } g > g_+. \tag{10.57}$$

For negative z and negative g, the same argument follows near λ_-. We realize that, while the limiting Stieltjes transform $\mathfrak{g}(z)$ loses all information about individual eigenvalues, its inverse function $\mathfrak{z}(g)$, or really the large N limit of the inverse of the function $g_N(z)$, retains information about the smallest and largest eigenvalues. In Chapter 14 we will study random matrices where a finite number of eigenvalues lie outside the support of ρ. In the large N limit, these eigenvalues do not change the density or $\mathfrak{g}(z)$ but they do show up in the inverse function $\mathfrak{z}(g)$.

Let us define $z_{\text{bulk}}(g)$, the inverse function of $\mathfrak{g}(z)$ without considering extreme eigenvalues or outliers. In the presence of outliers we have $\lambda_{\max} \geq \lambda_+$ and $g_{\max} := \mathfrak{g}(\lambda_{\max}) \leq \mathfrak{g}(\lambda_+)$ and similarly for $g_{\min}$. With arguments similar to those above we find the following result for the limit of the inverse of $g_N(z)$:

$$\mathfrak{z}(g) = \begin{cases} \lambda_{\min} & \text{for } g \leq g_{\min}, \\ z_{\text{bulk}}(g) & \text{for } g_{\min} < g < g_{\max}, \\ \lambda_{\max} & \text{for } g \geq g_{\max}. \end{cases} \tag{10.58}$$

In the absence of outliers the result still applies with max and min (extreme eigenvalues) replaced by $+$ and $-$ (edge of the spectrum), respectively.

10.4.4 Large t Behavior of I_t

Now that we understand the behavior of $\mathfrak{z}(g)$ for larger arguments we can go back to our study of the rank-1 HCIZ integral. There is indeed an apparent paradox in the result of our computation of $I_t(\mathbf{B})$. For a given matrix $\mathbf{B}$ there are two immediate bounds to $I_t(\mathbf{B}) = Z_t(\mathbf{B})/Z_t(0)$:

$$\exp\left(\frac{Nt\lambda_{\min}}{2}\right) \leq I_t(\mathbf{B}) \leq \exp\left(\frac{Nt\lambda_{\max}}{2}\right), \tag{10.59}$$

where $\lambda_{\min}$ and $\lambda_{\max}$ are the smallest and largest eigenvalues of $\mathbf{B}$, respectively. Focusing on the upper bound, we have

$$H_{\mathbf{B}}(t) \leq t\lambda_{\max}. \tag{10.60}$$

On the other hand, the anti-derivative of the R-transform for a unit Wigner matrix reads

$$R_{\mathbf{W}}(t) = t \longrightarrow H_{\mathbf{W}}(t) = \frac{t^2}{2}, \tag{10.61}$$

whereas $\lambda_{\max} \to \lambda_+ = 2$. One might thus think that the quadratic behavior of $H_{\mathbf{W}}(t)$ violates the bound (10.60) for $t > 4$. We should, however, remember that Eq. (10.42) is in fact only valid for $t < g_+$, the value at which $\mathfrak{g}(z)$ ceases to be invertible. In the absence of outliers, $g_+ = \mathfrak{g}(\lambda_+)$. For a unit Wigner this point is $g_+ = \mathfrak{g}(2) = 1$; the bound is not violated. For $t > g_+$, one can still compute $H_{\mathbf{B}}(t)$ but the result depends explicitly on $\lambda_{\max}$.

Now that we understand the behavior of $\mathfrak{z}(g)$ for larger arguments, including in the presence of outliers, we can extend our result for $H_{\mathbf{B}}(t)$ for large t's. We just need to use Eq. (10.58) into Eq. (10.41):

$$\frac{dH_{\mathbf{B}}(t)}{dt} = \begin{cases} R_{\mathbf{B}}(t) & \text{for } t \leq g_{\max} := \mathfrak{g}(\lambda_{\max}), \\ \lambda_{\max} - 1/t & \text{for } t > g_{\max}, \end{cases} \tag{10.62}$$

where the largest eigenvalue $\lambda_{\max}$ can be either the edge of the spectrum λ_+ or a true outlier. We will show in Section 13.3 how this result can also be derived for Wigner matrices using the replica method.

10.5 The Full-Rank HCIZ Integral

We have defined in Eq. (10.23) the HCIZ integral as a generalization of the Fourier transform for matrices, and have seen how to evaluate this integral in the limit $N \to \infty$ when one matrix is of low rank. A generalized HCIZ integral $I_\beta(\mathbf{A}, \mathbf{B})$ can be defined as

$$I_\beta(\mathbf{A},\mathbf{B}) = \int_{G(N)} \mathrm{d}\mathbf{U}\, \mathrm{e}^{\frac{\beta N}{2} \operatorname{Tr} \mathbf{AUBU}^\dagger}, \tag{10.63}$$

where the integral is over the (flat) Haar measure of the compact group $\mathbf{U} \in G(N) = O(N), U(N)$ or $Sp(N)$ in N dimensions and $\mathbf{A},\mathbf{B}$ are arbitrary $N \times N$ symmetric (resp. Hermitian or symplectic) matrices, with, correspondingly, $\beta = 1,2$ or 4. Note that by construction $I_\beta(\mathbf{A},\mathbf{B})$ can only depend on the eigenvalues of $\mathbf{A}$ and $\mathbf{B}$, since any change of basis on $\mathbf{B}$ (say) can be reabsorbed in $\mathbf{U}$, over which we integrate. Note also that the Haar measure is normalized, i.e. $\int_{G(N)} \mathrm{d}\mathbf{U} = 1$.

Interestingly, it turns out that in the unitary case $G(N) = U(N)$ ($\beta = 2$), the HCIZ integral can be expressed exactly, for all N, as the ratio of determinants that depend on $\mathbf{A},\mathbf{B}$ and additional N-dependent prefactors. This is the Harish-Chandra–Itzykson–Zuber celebrated result, which cannot be absent from a book on random matrices:

$$I_2(\mathbf{A},\mathbf{B}) = \frac{c_N}{N^{(N^2-N)/2}} \frac{\det\left(\mathrm{e}^{N\nu_i\lambda_j}\right)}{\Delta(\mathbf{A})\Delta(\mathbf{B})}, \tag{10.64}$$

with $\{\nu_i\}$, $\{\lambda_i\}$ the eigenvalues of $\mathbf{A}$ and $\mathbf{B}$, $\Delta(\mathbf{A})$, $\Delta(\mathbf{B})$ are the Vandermonde determinants of $\mathbf{A}$ and $\mathbf{B}$, and $c_N = \prod_\ell^{N-1} \ell!$.

Although this result is fully explicit for $\beta = 2$, the expression in terms of determinants is highly non-trivial and quite tricky. For example, the expression becomes degenerate (0/0) whenever two eigenvalues of $\mathbf{A}$ (or $\mathbf{B}$) coincide. Also, as is well known, determinants contain $N!$ terms of alternating signs, which makes their order of magnitude very hard to estimate *a priori*. The aim of this technical section is to discuss how the HCIZ result can be obtained using the Karlin–McGregor equation (9.69). We then use the mapping to the Dyson Brownian motion to derive a large N approximation for the full-rank HCIZ integral in the general case.

10.5.1 HCIZ and Karlin–McGregor

In order to understand the origin of Eq. (10.64), the basic idea is to interpret the HCIZ integrand in the unitary case, $\exp[N \operatorname{Tr} \mathbf{AUBU}^\dagger]$, as a part of the diffusion propagator in the space of Hermitian matrices, and use the Karlin–McGregor formula.

Indeed, adding to $\mathbf{A}$ a sequence of infinitesimal random Gaussian Hermitian matrices of variance $\mathrm{d}t/N$, the probability to end up with matrix $\mathbf{B}$ in a time $t = 1$ is given by

$$P(\mathbf{B}|\mathbf{A}) \propto N^{N^2/2}\, \mathrm{e}^{-N/2 \operatorname{Tr}(\mathbf{B}-\mathbf{A})^2}, \tag{10.65}$$

where we drop an overall normalization constant in our attempt to understand the structure of Eq. (10.64). The corresponding eigenvalues follow a Dyson Brownian motion, namely

$$\mathrm{d}x_i = \sqrt{\frac{1}{N}}\mathrm{d}B_i + \frac{1}{N}\sum_{\substack{j=1 \\ j\neq i}}^{N} \frac{\mathrm{d}t}{x_i - x_j}, \tag{10.66}$$

with $x_i(t=0) = \nu_i$ and $x_i(t=1) = \lambda_i$. Now, for $\beta = 2$ we can use the Karlin–McGregor equation (9.69) to derive the conditional distribution of the $\{\lambda_i\}$, given by

$$P(\{\lambda_i\}|\{\nu_i\}) = \frac{\Delta(\mathbf{B})}{\Delta(\mathbf{A})} P(\vec{\lambda}, t = 1|\vec{\nu}), \tag{10.67}$$

where $P(\vec{\lambda}, t = 1|\vec{\nu})$ is given by a determinant, Eq. (9.66). With the present normalization, this determinant reads

$$P(\vec{\lambda}, t = 1|\vec{\nu}) = \left(\frac{N}{2\pi}\right)^{N/2} \mathrm{e}^{-\frac{N}{2}(\mathrm{Tr}\,\mathbf{A}^2 + \mathrm{Tr}\,\mathbf{B}^2)} \det\left(\mathrm{e}^{N\nu_i\lambda_j}\right). \tag{10.68}$$

Now, the distribution of eigenvalues of $\mathbf{B}$ can be computed from Eq. (10.65). First we make $P(\mathbf{B}|\mathbf{A})$ unitary-invariant by integrating over $U(N)$:

$$\begin{aligned} P(\mathbf{B}|\mathbf{A}) \to \overline{P}(\mathbf{B}|\mathbf{A}) &= N^{N^2/2} \frac{\int_{U(N)} \mathrm{d}\mathbf{U}\, \mathrm{e}^{-N/2\, \mathrm{Tr}(\mathbf{U}\mathbf{B}\mathbf{U}^\dagger - \mathbf{A})^2}}{\Omega_N} \\ &= N^{N^2/2} \mathrm{e}^{-\frac{N}{2}(\mathrm{Tr}\,\mathbf{A}^2 + \mathrm{Tr}\,\mathbf{B}^2)} \frac{I_2(\mathbf{A}, \mathbf{B})}{\Omega_N}, \end{aligned} \tag{10.69}$$

where $\Omega_N = \int_{U(N)} \mathrm{d}\mathbf{U}$ is the "volume" of the unitary group $U(N)$. This new measure, by construction, only depends on $\{\lambda_i\}$, the eigenvalues of $\mathbf{B}$. Changing variables from $\mathbf{B}$ to $\{\lambda_i\}$ introduces a Jacobian, which in the unitary case is the square of the Vandermonde determinant of $\mathbf{B}$, $\Delta^2(\mathbf{B})$. We thus find a second expression for the distribution of the $\{\lambda_i\}$:

$$P(\{\lambda_i\}|\{\nu_i\}) \propto N^{N^2/2} \Delta^2(\mathbf{B}) \mathrm{e}^{-\frac{N}{2}(\mathrm{Tr}\,\mathbf{A}^2 + \mathrm{Tr}\,\mathbf{B}^2)} I_2(\mathbf{A}, \mathbf{B}). \tag{10.70}$$

Comparing with Eqs. (10.67) and (10.68) we thus find

$$I_2(\mathbf{A}, \mathbf{B}) \propto N^{(N - N^2)/2} \frac{\det\left(\mathrm{e}^{N\nu_i\lambda_j}\right)}{\Delta(\mathbf{A})\Delta(\mathbf{B})}, \tag{10.71}$$

which coincides with Eq. (10.64), up to an overall constant c_N which can be obtained by taking the limit $\mathbf{A} = \mathbf{1}$, i.e. when all the eigenvalues of $\mathbf{A}$ are equal to 1. The limit is singular but one can deal with it in a way similar to the one used by Brézin and Hikami to go from Eq. (6.65) to (6.67). In this limit, the right hand side of Eq. (10.71) reads $\exp(N\,\mathrm{Tr}\,\mathbf{B})/c_N$, while the left hand side is trivially equal to $\exp(N\,\mathrm{Tr}\,\mathbf{B})$. Hence a factor c_N is indeed missing in the right hand side of Eq. (10.71).

Equation (10.64) can also be used to obtain an exact formula for the rank-1 HCIZ integral (when $\beta = 2$). The trick is to have one of the eigenvalues of ν_i equal to some non-zero number t and let the $N - 1$ others go to zero. The limit can again be dealt with in the same way as Eq. (6.65). One finally finds

$$I_2(t, \mathbf{B}) = \frac{(N-1)!}{(Nt)^{N-1}} \sum_{j=1}^{N} \frac{\mathrm{e}^{Nt\lambda_j}}{\prod_{k \neq j}(\lambda_j - \lambda_k)}. \tag{10.72}$$

The above formula may look singular at $t = 0$, but we have $\lim_{t\to 0} I_2(t, \mathbf{B}) = 1$ as expected.

10.5.2 HCIZ at Large N: The Euler–Matytsin Equations

We now explain how $I_2(\mathbf{A}, \mathbf{B})$ can be estimated for large matrix size, using a Dyson Brownian representation of $\overline{P}(\mathbf{B}|\mathbf{A})$, Eq. (10.66). In terms of these interacting Brownian motions, the question is how to estimate the probability that the $x_i(t)$ start at $x_i(t = 0) =$

ν_i and end at $x_i(t=1) = \lambda_i$, when their trajectories are determined by Eq. (10.66), which we rewrite as

$$\mathrm{d}x_i = \sqrt{\frac{1}{N}}\mathrm{d}B_i - \partial_{x_i} V\,\mathrm{d}t, \qquad V(\{x_i\}) := -\frac{1}{N}\sum_{i<j} \ln|x_i - x_j|. \tag{10.73}$$

The probability of a given trajectory for the N Brownian motions between time $t = 0$ and time $t = 1$ is then given by[4]

$$P(\{x_i(t)\}) = Z^{-1}\exp -\left[\frac{N}{2}\int_0^1 \mathrm{d}t \sum_i (\dot{x}_i + \partial_{x_i} V)^2\right] := Z^{-1}\mathrm{e}^{-N^2 S}, \tag{10.74}$$

where Z is a normalization factor that we will not need explicitly. Expanding the square as $\dot{x}_i^2 + 2\partial_{x_i} V\dot{x}_i + (\partial_{x_i} V)^2$, one can decompose $S = S_1 + S_2$ into a total derivative term equal, in the continuum limit, to boundary terms, i.e.

$$S_1 = -\frac{1}{2}\left[\int \mathrm{d}x\mathrm{d}y\rho_C(x)\rho_C(y)\ln|x-y|\right]_{C=A}^{C=B} \tag{10.75}$$

and

$$S_2 := \frac{1}{2N}\int_0^1 \mathrm{d}t \sum_{i=1}^N \left[\dot{x}_i^2 + (\partial_{x_i} V)^2\right]. \tag{10.76}$$

We now look for the "instanton" trajectory that contributes most to the probability P for large N, in other words the trajectory that minimizes S_2. This extremal trajectory is such that the functional derivative of S_2 with respect to all $x_i(t)$ is zero:

$$-2\frac{\mathrm{d}^2 x_i}{\mathrm{d}t^2} + 2\sum_{\ell=1}^N \partial^2_{x_i,x_\ell} V\partial_{x_\ell} V = 0, \tag{10.77}$$

which leads, after a few algebraic manipulations, to

$$\frac{\mathrm{d}^2 x_i}{\mathrm{d}t^2} = -\frac{2}{N^2}\sum_{\ell\neq i}\frac{1}{(x_i - x_\ell)^3}. \tag{10.78}$$

This can be interpreted as the motion of unit mass particles, accelerated by an *attractive* force that derives from an effective two-body potential $\phi(r) = -(Nr)^{-2}$. The hydrodynamical description of such a fluid, justified when $N \to \infty$, is given by the Euler equations for the density field $\rho(x,t)$ and the velocity field $v(x,t)$:

$$\partial_t \rho(x,t) + \partial_x[\rho(x,t)v(x,t)] = 0 \tag{10.79}$$

and

$$\partial_t v(x,t) + v(x,t)\partial_x v(x,t) = -\frac{1}{\rho(x,t)}\partial_x \Pi(x,t), \tag{10.80}$$

where $\Pi(x,t)$ is the pressure field, which reads, from the "virial" formula for an interacting fluid at temperature T,[5]

[4] We neglect here a Jacobian which is not relevant to compute the leading term of $I_2(\mathbf{A},\mathbf{B})$ in the large N limit.

[5] See e.g. Le Bellac et al. [2004], p. 138.

$$\Pi = \rho T - \frac{1}{2}\rho \sum_{\ell \neq i} |x_i - x_\ell| \phi'(x_i - x_\ell) \approx -\frac{\rho}{N^2} \sum_{\ell \neq i} \frac{1}{(x_i - x_\ell)^2}, \tag{10.81}$$

because the fluid describing the instanton is at zero temperature, $T = 0$. Now, writing $x_i - x_\ell \approx (i - \ell)/(N\rho)$ and $\sum_{n=1}^{\infty} n^{-2} = \frac{\pi^2}{6}$, one finally finds

$$\Pi(x,t) = -\frac{\pi^2}{3}\rho(x,t)^3. \tag{10.82}$$

Equations (10.79) and (10.80) for ρ and v with Π given by (10.82) are called the Euler–Matytsin equations. They should be solved with the following boundary conditions:

$$\rho(x,t=0) = \rho_{\mathbf{A}}(x); \qquad \rho(x,t) = \rho_{\mathbf{B}}(x); \tag{10.83}$$

the velocity field $v(x,t=0)$ should be chosen such that these boundary conditions are fulfilled.

Expressing S_2 in terms of the solution of the Euler–Matytsin equations gives, in the continuum limit,

$$S_2(\mathbf{A},\mathbf{B}) \approx \frac{1}{2}\int \mathrm{d}x\rho(x,t)\left[v^2(x,t) + \frac{\pi^2}{3}\rho^2(x,t)\right]. \tag{10.84}$$

Hence, the probability $P(\{\lambda_i\}|\{\nu_i\})$ to observe the set of eigenvalues $\{\lambda_i\}$ of $\mathbf{B}$ for a given set of eigenvalues ν_i for $\mathbf{A}$ is, in the large N limit, proportional to $\exp[-N^2(S_1 + S_2)]$. Comparing with Eq. (10.70), we get as a final expression for $F_2(\mathbf{A},\mathbf{B}) := -\lim_{N\to\infty} N^{-2} \ln I_2(\mathbf{A},\mathbf{B})$:

$$F_2(\mathbf{A},\mathbf{B}) = \frac{3}{4} + S_2(\mathbf{A},\mathbf{B}) - \frac{1}{2}\int \mathrm{d}x\, x^2(\rho_{\mathbf{A}}(x) + \rho_{\mathbf{B}}(x))$$
$$+ \frac{1}{2}\int \mathrm{d}x\mathrm{d}y\, [\rho_{\mathbf{A}}(x)\rho_{\mathbf{A}}(y) + \rho_{\mathbf{B}}(x)\rho_{\mathbf{B}}(y)] \ln|x-y|. \tag{10.85}$$

This result was first derived in Matytsin [1994], and proven rigorously in Guionnet and Zeitouni [2002]. Note that this expression is symmetric in $\mathbf{A},\mathbf{B}$, as it should be, because the solution of the Euler–Matytsin equations for the time reversed path from $\rho_{\mathbf{B}}$ to $\rho_{\mathbf{A}}$ are simply obtained from $\rho(x,t) \to \rho(x,1-t)$ and $v(x,t) \to -v(x,1-t)$, which leaves $S_2(\mathbf{A},\mathbf{B})$ unchanged.

The whole calculation above can be repeated for the $\beta = 1$ (orthogonal group) or $\beta = 4$ (symplectic group) with the final (simple) result $F_\beta(\mathbf{A},\mathbf{B}) = \beta F_2(\mathbf{A},\mathbf{B})/2$.

Bibliographical Notes

- The Burgers' equation in the context of random matrices:

  - L. C. G. Rodgers and Z. Shi. Interacting Brownian particles and the Wigner law. *Probability Theory and Related Fields*, 95:555–570, 1993,
  - J.-P. Blaizot and M. A. Nowak. Universal shocks in random matrix theory. *Physical Review E*, 82:051115, 2010,
  - G. Menon. Lesser known miracles of Burgers equation. *Acta Mathematica Scientia*, 32(1):281–94, 2012.

- The HCIZ integral: Historical papers:
  - Harish-Chandra. Differential operators on a semisimple Lie algebra. *American Journal of Mathematics*, 79:87–120, 1957,
  - C. Itzykson and J.-B. Zuber. The planar approximation. II. *Journal of Mathematical Physics*, 21:411–421, 1980,

  for a particularly insightful introduction, see T. Tao, http://terrytao.wordpress.com/2013/02/08/theharish-chandra-itzykson-zuber-integral-formula/.
- The low-rank HCIZ integral:
  - E. Marinari, G. Parisi, and F. Ritort. Replica field theory for deterministic models. II. A non-random spin glass with glassy behaviour. *Journal of Physics A: Mathematical and General*, 27(23):7647, 1994,
  - A. Guionnet and M. Maïda. A Fourier view on the R-transform and related asymptotics of spherical integrals. *Journal of Functional Analysis*, 222(2):435–490, 2005.
- The HCIZ integral: Large N limit:
  - A. Matytsin. On the large-N limit of the Itzykson-Zuber integral. *Nuclear Physics B*, 411:805–820, 1994,
  - A. Guionnet and O. Zeitouni. Large deviations asymptotics for spherical integrals. *Journal of Functional Analysis*, 188(2):461–515, 2002,
  - B. Collins, A. Guionnet, and E. Maurel-Segala. Asymptotics of unitary and orthogonal matrix integrals. *Advances in Mathematics*, 222(1):172–215, 2009,
  - J. Bun, J. P. Bouchaud, S. N. Majumdar, and M. Potters. Instanton approach to large N Harish-Chandra-Itzykson-Zuber integrals. *Physical Review Letters*, 113:070201, 2014,
  - G. Menon. The complex Burgers equation, the HCIZ integral and the Calogero-Moser system, unpublished, 2017, available at: https://www.dam.brown.edu/people/menon/talks/cmsa.pdf.
- On the classical virial theorem, see
  - M. Le Bellac, F. Mortessagne, and G. G. Batrouni. *Equilibrium and Non-Equilibrium Statistical Thermodynamics*. Cambridge University Press, Cambridge, 2004.

11
Free Probabilities

In the previous chapter we saw how to compute the spectrum of the sum of two large random matrices, first when one of them is a Wigner and later when one is "rotationally invariant" with respect to the other. In this chapter, we would like to formalize the notion of relative rotational invariance, which leads to the abstract concept of freeness.

The idea is as follows. In standard probability theory, one can work abstractly by defining expectation values (moments) of random variables. The concept of independence is then equivalent to the factorization of moments (e.g. $\mathbb{E}[A^3 B^2] = \mathbb{E}[A^3]\mathbb{E}[B^2]$ when A and B are independent).

However, random matrices do not commute in general and the concept of factorization of moments is not powerful enough to deal with non-commuting random objects. Following von Neumann, Voiculescu extended the concept of independence to non-commuting objects and called this property *freeness*. He then showed how to characterize the sum and the product of *free* variables. It was later realized that large rotationally invariant matrices provide an explicit example of (asymptotically) free random variables. In other words, free probabilities gave us very powerful tools to compute sums and products of large random matrices. We have already encountered the free addition; the free product will allow us to study sample covariance matrices in the presence of non-trivial true correlations.

This chapter may seem too dry and abstract for someone looking for applications. Bear with us, it is in fact not that complicated and we will keep the jargon to a minimum. The reward will be one of the most powerful and beautiful recent developments in random matrix theory, which we will expand upon in Chapter 12.

11.1 Algebraic Probabilities: Some Definitions

The ingredients we will need in this chapter are as follows:[1]

- A **ring** $\mathcal{R}$ of random variables, which can be non-commutative with respect to the multiplication.[2]

[1] In mathematical language, the first three items give a *-algebra, while τ gives a tracial state on this algebra.
[2] Recall that a ring is a set equipped with two binary operations that generalize the arithmetic operations of addition and multiplication.

- A **field** of scalars, which is usually taken to be $\mathbb{C}$. The scalars commute with everything.
- An operation $*$, called involution. For instance, $*$ denotes the conjugate for complex numbers, the transpose for real matrices, and the conjugate transpose for complex matrices.
- A positive linear functional $\tau(.)$ ($\mathcal{R} \to \mathbb{C}$) that satisfies $\tau(AB) = \tau(BA)$ for $A, B \in \mathcal{R}$. By positive we mean $\tau(AA^*)$ is real non-negative. We also require that τ be *faithful*, in the sense that $\tau(AA^*) = 0$ implies $A = 0$. For instance, τ can be the expectation operator $\mathbb{E}[.]$ for standard probability theory, or the normalized trace operator $\frac{1}{N}\operatorname{Tr}(.)$ for a ring of matrices, or the combined operation $\frac{1}{N}\mathbb{E}[\operatorname{Tr}(.)]$.

We will call the elements in $\mathcal{R}$ the random variables and denote them by capital letters. For any $A \in \mathcal{R}$ and $k \in \mathbb{N}$, we call $\tau(A^k)$ the kth moment of A and we assume in what follows that $\tau(A^k)$ is finite for all k. In particular, we call $\tau(A)$ the mean of A and $\tau(A^2) - \tau(A)^2$ the variance of A. We will say that two elements A and B have the same distribution if they have the same moments of all orders.[3]

The ring of variables must have an element called $\mathbf{1}$ such that $A\mathbf{1} = \mathbf{1}A = A$ for every A. It satisfies $\tau(\mathbf{1}) = 1$. We will call $\mathbf{1}$ and its multiples $\alpha\mathbf{1}$ *constants*. Adding a constant simply shifts the mean as

$$\tau(A + \alpha\mathbf{1}) = \tau(A) + \alpha. \tag{11.1}$$

11.2 Addition of Commuting Variables

In this section, we recall some well-known properties of commuting random variables, i.e. such that

$$AB = BA, \quad \forall A, B \in \mathcal{R}. \tag{11.2}$$

Note that A is not necessarily a real (or complex) number but can be an element of a more abstract ring. We will say that A and B are independent, if $\tau(p(A)q(B)) = \tau(p(A))\tau(q(B))$ for any polynomial p, q. This condition is equivalent to the factorization of moments.

From a scalar α we can build the constant $\alpha\mathbf{1}$ and write $A+\alpha$ to mean $A+\alpha\mathbf{1}$. Constants of the ring are independent of all other random variables, so if A and B are independent, $A + \alpha$ and B are also independent. This setting recovers the classical probability theory of commutative random variables (with finite moments to every order).

11.2.1 Moments

Now let us study the moments of the sum of independent random variables $A + B$. First we trivially have by linearity

$$\tau(A + B) = \tau(A) + \tau(B). \tag{11.3}$$

[3] This is of course not correct for standard commuting random variables: some distributions are not uniquely determined by their moments.

From now on we will assume $\tau(A) = \tau(B) = 0$, i.e. A, B have mean zero, unless stated otherwise. For a non-zero mean variable $\tilde{A}$, we write $A = \tilde{A} - \tau(\tilde{A})$ such that $\tau(A) = 0$. One can recover the formulas for moments and cumulants of $\tilde{A}$ simply by substituting $A \to \tilde{A} - \tau(\tilde{A})$ in all formulas written for zero mean A. The procedure is straightforward but leads to rather cumbersome results.

For the second moment,

$$\begin{aligned}\tau\left((A+B)^2\right) &= \tau(A^2) + \tau(B^2) + 2\tau(AB) \\ &= \tau(A^2) + \tau(B^2) + 2\tau(A)\tau(B) = \tau(A^2) + \tau(B^2),\end{aligned} \tag{11.4}$$

i.e. the variance is also additive. For the third moment, we have

$$\tau\left((A+B)^3\right) = \tau(A^3) + \tau(B^3) + 3\tau(A)\tau(B^2) + 3\tau(B)\tau(A^2) = \tau(A^3) + \tau(B^3), \tag{11.5}$$

which is also additive. However, the fourth and higher moments are not additive anymore. For example we get, expanding $(A+B)^4$,

$$\tau\left((A+B)^4\right) = \tau(A^4) + \tau(B^4) + 6\tau(A^2)\tau(B^2). \tag{11.6}$$

11.2.2 Cumulants

For zero mean variables the first three moments are additive but not the higher ones. Nevertheless, certain combinations of higher moments are additive; we call them cumulants and denote them as κ_n for the nth cumulant. Note that for a variable with non-zero mean $\tilde{A}$, the second and third cumulants are the second and third moments of $A := \tilde{A} - \tau(\tilde{A})$:

$$\begin{aligned}\kappa_1(\tilde{A}) &= \tau(\tilde{A}), \\ \kappa_2(\tilde{A}) &= \tau(A^2) = \tau(\tilde{A}^2) - \tau(\tilde{A})^2, \\ \kappa_3(\tilde{A}) &= \tau(A^3) = \tau(\tilde{A}^3) - 3\tau(\tilde{A}^2)\tau(\tilde{A}) + 2\tau(\tilde{A})^3.\end{aligned} \tag{11.7}$$

For the fourth cumulant, let us consider for simplicity zero mean variables A and B, and define κ_4 as

$$\kappa_4(A) := \tau(A^4) - 3\tau(A^2)^2. \tag{11.8}$$

Then we can verify that

$$\begin{aligned}\kappa_4(A+B) &= \tau\left((A+B)^4\right) - 3\tau\left((A+B)^2\right)^2 \\ &= \tau(A^4) + \tau(B^4) + 6\tau(A^2)\tau(B^2) - 3\left(\tau(A^2) + \tau(B^2)\right)^2 \\ &= \tau(A^4) - 3\tau(A^2)^2 + \tau(B^4) - 3\tau(B^2)^2 = \kappa_4(A) + \kappa_4(B),\end{aligned} \tag{11.9}$$

which is additive again.

In general, $\tau((A+B)^n)$ will be of the form $\tau(A^n) + \tau(B^n)$ plus some homogeneous mix of lower order terms. We can then define the nth cumulant κ_n iteratively such that

$$\kappa_n(A+B) = \kappa_n(A) + \kappa_n(B), \tag{11.10}$$

where

$$\kappa_n(A) = \tau(A^n) + \text{lower order terms moments.} \tag{11.11}$$

In order to have a compact definition of cumulants, recall that we are looking for quantities that are additive for independent variables. But we already know that the log-characteristic function introduced in Eq. (8.4) is additive. In the present context, we define the characteristic function as[4]

$$\varphi_A(t) = \tau\left(\mathrm{e}^{\mathrm{i}tA}\right), \tag{11.12}$$

where the exponential function is formally defined through its power series:

$$\tau(\mathrm{e}^{\mathrm{i}tA}) = \sum_{\ell=0}^{\infty} \frac{(\mathrm{i}t)^\ell}{\ell!}\tau(A^\ell), \tag{11.13}$$

hence the characteristic function is also the moment generating function. Now, from the formal definition of the exponential and the factorization of moments one can easily show that for independent, commuting A, B,

$$\varphi_{A+B}(t) = \varphi_A(t)\varphi_B(t). \tag{11.14}$$

Here is an *algebraic* proof. For each k,

$$\tau((A+B)^k) = \sum_{i=0}^{k} \binom{k}{i} \tau(A^i)\tau(B^{k-i}), \tag{11.15}$$

with which we get

$$\begin{aligned} \varphi_{A+B}(t) &= \sum_{k=0}^{\infty}\sum_{i\le k} \frac{(\mathrm{i}t)^k}{k!}\binom{k}{i}\tau(A^i)\tau(B^{k-i}) = \sum_{i\le k} \frac{(\mathrm{i}t)^{k-i}\tau(B^{k-i})}{(k-i)!}\frac{(\mathrm{i}t)^i\tau(A^i)}{i!} \\ &= \left(\sum_i \frac{(\mathrm{i}t)^i\tau(A^i)}{i!}\right)\left(\sum_j \frac{(\mathrm{i}t)^j\tau(B^j)}{j!}\right) = \varphi_A(t)\varphi_B(t). \end{aligned} \tag{11.16}$$

We now define $H_A(t) := \log \varphi_A(t)$. Then, for independent, commuting A, B, we have

$$H_{A+B}(t) = H_A(t) + H_B(t). \tag{11.17}$$

We can expand $H_A(t)$ as a power series of t and call the corresponding coefficients the cumulants, i.e.

$$H_A(t) = \log \tau\left(\mathrm{e}^{\mathrm{i}tA}\right) := \sum_{n=1}^{\infty} \frac{\kappa_n(A)}{n!}(\mathrm{i}t)^n. \tag{11.18}$$

[4] The factor i in the definition is not necessary in this setting as the formal power series of the exponential and the logarithm do not need to converge. We nevertheless include it by analogy with the Fourier transform.

From the additive property of H, the cumulants defined in the above way are automatically additive. In fact, using the power series for $\log(1+x)$, we have

$$H_A(t) = \sum_{n=1}^{\infty} \frac{(-1)^{n-1}}{n} \left(\sum_{k=1}^{\infty} \frac{(it)^k}{k!} \tau(A^k) \right)^n \equiv \sum_{n=1}^{\infty} \frac{\kappa_n(A)}{n!} (it)^n. \tag{11.19}$$

Matching powers of (it) we obtain an expression for κ_n for any n. We can work out by hand the first few cumulants. For example, for $n = 1$, Eq. (11.19) readily yields $\kappa_1(A) = \tau(A)$. We now assume A has mean zero, i.e. $\tau(A) = 0$. Then

$$\tau\left(e^{itA}\right) = 1 + \frac{(it)^2}{2}\tau(A^2) + \frac{(it)^3}{6}\tau(A^3) + \frac{(it)^4}{24}\tau(A^4) + \cdots, \tag{11.20}$$

whereas the first few terms in the expansion of (11.19) are

$$H_A(t) = \frac{(it)^2}{2}\tau(A^2) + \frac{(it)^3}{6}\tau(A^3) + (it)^4\left(\frac{\tau(A^4)}{24} - \frac{\tau(A^2)^2}{8}\right) + \cdots, \tag{11.21}$$

from which we recover the first four cumulants defined above:

$$\kappa_1(A) = 0, \quad \kappa_2(A) = \tau(A^2), \quad \kappa_3(A) = \tau(A^3), \quad \kappa_4(A) = \tau(A^4) - 3\tau(A^2)^2. \tag{11.22}$$

The expression for κ_n soon becomes very cumbersome for larger n. Nevertheless, by exponentiating Eq. (11.18) and matching with Eq. (11.13), one can extract the following moment–cumulant relation for commuting variables:

$$\begin{aligned} \tau(A^n) &= \sum_{\substack{r_1,r_2,\ldots,r_n \geq 0 \\ r_1+2r_2+\cdots+nr_n=n}} \frac{n!\,\kappa_1^{r_1}\kappa_2^{r_2}\cdots\kappa_n^{r_n}}{(1!)^{r_1}(2!)^{r_2}\cdots(n!)^{r_n} r_1!\,r_2!\cdots r_n!} \\ &= \kappa_n + \text{products of lower order terms} + \kappa_1^n. \end{aligned} \tag{11.23}$$

In particular, the scaling properties for the moments and cumulants (see (11.28) below) are consistent due to the relation $r_1 + 2r_2 + \cdots + nr_n = n$.

Exercise 11.2.1 Cumulants of a constant

Show that a constant $\alpha\mathbf{1}$ has $\kappa_1 = \alpha$ and $\kappa_n = 0$ for $n \geq 2$. (Hint: compute $H_{\alpha\mathbf{1}}(k) = \log\left(\tau\left(e^{ik\alpha\mathbf{1}}\right)\right)$.)

11.2.3 Scaling of Moments and Cumulants

Moments and cumulants obey simple transformation rules under scalar addition and multiplication. For example, when adding a constant to a variable, $\tilde{A} := A + \alpha$, where $\tau(A) = 0$, we only change the first cumulant:

$$\kappa_1(\tilde{A}) = \alpha \quad \text{and} \quad \kappa_n(\tilde{A}) = \kappa_n(A) \quad \text{for} \quad n \geq 2. \tag{11.24}$$

For the case of multiplication by an arbitrary scalar α, by commutativity of scalars and linearity of τ we have

$$\tau\left((\alpha A)^k\right) = \alpha^k \tau\left(A^k\right). \tag{11.25}$$

For the cumulant, we first look at the scaling of the log-characteristic function:

$$H_{\alpha A}(t) = \log\left(\tau\left(e^{it\alpha A}\right)\right) = H_A(\alpha t). \tag{11.26}$$

And by (11.18), we have

$$H_{\alpha A}(t) = H_A(\alpha t) = \sum_{n=1}^{\infty} \frac{\alpha^n \kappa_n(A)}{n!}(it)^n. \tag{11.27}$$

Thus we have the simple scaling property

$$\kappa_n(\alpha A) = \alpha^n \kappa_n(A). \tag{11.28}$$

11.2.4 Law of Large Numbers and Central Limit Theorem

Continuing our study of algebraic probabilities, we would like to recover two very important theorems in probability theory, namely the law of large numbers (LLN) and the central limit theorem (CLT). The first states that the sample average converges to the mean (a constant) as the number of observations $N \to \infty$, and the second that a large sum of properly centered and rescaled random variables converge to a Gaussian.

First we need to define in our context what we mean by a constant and a Gaussian. For simplicity, we can think of the variables in this section as standard random variables. We will later introduce non-commutating cumulants. The arguments of this section apply in the non-commutative case with independence replaced by freeness.

We have defined the constant variable $A = \alpha\mathbf{1}$, which satisfies

$$\kappa_1(A) = \alpha, \quad \kappa_\ell(A) = 0, \ \forall \ell > 1. \tag{11.29}$$

Then we define the "Gaussian" random variable as an element A that satisfies

$$\kappa_2(A) \neq 0, \quad \kappa_\ell(A) = 0, \ \forall \ell > 2. \tag{11.30}$$

Note that this definition (in the scalar case) is equivalent to the standard Gaussian random variable with density

$$P_{\mu,\sigma^2}(x) = \frac{1}{\sqrt{2\pi\sigma^2}} \exp\left(-\frac{(x-\mu)^2}{2\sigma^2}\right), \tag{11.31}$$

with $\kappa_1 = \mu$ and $\kappa_2 = \sigma^2$.

By extension, we call $\kappa_1(A)$ the mean, and $\kappa_2(A)$ the variance. Now we can give a simple proof for the LLN and CLT within our algebraic setting. Let

$$S_K := \frac{1}{K}\sum_{i=1}^{K} A_i, \tag{11.32}$$

where A_i are K IID variables.[5] Then by (11.28) and the additive property of cumulants, we get that

$$\kappa_\ell(S_K) = \frac{K}{K^\ell}\kappa_\ell(A) \overset{K\to\infty}{\to} \begin{cases} \kappa_1(A), & \text{if } \ell = 1,\\ 0, & \text{if } \ell > 1. \end{cases} \tag{11.33}$$

In other words, S_K converges to a constant in the sense of cumulants.

Assume now that $\kappa_1(A) = 0$ and consider

$$\widehat{S}_K := \frac{1}{\sqrt{K}}\sum_{i=1}^{K} A_i. \tag{11.34}$$

Then it is easy to see that

$$\kappa_\ell(\widehat{S}_K) = \frac{K}{K^{\ell/2}}\kappa_\ell(A) \to \begin{cases} 0, & \text{if } \ell = 1,\\ \kappa_2(A), & \text{if } \ell = 2,\\ 0, & \text{if } \ell > 2. \end{cases} \tag{11.35}$$

In other words, $\widehat{S}_K$ converges to a "Gaussian" random variable with variance $\kappa_2(A) = \tau(A^2)$, in the sense of cumulants.

In our algebraic probability setting we have made the assumption that the variables we consider have finite moments of all orders. This is a very strong assumption. In particular it excludes any variable whose probability decays as a power law. If we relaxed this assumption we would find that some sums of power-law distributed variables converge not to a Gaussian but to a Lévy-stable distribution. A similar concept exists in the non-commutative case, but it is beyond the scope of this book.

11.3 Non-Commuting Variables

We now return to our original goal of developing an extension of standard probabilities for non-commuting objects. One of the goals is to generalize the law of addition of independent variables. We consider a variable equal to $A + B$ where A and B are now non-commutative objects such as large random matrices. If we compute the first three moments of $A + B$, no particular problems arise thanks to the tracial property of τ, and they behave as in the commutative case. For example, consider the third moment

[5] IID copies are variables A_i that have exactly the same cumulants and are all independent. We did not define independence for more than two variables but the factorization of moments can be easily extended to more variables. Note that pairwise independence is not enough to assure independence as a group. For example if x_1, x_2 and x_3 are IID Gaussians, the Gaussian variables x_1, x_2 and $y_3 = \text{sign}(x_1x_2)|x_3|$ are pairwise independent but not all independent as $\mathbb{E}[x_1x_2y_3] > 0$ whereas $\mathbb{E}[x_1]\mathbb{E}[x_2]\mathbb{E}[y_3] = 0$.

$$\tau\big((A+B)^3\big) = \tau(A^3) + \tau(A^2B) + \tau(ABA) + \tau(BA^2) \\ + \tau(AB^2) + \tau(BAB) + \tau(B^2A) + \tau(B^3). \tag{11.36}$$

But since $\tau(A^2B) = \tau(ABA) = \tau(BA^2)$ (and similarly when A appears once and B twice), the classic independence property

$$\tau(A^2B) = \tau(A^2)\tau(B) = 0 \tag{11.37}$$

appears to be sufficient. Things become more interesting for the fourth moment. Indeed,

$$\begin{aligned}\tau\big((A+B)^4\big) &= \tau(A^4) + \tau(A^3B) + \tau(A^2BA) + \tau(ABA^2) + \tau(BA^3) \\ &\quad + \tau(A^2B^2) + \tau(ABAB) + \tau(BA^2B) + \tau(AB^2A) + \tau(BABA) \\ &\quad + \tau(B^2A^2) + \tau(B^3A) + \tau(B^2AB) + \tau(BAB^2) + \tau(AB^3) + \tau(B^4) \\ &= \tau(A^4) + 4\tau(A^3B) + 4\tau(A^2B^2) + 2\tau(ABAB) + 4\tau(AB^3) + \tau(B^4),\end{aligned} \tag{11.38}$$

where in the second step we again used the tracial property of τ. For commutative random variables, independence of A, B means $\tau(A^2B^2) = \tau(A^2)\tau(B^2)$, and this is enough to treat all the terms above. In the non-commutative case, we also need to handle the term $\tau(ABAB)$. In general $ABAB$ is not equal to A^2B^2. "Independence" is therefore not enough to deal with this term, so we need a new concept. A radical solution would be to postulate that $\tau(ABAB)$ is zero whenever $\tau(A) = \tau(B) = 0$. As we compute higher moments of $A+B$ we will encounter more and more complicated similar mixed moments. The concept of *freeness* deals with all such terms at once.

11.3.1 Freeness

Given two random variables A, B, we say they are **free** if for any polynomials $p_1, \dots, p_n$ and $q_1, \dots, q_n$ such that

$$\tau(p_k(A)) = 0, \quad \tau(q_k(B)) = 0, \quad \forall k, \tag{11.39}$$

we have

$$\tau\left(p_1(A)q_1(B)p_2(A)q_2(B)\cdots p_n(A)q_n(B)\right) = 0. \tag{11.40}$$

We will call a polynomial (or a variable) traceless if $\tau(p(A)) = 0$. Note that $\alpha\mathbf{1}$ is free with respect to any $A \in \mathcal{R}$ because $\tau(p(\alpha\mathbf{1})) \equiv p(\alpha\mathbf{1})$, from the definition of $\mathbf{1}$. Hence,

$$\tau(p(\alpha\mathbf{1})) = 0 \Leftrightarrow p(\alpha\mathbf{1}) = 0. \tag{11.41}$$

Moreover, it is easy to see that if A, B are free, then $p(A), q(B)$ are free for any polynomials p, q. By extension, $F(A)$ and $G(B)$ are also free for any function F and G defined by their power series.

The freeness is non-trivial only in the non-commutative case. For the commutative case, it is easy to check that A, B are free if and only if either A or B is a constant. Free random variables are "maximally" non-commuting, in a sense made more precise for the example of free random matrices in the next chapter. For example, for free and mean zero variables A and B, we have $\tau(ABAB) = 0$ whereas $\tau(A^2B^2) = \tau((A^2 - \tau(A^2))B^2) + \tau(A^2)\tau(B^2) = \tau(A^2)\tau(B^2)$.

Assuming A, B are free with $\tau(A) = \tau(B) = 0$, we can compute the moments of the free addition $A + B$. The second moment is easy:

$$\tau\big((A+B)^2\big) = \tau(A^2) + \tau(B^2) + 2\tau(AB) = \tau(A^2) + \tau(B^2), \tag{11.42}$$

because both $\tau(A), \tau(B)$ are zero. For the third and higher moments the trick is, as just above, to add and subtract quantities such that, in each term, at least one object of the form $(C - \tau(C))$ is present:

$$\begin{aligned}\tau\big((A+B)^3\big) &= \tau(A^3) + \tau(B^3) + 3\tau(A^2B) + 3\tau(AB^2)\\ &= \tau(A^3) + \tau(B^3) + 3\tau((A^2 - \tau(A^2))B) + 3\tau(A^2)\tau(B)\\ &\quad + 3\tau(A(B^2 - \tau(B^2))) + 3\tau(A)\tau(B^2)\\ &= \tau(A^3) + \tau(B^3),\end{aligned} \tag{11.43}$$

and for the fourth moment:

$$\begin{aligned}\tau\big((A+B)^4\big) &= \tau(A^4) + 4\tau(A^3B) + 4\tau(A^2B^2) + 2\tau(ABAB) + 4\tau(AB^3) + \tau(B^4)\\ &= \tau(A^4) + 4\tau((A^3 - \tau(A^3))B) + 4\tau(A^3)\tau(B)\\ &\quad + 4\tau\big((A^2 - \tau(A^2))(B^2 - \tau(B^2))\big) + 4\tau(A^2)\tau(B^2)\\ &\quad + 4\tau\big(A\big(B^3 - \tau(B^3)\big)\big) + 4\tau(A)\tau(B^3) + \tau(B^4)\\ &= \tau(A^4) + \tau(B^4) + 4\tau(A^2)\tau(B^2).\end{aligned} \tag{11.44}$$

In particular, we find

$$\tau((A+B)^4) - 2\tau\big((A+B)^2\big)^2 = \tau(A^4) + \tau(B^4) - 2\tau(A^2)^2 - 2\tau(B^2)^2. \tag{11.45}$$

11.3.2 Free Cumulants

Let us define the cumulants of A as

$$\kappa_1(A) = \tau(A), \quad \kappa_2(A) = \tau(A_0^2), \quad \kappa_3(A) = \tau(A_0^3), \quad \kappa_4(A) = \tau(A_0^4) - 2\tau(A_0^2)^2, \tag{11.46}$$

where A_0 is a short-hand for $A - \tau(A)\mathbf{1}$. Then these objects are additive for free random variables. The first three are the same as the commutative ones. But for the fourth cumulant, the coefficient in front of $\tau(A_0^2)^2$ is now 2 instead of 3. Higher cumulants all differ from their commutative counterparts.

As in the commutative case, we can define the kth free cumulant iteratively as

$$\kappa_k(A) = \tau(A^k) + \text{homogeneous products of lower order moments}, \tag{11.47}$$

such that

$$\kappa_k(A + B) = \kappa_k(A) + \kappa_k(B), \quad \forall k, \tag{11.48}$$

whenever A, B are free.

An important example of non-commutative free random variables is two independent large random matrices where one of them is rotational invariant – see next chapter.[6]

11.3.3 Additivity of the R-Transform

In the previous chapter, we saw that the R-transform is additive for large rotationally invariant matrices. We will show here that we can define the R-transform in our abstract algebraic probability setting and that this R-transform is also additive for free variables. In the next chapter, we will dwell on why large rotationally invariant matrices are free.

First we define the Stieltjes transform as a moment generating function as in (2.22); we can define $\mathfrak{g}_A(z)$ for large z as

$$\mathfrak{g}_A(z) = \sum_{k=0}^{\infty} \frac{1}{z^{k+1}} \tau(A^k). \tag{11.49}$$

Then we can also define the R-transform as before:

$$R_A(g) := \mathfrak{z}_A(g) - \frac{1}{g} \tag{11.50}$$

for small enough g. Here the inverse function $\mathfrak{z}_A(g)$ is defined as the formal power series that satisfies $\mathfrak{g}_A(\mathfrak{z}_A(g)) = g$ to all orders.

We now claim that the R-transform is additive for free random variables, i.e.

$$R_{A+B}(g) = R_A(g) + R_B(g), \tag{11.51}$$

whenever A, B are free.

We let $\mathfrak{z}_A(g)$ be the inverse function of

$$\mathfrak{g}_A(z) = \tau\left[(z - A)^{-1}\right], \tag{11.52}$$

whose power series is actually given by (11.49). Consider a fixed scalar g. By construction

$$\tau(g\mathbf{1}) = g = \mathfrak{g}_A(\mathfrak{z}_A(g)) = \tau\left[(\mathfrak{z}_A(g) - A)^{-1}\right]. \tag{11.53}$$

The arguments of $\tau(.)$ on the left and on the right of the above equation have the same mean but they are in general different, so let us define their difference as gX_A via

$$gX_A := (z_A - A)^{-1} - g\mathbf{1}, \tag{11.54}$$

where $z_A := \mathfrak{z}_A(g)$. From its very definition, we have $\tau(X_A) = 0$.

[6] Freeness is only exact in the large N limit of random matrices.

We can invert Eq. (11.54) and find

$$A - z_A = -\frac{1}{g}(1 + X_A)^{-1}. \tag{11.55}$$

Consider another variable B, free from A. For the same fixed g we can find the scalar $z_B := \mathfrak{z}_B(g)$ and define X_B with $\tau(X_B) = 0$ as for A, to find

$$B - z_B = -\frac{1}{g}(1 + X_B)^{-1}. \tag{11.56}$$

Since X_A and X_B are functions of A and B, X_A and X_B are also free. Now,

$$\begin{aligned} A + B - z_A - z_B &= -\frac{1}{g}(1 + X_A)^{-1} - \frac{1}{g}(1 + X_B)^{-1} \\ &= -\frac{1}{g}(1 + X_A)^{-1}(2 + X_A + X_B)(1 + X_B)^{-1}. \end{aligned} \tag{11.57}$$

Hence, noting that $(1 + X_A)(1 + X_B) + 1 - X_A X_B = 2 + X_A + X_B$,

$$A + B - z_A - z_B + \frac{1}{g} = -\frac{1}{g}(1 + X_A)^{-1}(1 - X_A X_B)(1 + X_B)^{-1},$$

$$\left[A + B - \left(z_A + z_B - \frac{1}{g}\right)\right]^{-1} = -g(1 + X_B)(1 - X_A X_B)^{-1}(1 + X_A). \tag{11.58}$$

Using the identity

$$(1 - X_A X_B)^{-1} = \sum_{n=0}^{\infty} (X_A X_B)^n, \tag{11.59}$$

we can expand the expression

$$\tau\left[(1 + X_B)(1 - X_A X_B)^{-1}(1 + X_A)\right], \tag{11.60}$$

which will contain 1 plus terms of the form $\tau(X_A X_B X_A X_B \dots X_B)$ where the initial and final factor might be either X_A or X_B but the important point is that X_A and X_B always alternate. By the freeness and zero mean of X_A and X_B, all these terms are thus zero. Hence we get

$$\tau\left\{\left[A + B - \left(z_A + z_B - \frac{1}{g}\right)\right]^{-1}\right\} = -g \Rightarrow \mathfrak{g}_{A+B}(z_A + z_B - g^{-1}) = g, \tag{11.61}$$

finally leading to the announced result:[7]

$$\mathfrak{z}_{A+B} = z_A + z_B - g^{-1} \Rightarrow R_{A+B} = R_A + R_B. \tag{11.62}$$

[7] The above compact proof is taken from Tao [2012].

11.3.4 R-Transform and Cumulants

The R-transform is defined as a power series in g. We claim that the coefficients of this power series are in fact exactly the non-commutative cumulants defined earlier. In other words, $R_A(g)$ is the cumulant generating function:

$$R_A(g) := \sum_{k=1}^{\infty} \kappa_k(A) g^{k-1}. \tag{11.63}$$

To show that these coefficients are indeed the cumulants we first realize that the general equality $R(g) = \mathfrak{z}(g) - 1/g$ is equivalent to

$$z\mathfrak{g}_A(z) - 1 = \mathfrak{g}_A(z) R_A(\mathfrak{g}_A(z)). \tag{11.64}$$

We can compute the power series of the two sides of this equality:

$$z\mathfrak{g}_A(z) - 1 = \sum_{k=1}^{\infty} \frac{m_k}{z^k}, \tag{11.65}$$

where $m_k := \tau(A^k)$ denotes the kth moment, and

$$\mathfrak{g}_A(z) R_A(\mathfrak{g}_A(z)) = \sum_{k=1}^{\infty} \kappa_k \left(\frac{1}{z} + \sum_{\ell=1}^{\infty} \frac{m_l}{z^{\ell+1}} \right)^k. \tag{11.66}$$

Equating the right hand sides of Eqs. (11.65) and (11.66) and matching powers of $1/z$ we get recursive relations between moments (m_k) and cumulants (κ_k):

$$\begin{aligned}
m_1 &= \kappa_1 && \Rightarrow \quad m_1 = \kappa_1, \\
m_2 &= \kappa_2 + \kappa_1 m_1 && \Rightarrow \quad m_2 = \kappa_2 + \kappa_1^2, \\
m_3 &= \kappa_3 + 2\kappa_2 m_1 + \kappa_1 m_2 && \Rightarrow \quad m_3 = \kappa_3 + 3\kappa_2\kappa_1 + \kappa_1^3, \\
m_4 &= \kappa_4 + 4\kappa_3 m_1 + \kappa_2[2m_2 + m_1^2] + \kappa_1 m_3 && \Rightarrow \quad m_4 = \kappa_4 + 6\kappa_2\kappa_1^2 + 2\kappa_2^2 + 4\kappa_3\kappa_1 + \kappa_1^4.
\end{aligned} \tag{11.67}$$

By looking at the z^{-k} term coming from the $[1/z + \cdots]^k$ term in Eq. (11.66) we realize that $m_k = \kappa_k + \cdots$ where "$\cdots$" are homogeneous combinations of lower order κ_k and m_k. Since the coefficients of the power series Eq. (11.63) are additive under addition of free variables and obey the property

$$\kappa_k(A) = \tau(A^k) + \text{homogeneous products of lower order moments}, \tag{11.68}$$

they are therefore the cumulants defined in Section 11.3.2.

11.3.5 Cumulants and Non-Crossing Partitions

We saw that Eq. (11.64) can be used to compute cumulants iteratively. Actually that equation can be translated into a systematic relation between moments and cumulants:

$$m_n = \sum_{\pi \in \mathrm{NC}(n)} \kappa_{\pi_1} \cdots \kappa_{\pi_{\ell_\pi}}, \tag{11.69}$$

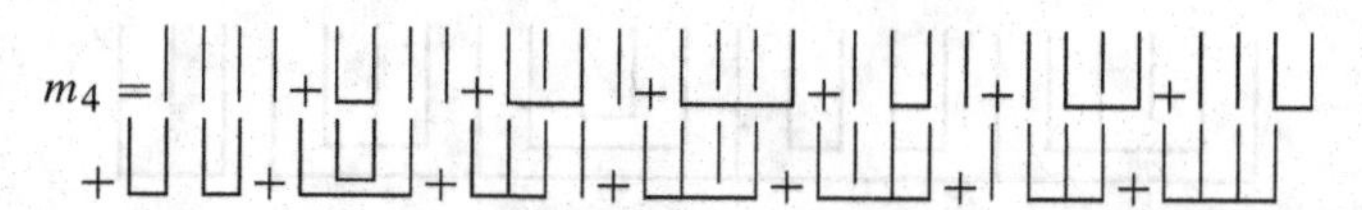

Figure 11.1 List of all non-crossing partitions of four elements. In Eq. (11.69) for m_4, the first partition contributes $\kappa_1\kappa_1\kappa_1\kappa_1 = \kappa_1^4$. The next six all contribute $\kappa_2\kappa_1^2$ and so forth. We read $m_4 = \kappa_1^4 + 6\kappa_2\kappa_1^2 + 2\kappa_2^2 + 4\kappa_3\kappa_1 + \kappa_4$.

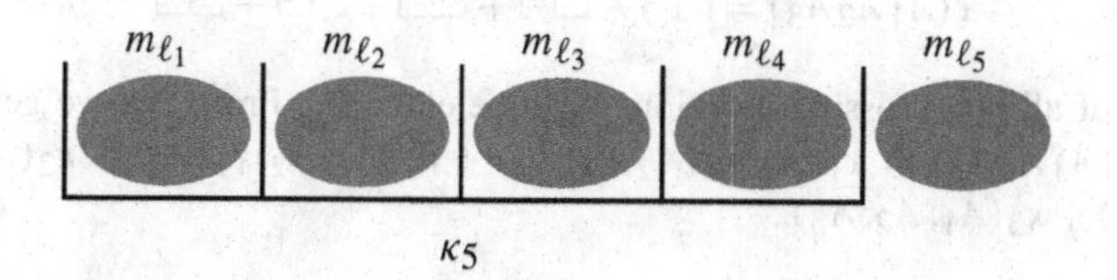

Figure 11.2 A typical non-crossing diagram contributing to a large moment m_n. In this example the first element is connected to four others (giving a factor of κ_5) which breaks the diagram into five disjoint non-crossing diagrams contributing a factor $m_{\ell_1}m_{\ell_2}m_{\ell_3}m_{\ell_4}m_{\ell_5}$. Note that we must have $\ell_1 + \ell_2 + \ell_3 + \ell_4 + \ell_5 = n$.

where $\pi \in \mathrm{NC}(n)$ indicates that the sum is over all possible non-crossing partitions of n elements. For any such partition π the integers $\{\pi_1, \pi_2, \ldots, \pi_{\ell_\pi}\}$ $(1 \leq \ell_\pi \leq n)$ equal the number elements in each group (see Fig. 11.3). They satisfy

$$n = \sum_{k=1}^{\ell_\pi} \pi_k. \tag{11.70}$$

We will show that, provided we define cumulants by Eq. (11.69), we recover Eq. (11.64). But before we do so, let us first show this relation on a simple example. Figure 11.1 shows the computation of the fourth moment in terms of the cumulants.

The argument is very similar to the recursion relation obtained for Catalan numbers where we considered non-crossing pair partitions (see Section 3.2.3). Here the argument is slightly more complicated as we have partitions of all possible sizes. We consider the moment m_n for $n \geq 1$. We break down the sum over all non-crossing partitions of n elements by looking at ℓ, the size of the set containing the first element (for example in Fig. 11.3, the first element belongs to a set of size $\ell = 5$). The size of this first set can be $1 \leq \ell \leq n$. This initial set breaks the partition into ℓ (possibly empty) disjoint smaller partitions. They must be disjoint, otherwise there would be a crossing. In Figure 11.2 we show how an initial 5-set breaks the full partition into 5 blocks. In each of these blocks, every non-crossing partition is possible, the only constraint is that the total size of the partition must be n. The sum over all possible non-crossing partitions of size k of the relevant κ's is the moment m_k. Note that the empty partition contributes a multiplicative factor 1, so we define $m_0 \equiv 1$. Putting everything together we obtain the following recursion relation for m_n:

$$m_n = \sum_{\ell=1}^{n} \kappa_\ell \prod_{\substack{k_1,k_2,\ldots,k_\ell \geq 0 \\ k_1+k_2+\cdots+k_\ell = n-\ell}} m_{k_1} m_{k_2} \ldots m_{k_\ell}. \tag{11.71}$$

Let us multiply both sides of this equation by z^{-n} and sum over n from 1 to ∞. The left hand side gives $zg(z) - 1$, by definition of $g(z)$. The right hand side reads

Figure 11.3 Generic non-crossing partition of 23 elements with two singletons, five doublets, two triplets, and one quintent, such that $23 = 5+2\cdot 3+5\cdot 2+2\cdot 1$. In Eq. (11.69), this particular partition appears for m_{23} and contributes $\kappa_5\kappa_3^2\kappa_2^5\kappa_1^2$.

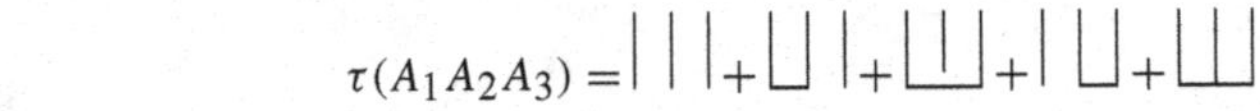

$\tau(A_1A_2A_3) =$ + + + +

Figure 11.4 List of all non-crossing partitions of three elements. From this we get Eq. (11.74) for three elements: $\tau(A_1A_2A_3) = \kappa_1(A_1)\kappa_1(A_2)\kappa_1(A_3) + \kappa_2(A_1,A_2)\kappa_1(A_3) + \kappa_2(A_1,A_3)\kappa_1(A_2) + \kappa_2(A_2,A_3)\kappa_1(A_3) + \kappa_3(A_1,A_2,A_3)$.

$$\sum_{n=1}^{\infty}\sum_{\ell=1}^{n}\kappa_\ell \prod_{k_1,k_2,\ldots,k_\ell\geq 0} \delta_{k_1+k_2+\cdots+k_\ell=n-\ell}\frac{m_{k_1}m_{k_2}\ldots m_{k_\ell}}{z^{1+k_1}z^{1+k_2}\ldots z^{1+k_\ell}}, \tag{11.72}$$

which can be transformed into

$$\sum_{\ell=1}^{\infty}\kappa_\ell\left[\sum_{k_1=0}^{\infty}\frac{m_{k_1}}{z^{1+k_1}}\right]^\ell = g(z)R(g(z)), \tag{11.73}$$

where we have used Eq. (11.63). We thus recover exactly Eq. (11.64), showing that the relation (11.69) is equivalent to our previous definition of the free cumulants.

It is interesting to contrast the moment–cumulant relation in the standard (commutative) case (Eq. (11.23)) and the free (non-commutative) case (Eq. (11.69)). Both can be written as a sum over all partitions on n elements; in the standard case all partitions are allowed, while in the free case the sum is only over non-crossing partitions.

11.3.6 Freeness as the Vanishing of Mixed Cumulants

We have defined freeness in Section 11.3.1 as the property of two variables A and B such that the trace of any mixed combination of traceless polynomials in A and in B vanishes. There exists another equivalent definition of freeness, namely that every mixed cumulant of A and B vanish. To make sense of this definition we first need to introduce cumulants of several variables. They are defined recursively by

$$\tau(A_1A_2\ldots A_n) = \sum_{\pi\in \mathrm{NC}(n)}\kappa_\pi(A_1A_2\ldots A_n), \tag{11.74}$$

where the A_i's are not necessarily distinct and $\mathrm{NC}(n)$ is the set of all non-crossing partitions of n elements. Here

$$\kappa_\pi(A_1A_2\ldots A_n) = \kappa_{\pi_1}(\ldots)\ldots\kappa_{\pi_\ell}(\ldots) \tag{11.75}$$

are the products of cumulants of variables belonging to the same group of the corresponding partition – see Figure 11.4 for an illustration. We also call these generalized cumulants the free cumulants.

When all the variables in Eq. (11.74) are the same ($A_i = A$) we recover the previous definition of cumulants with a slightly different notation (e.g. $\kappa_3(A,A,A) \equiv \kappa_3(A)$).

Cumulants with more than one variable are called mixed cumulants (e.g. $\kappa_4(A,A,B,A)$). By applying Eq. (11.74) we find for the low generalized cumulants of two variables

$$\begin{aligned} m_1(A) &= \kappa_1(A), \\ m_2(A,B) &= \kappa_1(B)\kappa_1(B) + \kappa_2(A,B), \\ m_3(A,A,B) &= \kappa_1(A)^2\kappa_1(B) + \kappa_2(A,A)\kappa_1(B) + 2\kappa_2(A,B)\kappa_1(A) + \kappa_3(A,A,B). \end{aligned} \tag{11.76}$$

We can now state more precisely the alternative definition of freeness: a set of variables is free if and only if all their mixed cumulants vanish. For example, in the low cumulants listed above, freeness of A and B implies that $\kappa_2(A,B) = \kappa_3(A,A,B) = 0$.

This definition of freeness is easy to generalize to a collection of variables, i.e. a collection of variables is free if all its mixed cumulants are zero. As noted at the end of Section 11.3.7 below, pairwise freeness is not enough to ensure that a collection is free.

We remark that vanishing of mixed cumulants implies that free cumulants are additive. In Speicher's notation, $\kappa_k(A,B,C,\ldots)$ is a multi-linear function in each of its arguments, where k gives the number of variables. Thus we have

$$\begin{aligned} \kappa_k(A+B,A+B,\ldots) &= \kappa_k(A,A,\ldots) + \kappa_k(B,B,\ldots) + \text{mixed cumulants} \\ &= \kappa_k(A,A,\ldots) + \kappa_k(B,B,\ldots), \end{aligned} \tag{11.77}$$

i.e. κ_k is additive.

We will give a concrete application of the formalism of free mixed cumulants in Section 12.2.

11.3.7 The Central Limit Theorem for Free Variables

We can now go back and re-read Section 11.2.4. We can replace every occurrence of the word *independent* with *free*, and *cumulant* with *free cumulant*. The LLN now states that the sum of K free identically distributed (FID) variables normalized by K^{-1} converges to a constant (also called a scalar) with the same mean.

Let us define a free Wigner variable as a variable with second free cumulant $\kappa_2 > 0$ and all other free cumulants equal to zero. In other words, a free Wigner variable is such that $R_W(x) = \kappa_2 x$. The CLT then states that the sum of K zero-mean free identical variables normalized by $K^{-1/2}$ converges to a free Wigner variable with the same second cumulant.

In the case where our free random variables are large symmetric random matrices, the Wigner defined here by its cumulant coincides with the Wigner matrices defined in Chapter 2. We indeed saw that the R-transform of a Wigner matrix is given by $R(x) = \sigma^2 x$, i.e. the cumulant generating function has a single term corresponding to $\kappa_2 = \sigma^2$.

Alternatively, we note that the moments of a Wigner are given by the sum over non-crossing *pair* partitions (Eq. (3.26)). Comparing with Eq. (11.69), we realize that partitions containing anything other than pairs must contribute zero, hence only the second cumulant of the Wigner is non-zero.

The LLN and the CLT require variables to be collectively free, in the sense that all mixed cumulants are zero. As is the case with *independence*, pairwise freeness is not enough to ensure freeness as a collection (see footnote on page 161). Indeed, in Section 12.5 we will encounter variables that are pairwise free but not free as a collection. One can have A and B mutually free and both free with respect to C but $A + B$ is not free with respect to C. This does not happen for rotationally invariant large matrices but can arise in other constructions. The definition of a *free collection* is just an extension of definition (11.40) including traceless polynomials in all variables in the collection. With this definition, sums of variables in the collection are free from those not included in the sum (e.g. $A + B$ is free from C).

11.3.8 Subordination Relation for Addition of Free Variables

We now introduce the subordination relation for free addition, which is just a rewriting of the addition of R-transforms. For free A and B, we have

$$R_A(g) + R_B(g) = R_{A+B}(g) \Rightarrow \mathfrak{z}_A(g) + R_B(g) = \mathfrak{z}_{A+B}(g), \tag{11.78}$$

where

$$\mathfrak{g}_{A+B}(\mathfrak{z}_{A+B}) = g = \mathfrak{g}_A(z_A) = \mathfrak{g}_A\left(\mathfrak{z}_{A+B} - R_B(g)\right). \tag{11.79}$$

We call $z := \mathfrak{z}_{A+B}(g)$, then the above relations give

$$\mathfrak{g}_{A+B}(z) = \mathfrak{g}_A\left(z - R_B(\mathfrak{g}_{A+B}(z))\right), \tag{11.80}$$

which is called a subordination relation (compare with Eq. (10.2)).

11.4 Free Product

In the previous section, we have studied the property of the sum of free random variables. In the case of commuting variables, the question of studying the product of (positive) random variables is trivial, since taking the logarithm of this product we are back to the problem of sums again. In the case of non-commuting variables, things are more interesting. We will see below that one needs to introduce the so-called S-transform, which is the counterpart of the R-transform for products of free variables.

We start by noticing that the free product of traceless variables is trivial. If A, B are free and $\tau(A) = \tau(B) = 0$, we have

$$\tau((AB)^k) = \tau(ABAB\ldots AB) = 0. \tag{11.81}$$

11.4.1 Low Moments of Free Products

We will now compute the first few moments of the free products of two variables with a non-zero trace: $C := AB$ where A, B are free and $\tau(A) \neq 0$, $\tau(B) \neq 0$. Without loss of generality, we can assume that $\tau(A) = \tau(B) = 1$ by rescaling A and B. Then

$$\tau(C) = \tau\left((A - \tau(A))(B - \tau(B))\right) + \tau(A)\tau(B) = \tau(A)\tau(B) = 1. \tag{11.82}$$

We can also use (11.74) to get

$$\tau(C) = \kappa_2(A, B) + \kappa_1(A)\kappa_1(B) = \kappa_1(A)\kappa_1(B) = 1, \tag{11.83}$$

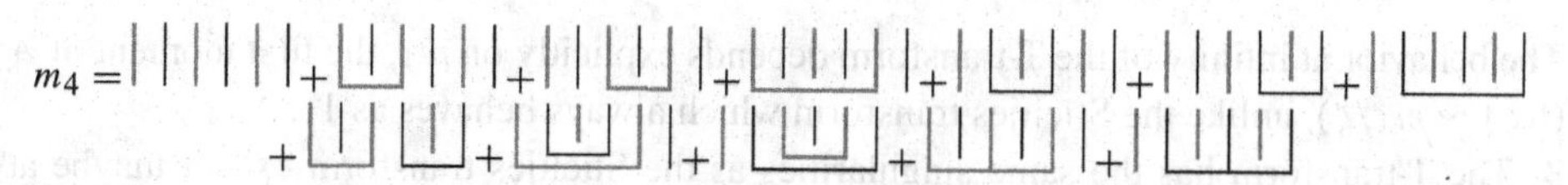

Figure 11.5 List of all non-crossing partitions of six elements contributing to $\tau(ABABAB)$ excluding mixed AB terms. Terms involving A are in thick gray and B in black. Equation (11.86) can be read off from these diagrams. Note that $\kappa_1(A) = \kappa_2(B) = 1$.

since mixed cumulants are zero for mutually free variables. Similarly, using Eq. (11.74) we can get that (see Fig. 11.1 for the non-crossing partitions of four elements)

$$\begin{aligned}\tau(C^2) = \tau(ABAB) &= \kappa_1(A)^2\kappa_1(B)^2 + \kappa_2(A)\kappa_1(B)^2 + \kappa_1(A)^2\kappa_2(B) \\ &= 1 + \kappa_2(A) + \kappa_2(B),\end{aligned} \tag{11.84}$$

which gives

$$\kappa_2(C) := \tau(C^2) - \tau(C)^2 = \kappa_2(A) + \kappa_2(B). \tag{11.85}$$

For the third moment of $C = AB$, we can follow Figure 11.5 and get

$$\begin{aligned}\tau(C^3) &= \tau(ABABAB) \\ &= 1 + 3\kappa_2(A) + 3\kappa_2(B) + 3\kappa_2(A)\kappa_2(B) + \kappa_3(A) + \kappa_3(B),\end{aligned} \tag{11.86}$$

leading to

$$\begin{aligned}\kappa_3(C) :&= \tau(C^3) - 3\tau(C^2)\tau(C) + 2\tau(C)^3 \\ &= \kappa_3(A) + \kappa_3(B) + 3\kappa_2(A)\kappa_2(B).\end{aligned} \tag{11.87}$$

Under free products of unit-trace variables, the mean remains equal to one and the variance is additive. The third cumulant is not additive; it is strictly greater than the sum of the third cumulants unless one of the two variables is the identity (unit scalar).

11.4.2 Definition of the S-Transform

We will now show that the above relations can be encoded into the S-transform $S(t)$ which is multiplicative for products of free variables:

$$S_{AB}(t) = S_A(t)S_B(t) \tag{11.88}$$

for A and B free. To define the S-transform, we first introduce the T-transform as

$$\begin{aligned}t_A(\zeta) &= \tau\left[(1 - \zeta^{-1}A)^{-1}\right] - 1 \\ &= \zeta g_A(\zeta) - 1 \\ &= \sum_{k=1}^{\infty} \frac{m_k}{\zeta^k}.\end{aligned} \tag{11.89}$$

The behavior at infinity of the T-transform depends explicitly on m_1, the first moment of A ($\mathfrak{t}(\zeta) \sim m_1/\zeta$), unlike the Stieltjes transform which always behaves as $1/z$.

The T-transform has the same singularities as the Stieltjes transform except maybe at zero. When A is a matrix, one can recover the continuous part of its eigenvalue density $\rho(x)$ using the following T-version of the Sokhotski–Plemelj formula:

$$\lim_{\eta \to 0+} \operatorname{Im} \mathfrak{t}(x - i\eta) = \pi x \rho(x). \tag{11.90}$$

Poles in the T-transform indicate Dirac masses. If $\mathfrak{t}(\zeta) \sim A/(\zeta - \lambda_0)$ near λ_0 then the density is a Dirac mass of amplitude A/λ_0 at λ_0. The behavior at zero of the T-transform is a bit different from that of the Stieltjes transform. A regular density at zero gives a regular Stieltjes and hence $\mathfrak{t}(0) = -1$. Deviations from this value indicate a Dirac mass at zero, hence when $\mathfrak{t}(0) \neq -1$, the density has a Dirac at zero of amplitude $\mathfrak{t}(0) + 1$.

The T-transform can also be written as

$$\mathfrak{t}_A(\zeta) = \tau\left[A\,(\zeta - A)^{-1}\right]. \tag{11.91}$$

We define $\zeta_A(t)$ to be the inverse function of $\mathfrak{t}_A(\zeta)$. When $m_1 \neq 0$, $\mathfrak{t}_A$ is invertible for large ζ, and hence ζ_A exists for small enough t. We then define the S-transform as[8]

$$S_A(t) := \frac{t+1}{t\zeta_A(t)}, \tag{11.92}$$

for variables A such that $\tau(A) \neq 0$.

Let us compute the S-transform of the identity $S_1(t)$:

$$\mathfrak{t}_1(\zeta) = \frac{1}{\zeta - 1} \Rightarrow \zeta_1(t) = \frac{t+1}{t} \Rightarrow S_1(t) = 1, \tag{11.93}$$

as expected as the identity is free with respect to any variable. The S-transform scales in a simple way with the variable A. To find its scaling we first note that

$$\mathfrak{t}_{\alpha A}(\zeta) = \tau\left[(1 - (\alpha^{-1}\zeta)^{-1}A)^{-1}\right] - 1 = \mathfrak{t}_A(\zeta/\alpha), \tag{11.94}$$

which gives

$$\zeta_{\alpha A}(t) = \alpha\zeta_A(t). \tag{11.95}$$

Then, using (11.92), we get that

$$S_{\alpha A}(t) = \alpha^{-1} S_A(t). \tag{11.96}$$

The above scaling is slightly counterintuitive but it is consistent with the fact that $S_A(0) = 1/\tau(A)$. We will be focusing on unit trace objects such that $S(0) = 1$.

[8] Most authors prefer to define the S-transform in terms of the moment generating function $\psi(z) := \mathfrak{t}(1/z)$. The definition $S(t) = \psi^{-1}(t)(t+1)/t$ is equivalent to ours ($\psi^{-1}(t)$ is the functional inverse of $\psi(z)$). We prefer to work with the T-transform as the function $t(\zeta)$ has an analytic structure very similar to that of $g(z)$. The function $\psi(z)$ is analytic near zero and singular for large values of z corresponding to the reciprocal of the eigenvalues.

The construction of the S-transform relies on the properties of mixed moments of free variables. In that respect it is closely related to the R-transform. Using the relation $t_A(\zeta) = \zeta g_A(\zeta) - 1$, one can get the following relationships between R_A and S_A:

$$S_A(t) = \frac{1}{R_A(tS_A(t))}, \quad R_A(g) = \frac{1}{S_A(gR_A(g))}. \tag{11.97}$$

11.4.3 Multiplicativity of the S-Transform

We can now show the multiplicative property (11.88). The proof is similar to the one given for the additive case and is adapted from it.

We fix t and let ζ_A and ζ_B be the inverse T-transforms of t_A and t_B. Then we define E_A through

$$1 + t + E_A = (1 - A/\zeta_A)^{-1}, \tag{11.98}$$

and similarly for E_B. We have $\tau(E_A) = 0$, $\tau(E_B) = 0$, and, since A and B are free, E_A, E_B are also free. Then we have

$$\frac{A}{\zeta_A} = 1 - (1 + t + E_A)^{-1}, \tag{11.99}$$

which gives

$$\begin{aligned}\frac{AB}{\zeta_A\zeta_B} &= \left[1 - (1+t+E_A)^{-1}\right]\left[1 - (1+t+E_B)^{-1}\right] \\ &= (1+t+E_A)^{-1}\left[(t+E_A)(t+E_B)\right](1+t+E_B)^{-1}.\end{aligned} \tag{11.100}$$

Using the identity

$$t(E_A + E_B) = \frac{t}{1+t}\left[(1+t+E_A)(1+t+E_B) - (1+t)^2 - E_AE_B\right], \tag{11.101}$$

we can rewrite the above expression as

$$\begin{aligned}\frac{AB}{\zeta_A\zeta_B} &= \frac{t}{1+t} + (1+t+E_A)^{-1}\left[-t + \frac{E_AE_B}{1+t}\right](1+t+E_B)^{-1} \\ \Rightarrow 1 - \frac{1+t}{t}\frac{AB}{\zeta_A\zeta_B} &= (1+t)(1+t+E_A)^{-1}\left[1 - \frac{E_AE_B}{t(1+t)}\right](1+t+E_B)^{-1} \\ \Rightarrow \left[1 - \frac{1+t}{t}\frac{AB}{\zeta_A\zeta_B}\right]^{-1} &= \frac{1}{1+t}(1+t+E_B)\left[1 - \frac{E_AE_B}{t(1+t)}\right]^{-1}(1+t+E_A).\end{aligned} \tag{11.102}$$

Using the expansion

$$\left[1 - \frac{E_AE_B}{t(1+t)}\right]^{-1} = \sum_{n=0}\left(\frac{E_AE_B}{t(1+t)}\right)^n, \tag{11.103}$$

one can check that

$$\tau\left[(1+t+E_B)\left[1 - \frac{E_AE_B}{t(1+t)}\right]^{-1}(1+t+E_A)\right] = (1+t)^2, \tag{11.104}$$

where we used the freeness condition for E_A and E_B. Thus we get that

$$\tau\left\{\left[1-\frac{1+t}{t}\frac{AB}{\zeta_A\zeta_B}\right]^{-1}\right\}=1+t \quad \Rightarrow \quad \mathfrak{t}_{AB}\left(\frac{t\zeta_A\zeta_B}{1+t}\right)=t, \tag{11.105}$$

which gives that

$$S_{AB}(t)=S_A(t)S_B(t) \tag{11.106}$$

thanks to the definition (11.92).

11.4.4 Subordination Relation for the Free Product

We next derive a subordination relation for the free product using (11.88) and (11.92):

$$S_{AB}(t)=S_A(t)S_B(t) \quad \Rightarrow \quad \zeta_{AB}(t)=\frac{\zeta_A(t)}{S_B(t)}, \tag{11.107}$$

where

$$\mathfrak{t}_{AB}(\zeta_{AB}(t))=t=\mathfrak{t}_A(\zeta_A(t))=\mathfrak{t}_A(\zeta_{AB}(t)S_B(t)). \tag{11.108}$$

We call $\zeta := \zeta_{AB}(t)$, then the above relations give

$$\mathfrak{t}_{AB}(\zeta)=\mathfrak{t}_A\left(\zeta S_B(\mathfrak{t}_{AB}(\zeta))\right), \tag{11.109}$$

which is the subordination relation for the free product. In fact, the above is true even when S_A does not exist, e.g. when $\tau(A)=0$.

When applied to free random matrices, the form AB is not very useful since it is not necessarily symmetric even if A and B are. But if $A \succeq 0$ (i.e. A is positive semi-definite symmetric) and B is symmetric, then $A^{\frac{1}{2}}BA^{\frac{1}{2}}$ has the same moments as AB and is also symmetric. In our applications below we will always encounter the case $A \succeq 0$ and call $A^{\frac{1}{2}}BA^{\frac{1}{2}}$ the free product of A and B.

Exercise 11.4.1 Properties of the S-transform

(a) Using Eq. (11.92), show that

$$R(x)=\frac{1}{S(xR(x))}. \tag{11.110}$$

Hint: define $t=xR(x)=zg-1$ and identify x as g.

(b) For a variable such that $\tau(M)=\kappa_1=1$, write $S(t)$ as a power series in t, compute the first few terms of the powers series, up to (and including) the t^2 term, using Eq. (11.110) and Eq. (11.63). You should find

$$S(t)=1-\kappa_2 t+(2\kappa_2^2-\kappa_3)t^2+O(t^3). \tag{11.111}$$

(c) We have shown that, when **A** and **B** are mutually free with unit trace,

$$\tau(\mathbf{AB}) = 1, \tag{11.112}$$

$$\tau(\mathbf{ABAB}) - 1 = \kappa_2(\mathbf{A}) + \kappa_2(\mathbf{B}), \tag{11.113}$$

$$\tau(\mathbf{ABABAB}) = \kappa_3(\mathbf{A}) + \kappa_3(\mathbf{B}) + 3\kappa_2(\mathbf{A})\kappa_2(\mathbf{B}) + 3(\kappa_2(\mathbf{A}) + \kappa_2(\mathbf{B})) + 1. \tag{11.114}$$

Show that these relations are compatible with $S_{\mathbf{AB}}(t) = S_{\mathbf{A}}(t)S_{\mathbf{B}}(t)$ and the first few terms of your power series in (b).

(d) Consider $\mathbf{M}_1 = \mathbf{1} + \sigma_1 \mathbf{W}_1$ and $\mathbf{M}_2 = \mathbf{1} + \sigma_2 \mathbf{W}_2$ where $\mathbf{W}_1$ and $\mathbf{W}_2$ are two different (free) unit Wigner matrices and both σ's are less than $1/2$. $\mathbf{M}_1$ and $\mathbf{M}_2$ have $\kappa_3 = 0$ and are positive definite in the large N limit. What is $\kappa_3(\mathbf{M}_1\mathbf{M}_2)$?

Exercise 11.4.2 S-transform of the matrix inverse

(a) Consider **M** an invertible symmetric random matrix and $\mathbf{M}^{-1}$ its inverse. Using Eq. (11.89), show that

$$\mathfrak{t}_{\mathbf{M}}(\zeta) + \mathfrak{t}_{\mathbf{M}^{-1}}\left(\frac{1}{\zeta}\right) + 1 = 0. \tag{11.115}$$

(b) Using Eq. (11.115), show that

$$S_{\mathbf{M}^{-1}}(x) = \frac{1}{S_{\mathbf{M}}(-x-1)}. \tag{11.116}$$

Hint: write $u(x) = 1/\zeta(t)$ where $u(x)$ is such that $x = \mathfrak{t}_{\mathbf{M}^{-1}}(u(x))$. Equation (11.115) is then equivalent to $x = -1 - t$.

Bibliographical Notes

- The concept of freeness was introduced in
  - D. Voiculescu. Symmetries of some reduced free product C*-algebras. In H. Araki et al. (eds.), *Operator Algebras and Their Connections with Topology and Ergodic Theory*. Lecture Notes in Mathematics, volume 1132. Springer, Berlin, Heidelberg, 1985,
  - D. Voiculescu. Limit laws for random matrices and free products. *Inventiones mathematicae*, 104(1):201–220, 1991,

  with the second reference making the link to RMT.
- For general introductions to freeness, see
  - J. A. Mingo and R. Speicher. *Free Probability and Random Matrices*. Springer, New York, 2017,

- J. Novak. Three lectures on free probability. In *Random Matrix Theory, Interacting Particle Systems, and Integrable Systems*. Cambridge University Press, Cambridge, 2014.

- For a more operational approach to freeness, see

- A. M. Tulino and S. Verdú. *Random Matrix Theory and Wireless Communications*. Now publishers, Hanover, Mass., 2004,.

- The proof of the additivity of the R-transform presented here is from

- T. Tao. *Topics in Random Matrix Theory*. American Mathematical Society, Providence, Rhode Island, 2012,

see also

- P. Zinn-Justin. Adding and multiplying random matrices: A generalization of Voiculescu's formulas. *Physical Review E*, 59:4884–4888, 1999,

for an alternative point of view.

- Central limit theorems and Lévy-stable free variables:

- H. Bercovici, V. Pata, and P. Biane. Stable laws and domains of attraction in free probability theory. *Annals of Mathematics*, 149(3):1023–1060, 1999,
- Z. Burda, J. Jurkiewicz, M. A. Nowak, G. Papp, and I. Zahed. Free random Lévy and Wigner-Lévy matrices. *Physical Review E*, 75:051126, 2007.

12
Free Random Matrices

In the last chapter, we introduced the concept of freeness rather abstractly, as the proper non-commutative generalization of *independence* for usual random variables. In the present chapter, we explain why large, randomly rotated matrices behave as free random variables. This justifies the use of R-transforms and S-transforms to deal with the spectrum of sums and products of large random matrices. We also revisit the abstract central limit theorem of the previous chapter (Section 11.2.4) in the more concrete case of sums of randomly rotated matrices.

12.1 Random Rotations and Freeness

12.1.1 Statement of the Main Result

Recall the definition of freeness. A and B are free if for any set of traceless polynomials $p_1, \ldots, p_n$ and $q_1, \ldots, q_n$ the following equality holds:

$$\tau \left(p_1(A)q_1(B)p_2(A)q_2(B) \ldots p_n(A)q_n(B)\right) = 0. \tag{12.1}$$

In order to make the link with large matrices we will consider $\mathbf{A}$ and $\mathbf{B}$ to be large symmetric matrices and $\tau(\mathbf{M}) := 1/N\,\mathrm{Tr}(\mathbf{M})$. The matrix $\mathbf{A}$ can be diagonalized as $\mathbf{U}\Lambda\mathbf{U}^T$ and $\mathbf{B}$ as $\mathbf{V}\Lambda'\mathbf{V}^T$. A traceless polynomial $p_i(\mathbf{A})$ can be diagonalized as $\mathbf{U}\Lambda_i\mathbf{U}^T$, where $\mathbf{U}$ is the same orthogonal matrix as for $\mathbf{A}$ itself and $\Lambda_i = p_i(\Lambda)$ is some traceless diagonal matrix, and similarly for $q_i(\mathbf{B})$. Equation (12.1) then becomes

$$\tau \left(\Lambda_1\mathbf{O}\Lambda_1'\mathbf{O}^T\Lambda_2\mathbf{O}\Lambda_2'\mathbf{O}^T \ldots \Lambda_n\mathbf{O}\Lambda_n'\mathbf{O}^T\right) = 0, \tag{12.2}$$

where we have introduced $\mathbf{O} = \mathbf{U}^T\mathbf{V}$ as the orthogonal matrix of basis change rotating the eigenvectors of $\mathbf{A}$ into those of $\mathbf{B}$.

As we argue below, in the large N limit Eq. (12.2) always holds true when averaged over the orthogonal matrix $\mathbf{O}$ and whenever matrices Λ_i and Λ_i' are traceless. We also expect that in the large N limit Eq. (12.2) becomes self-averaging, so a single matrix $\mathbf{O}$ behaves as the average over all such matrices. Hence, two large symmetric matrices whose eigenbases are randomly rotated with respect to one another are essentially free. For example, Wigner matrices $\mathbf{X}$ and white Wishart matrices $\mathbf{W}$ are rotationally invariant, meaning that the matrices of their eigenvectors are random orthogonal matrices. We conclude that for N large, both $\mathbf{X}$ and $\mathbf{W}$ are free with respect to any matrix independent from them, in particular they are free from any deterministic matrix.

12.1.2 Integration over the Orthogonal Group

We now come back to the central statement that in the large N limit the average over $\mathbf{O}$ of Eq. (12.2) is zero for traceless matrices Λ_i and Λ'_i. In order to compute quantities like

$$\left\langle \tau\left(\Lambda_1 \mathbf{O} \Lambda'_1 \mathbf{O}^T \Lambda_2 \mathbf{O} \Lambda'_2 \mathbf{O}^T \dots \Lambda_n \mathbf{O} \Lambda'_n \mathbf{O}^T\right)\right\rangle_{\mathbf{O}}, \tag{12.3}$$

one needs to understand how to compute the following moments of rotation matrices, averaged over the Haar (flat) measure over the orthogonal group $O(N)$:

$$I(\mathbf{i},\mathbf{j},n) := \left\langle \mathbf{O}_{i_1 j_1} \mathbf{O}_{i_2 j_2} \dots \mathbf{O}_{i_{2n} j_{2n}} \right\rangle_{\mathbf{O}}. \tag{12.4}$$

The general formula has been worked out quite recently and involves the *Weingarten functions*. A full discussion of these functions is beyond the scope of this book, but we want to give here a brief account of the structure of the result. When $N \to \infty$, the leading term is quite simple: one recovers the Wick's contraction rules, as if $\mathbf{O}_{i_1 j_1}$ were independent random Gaussian variables with variance $1/N$. Namely,

$$I(\mathbf{i},\mathbf{j},n) = N^{-n} \sum_{\text{pairings } \pi} \delta_{i_{\pi(1)} i_{\pi(2)}} \delta_{j_{\pi(1)} j_{\pi(2)}} \dots \delta_{i_{\pi(2n-1)} i_{\pi(2n)}} \delta_{j_{\pi(2n-1)} j_{\pi(2n)}} + O(N^{-n-1}). \tag{12.5}$$

Note that all pairings of the i-indices are the same as those of the j-indices. For example, for $n = 1$ and $n = 2$ one has explicitly, for $N \to \infty$,

$$N \left\langle \mathbf{O}_{i_1 j_1} \mathbf{O}_{i_2 j_2} \right\rangle_{\mathbf{O}} = \delta_{i_1 i_2} \delta_{j_1 j_2} \tag{12.6}$$

and

$$\begin{aligned} N^2 \left\langle \mathbf{O}_{i_1 j_1} \mathbf{O}_{i_2 j_2} \mathbf{O}_{i_3 j_3} \mathbf{O}_{i_4 j_4} \right\rangle_{\mathbf{O}} &= \delta_{i_1 i_2} \delta_{j_1 j_2} \delta_{i_3 i_4} \delta_{j_3 j_4} \\ &+ \delta_{i_1 i_3} \delta_{j_1 j_3} \delta_{i_2 i_4} \delta_{j_2 j_4} \\ &+ \delta_{i_1 i_4} \delta_{j_1 j_4} \delta_{i_2 i_3} \delta_{j_2 j_3}. \end{aligned} \tag{12.7}$$

The case $n = 1$ is exact and has no subleading corrections in N, so we can use it to compute

$$\begin{aligned} \left\langle \tau(\Lambda_1 \mathbf{O} \Lambda'_1 \mathbf{O}^T) \right\rangle_{\mathbf{O}} &= N^{-1} \sum_{i,j=1}^{N} (\Lambda_1)_i \left\langle \mathbf{O}_{ij} (\Lambda'_1)_j \mathbf{O}^T_{ji} \right\rangle_{\mathbf{O}} \\ &= N^{-2} \sum_{i,j=1}^{N} (\Lambda_1)_i (\Lambda'_1)_j \\ &= \tau(\Lambda_1)\tau(\Lambda'_1). \end{aligned} \tag{12.8}$$

(Recall that $\tau(\mathbf{A})$ is equal to $N^{-1}\operatorname{Tr}\mathbf{A}$.) Clearly the result is zero when $\tau(\Lambda_1) = \tau(\Lambda'_1) = 0$, as required by the freeness condition.

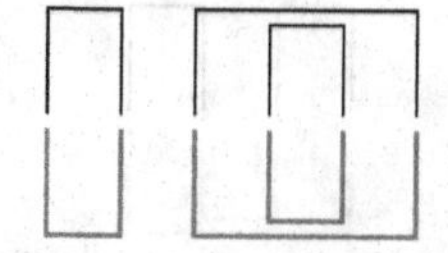

Figure 12.1 Number of loops $\ell(\pi,\pi)$ for $\pi = (1,2)(3,6)(4,5)$. The bottom part of the diagram (thick gray) corresponds to the first partition π and the top part (black) to the second partition, here also equal to π. In this example there are three loops each of size 2.

Now, using only the leading Wick terms for $n = 2$ and after some index contractions and manipulations, one would obtain

$$\lim_{N\to\infty} \langle \tau(\Lambda_1 \mathbf{O}\Lambda_1' \mathbf{O}^T \Lambda_2 \mathbf{O}\Lambda_2' \mathbf{O}^T)\rangle_{\mathbf{O}} = \tau(\Lambda_1\Lambda_2)\tau(\Lambda_1')\tau(\Lambda_2') + \tau(\Lambda_1)\tau(\Lambda_2)\tau(\Lambda_1'\Lambda_2'). \tag{12.9}$$

However, this cannot be correct. Take for example $\Lambda_1 = \Lambda_2 = \mathbf{1}$, for which $\tau(\Lambda_1 \mathbf{O}\Lambda_1' \mathbf{O}^T \Lambda_2 \mathbf{O}\Lambda_2' \mathbf{O}^T) = \tau(\Lambda_1'\Lambda_2')$ exactly, whereas the formula above adds an extra term $\tau(\Lambda_1')\tau(\Lambda_2')$. The correct formula actually reads

$$\lim_{N\to\infty} \langle \tau(\Lambda_1 \mathbf{O}\Lambda_1' \mathbf{O}^T \Lambda_2 \mathbf{O}\Lambda_2' \mathbf{O}^T)\rangle_{\mathbf{O}} = \tau(\Lambda_1\Lambda_2)\tau(\Lambda_1')\tau(\Lambda_2') + \tau(\Lambda_1)\tau(\Lambda_2)\tau(\Lambda_1'\Lambda_2') - \tau(\Lambda_1)\tau(\Lambda_2)\tau(\Lambda_1')\tau(\Lambda_2'), \tag{12.10}$$

which again is zero whenever all individual traces are zero (i.e. the freeness condition).

12.1.3 Beyond Wick Contractions: Weingarten Functions

Where does the last term in Eq. (12.10) come from? The solution to this puzzle lies in the fact that some subleading corrections to Eq. (12.7) also contribute to the trace we are computing: summing over indices from 1 to N can prop up some subdominant terms and make them contribute to the final result. Hence we need to know a little more about the Weingarten functions. This will allow us to conclude that the freeness condition holds for arbitrary n. The general Weingarten formula reads

$$I(\mathbf{i},\mathbf{j},n) = \sum_{\text{pairings } \pi,\sigma} W_n(\pi,\sigma)\delta_{i_{\pi(1)}i_{\pi(2)}}\delta_{j_{\sigma(1)}j_{\sigma(2)}}\dots\delta_{i_{\pi(2n-1)}i_{\pi(2n)}}\delta_{j_{\sigma(2n-1)}j_{\sigma(2n)}}, \tag{12.11}$$

where now the pairings π of i's and σ of j's do not need to coincide. The Weingarten functions $W_n(\pi,\sigma)$ can be thought of as matrices with pairings as indices. They are given by the pseudo-inverse[1] of the matrices $M_n(\pi,\sigma) := N^{\ell(\pi,\sigma)}$, where $\ell(\pi,\sigma)$ is the number of loops obtained when superposing π and σ. For example, when $\pi = \sigma$ one finds n loops, each of length 2, see Figure 12.1. While when $\pi \neq \sigma$ the number of loops is always less than n ($\ell(\pi,\sigma) < n$), see Figure 12.2. At large N, the diagonal of the matrix M_n dominates and the matrix is always invertible. By expanding in powers of $1/N$, we see that its inverse W_n, whose elements are the Weingarten functions, has an N^{-n} behavior

[1] The pseudo-inverse of M is such that $WMW = W$ and $MWM = M$. When M is invertible, $W = M^{-1}$. If M is diagonalizable, the eigenvalues of W are the reciprocal of those of M with the rule $1/0 \to 0$.

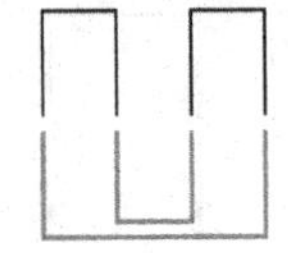

Figure 12.2 Number of loops $\ell(\pi,\sigma)$ for $\pi = (1,4)(2,3)$ and $\sigma = (1,2)(3,4)$. In this example there is only one loop.

on the diagonal and off-diagonal terms are at most N^{-n-1}. More generally, one has the following beautiful expansion:

$$W_n(\pi,\sigma) = N^{\ell(\pi,\sigma)-2n} \sum_{g=0}^{\infty} \Omega_g(\pi,\sigma) N^{-g}, \tag{12.12}$$

where the coefficients $\Omega_g(\pi,\sigma)$ depend on properties of certain "geodesic paths" in the space of partitions. The bottom line of this general expansion formula is that non-Wick contractions are of higher order in N^{-1}.

As an illustration, let us come back to the missing term in Eq. (12.9) when one restricts to Wick contractions, for which the number of loops $\ell(\pi,\pi) = 2$. The next term corresponds to $\ell(\pi,\sigma) = 1$ for which $W_2(\pi,\sigma) \sim -N^{-3}$ (see Exercise 12.1.1). Consider the pairings $i_1 = i_4, i_2 = i_3$ and $j_1 = j_2, j_3 = j_4$, for which $\ell(\pi,\sigma) = 1$ (see Fig. 12.2). Such pairings do not add any constraint on the $2n = 4$ free indices that appear in Eq. (12.3); summing over them thus yields $\tau(\Lambda_1)\tau(\Lambda_2)\tau(\Lambda_1')\tau(\Lambda_2')$, with a -1 coming from the Weingarten function $W_2(\pi,\sigma)$. Hence we recover the last term of Eq. (12.10).

Exercise 12.1.1 Exact Weingarten functions at $n = 2$

There are three possible pair partitions of four elements. They are shown on Figure 3.2. If we number these π_1,π_2 and π_3, M_2 is a 3×3 matrix whose elements are equal to $N^{\ell(\pi_i,\pi_j)}$.

(a) By trying a few combinations of the three pairings, convince yourself that for $n = 2$, $\ell(\pi_i,\pi_j) = 2$ if $i = j$ and 1 otherwise.

(b) Build the matrix M_2. For $N > 1$ it is invertible, find its inverse W_2. Hint: use an ansatz for W_2 with a variable a on the diagonal and b off-diagonal. For $N = 1$ find the pseudo-inverse of M_2.

(c) Finally show that (when $N > 1$) of the nine pairs of pairings, the three Wick contractions (diagonal elements) have

$$W_2^{\text{Wick}} = \frac{N+1}{N^3+N^2-2N} \overset{N\to\infty}{\to} N^{-2}, \tag{12.13}$$

and the six non-Wick parings (off-diagonal) have

$$W_2^{\text{non-Wick}} = -\frac{1}{N^3+N^2-2N} \overset{N\to\infty}{\to} -N^{-3}. \tag{12.14}$$

For $N = 1$, all nine Weingarten functions are equal to $1/9$.

(d) The expression $\langle\tau(\mathbf{O}\mathbf{O}^T\mathbf{O}\mathbf{O}^T)\rangle$ is always equal to 1. Write it as a sum over four indices and expand the expectation value over orthogonal matrices as nine terms each containing two Dirac deltas multiplied by a Weingarten function. Each sum of delta terms gives a power of N; find these for all nine terms and using your result from (c), show that indeed the Weingarten functions give $\langle\tau(\mathbf{O}\mathbf{O}^T\mathbf{O}\mathbf{O}^T)\rangle = 1$ for all N.

12.1.4 Freeness of Large Matrices

We are now ready to show that all expectations of the form (12.3) are zero. Let us look at them more closely:

$$\begin{aligned}&\langle\tau\left(\Lambda_1\mathbf{O}\Lambda_1'\mathbf{O}^T\Lambda_2\mathbf{O}\Lambda_2'\mathbf{O}^T\ldots\Lambda_n\mathbf{O}\Lambda_n'\mathbf{O}^T\right)\rangle_{\mathbf{O}}\\&=\frac{1}{N}\sum_{\mathbf{ij}}I(\mathbf{i},\mathbf{j},n)[\Lambda_1]_{i_{2n}i_1}[\Lambda_2]_{i_2i_3}\ldots[\Lambda_n]_{i_{2n-2}i_{2n-1}}[\Lambda_1']_{j_1j_2}[\Lambda_2']_{j_3j_4}\ldots[\Lambda_n']_{j_{2n-1}j_{2n}}.\end{aligned} \tag{12.15}$$

The object $I(\mathbf{i},\mathbf{j},n)$ contains all possible pairings of the i indices and all those of the j indices with a Weingarten function as its prefactor. The i and j indices never mix. We concentrate on i pairings. Each pairing will give rise to a product of normalized traces of Λ_i matrices. For example, the term

$$\tau(\Lambda_5)\tau(\Lambda_4\Lambda_1\Lambda_2)\tau(\Lambda_3\Lambda_6) \tag{12.16}$$

would appear for $n = 6$. Each normalized trace introduces a factor of N when going from Tr(.) to $\tau(.)$. Since by hypothesis $\tau(\Lambda_i) = 0$, the maximum number of non-zero traces is $\lfloor n/2\rfloor$, e.g.

$$\tau(\Lambda_1\Lambda_3)\tau(\Lambda_5\Lambda_6)\tau(\Lambda_2\Lambda_4). \tag{12.17}$$

The maximum factor of N that can be generated is therefore $N^{\lfloor n/2\rfloor}$. Applying the same reasoning to the j pairing, and using the fact that the Weingarten function is at most $O(N^{-n})$, we find

$$\left|\langle\tau\left(\Lambda_1\mathbf{O}\Lambda_1'\mathbf{O}^T\Lambda_2\mathbf{O}\Lambda_2'\mathbf{O}^T\ldots\Lambda_n\mathbf{O}\Lambda_n'\mathbf{O}^T\right)\rangle_{\mathbf{O}}\right| \le O\left(N^{-1+2\lfloor n/2\rfloor-n}\right) \overset{N\to\infty}{\to} 0. \tag{12.18}$$

Using the same arguments one can shown that large unitary invariant complex Hermitian random matrices are free. In this case one should consider an integral of unitary matrices in Eq. (12.4). The result is also given by Eq. (12.11) where the functions $W_n(\pi,\sigma)$ are now unitary Weingarten functions. They are different than the orthogonal Weingarten functions presented above but they share an important property in the large N limit, namely

$$W_n(\pi,\sigma) = \begin{cases}O(N^{-n}) \text{ if } \pi = \sigma,\\ O(N^{-n-k}) \text{ if } \pi \ne \sigma \text{ for some } k \ge 1,\end{cases} \tag{12.19}$$

which was the property needed in our proof of freeness of large rotationally invariant symmetric matrices.

12.2 R-Transforms and Resummed Perturbation Theory

In this section, we want to explore yet another route to obtain the additivity of R-transforms, which makes use of perturbation theory and of the mixed cumulant calculus introduced in the last chapter, Section 11.3.6, exploited in a concrete case.

We want to study the average Stieltjes transform of $\mathbf{A} + \mathbf{B}^{\mathrm{R}}$ where $\mathbf{B}^{\mathrm{R}}$ is a randomly rotated matrix $\mathbf{B}$: $\mathbf{B}^{\mathrm{R}} := \mathbf{O}\mathbf{B}\mathbf{O}^T$. We thus write

$$\mathfrak{g}(z) := \left\langle\tau\left((z\mathbf{1} - \mathbf{A} - \mathbf{B}^{\mathrm{R}})^{-1}\right)\right\rangle_{\mathbf{O}} := \tau_{\mathrm{R}}\left((z\mathbf{1} - \mathbf{A} - \mathbf{B}^{\mathrm{R}})^{-1}\right), \tag{12.20}$$

where τ_{R} is meant for both the normalized trace τ and the average over the Haar measure of the rotation group. We now formally expand the resolvent in powers of $\mathbf{B}^{\mathrm{R}}$. Introducing $\mathbf{G}_A = (z\mathbf{1} - \mathbf{A})^{-1}$, one has

$$\mathfrak{g}(z) = \tau_{\mathrm{R}}(\mathbf{G}_A) + \tau_{\mathrm{R}}(\mathbf{G}_A\mathbf{B}^{\mathrm{R}}\mathbf{G}_A) + \tau_{\mathrm{R}}(\mathbf{G}_A\mathbf{B}^{\mathrm{R}}\mathbf{G}_A\mathbf{B}^{\mathrm{R}}\mathbf{G}_A) + \cdots. \tag{12.21}$$

Now, since in the large N limit $\mathbf{G}_A$ and $\mathbf{B}^{\mathrm{R}}$ are free, we can use the general tracial formula, Eq. (11.74), noting that all mixed cumulants (containing both $\mathbf{G}_A$ and $\mathbf{B}^{\mathrm{R}}$) are identically zero.

In order to proceed, one needs to introduce three types of mixed moments where $\mathbf{B}^{\mathrm{R}}$ appears exactly n times:

$$m_n^{(1)} := \tau_{\mathrm{R}}(\mathbf{G}_A\mathbf{B}^{\mathrm{R}}\mathbf{G}_A \ldots \mathbf{G}_A\mathbf{B}^{\mathrm{R}}\mathbf{G}_A), \qquad m_n^{(2)} := \tau_{\mathrm{R}}(\mathbf{B}^{\mathrm{R}}\mathbf{G}_A \ldots \mathbf{G}_A\mathbf{B}^{\mathrm{R}}) \tag{12.22}$$

and

$$m_n^{(3)} := \tau_{\mathrm{R}}(\mathbf{B}^{\mathrm{R}}\mathbf{G}_A \ldots \mathbf{B}^{\mathrm{R}}\mathbf{G}_A) = \tau_{\mathrm{R}}(\mathbf{G}_A\mathbf{B}^{\mathrm{R}} \ldots \mathbf{G}_A\mathbf{B}^{\mathrm{R}}). \tag{12.23}$$

Note that $m_0^{(1)} \equiv \mathfrak{g}_A(z)$ and $m_0^{(2)} = m_0^{(3)} = 0$. We also introduce, for full generality, the corresponding generating functions:

$$\tilde{M}^{(a)}(u) = \sum_{n=0}^{\infty} m_n^{(a)} u^n, \qquad a = 1, 2, 3. \tag{12.24}$$

Note however that we will only be interested here in $\mathfrak{g}(z) := \tilde{M}^{(1)}(u = 1)$ (cf. Eq. (12.21)).

Let us compute $m_n^{(1)}$ using the same method as in Section 11.3.5, i.e. expanding in the size ℓ of the group to which the first $\mathbf{G}_A$ belongs (see Eq. (11.71)):

$$m_n^{(1)} = \sum_{\ell=1}^{n+1} \kappa_{G_A,\ell} \prod_{\substack{k_1,k_2,\ldots,k_\ell \geq 0 \\ k_1+k_2+\cdots+k_\ell = n-\ell}} m_{k_1}^{(2)} m_{k_2}^{(2)} \ldots m_{k_{\ell-1}}^{(2)} m_{k_\ell}^{(3)}, \tag{12.25}$$

where $n \geq 1$ and $\kappa_{G_A,\ell}$ are the free cumulants of $\mathbf{G}_A$. Similarly,

$$m_n^{(2)} = \sum_{\ell=1}^{n} \kappa_{B,\ell} \prod_{\substack{k_1,k_2,\ldots,k_\ell \geq 0 \\ k_1+k_2+\cdots+k_\ell = n-\ell}} m_{k_1}^{(1)} m_{k_2}^{(1)} \ldots m_{k_{\ell-1}}^{(1)} m_{k_\ell}^{(3)} \tag{12.26}$$

and

$$m_n^{(3)} = \sum_{\ell=1}^{n+1} \kappa_{B,\ell} \prod_{\substack{k_1,k_2,\ldots,k_\ell \geq 0 \\ k_1+k_2+\cdots+k_\ell = n-\ell}} m_{k_1}^{(1)} m_{k_2}^{(1)} \ldots m_{k_\ell}^{(1)}, \tag{12.27}$$

where $\kappa_{B,\ell}$ are the free cumulants of $\mathbf{B}$. Multiplying both sides of these equations by u^n and summing over n leads to, respectively,

$$\tilde{M}^{(1)}(u) = \mathfrak{g}_A(z) + \sum_{\ell=1}^{\infty} \kappa_{G_A,\ell} u^\ell [\tilde{M}^{(2)}(u)]^{\ell-1} \tilde{M}^{(3)}(u), \tag{12.28}$$

and

$$\tilde{M}^{(2)}(u) = \sum_{\ell=1}^{\infty} \kappa_{B,\ell} u^{\ell} [\tilde{M}^{(1)}(u)]^{\ell-1} \tilde{M}^{(3)}(u), \qquad \tilde{M}^{(3)}(u) = \sum_{\ell=1}^{\infty} \kappa_{B,\ell} u^{\ell} [\tilde{M}^{(1)}(u)]^{\ell}. \tag{12.29}$$

Recalling the definition of R-transforms as a power series of cumulants, we thus get

$$\tilde{M}^{(1)}(u) = \mathfrak{g}_A(z) + u\tilde{M}^{(3)}(u) R_{G_A}\left(u\tilde{M}^{(2)}(u)\right) \tag{12.30}$$

and

$$\tilde{M}^{(2)}(u) = u\tilde{M}^{(3)}(u) R_B\left(u\tilde{M}^{(1)}(u)\right), \qquad \tilde{M}^{(3)}(u) = u\tilde{M}^{(1)}(u) R_B\left(u\tilde{M}^{(1)}(u)\right). \tag{12.31}$$

Eliminating $\tilde{M}^{(2)}(u)$ and $\tilde{M}^{(3)}(u)$ and setting $u = 1$ then yields the following relation:

$$\mathfrak{g}(z) = \mathfrak{g}_A(z) + \mathfrak{g}(z)\, R_B(\mathfrak{g}(z))\, R_{G_A}\left(\mathfrak{g}(z) R_B^2(\mathfrak{g}(z))\right). \tag{12.32}$$

In order to rewrite this result in more familiar terms, let us consider the case where $\mathbf{B} = b\mathbf{1}$, in which case $R_B(z) \equiv b$ and, since $\mathbf{A}+\mathbf{B} = \mathbf{A}+b\mathbf{1}$, $\mathfrak{g}(z) \equiv \mathfrak{g}_A(z-b)$. Hence, for arbitrary b, R_{G_A} must obey the relation

$$R_{G_A}(b^2\mathfrak{g}(z)) = \frac{\mathfrak{g}(z) - \mathfrak{g}(z+b)}{b\mathfrak{g}(z)}. \tag{12.33}$$

Now, if for a fixed z we set $b = R_B(\mathfrak{g}(z))$, we find that Eq. (12.32) is obeyed provided $\mathfrak{g}(z) = \mathfrak{g}_A(z - R_B(\mathfrak{g}(z)))$, i.e. precisely the subordination relation Eq. (11.80).

12.3 The Central Limit Theorem for Matrices

In the last chapter, we briefly discussed the extension of the CLT for non-commuting variables. We now restate the result in the context of random matrices, with a special focus on the preasymptotic (cumulant) corrections to the Wigner distribution.

Let us consider the following sum of K large, randomly rotated matrices, all assumed to be traceless:

$$\mathbf{M}_K := \frac{1}{\sqrt{K}} \sum_{i=1}^{K} \mathbf{O}_i \mathbf{A}_i \mathbf{O}_i^T, \qquad \operatorname{Tr} \mathbf{A}_i = 0, \tag{12.34}$$

where $\mathbf{O}_i$ are independent, random rotation matrices, chosen with a flat measure over the orthogonal group $O(N)$. We also assume, for simplicity, that all $\mathbf{A}_i$ have the same (arbitrary) moments:

$$\tau(\mathbf{A}_i^{\ell}) \equiv m_{\ell}, \qquad \forall i. \tag{12.35}$$

This means in particular that all $\mathbf{A}_i$'s share the same R-transform:

$$R_{\mathbf{A}_i}(z) \equiv \sum_{\ell=2}^{\infty} \kappa_{\ell} z^{\ell-1}. \tag{12.36}$$

Using the fact that R-transforms of randomly rotated matrices simply add, together with $R_{\alpha\mathbf{M}}(z) = R_{\mathbf{M}}(\alpha z)$, one finds

$$R_{\mathbf{M}_K}(z) = \sum_{\ell=2}^{\infty} K^{1-\ell/2}\kappa_\ell z^{\ell-1}. \tag{12.37}$$

We thus see that, as K becomes large, all free cumulants except the second one tend to zero, which implies that the limit of $\mathbf{M}_K$ when K goes to infinity is a Wigner matrix, with a semi-circle eigenvalue spectrum.

It is interesting to study the finite K corrections to this result. First, assume that the spectrum of $\mathbf{A}_i$ is not symmetric around zero, such that the skewness $m_3 = \kappa_3 \neq 0$. For large K, the R-transform of $\mathbf{M}_K$ can be approximated as

$$R_{\mathbf{M}_K}(z) \approx \sigma^2 z + \frac{\kappa_3}{\sqrt{K}} z^2 + \cdots. \tag{12.38}$$

In order to derive the corrections to the semi-circle induced by skewness, we posit that the Stieltjes transform can be expanded around the Wigner result $\mathfrak{g}_{\mathbf{X}}(z)$ as

$$\mathfrak{g}_{\mathbf{M}_K}(z) = \mathfrak{g}_{\mathbf{X}}(z) + \frac{\kappa_3}{\sqrt{K}}\mathfrak{g}_3(z) + \cdots \tag{12.39}$$

and assume the second term to be very small. The R-transform, $R_{\mathbf{M}_K}(z)$, can be similarly expanded, yielding

$$R_{\mathbf{X}}\left(\mathfrak{g}_{\mathbf{X}}(z) + \frac{\kappa_3}{\sqrt{K}}\mathfrak{g}_3(z)\right) + \frac{\kappa_3}{\sqrt{K}} R_3(\mathfrak{g}_{\mathbf{X}}(z)) = z \Rightarrow R_3(\mathfrak{g}_{\mathbf{X}}(z)) = -\frac{\mathfrak{g}_3(z)}{\mathfrak{g}'_{\mathbf{X}}(z)} \tag{12.40}$$

or, equivalently,

$$\mathfrak{g}_3(z) = -\mathfrak{g}'_{\mathbf{X}}(z) R_3(\mathfrak{g}_{\mathbf{X}}(z)) = -\mathfrak{g}'_{\mathbf{X}}(z)\mathfrak{g}^2_{\mathbf{X}}(z). \tag{12.41}$$

For simplicity, we normalize the Wigner semi-circle such that $\sigma^2 = 1$, and hence

$$\mathfrak{g}_{\mathbf{X}}(z) = \frac{1}{2}\left(z - \sqrt[\oplus]{\Delta}\right); \qquad \Delta := z^2 - 4. \tag{12.42}$$

One then finds

$$\mathfrak{g}_3(z) = -\frac{1}{4}\left(1 - \frac{z\sqrt[\oplus]{\Delta}}{\Delta}\right)\left(z^2 - 2 - z\sqrt[\oplus]{\Delta}\right). \tag{12.43}$$

The imaginary part of this expression when $z \to \lambda + \mathrm{i}0$ gives the correction to the semi-circle eigenvalue spectrum, and reads, for $\lambda \in [-2,2]$,

$$\delta\rho(\lambda) = \frac{\kappa_3}{2\pi\sqrt{K}} \frac{\lambda(\lambda^2 - 3)}{\sqrt{4 - \lambda^2}}. \tag{12.44}$$

This correction is plotted in Figure 12.3. Note that it is odd in λ, as expected, and diverges near the edges of the spectrum, around which the above perturbation approach breaks down

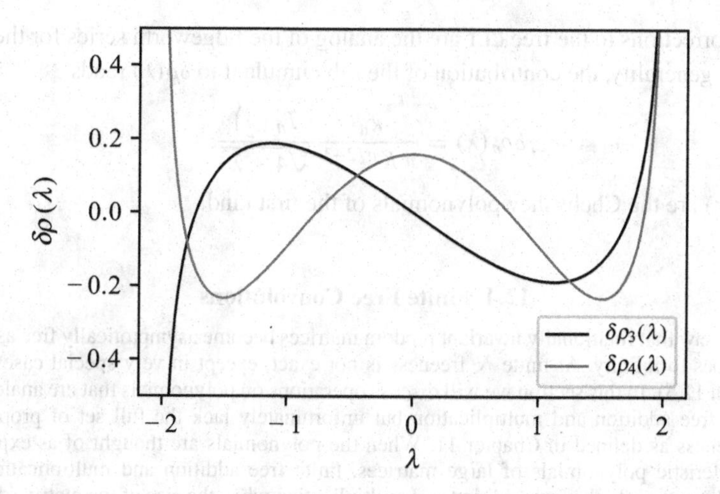

Figure 12.3 First order correction to the Wigner density of eigenvalues for non-zero skewness ($\delta\rho_3(\lambda)$) and kurtosis ($\delta\rho_4(\lambda)$), Eqs. (12.44) and (12.47), respectively.

(because the density of eigenvalues of the Wigner matrix goes to zero). Note that the excess skewness is computed to be

$$\int_{-2}^{2} \lambda^3 \delta\rho(\lambda) d\lambda = \frac{\kappa_3}{\sqrt{K}}, \tag{12.45}$$

as expected.

One can obtain the corresponding correction when $\mathbf{A}_i$ is symmetric around zero, in which case the first correction term comes from the kurtosis. For large K, the R-transform of $\mathbf{M}_K$ can now be approximated as

$$R_{\mathbf{M}_K}(z) \approx \sigma^2 z + \frac{\kappa_4}{K} z^3 + \cdots. \tag{12.46}$$

Following the same path as above, one finally derives the correction to the eigenvalue spectrum, which in this case reads, for $\lambda \in [-2, 2]$,

$$\delta\rho(\lambda) = \frac{\kappa_4}{2\pi K} \frac{\lambda^4 - 4\lambda^2 + 2}{\sqrt{4 - \lambda^2}} \tag{12.47}$$

(see Fig. 12.3). The correction is now even in λ and one can check that

$$\int_{-2}^{2} \delta\rho(\lambda) d\lambda = 0, \tag{12.48}$$

as it should be, since all the mass is carried by the semi-circle. Another check that this result is correct is to compute the excess free kurtosis, given by

$$\int_{-2}^{2} (\lambda^4 - 2\lambda^2) \delta\rho(\lambda) d\lambda = \frac{\kappa_4}{K}, \tag{12.49}$$

again as expected.

These corrections to the free CLT are the analog of the Edgeworth series for the classical CLT. In full generality, the contribution of the nth cumulant to $\delta\rho(\lambda)$ reads

$$\delta\rho_n(\lambda) = \frac{\kappa_n}{\pi K^{n/2-1}} \frac{T_n\left(\frac{\lambda}{2}\right)}{\sqrt{4-\lambda^2}}, \tag{12.50}$$

where $T_n(x)$ are the Chebyshev polynomials of the first kind.

12.4 Finite Free Convolutions

We saw that rotationally invariant random matrices become asymptotically free as their size goes to infinity. At finite N freeness is not exact, except in very special cases (see Section 12.5). In this section we will discuss operations on polynomials that are analogous to the free addition and multiplication but unfortunately lack the full set of properties of freeness as defined in Chapter 11. When the polynomials are thought of as expected characteristic polynomials of large matrices, finite free addition and multiplication do indeed converge to the free addition and multiplication when the size of the matrix (degree of the polynomial) goes to infinity.

12.4.1 Notations: Roots and Coefficients

Let $p(z)$ be a polynomial of degree N. By the fundamental theorem of algebra, this polynomial will have exactly N roots (when counted with their multiplicity) and can be written as

$$p(z) = a_0 \prod_{i=1}^{N} (z - \lambda_i). \tag{12.51}$$

If $a_0 = 1$ we say that $p(z)$ is monic and if all the λ_i's are real the polynomial is called *real-rooted*. In this section we will only consider real-rooted monic polynomials. Such a polynomial can always be viewed as the characteristic polynomial of the diagonal matrix Λ containing its roots, i.e.

$$p(z) = \det\left(z\mathbf{1} - \Lambda\right). \tag{12.52}$$

We can expand the product (12.51) as

$$p(z) = \sum_{k=0}^{N} (-1)^k a_k z^{N-k}. \tag{12.53}$$

Note that we have defined the coefficient a_k as the coefficient of z^{N-k} and *not* that of z^k and that we have included alternating signs $(-1)^k$ in its definition. The reason is that we want a simple link between the coefficients a_k and the roots λ_i. We have

$$a_k = \sum_{\substack{\text{ordered} \\ k\text{-tuples } \mathbf{i}}}^{N} \lambda_{i_1}\lambda_{i_2}\ldots\lambda_{i_k}, \tag{12.54}$$

$$a_0 = 1, \quad a_1 = \sum_{i=1}^{N} \lambda_i, \quad a_2 = \sum_{\substack{i=1 \\ j=i+1}}^{N} \lambda_i\lambda_j, \quad \ldots, \quad a_N = \prod_{i=1}^{N} \lambda_i. \tag{12.55}$$

Note that the coefficient a_k is homogeneous to λ^k. From the coefficients a_k we can compute the sample moments[2] of the set of λ_i's. In particular we have

$$\mu(\{\lambda_i\}) = \frac{a_1}{N} \quad \text{and} \quad \sigma^2(\{\lambda_i\}) = \frac{a_1^2}{N} - \frac{2a_2}{N-1}. \tag{12.56}$$

The polynomial $p(z)$ will often be the expected characteristic polynomial of some random matrix $\mathbf{M}$ of size N with joint distribution of eigenvalues $P(\{\mu_i\})$. The coefficients a_i are then multi-linear moments of this joint distribution. If the random eigenvalues μ_i do not fluctuate, we have $\lambda_i = \mu_i$ but in general we should think of the λ_i's as deterministic numbers fixed by the random ensemble considered. The case of independent eigenvalues gives a trivial expected characteristic polynomial, i.e. $p(z) = (z - \mathbb{E}(\mu))^N$ or else $\lambda_i = \mathbb{E}(\mu)\ \forall i$.

Shifts and scaling of the matrix $\mathbf{M}$ can be mapped onto operations on the polynomial $p_{\mathbf{M}}(z)$. For a shift, we have

$$p_{\mathbf{M}+\alpha\mathbf{1}}(z) = p_{\mathbf{M}}(z - \alpha). \tag{12.57}$$

Multiplication by a scalar gives

$$p_{\alpha\mathbf{M}}(z) = \alpha^N p_{\mathbf{M}}(\alpha^{-1} z) \Longleftrightarrow a_k = \alpha^k a_k. \tag{12.58}$$

Finally there is a formula for matrix inversion which is only valid when the eigenvalues are deterministic (e.g. $\mathbf{M} = \mathbf{O}\Lambda\mathbf{O}^T$ with fixed Λ):

$$p_{\mathbf{M}^{-1}}(z) = \frac{z^N}{a_N} p_{\mathbf{M}}(1/z) \Longleftrightarrow a_k = \frac{a_{N-k}}{a_N}. \tag{12.59}$$

A degree N polynomial can always be written as a degree N polynomial of the derivative operator acting on the monomial z^N. We introduce the notation $\hat{p}$ as

$$p(z) =: \hat{p}\,(\mathrm{d}_z)\, z^N \qquad \text{the coefficients of } \hat{p} \text{ are} \quad \hat{a}_k = (-1)^N \frac{k!}{N!} a_{N-k}, \tag{12.60}$$

where d_z is a shorthand notation for $\mathrm{d}/\mathrm{d}z$. It will prove useful to compute finite free convolutions.

12.4.2 Finite Free Addition

The equivalent of the free addition for two monic polynomials $p_1(x)$ and $p_2(x)$ of the same degree N is the finite free additive convolution defined as

$$p_1 \boxplus p_2(z) = \left\langle \det[z\mathbf{1} - \Lambda_1 - \mathbf{O}\Lambda_2\mathbf{O}^T] \right\rangle_{\mathbf{O}}, \tag{12.61}$$

where the diagonal matrices $\Lambda_{1,2}$ contain the roots of $p_{1,2}(z)$ all assumed to be real. The averaging over the orthogonal matrix $\mathbf{O}$ is as usual done over the flat (Haar) measure. We could have chosen to integrate $\mathbf{O}$ over unitary matrices or even permutation matrices; the final result would be the same.

As we will show in Section 12.4.5, the additive convolution can be expressed in a very concise form using the polynomials $\hat{p}_{1,2}(x)$:

$$p_1 \boxplus p_2(z) = \hat{p}_1\,(\mathrm{d}_z)\, p_2(z) = \hat{p}_2\,(\mathrm{d}_z)\, p_1(z) = \hat{p}_1\,(\mathrm{d}_z)\, \hat{p}_2\,(\mathrm{d}_z)\, z^N. \tag{12.62}$$

[2] Here we have chosen to normalize the sample variance with a factor $(N-1)^{-1}$. It may seem odd to use the formula suited to when the mean is unknown for computing the variance of deterministic numbers, but this definition will later match that of the finite free cumulant and give the Hermite polynomial the variance of the corresponding Wigner matrix.

It is easy to see that when $p_s(z) = p_1 \boxplus p_2(z)$, $p_s(z)$ is again a monic polynomial of degree N. What is less obvious, but true, is that $p_s(x)$ is also real-rooted. A proof of this is beyond the scope of this book. The additive convolution is bilinear in the coefficients of $p_1(z)$ and $p_2(z)$, which means that the operation commutes with the expectation value. If $p_{1,2}(z)$ are independent random polynomials (for example, characteristic polynomials of independent random matrices) we have a relation for their expected value:

$$\mathbb{E}[p_1 \boxplus p_2(z)] = \mathbb{E}[p_1(z)] \boxplus \mathbb{E}[p_2(z)]. \tag{12.63}$$

The finite free addition can also be written as a relation between the coefficients $a_k^{(s)}$ of $p_s(z)$ and those of $p_{1,2}(z)$:

$$a_k^{(s)} = \sum_{i+j=k} \frac{(N-i)!\,(N-j)!}{N!\,(N-k)!} a_i^{(1)} a_j^{(2)}. \tag{12.64}$$

More explicitly, the first three coefficients are given by

$$\begin{aligned} a_0^{(s)} &= 1, \\ a_1^{(s)} &= a_1^{(1)} + a_1^{(2)}, \\ a_2^{(s)} &= a_2^{(1)} + a_2^{(2)} + \frac{N-1}{N} a_1^{(2)} a_1^{(2)}. \end{aligned} \tag{12.65}$$

From which we can verify that both the sample mean and the variance (Eq. (12.56)) are additive under the finite free addition.

If we call $p_\mu(z) = (z-\mu)^N$ the polynomial with a single root μ so $p_0(z) = z^N$ is the trivial monic polynomial, we have that, under additive convolution with any $p(z)$, $p_0(z)$ acts as a null element and $p_\mu(z)$ acts as a shift:

$$p \boxplus p_0(z) = p(z) \quad \text{and} \quad p \boxplus p_\mu(z) = p(z-\mu). \tag{12.66}$$

Hermite polynomials are stable under this addition:

$$H_N \boxplus H_N(z) = 2^{N/2} H_N(2^{-N/2} z), \tag{12.67}$$

where the factors $2^{N/2}$ compensate the doubling of the sample variance.

The average characteristic polynomials of Wishart matrices can easily be understood in terms the finite free sum. Consider an N-dimensional rank-1 matrix $\mathbf{M} = \mathbf{x}\mathbf{x}^T$ where $\mathbf{x}$ is a vector of IID unit variance numbers. It has one eigenvalue equal to N (on average) and all others are zero, so its average characteristic polynomial is

$$p(z) = z^{N-1}(z - N) = (1 - \mathrm{d}_z)\, z^N, \tag{12.68}$$

from which we read that $\hat{p}(\mathrm{d}_z) = (1 - \mathrm{d}_z)$. But since an unnormalized Wishart matrix of parameter T is just the free sum of T such projectors, Eq. (12.62) immediately leads to $p_T(z) = (1 - \mathrm{d}_z)^T z^N$, which coincides with Eq. (6.41).

12.4.3 A Finite R-Transform

If we look back at Eq. (12.62), we notice that the polynomial $\hat{p}(\mathrm{d}_z)$ behaves like the Fourier transform under free addition. Its logarithm is therefore additive. We need to be a bit careful of what we mean by the logarithm of a polynomial. The function $\hat{p}(\mathrm{d}_z)$ has two important properties: first as a power series in d_z it always starts $1 + O(\mathrm{d}_z)$; second it is defined by its action on Nth order polynomials in z, so only its $N+1$ first terms in its Taylor series matter. We will say that the polynomial is defined modulo d_z^{N+1}, meaning that higher order terms are set to zero. When we take the logarithm of $\hat{p}(u)$ we thus mean: apply the Taylor series of the logarithm around 1 and expand up to the power u^N. The resulting function

$$L(u) := -\left[\log \hat{p}(u)\right] \mod u^{N+1} \tag{12.69}$$

is then additive. For average characteristic polynomials of random matrices, it should be related to the R-transform in the large N limit.

Let us examine more closely $L(u)$ in three simple cases: identity, Wigner and Wishart.

- For the identity matrix of size N we have

$$p_{\mathbf{1}}(z) = (z-1)^N = \sum_{k=0}^{N} \binom{N}{k} (-1)^k z^{N-k} = \sum_{k=0}^{N} \frac{(-\mathrm{d}_z)^k}{k!} z^N = \exp(-\mathrm{d}_z) z^N. \tag{12.70}$$

  So we find

$$\hat{p}_{\mathbf{1}}(u) = \left[\exp(-u)\right] \mod u^{N+1} \quad \Rightarrow \quad L_{\mathbf{1}}(u) = u. \tag{12.71}$$

- For Wigner matrices, we saw in Chapter 6 that the expected characteristic polynomial of a unit Wigner matrix is given by a Hermite polynomial normalized as

$$p_{\mathbf{X}}(z) = N^{-N/2} H_N(\sqrt{N} z) = \exp\left[-\frac{1}{2N}\left(\frac{\mathrm{d}}{\mathrm{d}z}\right)^2\right] z^N, \tag{12.72}$$

  where the right hand side comes from Eq. (6.8). We then have

$$\hat{p}_{\mathbf{X}}(u) = \left[\exp\left(-\frac{u^2}{2N}\right)\right] \mod u^{N+1} \quad \Rightarrow \quad L_{\mathbf{X}}(u) = \frac{u^2}{2N}. \tag{12.73}$$

- For Wishart matrices, Eq. (6.41) expresses the expected characteristic polynomial as a derivative operator acting on the monomial z^N. For a correctly normalized Wishart matrix, we then find the monic Laguerre polynomial:

$$p_{\mathbf{W}}(z) = \left(1 - \frac{1}{T}\frac{\mathrm{d}}{\mathrm{d}z}\right)^T z^N, \tag{12.74}$$

  from which we can immediately read off the polynomial $\hat{p}(z)$:

$$\begin{aligned} \hat{p}_{\mathbf{W}}(u) &= \left[\left(1 - \frac{qu}{N}\right)^{N/q}\right] \mod u^{N+1} \\ \Rightarrow L_{\mathbf{W}}(u) &= -\frac{N}{q}\left[\log\left(1 - \frac{qu}{N}\right)\right] \mod u^{N+1}. \end{aligned} \tag{12.75}$$

We notice that in these three cases, the L function is related to the corresponding limiting R-transform of the infinite size matrices as

$$L'(u) = \left[R(u/N)\right] \mod u^{N+1}. \tag{12.76}$$

The equality at finite N holds for these simple cases but not in the general case. But the equality holds in general in the limiting case:

$$\lim_{N\to\infty} L'(Nx) = R(x). \tag{12.77}$$

12.4.4 Finite Free Product

The free product can also be generalized to an operation on a real-rooted monic polynomial of the same degree. We define

$$p_1 \boxtimes p_2(z) = \left\langle \det[z\mathbf{1} - \Lambda_1 \mathbf{O} \Lambda_2 \mathbf{O}^T] \right\rangle_{\mathbf{O}}, \tag{12.78}$$

where as usual the diagonal matrices $\Lambda_{1,2}$ contain the roots of $p_{1,2}(z)$. As in the additive case, averaging the matrix $\mathbf{O}$ over the permutation, orthogonal or unitary group gives the same result.

We will show in Section 12.4.5 that the result of the finite free product has a simple expression in terms of the coefficient a_k defined by (12.53). When $p_\mathrm{m}(z) = p_1 \boxtimes p_2(z)$ we have

$$a_k^{(\mathrm{m})} = \left[\binom{N}{k}\right]^{-1} a_k^{(1)} a_k^{(2)}. \tag{12.79}$$

Note that if $\Lambda_1 = \alpha \mathbf{1}$ is a multiple of the identity with $\alpha \neq 0$ we have

$$p_{\alpha \mathbf{1}} = (z - \alpha)^N \quad \Rightarrow \quad a_k^{(\alpha \mathbf{1})} = \binom{N}{k} \alpha^k. \tag{12.80}$$

Plugging this into Eq. (12.79), we see that the free product with a multiple of the identity multiplies each a_k by α^k which is equivalent to multiplying each root by α. In particular the identity matrix ($\alpha = 1$) is the neutral element for that convolution.

When $p_\mathrm{m}(z) = p_1 \boxtimes p_2(z)$, the sample mean of the roots of $p_\mathrm{m}(z)$ (Eq. (12.56)) behaves as

$$\mu_{(\mathrm{m})} = \mu_{(1)} \mu_{(2)}.$$

When both means are unity, we have for the sample variance

$$\sigma^2_{(\mathrm{m})} = \sigma^2_{(1)} + \sigma^2_{(2)} - \frac{\sigma^2_{(1)} \sigma^2_{(2)}}{N}. \tag{12.81}$$

At large N the last term becomes negligible and we recover the additivity of the variance for the product of unit-trace free variables (see Eq. (11.85)).

Exercise 12.4.1 Free product of polynomials with roots 0 and 1

Consider an even degree $N = 2M$ polynomial with roots 0 and 1 both with multiplicity M: $p(z) = z^M (z-1)^M$. We will study the finite free product of this polynomial with itself $p_\mathrm{m}(z) = p \boxtimes p(z)$. We will study the large N limit of this problem in Section 15.4.2.

(a) Expand the polynomial and write the coefficients a_k in terms of binomial coefficients.

(b) Using Eq. (12.79) show that the coefficients of $p_\mathrm{m}(z)$ are given by

$$a_k^{(\mathrm{m})} = \begin{cases} \binom{M}{k}^2 \left[\binom{N}{k}\right]^{-1} & \text{when } k \leq M, \\ 0 & \text{otherwise.} \end{cases} \tag{12.82}$$

(c) The polynomial $p_\mathrm{m}(z)$ always has zero as a root. What is its multiplicity?

(d) The degree M polynomial $q(z) = z^{-M} p_{\rm m}(z)$ only has non-zero roots. What is its average root?

(e) For $M = 2$, show that the two roots of $q(z)$ are $1/2 \pm 1/\sqrt{12}$.

(f) The case $M = 4$ can be solved by noticing that $q(z)$ is a symmetric function of $(z - 1/2)$. Show that the four roots are given by

$$\lambda_{\pm\pm} = \frac{1}{2} \pm \frac{1}{2}\sqrt{\frac{15 \pm 2\sqrt{30}}{35}}. \tag{12.83}$$

12.4.5 Finite Free Convolutions: Derivation of the Results

We will first study the definition (12.61) when the matrix $\mathbf{O}$ is averaged over the permutation group S_N. In this case we can write

$$p_{\rm s}(z) := p_1 \boxplus p_2(z) = \frac{1}{N!} \sum_{\substack{\text{permutations}\\ \sigma}} \prod_{i=1}^{N} \left(z - \lambda_i^{(1)} - \lambda_{\sigma(i)}^{(2)}\right). \tag{12.84}$$

The coefficients $a_k^{(\rm s)}$ are given by the average over the permutation group of the coefficients of the polynomial with roots $\{\lambda_i^{(1)} + \lambda_{\sigma(i)}^{(2)}\}$. For example for $a_1^{(\rm s)}$ we have

$$\begin{aligned} a_1^{(\rm s)} &= \frac{1}{N!} \sum_{\substack{\text{permutations}\\ \sigma}} \sum_{i}^{N} \left(\lambda_i^{(1)} + \lambda_{\sigma(i)}^{(2)}\right) \\ &= \sum_{i}^{N} \left(\lambda_i^{(1)} + \lambda_i^{(2)}\right) = a_1^{(1)} + a_1^{(2)}. \end{aligned} \tag{12.85}$$

For the other coefficients a_k, the combinatorics is a bit more involved. Let us first figure out the structure of the result. For each permutation we can expand the product

$$\left(z - \lambda_1^{(1)} - \lambda_{\sigma(1)}^{(2)}\right)\left(z - \lambda_2^{(1)} - \lambda_{\sigma(2)}^{(2)}\right) \cdots \left(z - \lambda_N^{(1)} - \lambda_{\sigma(N)}^{(2)}\right). \tag{12.86}$$

To get a contribution to a_k we need to choose the variable z $(N - k)$ times, we can then choose a $\lambda^{(1)}$ i times and a $\lambda^{(2)}$ $(k - i)$ times. For a given k and for each choice of i, once averaged over all permutations, the product of the $\lambda^{(1)}$ and that of the $\lambda^{(2)}$ must both be completely symmetric and therefore proportional to $a_i^{(1)} a_{k-i}^{(2)}$. We thus have

$$a_k^{(\rm s)} = \sum_{i=0}^{k} C(i,k,N) a_i^{(1)} a_{k-i}^{(2)}, \tag{12.87}$$

where $C(i,k,N)$ are combinatorial coefficients that we still need to determine. There is an easy way to get these coefficients: we can use a case that we can compute directly and match the coefficients. If $\Lambda_2 = \mathbf{1}$, the identity matrix, we have

$$p_1(z) = (z-1)^N \quad \Rightarrow \quad a_k^{(1)} = \binom{N}{k}. \tag{12.88}$$

For a generic polynomial $p(z)$, the free sum with $p_1(z)$ is given by a simple shift in the argument z by 1:

$$p_s(z) = p(z-1) = \sum_{k=0}^{N} (-1)^k a_k^{(1)} (z-1)^{N-k} \quad \Rightarrow \quad a_k^{(s)} = \sum_{i=0}^{k} \binom{N-i}{N-k} a_i^{(1)}. \tag{12.89}$$

Combining Eqs. (12.87) and (12.88) and matching the coefficient to (12.89), we arrive at

$$a_k^{(s)} = \sum_{i=0}^{k} \frac{(N-i)!\,(N-k+i)!}{N!\,(N-k)!} a_i^{(1)} a_{k-i}^{(2)}, \tag{12.90}$$

which is equivalent to Eq. (12.64). For the equivalence with Eq. (12.62) see Exercise 12.4.2.

Now suppose we want to average Eq. (12.61) with respect to the orthogonal or unitary group (O(N) or U(N)). For a given rotation matrix $\mathbf{O}$, we can expand the determinant, keeping track of powers of z and of the various $\lambda^{(1)}$ that appear in products containing at most one of each $\lambda_i^{(1)}$. After averaging over the group, the combinations of $\lambda^{(1)}$ must be permutation invariant, i.e. proportional to $a_i^{(1)}$; we then get for the coefficient $a_k^{(s)}$,

$$a_k^{(s)} = \sum_{i=0}^{k} C\left(i,k,N,\left\{\lambda_j^{(2)}\right\}\right) a_i^{(1)}, \tag{12.91}$$

where the coefficients $C\left(i,k,N,\left\{\lambda_j^{(2)}\right\}\right)$ depend on the roots $\lambda_j^{(2)}$. By dimensional analysis, they must be homogeneous to $(\lambda^{(2)})^{k-i}$. Since the expression must be symmetrical in (1) $\leftrightarrow$ (2), it must be of the form (12.87). And since the free addition with the unit matrix is the same for all three groups, Eq. (12.64) must be true in all three cases (S_N, O(N) and U(N)).

We now turn to the proof of Eq. (12.79) for the finite free product. Consider Eq. (12.78), where the matrix $\mathbf{O}$ is averaged over the permutation group S_N. For $p_m(z) = p_1 \boxtimes p_2(z)$ we have

$$p_m(z) = \frac{1}{N!} \sum_{\substack{\text{permutations}\\ \sigma}} \prod_{i=1}^{N} \left(z - \lambda_i^{(1)} \lambda_{\sigma(i)}^{(2)}\right) := \frac{1}{N!} \sum_{\substack{\text{permutations}\\ \sigma}} p_\sigma(z). \tag{12.92}$$

For a given permutation σ, the coefficients a_k^σ are given by

$$a_k^\sigma = \sum_{\substack{\text{ordered}\\ k\text{-tuples } \mathbf{i}}}^{N} \lambda_{i_1}^{(1)} \lambda_{i_2}^{(1)} \cdots \lambda_{i_k}^{(1)} \lambda_{\sigma(i_1)}^{(2)} \lambda_{\sigma(i_2)}^{(2)} \cdots \lambda_{\sigma(i_k)}^{(2)}. \tag{12.93}$$

After averaging over the permutations σ, we must have that $a_k^{(s)} \propto a_k^{(1)} a_k^{(2)}$. By counting the number of terms in the sum defining each a_k, we realize that the proportionality constant must be one over this number. We then have

$$a_k^{(m)} = \left[\binom{N}{k}\right]^{-1} a_k^{(1)} a_k^{(2)}. \tag{12.94}$$

We could have derived the proportionality constant by requiring that the polynomial $p_1(z) = (z-1)^N$ is the neutral element of this convolution.

Exercise 12.4.2 Equivalence of finite free addition formulas

For a real-rooted monic polynomial $p_1(z)$ of degree N, we define the polynomial $\hat{p}_1(u)$ as

$$\hat{p}_1(u) = \sum_{k=0}^{N} \frac{(N-k)!}{N!} a_k^{(1)} (-1)^k u^k, \tag{12.95}$$

where the coefficients $a_k^{(1)}$ are given by Eq. (12.53).

(a) Show that

$$p_1(z) = \hat{p}_1\left(\frac{\mathrm{d}}{\mathrm{d}z}\right) z^N. \tag{12.96}$$

(b) For another polynomial $p_2(z)$, show that

$$\hat{p}_1\left(\frac{\mathrm{d}}{\mathrm{d}z}\right) p_2(z) = \sum_{k=0}^{N} (-1)^k \sum_{i=0}^{k} \frac{(N-i)!\,(N-k+i)!}{N!\,(N-k)!} a_i^{(1)} a_{k-i}^{(2)} z^{N-k}, \tag{12.97}$$

which shows the equivalence of Eqs. (12.62) and (12.64).

12.5 Freeness for 2 × 2 Matrices

Certain low-dimensional random matrices can be free. For $N = 1$ (1×1 matrices) all matrices commute and thus behave as classical random numbers. As mentioned in Chapter 11, freeness is trivial for commuting variables as only constants (deterministic variables) can be free with respect to non-constant random variables.

For $N = 2$, there exist non-trivial matrices that can be mutually free. To be more precise, we consider the space of 2×2 symmetric random matrices and define the operator[3]

$$\tau(\mathbf{A}) = \frac{1}{2}\mathbb{E}[\operatorname{Tr} \mathbf{A}]. \tag{12.98}$$

We now consider matrices that have deterministic eigenvalues but random, rotationally invariant eigenvectors. We will see that any two such matrices are free. Since 2×2 matrices only have two eigenvalues we can write these matrices as

$$\mathbf{A} = a\mathbf{1} + \sigma\, \mathbf{O} \begin{pmatrix} 1 & 0 \\ 0 & -1 \end{pmatrix} \mathbf{O}^T, \tag{12.99}$$

[3] More formally, we need to consider 2×2 symmetric random matrices with finite moments to all orders. This space is closed under addition and multiplication and forms a ring satisfying all the axioms described in Section 11.1.

where a is the mean of the two eigenvalues and σ their half-difference. The matrix $\mathbf{O}$ is a random rotation matrix which for $N = 2$ only has one degree of freedom and can always be written as

$$\mathbf{O} = \begin{pmatrix} \cos\theta & \sin\theta \\ -\sin\theta & \cos\theta \end{pmatrix}, \tag{12.100}$$

where the angle θ is uniformly distributed in $[0, 2\pi]$. Note again that we are considering matrices for which a and σ are non-random. If we perform the matrix multiplications and use some standard trigonometric identities we find

$$\mathbf{A} = a\mathbf{1} + \sigma \begin{pmatrix} \cos 2\theta & \sin 2\theta \\ \sin 2\theta & -\cos 2\theta \end{pmatrix}. \tag{12.101}$$

12.5.1 Freeness of Matrices with Deterministic Eigenvalues

We can now show that any two such matrices $\mathbf{A}$ and $\mathbf{B}$ are free. If we put ourselves in the basis where $\mathbf{A}$ is diagonal, we see that traceless polynomials $p_k(\mathbf{A})$ and $q_k(\mathbf{B})$ are necessarily of the form

$$p_k(\mathbf{A}) = a_k \begin{pmatrix} 1 & 0 \\ 0 & -1 \end{pmatrix} \quad \text{and} \quad q_k(\mathbf{B}) = b_k \begin{pmatrix} \cos 2\theta & \sin 2\theta \\ \sin 2\theta & -\cos 2\theta \end{pmatrix}, \tag{12.102}$$

for some deterministic numbers a_k and b_k and where θ is the random angle between the eigenvectors of $\mathbf{A}$ and $\mathbf{B}$. We can now compute the expectation value of the trace of products of such polynomials:

$$\tau\left[\prod_{k=1}^{n} p_k(\mathbf{A})q_k(\mathbf{B})\right] = \frac{1}{2}\left(\prod_{k=1}^{n} a_k b_k\right) \mathbb{E}\,\mathrm{Tr}\begin{pmatrix} \cos 2\theta & \sin 2\theta \\ -\sin 2\theta & \cos 2\theta \end{pmatrix}^n. \tag{12.103}$$

We notice that the matrix on the right hand side is a rotation matrix (of angle 2θ) raised to the power n and therefore itself a rotation matrix (of angle $2n\theta$). The average of such a matrix is zero, thus finishing our proof that two rotationally invariant 2×2 matrices with deterministic eigenvalues are free.

As a consequence, we can use the R-transform to compute the average eigenvalue spectrum of $\mathbf{A} + \mathbf{OBO}^T$ when $\mathbf{A}$ and $\mathbf{B}$ have deterministic eigenvalues and $\mathbf{O}$ is a random rotation matrix. In particular if $\mathbf{A}$ and $\mathbf{B}$ have the same variance (σ) then this spectrum is given by the arcsine law that we encountered in Section 7.1.3. For positive definite $\mathbf{A}$ we can also compute the spectrum of $\sqrt{\mathbf{A}}\mathbf{OBO}^T\sqrt{\mathbf{A}}$ using the S-transform (see Exercises 12.5.1 and 12.5.2).

Exercise 12.5.1 Sum of two free 2 × 2 matrices

(a) Consider $\mathbf{A}_1$ a traceless rotationally invariant 2×2 matrix with deterministic eigenvalues $\lambda_\pm = \pm\sigma$. Show that

$$\mathfrak{g}_{\mathbf{A}}(z) = \frac{z}{z^2 - \sigma^2} \quad \text{and} \quad R(g) = \frac{\sqrt{1 + 4\sigma^2 g^2} - 1}{2g}. \tag{12.104}$$

(b) Two such matrices $\mathbf{A}_1$ and $\mathbf{A}_2$ (with eigenvalues $\pm\sigma_1$ and $\pm\sigma_2$ respectively) are free, so we can sum their R-transforms to find the spectrum of their sum. Show that $\mathfrak{g}_{\mathbf{A}_1+\mathbf{A}_2}(z)$ is given by one of the roots of

$$\mathfrak{g}_{\mathbf{A}_1+\mathbf{A}_2}(z) = \frac{\pm z}{\sqrt{(\sigma_1^4 + \sigma_2^4) - 2(\sigma_1^2 + \sigma_2^2)z^2 + z^4}}. \tag{12.105}$$

(c) In the basis where $\mathbf{A}_1$ is diagonal $\mathbf{A}_1 + \mathbf{A}_2$ has the form

$$\mathbf{A} = \begin{pmatrix} \sigma_1 + \sigma_2 \cos 2\theta & \sigma_2 \sin 2\theta \\ \sigma_2 \sin 2\theta & -\sigma_1 - \sigma_2 \cos 2\theta \end{pmatrix}, \tag{12.106}$$

for a random angle θ uniformly distributed between $[0, 2\pi]$. Show that the eigenvalues of $\mathbf{A}_1 + \mathbf{A}_2$ are given by

$$\lambda_\pm = \pm\sqrt{\sigma_1^2 + \sigma_2^2 + 2\sigma_1\sigma_2 \cos 2\theta}. \tag{12.107}$$

(d) Show that the densities implied by (b) and (c) are the same. For $\sigma_1 = \sigma_2$, it is called the arcsine law (see Section 7.1.3).

Exercise 12.5.2 Product of two free 2 × 2 matrices

A rotationally invariant 2×2 matrix with deterministic eigenvalues 0 and $a_1 \geq 0$ has the form

$$A_1 = \mathbf{O} \begin{pmatrix} 0 & 0 \\ 0 & a_1 \end{pmatrix} \mathbf{O}^T. \tag{12.108}$$

Two such matrices are free so we can use the S-transform to compute the eigenvalue distribution of their product.

(a) Show that the T-transform and the S-transform of $\mathbf{A}_1$ are given by

$$\mathfrak{t}(\zeta) = \frac{a_1}{2(\zeta - a_1)} \quad \text{and} \quad S(t) = \frac{2}{a_1} \frac{t+1}{2t+1}. \tag{12.109}$$

(b) Consider another such matrix $\mathbf{A}_2$ with independent eigenvectors and non-zero eigenvalue a_2. Using the multiplicativity of the S-transform, show that the T-transform and the density of eigenvalues of $\sqrt{\mathbf{A}_1}\mathbf{A}_2\sqrt{\mathbf{A}_1}$ are given by

$$\mathfrak{t}_{\mathbf{A}_1\mathbf{A}_2} = \frac{1}{2} - \frac{1}{2}\sqrt{\frac{\zeta}{\zeta - a_1 a_2}} \tag{12.110}$$

and

$$\rho_{\mathbf{A}_1\mathbf{A}_2}(\lambda) = \frac{1}{2}\delta(\lambda) + \frac{1}{2\pi\sqrt{\lambda(a_1 a_2 - \lambda)}} \quad \text{for } 0 \leq \lambda \leq a_1 a_2, \tag{12.111}$$

where the delta function in the density indicates the fact that one eigenvalue is always zero.

(c) By directly computing the matrix product in the basis where $\mathbf{A}_1$ is diagonal, show that, in that basis,

$$\sqrt{\mathbf{A}_1}\mathbf{A}_2\sqrt{\mathbf{A}_1} = a_1 a_2 \begin{pmatrix} 0 & 0 \\ 0 & 1 - \cos 2\theta \end{pmatrix}, \tag{12.112}$$

where θ is a random angle uniformly distributed between 0 and 2π.

(d) Show that the distribution of the non-zero eigenvalue implied by (b) and (c) is the same. It is the shifted arcsine law.

12.5.2 Pairwise Freeness and Free Collections

For the 2×2 matrices $\mathbf{A}$ and $\mathbf{B}$ to be free, they need to have deterministic eigenvalues. The proof above does not work if the eigenvalues of one of these matrices are random. As an illustration, consider three matrices $\mathbf{A}$, $\mathbf{B}$ and $\mathbf{C}$ as above (2×2 symmetric rotationally invariant matrices with deterministic eigenvalues). Each pair of these matrices is free. On the other hand, the free sum $\mathbf{A} + \mathbf{B}$ has random eigenvalues and it is not necessarily free from $\mathbf{C}$. Actually we can show that it is *not* free with respect to $\mathbf{C}$.

First we show that $\mathbf{A}$, $\mathbf{B}$ and $\mathbf{C}$ do not form a free collection. For simplicity, we consider them traceless with $\sigma_\mathbf{A} = \sigma_\mathbf{B} = \sigma_\mathbf{C} = 1$. Then one can show that

$$\tau(\mathbf{ABCABC}) = \tau((\mathbf{ABC})^2) = 1, \tag{12.113}$$

violating the freeness condition for three variables. Indeed, let us compute explicitly $\mathbf{ABC}$ in the basis where $\mathbf{A}$ is diagonal. We find

$$\mathbf{ABC} = \begin{pmatrix} \cos 2\theta \cos 2\phi + \sin 2\theta \sin 2\phi & \cos 2\theta \sin 2\phi - \sin 2\theta \cos 2\phi \\ \cos 2\theta \sin 2\phi - \sin 2\theta \cos 2\phi & -\cos 2\theta \cos 2\phi - \sin 2\theta \sin 2\phi \end{pmatrix}, \tag{12.114}$$

where ϕ is the angle between the eigenvectors of $\mathbf{A}$ and those of $\mathbf{C}$. The matrix $\mathbf{ABC}$ is a non-zero symmetric matrix, so the trace of its square must be non-zero. Actually one finds $(\mathbf{ABC})^2 = \mathbf{1}$.

Another way to see that $\mathbf{A}$, $\mathbf{B}$ and $\mathbf{C}$ are not free as a group is to compute the mixed cumulant $\kappa_6(\mathbf{A},\mathbf{B},\mathbf{C},\mathbf{A},\mathbf{B},\mathbf{C})$, given that odd cumulants such as $\kappa_1(\mathbf{A})$ and $\kappa_3(\mathbf{A},\mathbf{B},\mathbf{C})$ are zero and that mixed cumulants involving two matrices are zero (they are pairwise free). The only non-zero term in the moment–cumulant relation for $\tau(\mathbf{ABCABC})$ (see Eq. (11.74)) is

$$\tau(\mathbf{ABCABC}) = \kappa_6(\mathbf{A},\mathbf{B},\mathbf{C},\mathbf{A},\mathbf{B},\mathbf{C}) = 1. \tag{12.115}$$

The matrices **A**, **B** and **C** have therefore at least one non-zero cross-cumulant and cannot be free as a collection.

Now, to show that $\mathbf{A}+\mathbf{B}$ is not free from **C**, we realize that if we expand the sixth cross-cumulant $\kappa_6\,(\mathbf{A}+\mathbf{B},\mathbf{A}+\mathbf{B},\mathbf{C},\mathbf{A}+\mathbf{B},\mathbf{A}+\mathbf{B},\mathbf{C})$, we will encounter the above non-zero cross-cumulant of **A**, **B** and **C**. Indeed, a cross-cumulant is linear in each of its arguments. Since all other terms in this expansion are zero, we find

$$\begin{aligned}&\kappa_6(\mathbf{A}+\mathbf{B},\mathbf{A}+\mathbf{B},\mathbf{C},\mathbf{A}+\mathbf{B},\mathbf{A}+\mathbf{B},\mathbf{C})\\&\quad=\kappa_6\,(\mathbf{A},\mathbf{B},\mathbf{C},\mathbf{A},\mathbf{B},\mathbf{C})+\kappa_6\,(\mathbf{B},\mathbf{A},\mathbf{C},\mathbf{A},\mathbf{B},\mathbf{C})\\&\quad\quad+\kappa_6\,(\mathbf{A},\mathbf{B},\mathbf{C},\mathbf{B},\mathbf{A},\mathbf{C})+\kappa_6\,(\mathbf{B},\mathbf{A},\mathbf{C},\mathbf{B},\mathbf{A},\mathbf{C})=4\neq 0.\end{aligned}\tag{12.116}$$

As a consequence, even though 2×2 symmetric rotationally invariant random matrices with deterministic eigenvalues are pairwise free, they do not satisfy the free CLT. If they did, it would imply that 2×2 matrices with a semi-circle spectrum would be stable under addition, which is not the case. Note that Gaussian 2×2 Wigner matrices (which *are* stable under addition) do not have a semi-circle spectrum (see Exercise 12.5.3).

Exercise 12.5.3 Eigenvalue spectrum of 2 × 2 Gaussian matrices

Real symmetric and complex Hermitian Gaussian 2×2 matrices are stable under addition but they are not free. In this exercise we see that their spectrum is not given by a semi-circle law.

(a) Use Eq. (5.22) and the Gaussian potential $V(x)=x^2/2$ to write the joint probability (up to a normalization) of λ_1 and λ_2, the two eigenvalues of a real symmetric Gaussian 2×2 matrix.

(b) To find the eigenvalue density, we need to compute

$$\rho(\lambda_1)=\int_{-\infty}^{\infty}\mathrm{d}\lambda_2 P(\lambda_1,\lambda_2).\tag{12.117}$$

This integral will involve an error function because of the absolute value in $P(\lambda_1,\lambda_2)$. If you have the courage compute $\rho(\lambda)$ leaving the normalization undetermined.

(c) It is easier to do the complex Hermitian case. Use Eq. (5.26) and the same potential to adapt your answer in (a) to the $\beta=2$ case.

(d) The absolute value has now disappeared and the integral in (b) is now much easier. Perform this integral and find the normalization constant. You should obtain

$$\rho(\lambda)=\frac{\lambda^2+\frac{1}{2}}{\sqrt{\pi}}\mathrm{e}^{-\lambda^2}.\tag{12.118}$$

Bibliographical Notes

- On integration over the orthogonal and unitary groups, see
  - B. Collins and P. Śniady. Integration with respect to the Haar measure on unitary, orthogonal and symplectic group. *Communications in Mathematical Physics*, 264(3):773–795, 2006,
  - B. Collins, A. Guionnet, and E. Maurel-Segala. Asymptotics of unitary and orthogonal matrix integrals. *Advances in Mathematics*, 222(1):172–215, 2009,
  - M. Bergère and B. Eynard. Some properties of angular integrals. *Journal of Physics A: Mathematical and Theoretical*, 42(26):265201, 2009.
- On the Weingarten coefficients, see
  - D. Weingarten. Asymptotic behavior of group integrals in the limit of infinite rank. *Journal of Mathematical Physics*, 19(5):999–1001, 1978,
  - P. W. Brouwer and C. W. J. Beenakker. Diagrammatic method of integration over the unitary group, with applications to quantum transport in mesoscopic systems. *Journal of Mathematical Physics*, 37(10):4904–4934, 1996,
  - B. Collins and S. Matsumoto. On some properties of orthogonal Weingarten functions. *Journal of Mathematical Physics*, 50(11):113516, 2009,
  - T. Banica. The orthogonal Weingarten formula in compact form. *Letters in Mathematical Physics*, 91(2):105–118, 2010.
- On the Edgeworth series, see e.g. https://en.wikipedia.org/wiki/Edgeworth_series and
  - J.-P. Bouchaud and M. Potters. *Theory of Financial Risk and Derivative Pricing: From Statistical Physics to Risk Management*. Cambridge University Press, Cambridge, 2nd edition, 2003.
- On the Edgeworth series for the free CLT, see
  - G. P. Chistyakov and F. Götze. Asymptotic expansions in the CLT in free probability. *Probability Theory and Related Fields*, 157:107–156, 2011.
- On finite N free matrices, see
  - A. Marcus. Polynomial convolutions and (finite) free probability, 2018: preprint available at https://web.math.princeton.edu/~amarcus/papers/ff_main.pdf,
  - A. Marcus, D. A. Spielman, and N. Srivastava. Finite free convolutions of polynomials. *preprint arXiv:1504.00350*, 2015.

13
The Replica Method*

In this chapter we will review another important tool to perform compact computations in random systems and in particular in random matrix theory, namely the "replica method". For example, one can use replicas to understand the R-transform addition rule when one adds two large, randomly rotated matrices.

Suppose that we want to compute the free energy of a large random system. The free energy is the logarithm of some partition function Z.[1] We expect that the free energy does not depend on the particular sample so we can average the free energy with respect to the randomness in the system to get the typical free energy of a given sample. Unfortunately, averaging the logarithm of a partition function is hard. What we can do more easily is to compute the partition function to some power n and later let $n \to 0$ using the "replica trick":

$$\log Z = \lim_{n \to 0} \frac{Z^n - 1}{n}. \tag{13.1}$$

The partition function Z^n is just the partition function of n non-interacting copies of the same system Z, these copies are called "replicas", hence the name of the technique. Averaging the logarithm is then equivalent to averaging Z^n and taking the limit $n \to 0$ as above. The averaging procedure will however couple the n copies and the resulting interacting system is in general hard to solve. In many interesting cases, the partition function can only be computed as the size of the system (say N) goes to infinity. Naturally one is tempted to interchange the limits ($n \to 0$ and $N \to \infty$) but there is no mathematical justification (yet) for doing so. Another problem is that we can hope to compute $\mathbb{E}[Z^n]$ for all integers n but is that really sufficient to do a proper $n \to 0$ limit?

For all these reasons, replica computations are not considered rigorous. Nevertheless, they are a precious source of intuition and they allow one to obtain results mathematicians would call conjectures, but that often turn out to be mathematically exact. Although a lot of progress has been made to understand why the replica trick works, there is still a halo of mystery and magic surrounding the method, and a nagging impression that an equivalent but more transparent formalism awaits revelation.

[1] See e.g. Section 13.4 for an explicit example.

In the present chapter, we will show how all the results obtained up to now can be rederived using replicas. We start by showing how the Stieltjes transform can be expressed using the replica method, and obtain once more the semi-circle law for Gaussian random matrices. We then discuss R-transforms and S-transforms in the language of replicas.

13.1 Stieltjes Transform

As we have shown in Chapter 2, the density of eigenvalues $\rho(\lambda)$ of a random matrix is encoded in the trace of the resolvent of that matrix, which defines the Stieltjes transform of $\rho(\lambda)$. Here we show how this quantity can be computed using the replica formalism.

13.1.1 General Set-Up

To use the replica trick in random matrix theory, we first need to express the Stieltjes transform as the average logarithm of a (possibly random) determinant. In the large N limit and for z sufficiently far from the real eigenvalues, the discrete Stieltjes transform $g_N(z)$ converges to a deterministic function $\mathfrak{g}(z)$. The replica trick will allow us to compute $\mathbb{E}[g_N(z)]$, which also converges to $\mathfrak{g}(z)$.

Using the definition Eq. (2.19) and dropping the N subscript, we have

$$\mathbb{E}[g_{\mathbf{A}}(z)] = \frac{1}{N}\mathbb{E}\left[\sum_{k=1}^{N} \frac{1}{z-\lambda_k}\right], \tag{13.2}$$

whereas the determinant of $z\mathbf{1} - \mathbf{A}$ is given by

$$\det(z\mathbf{1} - \mathbf{A}) = \prod_{k=1}^{N}(z-\lambda_k). \tag{13.3}$$

We can turn the product in the determinant into a sum by taking the logarithm and obtain $(z-\lambda_k)^{-1}$ from $\log(z-\lambda_k)$ by taking a derivative with respect to z. We then get

$$\mathbb{E}[g_{\mathbf{A}}(z)] = \frac{1}{N}\mathbb{E}\left[\frac{\mathrm{d}}{\mathrm{d}z}\log\det(z\mathbf{1} - \mathbf{A})\right]. \tag{13.4}$$

To compute the determinant we may use the multivariate Gaussian identity

$$\int \frac{\mathrm{d}^N \boldsymbol{\psi}}{(2\pi)^{N/2}} \exp\left(-\frac{\boldsymbol{\psi}^T \mathbf{M} \boldsymbol{\psi}}{2}\right) = \frac{1}{\sqrt{\det \mathbf{M}}}, \tag{13.5}$$

which is exact for any N as long as the matrix $\mathbf{M}$ is positive definite. For z larger than the top eigenvalue of $\mathbf{A}$, $(z\mathbf{1} - \mathbf{A})$ will be positive definite. The Gaussian formula allows us to compute the inverse of the square-root of the determinant, but we can neutralize the power $-1/2$ by introducing an extra factor -2 in front of the logarithm. Applying the replica trick (13.1) we thus get

$$\mathbb{E}[g_{\mathbf{A}}(z)] = -2\mathbb{E}\left[\frac{\mathrm{d}}{\mathrm{d}z}\lim_{n\to 0}\frac{Z^n - 1}{Nn}\right], \tag{13.6}$$

with

$$Z^n := \int \prod_{\alpha=1}^{n} \frac{\mathrm{d}^N \boldsymbol{\psi}_\alpha}{(2\pi)^{N/2}} \exp\left(-\sum_{\alpha=1}^{n} \frac{\boldsymbol{\psi}_\alpha^T (z\mathbf{1}-\mathbf{A})\boldsymbol{\psi}_\alpha}{2}\right), \tag{13.7}$$

where we have written Z^n as the product of n copies of the same Gaussian integral. This is all fine, except our Z^n is only defined for integer n and we need to take $n \to 0$. The limiting Stieltjes transform is defined as $\mathfrak{g}_{\mathbf{A}}(z) = \lim_{N\to\infty} \mathbb{E}[g_{\mathbf{A}}(z)]$, but in practice the replica trick will allow us to compute

$$\widehat{\mathfrak{g}}_{\mathbf{A}}(z) = -2\frac{\mathrm{d}}{\mathrm{d}z}\lim_{n\to 0}\lim_{N\to\infty}\frac{\mathbb{E}[Z^n] - 1}{Nn}, \tag{13.8}$$

and hope that the two limits $n \to 0$ and $N \to \infty$ commute, such that $\widehat{\mathfrak{g}}_{\mathbf{A}}(z) = \mathfrak{g}_{\mathbf{A}}(z)$. There is, however, no guarantee that these limits do commute.

13.1.2 The Wigner Case

As a detailed example of replica trick calculation, we now give all the steps necessary to compute the Stieltjes transform for the Wigner ensemble. We want to take the expectation value of Eq. (13.7) in the case where $\mathbf{A} = \mathbf{X}$, a symmetric Gaussian rotational invariant matrix:

$$\begin{aligned}
\mathbb{E}[Z^n] &= \int \prod_{\alpha=1}^{n} \frac{\mathrm{d}^N \boldsymbol{\psi}_\alpha}{(2\pi)^{N/2}} \mathbb{E}\left[\exp\left(-\frac{z}{2}\sum_{\alpha=1}^{n}\sum_{i=1}^{N}\psi_{\alpha i}^2 - \sum_{i<j}^{N}\mathbf{X}_{ij}\psi_{\alpha i}\psi_{\alpha j} - \frac{1}{2}\sum_{i}^{N}\mathbf{X}_{ii}\psi_{\alpha i}\psi_{\alpha i}\right)\right],\\
&= \int \prod_{\alpha=1}^{n} \frac{\mathrm{d}^N \boldsymbol{\psi}_\alpha}{(2\pi)^{N/2}} \exp\left(-\frac{z}{2}\sum_{\alpha=1}^{n}\sum_{i=1}^{N}\psi_{\alpha i}^2\right)\prod_{i<j}^{N}\mathbb{E}\left[\exp\left(-\mathbf{X}_{ij}\sum_{\alpha=1}^{n}\psi_{\alpha i}\psi_{\alpha j}\right)\right]\\
&\quad\times\prod_{i}^{N}\mathbb{E}\left[\exp\left(-\frac{1}{2}\mathbf{X}_{ii}\sum_{\alpha=1}^{n}\psi_{\alpha i}\psi_{\alpha i}\right)\right],
\end{aligned} \tag{13.9}$$

where we have isolated the products of expectation of independent terms and separated the diagonal and off-diagonal terms. We can evaluate the expectation values using the following identity: for a centered Gaussian variable x of variance σ^2, we have

$$\mathbb{E}[\mathrm{e}^{ax}] = \mathrm{e}^{\sigma^2 a^2/2}. \tag{13.10}$$

Using the fact that the diagonal and off-diagonal elements have a variance equal to, respectively, $2\sigma^2/N$ and σ^2/N (see Section 2.2.2), we get

$$\mathbb{E}[Z^n] = \int \prod_{\alpha=1}^{n} \frac{\mathrm{d}^N \boldsymbol{\psi}_\alpha}{(2\pi)^{N/2}} \exp\left(-\frac{z}{2}\sum_{\alpha=1}^{n}\sum_{i=1}^{N}\psi_{\alpha i}^2\right) \prod_{i<j}^{N} \exp\left(\frac{\sigma^2}{2N}\left(\sum_{\alpha=1}^{n}\psi_{\alpha i}\psi_{\alpha j}\right)^2\right)$$
$$\times \prod_{i}^{N} \exp\left(\frac{\sigma^2}{4N}\left(\sum_{\alpha=1}^{n}\psi_{\alpha i}\psi_{\alpha i}\right)^2\right). \tag{13.11}$$

We can now combine the last two sums in the exponential into a single sum over $\{ij\}$, which we can further transform into

$$\frac{\sigma^2}{4N}\sum_{i,j=1}^{N}\left(\sum_{\alpha=1}^{n}\psi_{\alpha i}\psi_{\alpha j}\right)^2 = \frac{\sigma^2}{4N}\sum_{\alpha,\beta=1}^{n}\left(\sum_{i=1}^{N}\psi_{\alpha i}\psi_{\beta i}\right)^2. \tag{13.12}$$

We would like to integrate over the variables $\psi_{\alpha i}$ but the argument of the exponential contains fourth order terms in the ψ's. To tame this term, one uses the Hubbard–Stratonovich identity:

$$\exp\left(\frac{ax^2}{2}\right) = \int \frac{\mathrm{d}q}{\sqrt{2\pi a}} \exp\left(-\frac{q^2}{2a} + xq\right). \tag{13.13}$$

Before we use Hubbard–Stratonovich, we need to regroup diagonal and off-diagonal terms in $\alpha\beta$:

$$\frac{\sigma^2}{4N}\sum_{\alpha,\beta=1}^{n}\left(\sum_{i=1}^{N}\psi_{\alpha i}\psi_{\beta i}\right)^2 = \frac{\sigma^2}{2N}\sum_{\alpha<\beta}^{n}\left(\sum_{i=1}^{N}\psi_{\alpha i}\psi_{\beta i}\right)^2 + \frac{\sigma^2}{N}\sum_{\alpha=1}^{n}\left(\sum_{i=1}^{N}\frac{\psi_{\alpha i}\psi_{\alpha i}}{2}\right)^2, \tag{13.14}$$

where in the diagonal terms we have pushed the factor $1/4$ in the squared quantity for later convenience. We can now use Eq. (13.13), introducing diagonal $q_{\alpha\alpha}$ and upper triangular $q_{\alpha\beta}$ to linearize the squared quantities. Writing the q's as a symmetric matrix $\mathbf{q}$ we have

$$\mathbb{E}[Z^n] \propto \int \mathrm{d}\mathbf{q} \int \prod_{\alpha=1}^{n} \frac{\mathrm{d}^N \boldsymbol{\psi}_\alpha}{(2\pi)^{N/2}} \exp -\left(\frac{N\,\mathrm{Tr}\,\mathbf{q}^2}{4\sigma^2} + \sum_{i=1}^{N}\sum_{\alpha,\beta=1}^{n}\frac{(z\delta_{\alpha\beta} - q_{\alpha\beta})\psi_{\alpha i}\psi_{\beta i}}{2}\right), \tag{13.15}$$

where $\mathrm{d}\mathbf{q}$ is the integration over the independent component of the $n \times n$ symmetric matrix $\mathbf{q}$; note that we have dropped z-independent constant factors. The integral of $\psi_{\alpha i}$ is now a multivariate Gaussian integral, actually N copies of the very same n-dimensional Gaussian integral:

$$\int \prod_{\alpha=1}^{n} \frac{\mathrm{d}\psi_\alpha}{\sqrt{2\pi}} \exp -\left(\sum_{\alpha,\beta=1}^{n}\frac{(z\delta_{\alpha\beta} - q_{\alpha\beta})\psi_\alpha\psi_\beta}{2}\right) = (\det(z\mathbf{1} - \mathbf{q}))^{-1/2}. \tag{13.16}$$

Raising this integral to the Nth power and using $\det \mathbf{M} = \exp \operatorname{Tr} \log \mathbf{M}$ we find

$$\mathbb{E}[Z^n] \propto \int \mathrm{d}\mathbf{q} \exp - \left[N \operatorname{Tr} \left(\frac{\mathbf{q}^2}{4\sigma^2} + \frac{1}{2} \log(z\mathbf{1} - \mathbf{q}) \right) \right] := \int \mathrm{d}\mathbf{q} \exp \left(-\frac{N}{2} F(\mathbf{q}) \right). \tag{13.17}$$

We now fix n and evaluate $\mathbb{E}[Z^n]$ for very large N, leaving the limit $n \to 0$ for later. In the large N limit, the integral over the matrix $\mathbf{q}$ can be done by the saddle point method. More precisely, we should find an extremum of $F(\mathbf{q})$ in the $n(n+1)/2$ elements of $\mathbf{q}$. Alternatively we can diagonalize $\mathbf{q}$, introducing the log of a Vandermonde determinant in the exponential (see Section 5.1.4).[2] In terms of the eigenvalues q_α of $\mathbf{q}$, one has

$$F(\{q_\alpha\}) = \sum_{\alpha=1}^{n} \frac{q_\alpha^2}{2\sigma^2} + \log(z - q_\alpha) - \frac{1}{N} \sum_{\alpha \neq \beta} \log |q_\alpha - q_\beta|. \tag{13.18}$$

To find the saddle point, we take the partial derivatives of $F\{q_\alpha\}$ with respect to the $\{q_\alpha\}$ and equate them to zero:

$$\frac{q_\alpha}{\sigma^2} - \frac{1}{z - q_\alpha} - \frac{1}{N} \sum_{\alpha \neq \beta} \frac{2}{q_\alpha - q_\beta} = 0. \tag{13.19}$$

The effect of the last term is to push the eigenvalues q_α away from one another, such that in equilibrium their relative distance is of order $1/N$ and the last term is of the same order as the first two. Since there are only n such eigenvalues, the total spread (from the largest to smallest) is of order n/N, which we will neglect when $N \to \infty$. Hence we can assume that all eigenvalues are identical and equal to $q^*(z)$, where $q^*(z)$ satisfies

$$z - q^* = \frac{\sigma^2}{q^*}. \tag{13.20}$$

We recognize the self-consistent equation for the Stieltjes transform of the Wigner (Eq. (2.35)) if we make the identification $q^*(z) = \sigma^2 \mathfrak{g}_{\mathbf{X}}(z)$. For N large and n small we indeed have

$$\mathbb{E}[Z^n] = \exp\left(-\frac{Nn}{2} F_1(z, q^*(z)) \right) \text{ with } F_1(z,q) = \frac{q^2}{2\sigma^2} + \log(z - q), \tag{13.21}$$

so

$$\lim_{n \to 0} \lim_{N \to \infty} \frac{\mathbb{E}[Z^n] - 1}{Nn} = -\frac{F_1(z, q^*(z))}{2}. \tag{13.22}$$

Finally, from Eq. (13.8), we should have

$$\mathfrak{g}_{\mathbf{X}}(z) = \frac{\mathrm{d}}{\mathrm{d}z} F_1(z, q^*(z)). \tag{13.23}$$

[2] A third method is used in spin-glass problems where the integrand has permutation symmetry but not necessarily rotational symmetry, see Section 13.4.

To finish the computation we need to take the derivative of $F_1(z,q^*(z))$ with respect to z, but since $q^*(z)$ is an extremum of F_1, the partial derivative of $F_1(z,q)$ with respect to q is zero at $q = q^*(z)$. Hence

$$\mathfrak{g}_{\mathbf{X}}(z) = \frac{\partial}{\partial z} F_1(z,q)\bigg|_{q=q^*(z)} = \frac{1}{(z-q^*)} = \frac{q^*(z)}{\sigma^2}. \tag{13.24}$$

We thus recover the usual solution of the self-consistent Wigner equation.

13.2 Resolvent Matrix

13.2.1 General Case

We saw that the replica trick can be used to compute the average Stieltjes transform of a random matrix. The Stieltjes transform is the normalized trace of the resolvent matrix $\mathbf{G}_{\mathbf{A}}(z) = (z\mathbf{1} - \mathbf{A})^{-1}$. In Chapter 19 we will need to know the average of the elements of the resolvent matrix for free addition and multiplication. These averages can also be computed using the replica trick. An element of an inverse matrix can indeed be written as a multivariate Gaussian integral:

$$\left[\mathbf{M}^{-1}\right]_{ij} \equiv \frac{1}{Z}\int \frac{d^N \boldsymbol{\psi}}{(2\pi)^{N/2}} \psi_i \psi_j \exp\left(-\frac{\boldsymbol{\psi}^T\mathbf{M}\boldsymbol{\psi}}{2}\right), \quad Z := \int \frac{d^N \boldsymbol{\psi}}{(2\pi)^{N/2}} \exp\left(-\frac{\boldsymbol{\psi}^T\mathbf{M}\boldsymbol{\psi}}{2}\right), \tag{13.25}$$

which we can rewrite as

$$\left[\mathbf{M}^{-1}\right]_{ij} = \lim_{m\to -1} Z^m \int d^N \boldsymbol{\psi}\, \psi_i \psi_j \exp\left(-\frac{\boldsymbol{\psi}^T\mathbf{M}\boldsymbol{\psi}}{2}\right). \tag{13.26}$$

If we express Z^m for $m \in \mathbb{N}^+$ as m Gaussian integrals and combine them with the integral with the $\psi_i\psi_j$ term (which we label number 1) we get, with $n = m+1$:

$$\left[\mathbf{M}^{-1}\right]_{ij} = \lim_{n\to 0} \int \prod_{\alpha=1}^{n} \frac{d^N \boldsymbol{\psi}_\alpha}{(2\pi)^{nN/2}} \psi_{1i}\psi_{1j} \exp\left(-\sum_{\alpha=1}^{n} \frac{\boldsymbol{\psi}_\alpha^T\mathbf{M}\boldsymbol{\psi}_\alpha}{2}\right) := \left\langle \psi_{1i}\psi_{1j}\right\rangle_{n=0}. \tag{13.27}$$

This equation can then be used to compute averages of elements of the resolvent matrix by using $\mathbf{M} = z\mathbf{1} - \mathbf{A}$ for the relevant random matrix $\mathbf{A}$. For example, in the case of Gaussian Wigner matrices, the correlation $\left\langle \psi_{1i}\psi_{1j}\right\rangle_{n=0}$ can be computed using the saddle point configuration of the ψ's. Since different i all decouple and play the same role, it is clear that

$$\mathbb{E}[\mathbf{G}_{\mathbf{X}}(z)]_{ij} = \left\langle \psi_{1i}\psi_{1j}\right\rangle_{n=0} = \delta_{ij}\,\mathfrak{g}_{\mathbf{X}}(z), \tag{13.28}$$

i.e. the average resolvent of a Gaussian matrix is simply the average Stieltjes transform times the identity matrix.

13.2.2 Free Addition

In this section we will show how to use Eq. (13.27) to compute the average of the full resolvent for the sum of two randomly rotated matrices. Since we know these matrices are free in the large N limit, we expect to recover the additivity of R-transforms, but we will in fact obtain a slightly richer result. Also, the replica method is very convenient to manipulate and resum the perturbation theory alluded to in Section 12.2.

Consider two symmetric matrices $\mathbf{A}$ and $\mathbf{B}$ and the new matrix $\mathbf{C} = \mathbf{A} + \mathbf{O}\mathbf{B}\mathbf{O}^T$ where $\mathbf{O}$ is a random orthogonal matrix. We want to compute

$$\mathbb{E}[\mathbf{G}_\mathbf{C}(z)] = \mathbb{E}\left[(z\mathbf{1} - \mathbf{A} - \mathbf{O}\mathbf{B}\mathbf{O}^T)^{-1}\right], \tag{13.29}$$

where the expectation value is over the orthogonal matrix $\mathbf{O}$. We can always choose $\mathbf{B}$ to be diagonal. If $\mathbf{B}$ is not diagonal to start with, we just absorb the orthogonal matrix that diagonalizes $\mathbf{B}$ in the matrix $\mathbf{O}$. Expressing $\mathbf{G}_\mathbf{C}(z)$ in the eigenbasis of $\mathbf{A}$ is equivalent to choosing $\mathbf{A}$ to be diagonal. In that basis, the off-diagonal elements of $\mathbb{E}[\mathbf{G}_\mathbf{C}(z)]$ must be zero by the following argument: since both $\mathbf{A}$ and $\mathbf{B}$ are diagonal, for every matrix $\mathbf{O}$ that contributes to an off-diagonal element of $\mathbb{E}[\mathbf{G}_\mathbf{C}(z)]$ there exists an equally probable matrix $\mathbf{O}'$ with the same contribution but opposite sign, hence the average must be zero. Note that while the average matrix $\mathbb{E}[\mathbf{G}_\mathbf{C}(z)]$ commutes with $\mathbf{A}$, a particular realization of the random matrix $\mathbf{G}_\mathbf{C}(z)$ (corresponding to a specific choice for $\mathbf{O}$) will not in general commute with $\mathbf{A}$.

Now, let us use the replica formalism to compute $\mathbb{E}[\mathbf{G}_\mathbf{C}(z)]$, i.e. start with

$$\begin{aligned}\mathbf{G}_\mathbf{C}(z)_{ij} = \lim_{n\to 0} \int \prod_{\alpha=1}^{n} \frac{d^N \boldsymbol{\psi}_\alpha}{(2\pi)^{N/2}} \psi_{1i}\psi_{1j} \exp\left(-\sum_{\alpha=1}^{n} \frac{\boldsymbol{\psi}_\alpha^T (z\mathbf{1} - \mathbf{A})\boldsymbol{\psi}_\alpha}{2}\right) \\ \times \mathbb{E}\left[\exp\left(\sum_{\alpha=1}^{n} \frac{\boldsymbol{\psi}_\alpha^T \mathbf{O}\mathbf{B}\mathbf{O}^T \boldsymbol{\psi}_\alpha}{2}\right)\right],\end{aligned} \tag{13.30}$$

where now we skip the i, j indices on $\boldsymbol{\psi}_\alpha$, treated as vectors.

The last term with the expectation value can be rewritten as

$$I = \mathbb{E}\left[\exp\left(\frac{N}{2}\sum_{\alpha=1}^{n} \operatorname{Tr} \mathbf{Q}_{\alpha,\alpha}\mathbf{O}\mathbf{B}\mathbf{O}^T\right)\right], \tag{13.31}$$

where $\mathbf{Q}_{\alpha,\beta} = \boldsymbol{\psi}_\alpha \boldsymbol{\psi}_\beta^T / N$ is an $n \times n$ symmetric matrix. We recognize the Harish-Chandra–Itzykson–Zuber integral discussed in Chapter 10, with one matrix ($\sum_{\alpha=1}^{n} \mathbf{Q}_{\alpha,\alpha}$) being at most of rank $n \ll N$, so we can use the low-rank formula Eq. (10.45). Our expectation value thus becomes

$$I = \exp\left(\frac{N}{2} \operatorname{Tr}_n H_\mathbf{B}(\mathbf{Q})\right), \tag{13.32}$$

where Tr_n denotes the trace of an $n \times n$ matrix and $H_\mathbf{B}$ is the anti-derivative of the R-transform of $\mathbf{B}$.

We now need to perform the integral of $\boldsymbol{\psi}_\alpha$ in Eq. (13.30). But in order to do so we must deal with $\operatorname{Tr} H_{\mathbf{B}}(\mathbf{Q})$, which is a non-linear function of the $\boldsymbol{\psi}_\alpha$. The trick is to make the matrix $\mathbf{Q}$ an integration variable that we fix to its definition using a delta function. The delta function is itself represented as an integral over another $(n \times n)$ symmetric matrix $\mathbf{Y}$ along the imaginary axis. In other words, we introduce the following representation of the delta function in Eq. (13.30):

$$\int_{-i\infty}^{i\infty} \frac{N^{n(n+1)/2} \mathrm{d}\mathbf{Y}}{2^{3n/2}\pi^{n/2}} \exp\left(-\frac{N}{2} \operatorname{Tr}_n \mathbf{QY} + \frac{1}{2} \sum_{\alpha,\beta=1}^{n} \mathbf{Y}_{\alpha,\beta} \boldsymbol{\psi}_\alpha \boldsymbol{\psi}_\beta^T \right), \tag{13.33}$$

where the integrals over $\mathrm{d}\mathbf{Q}$ and $\mathrm{d}\mathbf{Y}$ are over symmetric matrices. We have absorbed a factor of N in $\mathbf{Y}$ and a factor of 2 on its diagonal, hence the extra factors of 2 and N in front of $\mathrm{d}\mathbf{Y}$. We can now perform the following Gaussian integral over $\boldsymbol{\psi}_\alpha$:

$$J_{ij} = \int \prod_{k=1}^{N} \prod_{\alpha=1}^{n} \frac{\mathrm{d}\psi_{\alpha k}}{\sqrt{2\pi}} \psi_{1i}\psi_{1j} \exp\left(-\frac{1}{2} \sum_{k=1}^{N} \sum_{\alpha,\beta=1}^{n} \psi_{\alpha k}(z\delta_{\alpha,\beta} - a_k \delta_{\alpha,\beta} - \mathbf{Y}_{\alpha\beta})\psi_{\beta k} \right), \tag{13.34}$$

where we have written the vectors $\boldsymbol{\psi}_\alpha$ in terms of their components $\psi_{\alpha k}$, and where a_k are the eigenvalues of $\mathbf{A}$. We notice that the Gaussian integral is diagonal in the index k, so we have $N-1$ n-dimensional Gaussian integrands differing only by their value of a_k, and a last integral for $k = i = j$, where the term ψ_{1i}^2 is in front of the Gaussian integrand (the integral is zero if $i \neq j$, meaning that $\mathbf{G_C}(z)$ is diagonal, as expected). The result is then

$$J_{ij} = \delta_{ij} \left[((z - a_i)\mathbf{1}_n - \mathbf{Y})^{-1} \right]_{11} \prod_{k=1}^{N} (\det((z - a_k)\mathbf{1}_n - \mathbf{Y}))^{-1/2}, \tag{13.35}$$

where the first term is the 11 element of an $n \times n$ matrix, coming from the term ψ_{1i}^2.

Returning to our main expression Eq. (13.30) and dropping constants that are 1 as $n \to 0$,

$$\begin{aligned} \mathbb{E}[\mathbf{G_C}(z)]_{ij} = \lim_{n\to 0} \int_{-i\infty}^{i\infty} \mathrm{d}\mathbf{Q} \int \mathrm{d}\mathbf{Y} \delta_{ij} \left[((z - a_i)\mathbf{1}_n - \mathbf{Y})^{-1} \right]_{11} \\ \times \exp\left[\frac{N}{2} \left(-\operatorname{Tr}_n \mathbf{QY} + \operatorname{Tr}_n H_{\mathbf{B}}(\mathbf{Q}) - \frac{1}{N} \sum_{k=1}^{N} \operatorname{Tr}_n \log((z - a_k)\mathbf{1}_n - \mathbf{Y}) \right) \right]. \end{aligned} \tag{13.36}$$

For large N the integral over $\mathbf{Y}$ and $\mathbf{Q}$ is dominated by the saddle point, i.e. the extremum of the argument of the exponential. The inverse-matrix term in front of the exponential does not contain a power of N so it does not contribute to the determination of the saddle point. The extremum is over a function of two $n \times n$ symmetric matrices. Taking derivatives with respect to $\mathbf{Q}_{\alpha\beta}$ and equating it to zero gives

$$\mathbf{Y}_{\alpha\beta} = [R_{\mathbf{B}}(\mathbf{Q})]_{\alpha\beta}, \tag{13.37}$$

and similarly when taking derivatives with respect to $\mathbf{Y}_{\alpha\beta}$:

$$\mathbf{Q}_{\alpha\beta} = \frac{1}{N}\sum_{k=1}^{N}\left[((z-a_k)\mathbf{1}_n - \mathbf{Y})^{-1}\right]_{\alpha\beta}. \tag{13.38}$$

Let us look at these equations in a basis where $\mathbf{Q}$ is diagonal. The first equation shows that $\mathbf{Y}$ is also diagonal, so that the second equation reads

$$\mathbf{Q}_{\alpha\alpha} = \frac{1}{N}\sum_{k=1}^{N}\frac{1}{z - a_k - \mathbf{Y}_{\alpha\alpha}} \equiv g_{\mathbf{A}}(z - \mathbf{Y}_{\alpha\alpha}). \tag{13.39}$$

Hence all n diagonal elements $\mathbf{Q}_{\alpha\alpha}$ and $\mathbf{Y}_{\alpha\alpha}$, $\alpha = 1,\dots,n$ satisfy the same pair of equations. For large z, there is a unique solution to these equations, hence $\mathbf{Q}$ and $\mathbf{Y}$ must be multiples of the identity $\mathbf{Q} = q^*\mathbf{1}_n$ and $\mathbf{Y} = y^*\mathbf{1}_n$, as expected from the rotational symmetry of the argument of the exponential that we are maximizing. The quantities q^* and y^* are the unique solutions of

$$y = R_{\mathbf{B}}(q) \qquad \text{and} \qquad q = \mathfrak{g}_{\mathbf{A}}(z - y). \tag{13.40}$$

The saddle point for $\mathbf{Y}$ is real while our integral representation Eq. (13.33) was over purely imaginary matrices; but for large values of z, the solutions of Eqs. (13.40) give small values for q^* and y^*, and for such small values the integral contour can be deformed without encountering any singularities. We also justify the use of $R_{\mathbf{B}}(q^*) = H'_{\mathbf{B}}(q^*)$ as q^* can be made arbitrarily small by choosing a large enough z.

The expectation of the resolvent is thus given, for large enough z and N, by

$$\mathbb{E}[\mathbf{G_C}(z)_{ij}] \approx \lim_{n\to 0}\frac{\delta_{ij}}{(z - a_i - y^*)}\exp\left[\frac{nN}{2}\left(-q^*y^* + H_{\mathbf{B}}(q^*) - \frac{1}{N}\sum_{k=1}^{N}\log(z - a_k - y^*)\right)\right]. \tag{13.41}$$

As $n \to 0$ the exponential drops out and we obtain, in matrix form,

$$\mathbb{E}[\mathbf{G_C}(z)] = \mathbf{G_A}(z - R_{\mathbf{B}}(q^*)) \quad \text{with} \quad q^* = \mathfrak{g}_{\mathbf{A}}(z - R_{\mathbf{B}}(q^*)). \tag{13.42}$$

13.2.3 Resolvent Subordination for Addition and Multiplication

Equation (13.42) relates the average resolvent of $\mathbf{C}$ to that of $\mathbf{A}$. By taking the normalized trace on both sides we find

$$q^* = \mathfrak{g}_{\mathbf{C}}(z) = \mathfrak{g}_{\mathbf{A}}(z - R_{\mathbf{B}}(q^*)), \tag{13.43}$$

which is precisely the subordination relation for the Stieltjes transform of a free sum that we found in Section 11.3.8. We just have rederived this result once again with replicas.

But what is more interesting is that we have found a relationship for the average of a matrix element of the full resolvent matrix, namely

$$\mathbb{E}[\mathbf{G_C}(z)] = \mathbf{G_A}\left(z - R_{\mathbf{B}}(\mathfrak{g}_{\mathbf{C}}(z))\right). \tag{13.44}$$

This relation will give precious information on the overlap between the eigenvalues of **A** and **B** and those of **C**. Note that, by symmetry, one also has

$$\mathbb{E}[\mathbf{G_C}(z)] = \mathbf{G_B}\left(z - R_{\mathbf{A}}(\mathfrak{g}_{\mathbf{C}}(z))\right). \tag{13.45}$$

In the free product case, namely $\mathbf{C} = \mathbf{A}^{\frac{1}{2}}\mathbf{B}\mathbf{A}^{\frac{1}{2}}$ where **A** and **B** are large positive definite matrices whose eigenvectors are mutually random, a very similar replica computation gives a subordination relation for the average T matrix:

$$\mathbb{E}[\mathbf{T_C}(\zeta)] = \mathbf{T_A}[S_{\mathbf{B}}(\mathfrak{t}_{\mathbf{C}}(\zeta))\zeta], \tag{13.46}$$

with $S_{\mathbf{B}}(t)$ the S-transform of the matrix **B**. If we take the normalized trace on both sides, we recover the subordination relation Eq. (11.109). Equation (13.46) can then be turned into a subordination relation for the full resolvent:

$$\mathbb{E}[\mathbf{G_C}(z)] = S^*\mathbf{G_A}(zS^*) \quad \text{with} \quad S^* := S_{\mathbf{B}}(z\mathfrak{g}_{\mathbf{C}}(z) - 1). \tag{13.47}$$

13.2.4 *“Quenched” vs “Annealed”*

The replica trick is quite burdensome as one has to keep track of n copies of an integration vector $\boldsymbol{\psi}_n$ and these vectors interact through the averaging process. At large N one typically has to do a saddle point over one or several $n \times n$ matrices (e.g. **Q** and **Y** in the free addition computation of the previous section), and at the end take the $n \to 0$ limit. But in all computations of Stieltjes transforms so far, taking $n = 1$ instead of $n \to 0$ gives the correct saddle point and the correct final result. In other words, assuming that

$$\mathbb{E}[\log Z] \approx \log \mathbb{E}[Z] \tag{13.48}$$

leads to the correct result. For historical reasons coming from physics, $\mathbb{E}[\log Z]$ is called a *quenched average* whereas $\log \mathbb{E}[Z]$ is called an *annealed average*.

For example, if we go back to Eq. (13.21) we see that taking the logarithm of the $n = 1$ result gives the same result as the correct $n \to 0$ limit. The same is true for the Wishart case. For the free addition and multiplication one can also compute the Stieltjes transform using $n = 1$. This is a general result for bulk properties of random matrices. Most natural ensembles of random symmetric matrices (such as those from Chapter 5 and those arising from free addition and multiplication) feature a strong repulsion of eigenvalues. Because of this repulsion, eigenvalues do not fluctuate much around their *classical positions* – see the detailed discussion in Section 5.4.1. It is the absence of eigenvalue fluctuations on the global scale that makes the $n = 1$ and $n \to 0$ saddle points equivalent.

For the rank-1 HCIZ integral, on the other hand, things are more subtle. As we show in the next section, the annealed average $n = 1$ gives the right answer in some interval of parameters, when the integral is dominated by the bulk properties of eigenvalues. Outside this regime, fluctuations of the largest eigenvalue matter and the $n = 1$ result is no longer correct.

13.3 Rank-1 HCIZ and Replicas

In Chapter 10 we studied the rank-1 HCIZ integral and defined the function $H_{\mathbf{B}}(t)$ as

$$H_{\mathbf{B}}(t) := \lim_{N\to\infty} \frac{2}{N} \log \left\langle \exp\left(\frac{N}{2} \operatorname{Tr} \mathbf{TOBO}^T\right)\right\rangle_{\mathbf{O}}, \tag{13.49}$$

where the averaging is done over the orthogonal group for $\mathbf{O}$, $\mathbf{T}$ is a rank-1 matrix with eigenvalue t and $\mathbf{B}$ a fixed matrix. If $\mathbf{B}$ is a member of a random ensemble, such as the Wigner ensemble, the averaging over $\mathbf{O}$ should be done for a fixed $\mathbf{B}$ and only later the function $H_{\mathbf{B}}(t)$ can be averaged, if needed, over the randomness of $\mathbf{B}$ (quenched average). One could also do an annealed average over $\mathbf{B}$, defining another function $\hat{H}(t)$ as

$$\hat{H}(t) := \lim_{N\to\infty} \frac{2}{N} \log \left\langle \exp\left(\frac{N}{2} \operatorname{Tr} \mathbf{TOBO}^T\right)\right\rangle_{\mathbf{O},\mathbf{B}}. \tag{13.50}$$

It turns out that for small enough values of t, the two quantities are equal, i.e. $\hat{H}(t) = H_{\mathbf{B}}(t)$. For larger values of t, however, these two quantities differ. The aim of this section is to compute explicitly these quantities in the Wigner case using the replica trick, and show that there is a phase transition for a well-defined value $t = t_c$ beyond which quenched and annealed averages do not coincide.

13.3.1 Annealed Average

Let us compute directly the "annealed" average when $\mathbf{B} = \mathbf{X}$ is a Wigner matrix and $\mathbf{T} = t\,\mathrm{e}_1\mathrm{e}_1^T$, where t is the only non-zero eigenvalue of $\mathbf{T}$ and e_1 is the unit vector $(1,0,\dots,0)^T$. Then

$$\begin{aligned}\left\langle \exp\left(\frac{N}{2}\operatorname{Tr}\mathbf{TX}\right)\right\rangle_{\mathbf{X}} &= \left\langle \exp\left(\frac{Nt}{2}\mathrm{e}_1^T\mathbf{X}\mathrm{e}_1\right)\right\rangle_{\mathbf{X}} \\ &= \int \frac{\mathrm{d}X_{11}}{\sqrt{4\pi\sigma^2/N}} \exp\left(\frac{NtX_{11}}{2} - \frac{N}{4\sigma^2}X_{11}^2\right) \\ &= \exp\left(\frac{N}{2}\frac{t^2\sigma^2}{2}\right),\end{aligned} \tag{13.51}$$

so the annealed $\hat{H}_{\mathrm{wig}}(t)$ is given by

$$\hat{H}_{\mathrm{wig}}(t) = \frac{\sigma^2}{2}t^2, \tag{13.52}$$

which, at least superficially, coincides with the integral of the R-transform of a Wigner matrix, Eq. (10.61).

13.3.2 Quenched Average

The annealed average corresponds, in the replica language, to $n = 1$. Let us now turn to arbitrary (integer) n. To keep notation light we will set $\sigma^2 = 1$. As in Eq. (10.31) we define the partition function

$$Z_t(\mathbf{X}) = \int \frac{\mathrm{d}^N \boldsymbol{\psi}}{(2\pi)^{N/2}} \delta\left(\|\boldsymbol{\psi}\|^2 - Nt\right) \exp\left(\frac{1}{2}\boldsymbol{\psi}^T \mathbf{X} \boldsymbol{\psi}\right), \tag{13.53}$$

seeking to compute, at the end of the calculation,

$$\mathbb{E}[H_{\mathbf{X}}(t)] = \lim_{N\to\infty} \frac{2}{N} \lim_{n\to 0} \left(\frac{Z_t^n(\mathbf{X}) - 1}{n}\right) - 1 - \log t, \tag{13.54}$$

where $1 + \log t$ is the large N limit of $2/N \log Z_t(\mathbf{X} = 0)$, with $Z_t(0)$ given by Eq. (10.38). If we write $Z_t^n(\mathbf{X})$ as multiple copies of the same integral and express the Dirac deltas as Fourier integrals over z_α, we get

$$Z_t^n(\mathbf{X}) = \int_{-\mathrm{i}\infty}^{\mathrm{i}\infty} \prod_{\alpha=1}^{n} \mathrm{d}z_\alpha \int \prod_{\alpha=1}^{n} \frac{\mathrm{d}^N \boldsymbol{\psi}_\alpha}{(2\pi)^{N/2}} \exp\left(\frac{1}{2}\sum_{\alpha=1}^{n}(N z_\alpha t - z_\alpha \boldsymbol{\psi}_\alpha^T \boldsymbol{\psi}_\alpha) + \frac{1}{2}\sum_{\alpha=1}^{n} \boldsymbol{\psi}_\alpha^T \mathbf{X} \boldsymbol{\psi}_\alpha\right). \tag{13.55}$$

In order to take the expectation value over the Gaussian random matrix $\mathbf{X}$, we need as always to separate the diagonal and off-diagonal elements of $\mathbf{X}$. The steps are the same as those we took in Section 13.1.2:

$$\begin{aligned}\mathbb{E}\left[\exp\left(\frac{1}{2}\sum_{\alpha=1}^{n} \boldsymbol{\psi}_\alpha^T \mathbf{X} \boldsymbol{\psi}_\alpha\right)\right] &= \exp\left(\sum_{i,j=1}^{N} \frac{1}{4N}\left(\sum_{\alpha=1}^{n} \psi_{\alpha i}\psi_{\alpha j}\right)^2\right) \\ &= \exp\left(\frac{1}{4N}\sum_{\alpha,\beta=1}^{n}\left(\sum_{i=1}^{N} \psi_{\alpha i}\psi_{\beta i}\right)^2\right),\end{aligned} \tag{13.56}$$

which can be rewritten as a Hubbard–Stratonovich integral over an $n \times n$ matrix $\mathbf{q}$:

$$\mathbb{E}[\ldots] = C(n) \int \mathrm{d}\mathbf{q} \exp\left(-N\frac{\operatorname{Tr}\mathbf{q}^2}{4} + \sum_{i=1}^{N}\sum_{\alpha,\beta=1}^{n} \frac{q_{\alpha\beta}\psi_{\alpha i}\psi_{\beta i}}{2}\right), \tag{13.57}$$

where $C(n)$ is a numerical coefficient. After Gaussian integration, one thus finds

$$\mathbb{E}[Z_t^n] = \int_{-\mathrm{i}\infty}^{\mathrm{i}\infty} \prod_{\alpha=1}^{n} \mathrm{d}z_\alpha \int \mathrm{d}\mathbf{q} C(n) \exp\left[\frac{N}{2}\left(t \operatorname{Tr}\mathbf{z} - \frac{\operatorname{Tr}\mathbf{q}^2}{2} - \operatorname{Tr}\log(\mathbf{z} - \mathbf{q})\right)\right], \tag{13.58}$$

which makes sense provided that the real part of z is larger than all the eigenvalues of $\mathbf{q}$. We now define

$$F_n(\mathbf{q}, \mathbf{z}; t) = t \operatorname{Tr}\mathbf{z} - \frac{\operatorname{Tr}\mathbf{q}^2}{2} - \operatorname{Tr}\log(\mathbf{z} - \mathbf{q}), \tag{13.59}$$

where $\mathbf{z}$ is the vector of z_α treated as a diagonal matrix. For $n \geq 1$ we need to find a saddle point in the space of $n \times n$ matrices $\mathbf{z}$ and $\mathbf{q}$, i.e. a point in that space where the first derivatives of $F_t(\mathbf{q}, \mathbf{z})$ are zero with a negative Hessian.

As a check, for $n = 1$ we have at the saddle point

$$q^* = \frac{1}{z^* - q^*} \quad \text{and} \quad t = \frac{1}{z^* - q^*} \quad \Rightarrow \quad z^* = t + \frac{1}{t} \quad \text{and} \quad q^* = t. \tag{13.60}$$

Hence

$$F_1(q^*, z^*; t) = t^2 + 1 - \frac{t^2}{2} + \log t, \tag{13.61}$$

or

$$\hat{H}_{\text{wig}}(t) = \frac{t^2}{2}, \tag{13.62}$$

as it should be.

We now go back to the general n case. Using Eq. (1.37) we can take a matrix derivative of Eq. (13.59) with respect to $\mathbf{q}$ and $\mathbf{z}$:

$$\mathbf{q} = (\mathbf{z} - \mathbf{q})^{-1} \quad \text{and} \quad \left[(\mathbf{z} - \mathbf{q})^{-1}\right]_{\alpha\alpha} = t. \tag{13.63}$$

In the following technical part, we solve these equations for integer $n \geq 1$. We discuss the final result at the end of this subsection.

The second equation in (13.63) comes from the derivative with respect to z; remember $\mathbf{z}$ is only a diagonal matrix, the derivative with respect to $\mathbf{z}$ tells us only about the diagonal elements.

From this we can argue that $\mathbf{z}$ must be a multiple of the identity and $\mathbf{q}$ of the form[3]

$$\mathbf{q} = \begin{pmatrix} t & b & \dots & b \\ b & t & \dots & b \\ \vdots & \vdots & \ddots & b \\ b & b & \dots & t \end{pmatrix}, \tag{13.64}$$

for some b to be determined. To find an equation for b and z we need to express Eqs. (13.63) in terms of those quantities. To do so we first write the matrix $\mathbf{q}$ as a rank-1 perturbation of a multiple of the identity matrix:

$$\mathbf{q} = (t - b)\mathbf{1} + nb\mathbf{P}_1, \tag{13.65}$$

where $\mathbf{P}_1 = \mathbf{e}\mathbf{e}^T$ is the projector onto the normalized vector of all 1:

$$\mathbf{e} = \frac{1}{\sqrt{n}} \begin{pmatrix} 1 \\ 1 \\ \vdots \\ 1 \end{pmatrix}. \tag{13.66}$$

[3] More complicated, block diagonal structures for $\mathbf{q}$ sometimes need to be considered in the limit $n \to 0$. This is called "replica symmetry breaking", a phenomenon that occurs in many "complex" optimization problems, such as spin-glasses – see Section 13.4. Fortunately, in the present case, these complications are not present.

Note that the eigenvalues of the matrix $\mathbf{q}$ are $(t-b)+nb$ (with multiplicity 1) and $(t-b)$ (with multiplicity $(n-1)$). Since $\mathbf{z}$ is a multiple of the identity, the matrix $\mathbf{z}-\mathbf{q}$ is a rank-1 perturbation of a multiple of the identity and it can be inverted using the Sherman–Morrison formula, Eq. (1.28). The first of Eqs. (13.63) becomes

$$(t-b)\mathbf{1}+nb\mathbf{P}_1 = \frac{\mathbf{1}}{z-t+b} + \frac{nb\mathbf{P}_1}{(z-t+b)^2(1-nb(z-t+b)^{-1})}. \tag{13.67}$$

We can now equate the prefactors in front of the identity matrix $\mathbf{1}$ and of the projector $\mathbf{P}_1$ separately, to get two equations for our two unknowns (z and b). For the identity matrix we get

$$(t-b) = \frac{1}{z-t+b} \quad \Rightarrow \quad z = (t-b) + \frac{1}{t-b}. \tag{13.68}$$

For the second equation, we first replace $(z-t+b)^{-1}$ by $t-b$ and get

$$nb = \frac{(t-b)^2 nb}{1-nb(t-b)}. \tag{13.69}$$

We immediately find one solution: $b=0$. For this solution both $\mathbf{q}=q_0\mathbf{1}$ and $\mathbf{z}=z_0\mathbf{1}$ are multiples of the identity and we have $q_0=t$ and $z_0=t+t^{-1}$. This coincides with the (unique) solution we found in the annealed ($n=1$) case. For general n, there are potentially other solutions. Simplifying off nb, we find a quadratic equation for b:

$$1-nb(t-b) = (t-b)^2, \tag{13.70}$$

whose solutions we write as

$$b_\pm = \frac{(n-2)t \pm \sqrt{(n^2t^2-4(n-1))}}{2(n-1)}. \tag{13.71}$$

From the two solutions for b we can compute the corresponding values of z using Eq. (13.68). We get a term with a square-root on the denominator that we simplify using $(c\pm\sqrt{d})^{-1} \equiv (c\mp\sqrt{d})/(c^2-d)$. After further simplification we find

$$z_\pm = \frac{n^2t \pm (n-2)\sqrt{(n^2t^2-4(n-1))}}{2(n-1)}. \tag{13.72}$$

We now need to choose one of the three solutions z_0, z_+ and z_-. First, we consider integer $n \geq 1$ where the replica method is perfectly legitimate. We will later deal with the $n \to 0$ limit.

For $n=1$, z_- is ill defined while z_+ becomes identical to z_0. The only solution for all t is therefore z_0 and we recover the annealed result discussed in the previous subsection.

For $n \geq 2$, we first notice that the solutions $z_\pm$ do not exist for $t < t_s := 2\sqrt{n-1}/n$; they yield a complex result when the result must be real. So for $t < t_s$, z_0 is the solution. For larger values of t we should compare the values of $F_n(\mathbf{q}^*, \mathbf{z}^*; t)$ and choose the maximum one. For $t > t_s$, the z_+ solution always dominates z_-, so we only consider z_+ and z_0 henceforth.

For $n=2$, the analysis is easy, $t_s=1$ and $t_c=1$: at $t=1$, $z_+=z_0$ and for $t>1$ the z_+ solution dominates. For $n>2$, the situation is a bit more subtle. The z_+ solution appears at $t_s<1$ but at that point z_0 still dominates. At $t=1$, z_+ dominates. At some $t=t_c$, with $t_s<t_c<1$, we must have $F_n(\mathbf{q}_0, \mathbf{z}_0; t_c) = F_n(\mathbf{q}_+, \mathbf{z}_+; t_c)$. This point could in principle be shown analytically but it is easier numerically. In particular we do not have an analytical expression for $t_c(n)$ except the above bound $t_c(n>2)<1$ and the value $t_c(2)=1$ (see Fig. 13.1).

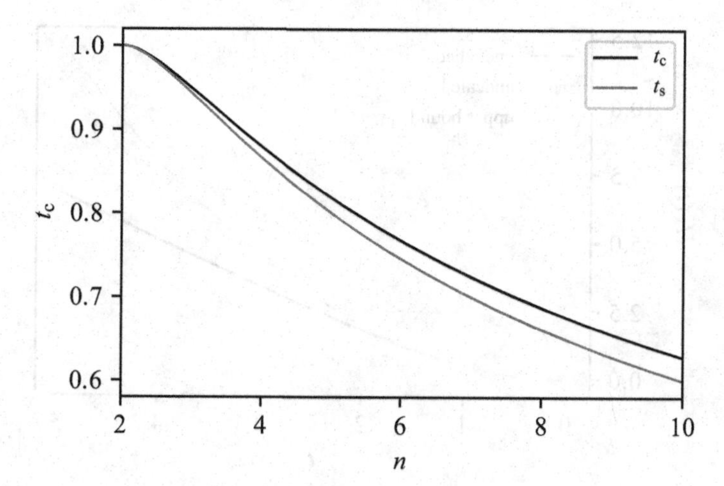

Figure 13.1 The point $t_c(n)$ where the $F_n(\mathbf{q}_0, \mathbf{z}_0; t_c) = F_n(\mathbf{q}_+, \mathbf{z}_+; t_c)$ and beyond which the z_+ solution dominates. Also shown is the point $t_s(n) = 2\sqrt{n-1}/n$ where the z_+ solution starts to exist. Note that $t_s \le t_c \le 1$ for all $n \ge 2$. Hence the transition appears below $t = 1$. Note also that $t_c \to 0$ as $n \to \infty$.

We can now put everything together but there is a trick to save computation effort. Given that our solutions cancel the partial derivatives of $F_t(\mathbf{q}, \mathbf{z}; t)$ with respect to $\mathbf{q}$ and $\mathbf{z}$, we can easily compute its derivative with respect to t:

$$\frac{\mathrm{d}}{\mathrm{d}t} F_n(\mathbf{q}^*, \mathbf{z}^*; t) = \frac{\partial}{\partial t} F_n(\mathbf{q}^*, \mathbf{z}^*; t) = \operatorname{Tr} \mathbf{z}^*(t) = n z^*(t). \tag{13.73}$$

Note that we can follow the value of $F_n(\mathbf{q}^*, \mathbf{z}^*; t)$ through the critical point t_c because $F_n(\mathbf{q}^*, \mathbf{z}^*; t)$ is continuous at that point (even if its derivative is not).

The above analysis therefore allows us to find, for $n \ge 1$,

$$\log \mathbb{E}[Z_t^n] \sim \frac{Nn}{2} \mathcal{F}_n(t), \tag{13.74}$$

where

$$\frac{\mathrm{d}}{\mathrm{d}t} \mathcal{F}_n(t) = \begin{cases} t + \frac{1}{t} & \text{for } t \le t_c(n), \\ \frac{n^2 t + (n-2)\sqrt{(n^2t^2 - 4(n-1))}}{2(n-1)} & \text{for } t > t_c(n), \end{cases} \tag{13.75}$$

with the boundary condition $\mathcal{F}_n(0) = 0$.

We can now analytically continue this solution down to $n \to 0$. The first regime, for small t, is easy as it does not depend on n. In the large t regime, the extrapolation of $z_+(t)$ to $n \to 0$ gives the very simple result (see Eq. (13.72)): $z_+ = 2$ for all t. The most tricky part is to find the critical point where one goes from the $z_0 = t + 1/t$ solution to the $z_+ = 2$ solution. We cannot analytically continue $t_c(n)$ to $n \to 0$, as we have no explicit formula for it. On the other hand, we can directly find the point $t_c(n = 0)$ at which the two solutions

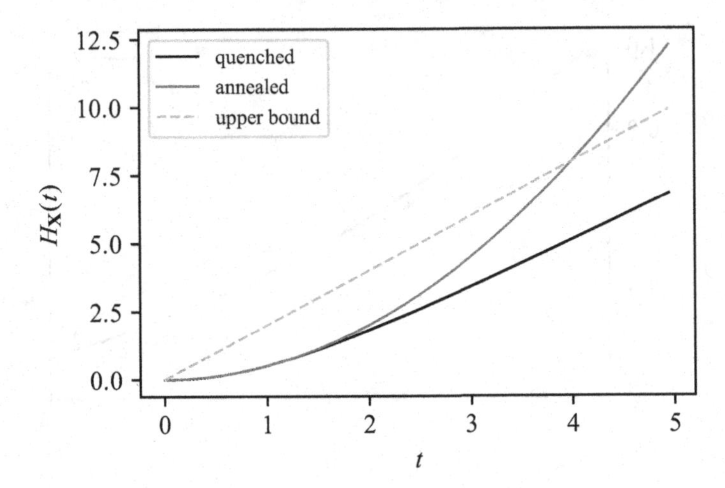

Figure 13.2 The function $H_{\mathbf{X}}(t)$ for a unit Wigner computed with a "quenched" average (HCIZ integral) and an "annealed" average. We also show the upper bound given by Eq. (10.59). The annealed and quenched averages are identical up to $t = t_c = 1$ and differ for larger t. The annealed average violates the bound, which is expected as in this case $\lambda_{\max}$ fluctuates and exceptionally large values of $\lambda_{\max}$ dominate the average exponential HCIZ integral.

lead to the same $\mathcal{F}_0$. It is relatively straightforward to show that this point is $t_c = 1$.[4] Correspondingly,

$$\frac{\mathrm{d}}{\mathrm{d}t}\mathcal{F}_0(t) = \begin{cases} t + \frac{1}{t} & \text{for } t \leq t_c(0) = 1, \\ 2 & \text{for } t > t_c(0) = 1. \end{cases} \tag{13.76}$$

We can now go back to the definition of the function $\mathbb{E}[H_{\mathbf{X}}(t)]$ (Eq. (13.54)). After taking the $n \to 0$ and $N \to \infty$ limits we find (see Fig. 13.2)

$$\frac{\mathrm{d}}{\mathrm{d}t}\mathbb{E}[H_{\mathbf{X}}(t)] = \begin{cases} t & \text{for } t \leq 1, \\ 2 - \frac{1}{t} & \text{for } t > 1, \end{cases} \tag{13.77}$$

with the condition $\mathbb{E}[H_{\mathbf{X}}(t = 0)] = 0$.

The upshot of this (rather complex) calculation is that, as announced in the introduction, for $t \leq t_c = 1$ the quenched and annealed results coincide, i.e. $\hat{H}_{\text{wig}}(t) = \mathbb{E}[H_{\mathbf{X}}(t)]$. For $t > t_c$, on the other hand, the two quantities are different. The reason is that for sufficiently large values of t, the average HCIZ integral becomes dominated by very rare Wigner matrices that happen to have a largest eigenvalue significantly larger than the Wigner semi-circle edge $\lambda = 2$. This allows $\hat{H}_{\text{wig}}(t)$ to continue its quadratic progression, while $\mathbb{E}[H_{\mathbf{X}}(t)]$ is dominated by the edge of the average spectrum and its growth with t is contained (see Fig. 13.2). When one computes higher and higher moments of the HCIZ

[4] Something peculiar happens as $n \to 0$, namely the minimum solution becomes the maximum one and vice versa. In other words for small t, where we know that z_0 is the right solution, we have $\mathcal{F}_0(z_0; t) < \mathcal{F}_0(z_+; t)$ and the opposite at large t. This paradox is always present within the replica method when $n \to 0$.

integral (i.e. as the number of replicas n increases), the dominance of extreme eigenvalues becomes more and more acute, leading to a smaller and smaller transition point $t_c(n)$.[5]

13.4 Spin-Glasses, Replicas and Low-Rank HCIZ

"Spin-glasses" are disordered magnetic materials exhibiting a freezing transition as the temperature is reduced. Typical examples are silver–manganese (or copper–manganese) alloys, where the manganese atoms carry a magnetic spin and are randomly dispersed in a non-magnetic matrix. Contrary to a ferromagnet (i.e. usual magnets like permalloy), where all microscopic spins agree to point more or less in the same direction when the temperature is below a certain transition temperature (872 K for permalloy), spins in spin-glasses freeze, but the configuration they adopt is disordered, "amorphous", with zero net magnetization.

A simple model to explain the phenomenon is the following. The energy of N spins $S_i = \pm 1$ is given by

$$\mathcal{H}(\{S\}) = \frac{1}{2} \sum_{i,j=1}^{N} \mathbf{J}_{ij} S_i S_j, \tag{13.78}$$

where $\mathbf{J}$ is a random matrix, which we take to be drawn from a rotational invariant ensemble, i.e. $\mathbf{J} = \mathbf{O}\Lambda\mathbf{O}^T$ where $\mathbf{O}$ is chosen according to the (flat) Haar measure over $O(N)$ and Λ is a certain fixed diagonal matrix with $\tau(\Lambda) = 0$, such that any pair of spins is as likely to want to point in the same direction or in opposite directions. The simplest case corresponds to $\mathbf{J} = \mathbf{X}$, a Wigner matrix, in which case the spectrum of Λ is the Wigner semi-circle. This case corresponds to the celebrated Sherrington–Kirkpatrick (SK) model, but other cases have been considered in the literature as well.

The physical properties of the system are encoded in the average free energy F, defined as

$$F := -T\mathbb{E}_{\mathbf{J}}\left[\log Z\right]; \qquad Z := \sum_{\{S\}} \exp\left(\frac{\mathcal{H}(\{S\})}{T}\right), \tag{13.79}$$

where the partition function Z is obtained as the sum over all 2^N configurations of the N spins, and T is the temperature. One of the difficulties of the theory of spin-glasses is to perform the average over the interaction matrix $\mathbf{J}$ of the *logarithm* of Z. Once again, one can try to use the replica trick to perform this average, to wit,

$$\mathbb{E}_{\mathbf{J}}\left[\log Z\right] = \frac{\partial}{\partial n}\mathbb{E}_{\mathbf{J}}\left[Z^n\right]\Big|_{n=0}. \tag{13.80}$$

One then computes the right hand side for integer n and hopes that the analytic continuation to $n \to 0$ makes sense. Introducing n replicas of the system, one has

$$\mathbb{E}_{\mathbf{J}}\left[Z^n\right] = \mathbb{E}_{\mathbf{J}}\left[\sum_{\{S^1,S^2,\ldots,S^n\}} \exp\left(\frac{\sum_{\alpha=1}^{n}\sum_{i,j=1}^{N} \mathbf{J}_{ij} S_i^\alpha S_j^\alpha}{2T}\right)\right]. \tag{13.81}$$

Now, for a fixed configuration of all nN spins $\{S^1, S^2, \ldots, S^n\}$, the $N \times N$ matrix $\mathbf{K}_{ij}^{(n)} := \sum_{\alpha=1}^{n} S_i^\alpha S_j^\alpha / N$ is at most of rank n, which is small compared to N which we will

[5] A very similar mechanism is at play in Derrida's random energy model, see Derrida [1981].

take to infinity. So we have to compute a low-rank HCIZ integral, which is given by Eq. (10.45):

$$\mathbb{E}_{\mathbf{J}}\left[\exp\left(\frac{N}{2T}\operatorname{Tr}\mathbf{O}\Lambda\mathbf{O}^T\mathbf{K}^{(n)}\right)\right] \approx \exp\left(\frac{N}{2}\operatorname{Tr}H_{\mathbf{J}}(\mathbf{K}^{(n)}/T)\right), \tag{13.82}$$

where $H_{\mathbf{J}}$ is the anti-derivative of the R-transform of $\mathbf{J}$ (or Λ). Now the non-zero eigenvalues of the $N \times N$ matrix $\mathbf{K}^{(n)}$ are the same as those of the $n \times n$ matrix $\mathbf{Q}_{\alpha\beta} = \sum_{i=1}^{N} S_i^\alpha S_i^\beta / N$, called the *overlap* matrix, because its elements measure the similarity of the configurations $\{S^a\}$ and $\{S^b\}$. Hence, we have to compute

$$\mathbb{E}_{\mathbf{J}}[Z^n] = \sum_{\{S^1, S^2, \ldots, S^n\}} \exp\left(\frac{N}{2}\operatorname{Tr}_n H_{\mathbf{J}}(\mathbf{Q}/T)\right). \tag{13.83}$$

It should be clear to the reader that all the steps above are very close to the ones followed in Section 13.2.2. We continue in the same vein, introducing a new $n \times n$ matrix for imposing the constraint $\mathbf{Q}_{\alpha\beta} = \sum_{i=1}^{N} S_i^\alpha S_i^\beta / N$:

$$1 = \int_{-i\infty}^{i\infty} \frac{N^{n(n+1)/2}\mathrm{d}\mathbf{Y}}{2^{3n/2}\pi^{n/2}} \int \mathrm{d}\mathbf{Q}\exp\left(-N\operatorname{Tr}_n \mathbf{Q}\mathbf{Y} + \sum_{\alpha,\beta=1}^{n} \mathbf{Y}_{\alpha,\beta}\sum_{i=1}^{N} S_i^\alpha S_i^\beta\right). \tag{13.84}$$

The nice thing about this representation is that sums over i become totally decoupled. So we get

$$\mathbb{E}_{\mathbf{J}}[Z^n] = C\int_{-i\infty}^{i\infty} \mathrm{d}\mathbf{Y}\int \mathrm{d}\mathbf{Q}\exp\left(\frac{N}{2}\operatorname{Tr}_n H_{\mathbf{J}}(\mathbf{Q}/T) - N\operatorname{Tr}_n \mathbf{Q}\mathbf{Y} + N\mathfrak{S}(\mathbf{Y})\right), \tag{13.85}$$

where C is an irrelevant constant and

$$\mathfrak{S}(\mathbf{Y}) := \log \mathcal{Z} \quad \text{with} \quad \mathcal{Z} := \left[\sum_{S} \mathrm{e}^{\sum_{\alpha,\beta=1}^{n} \mathbf{Y}_{\alpha,\beta} S^\alpha S^\beta}\right]. \tag{13.86}$$

In the large N limit, Eq. (13.85) can be estimated using a saddle point method over $\mathbf{Y}$ and $\mathbf{Q}$. As in Section 13.2.2 the first equation reads

$$\mathbf{Y}_{\alpha\beta} = \frac{1}{2T}[R_{\mathbf{J}}(\mathbf{Q}/T)]_{\alpha\beta}, \tag{13.87}$$

and taking derivatives with respect to $\mathbf{Y}_{\alpha\beta}$, we find

$$\mathbf{Q}_{\alpha\beta} = \frac{1}{\mathcal{Z}}\sum_{S} S^\alpha S^\beta \mathrm{e}^{\sum_{\alpha',\beta'=1}^{n} \mathbf{Y}_{\alpha',\beta'} S^{\alpha'} S^{\beta'}}, \tag{13.88}$$

which leads to the following self-consistent equation for $\mathbf{Q}$:

$$\mathbf{Q}_{\alpha\beta} = \frac{\sum_S S^\alpha S^\beta \mathrm{e}^{\frac{1}{2T}\sum_{\alpha',\beta'=1}^{n}[R_{\mathbf{J}}(\mathbf{Q}/T)]_{\alpha'\beta'} S^{\alpha'} S^{\beta'}}}{\sum_S \mathrm{e}^{\frac{1}{2T}\sum_{\alpha',\beta'=1}^{n}[R_{\mathbf{J}}(\mathbf{Q}/T)]_{\alpha'\beta'} S^{\alpha'} S^{\beta'}}} := \langle S^\alpha S^\beta \rangle_T. \tag{13.89}$$

At sufficiently high temperatures, one can expect the solution of these equations to be "replica symmetric", i.e.

$$\mathbf{Q}_{\alpha\beta} = \delta_{\alpha\beta}(1-q) + q. \tag{13.90}$$

This matrix has two eigenvalues, one non-degenerate equal to $1 + (n-1)q$ and another $(n-1)$-fold degenerate equal to $1 - q$. Correspondingly, $R_{\mathbf{J}}(\mathbf{Q})$ has eigenvalues $R_{\mathbf{J}}((1+(n-1)q)/T)$ and $R_{\mathbf{J}}((1-q)/T)$, from which we reconstruct the diagonal and off-diagonal elements of $R_{\mathbf{J}}(\mathbf{Q})$:

$$\begin{aligned} r &:= [R_{\mathbf{J}}(\mathbf{Q})]_{\alpha\beta} = \frac{1}{n}\left[R_{\mathbf{J}}((1+(n-1)q)/T) - R_{\mathbf{J}}((1-q)/T)\right], \\ r_d &:= [R_{\mathbf{J}}(\mathbf{Q})]_{\alpha\alpha} = R_{\mathbf{J}}((1-q)/T) + r. \end{aligned} \tag{13.91}$$

Injecting in the definition of $\mathcal{Z}$, we find

$$\mathcal{Z} = \sum_S \exp\left(\frac{1}{2T}\left[nr_d + r\sum_{\alpha\neq\beta=1}^{n} S^\alpha S^\beta\right]\right), \tag{13.92}$$

where we have used $S_\alpha^2 \equiv 1$. Writing $\sum_{\alpha\neq\beta=1}^n S^\alpha S^\beta = \left(\sum_\alpha^n S^\alpha\right)^2 - n$ and using a Hubbard–Stratonovich transformation, one gets

$$\mathcal{Z} = \sum_S e^{n\frac{r_d - r}{2T}} \int_{-\infty}^{+\infty} dx \exp\left(-\frac{x^2}{2} + x\sqrt{\frac{r}{T}}\sum_\alpha^n S^\alpha\right). \tag{13.93}$$

The sums of different S^α now again decouple, leading to

$$\mathcal{Z} = e^{n\frac{r_d - r}{2T} + n\log 2} \int_{-\infty}^{+\infty} dx \frac{e^{-\frac{x^2}{2}}}{\sqrt{2\pi}} \cosh^n\left[x\sqrt{\frac{r}{T}}\right]. \tag{13.94}$$

Now, one can notice that

$$\sum_{\alpha\neq\beta=1}^{n} \langle S^\alpha S^\beta \rangle_T = n(n-1)q = 2T\frac{\partial \log \mathcal{Z}}{\partial r}, \tag{13.95}$$

to get, in the limit $n \to 0$ and a few manipulations (including an integration by parts), an equation involving only q:

$$q = \int_{-\infty}^{+\infty} \frac{dx}{\sqrt{2\pi}} e^{-\frac{x^2}{2}} \tanh^2\left[x\sqrt{\frac{r}{T}}\right], \tag{13.96}$$

where, in the limit $n \to 0$,

$$r = \frac{q}{T} R_{\mathbf{J}}'\left(\frac{1-q}{T}\right). \tag{13.97}$$

Clearly, $q = 0$ is always a solution of this equation. The physical interpretation of q is the following: choose randomly two microscopic configurations of spins $\{S_i^\alpha\}$ and $\{S_i^\beta\}$, each with a weight given by $\exp\left(\frac{\mathcal{H}(\{S\})}{T}\right)/Z$. Then, the average overlap between these configurations, $\sum_{i=1}^N S_i^\alpha S_i^\beta / N$, is equal to q. When $q = 0$, these two configurations are thus uncorrelated. One expects this to be the case at high enough temperature, where the system explores randomly all configurations.

When the spins start freezing, on the other hand, one expects the system to strongly favor some (amorphous) configurations over others. Hence, one expects that $q > 0$

in the spin-glass phase. Expanding the right hand side of Eq. (13.96) for small q gives

$$\int_{-\infty}^{+\infty} \frac{\mathrm{d}x}{\sqrt{2\pi}} \mathrm{e}^{-\frac{x^2}{2}} \tanh^2\left[x\sqrt{\frac{r}{T}}\right]$$
$$= q\frac{R'_{\mathbf{J}}(1/T)}{T^2} - q^2\left(\frac{R''_{\mathbf{J}}(1/T)}{T^3} + 2\left(\frac{R'_{\mathbf{J}}(1/T)}{T^2}\right)^2\right) + O(q^3). \tag{13.98}$$

Assuming that the coefficient in front of the q^2 term is negative, we see that a non-zero q solution appears continuously below a critical temperature T_c given by

$$T_c^2 = R'_{\mathbf{J}}\left(\frac{1}{T_c}\right). \tag{13.99}$$

When $\mathbf{J}$ is a Wigner matrix, the spin-glass model is the one studied originally by Sherrington and Kirkpatrick in 1975. In this case, $R_{\mathbf{J}}(x) = x$ and therefore $T_c = 1$. There are cases, however, where the transition is discontinuous, i.e. where the overlap q jumps from zero for $T > T_c$ to a non-zero value at T_c. In these cases, the small q expansion is unwarranted and another method must be used to find the critical temperature. One example is the "random orthogonal model", where the coupling matrix $\mathbf{J}$ has $N/2$ eigenvalues equal to $+1$ and $N/2$ eigenvalues equal to -1.

The spin-glass phase $T < T_c$ is much more complicated to analyze, because the replica symmetric ansatz $\mathbf{Q}_{\alpha\beta} = \delta_{\alpha\beta}(1-q) + q$ is no longer valid. One speaks about "replica symmetry breaking", which encodes an exquisitely subtle, hierarchical organization of the phase space of these models. This is the physical content of the celebrated Parisi solution of the SK model, but is much beyond the scope of the present book, and we encourage the curious reader to learn more from the references given below.

Bibliographical Notes

- The replica trick was introduced by
  - R. Brout. Statistical mechanical theory of a random ferromagnetic system. *Physical Review*, 115:824–835, 1959,

  and popularized in
  - S. F. Edwards and P. W. Anderson. Theory of spin glasses. *Journal of Physics F: Metal Physics*, 5(5):965, 1975,

  for an authoritative book on the subject, see
  - M. Mézard, G. Parisi, and M. A. Virasoro. *Spin Glass Theory and Beyond*. World Scientific, Singapore, 1987,

  see also
  - M. Mézard and G. Parisi. Replica field theory for random manifolds. *Journal de Physique I, France*, 1(6):809–836, 1991.
- The replica method was first applied to random matrices by
  - S. F. Edwards and R. C. Jones. The eigenvalue spectrum of a large symmetric random matrix. *Journal of Physics A: Mathematical and General*, 9(10):1595, 1976,

  it is rarely discussed in random matrix theory books with the exception of

  - G. Livan, M. Novaes, and P. Vivo. *Introduction to Random Matrices: Theory and Practice*. Springer, New York, 2018.
- For a replica computation for the average resolvent for free addition and free product, see
  - J. Bun, R. Allez, J.-P. Bouchaud, and M. Potters. Rotational invariant estimator for general noisy matrices. *IEEE Transactions on Information Theory*, 62:7475–7490, 2016,
  - J. Bun, J.-P. Bouchaud, and M. Potters. Cleaning large correlation matrices: Tools from random matrix theory. *Physics Reports*, 666:1–109, 2017.
- For Derrida's random energy model where different moments of Z have different transition temperatures, see
  - B. Derrida. Random-energy model: An exactly solvable model of disordered systems. *Physical Review B*, 24:2613–2626, 1981.
- For an introduction to experimental spin-glasses, see
  - K. H. Fischer and J. A. Hertz. *Spin Glasses*. Cambridge University Press, Cambridge, 1991,
  - A. P. Young. *Spin Glasses and Random Fields*. World Scientific, Singapore, 1997,

  see also M. Mézard, G. Parisi, M. Virasoro, *op. cit.*
- For general orthogonal invariant spin-glasses, see
  - R. Cherrier, D. S. Dean, and A. Lefèvre. Role of the interaction matrix in mean-field spin glass models. *Physical Review E*, 67:046112, 2003.

14

Edge Eigenvalues and Outliers

In many instances, the eigenvalue spectrum of large random matrices is confined to a single interval of finite size. This is of course the case for Wigner matrices, where the correctly normalized eigenvalues fall between $\lambda_- = -2$ and $\lambda_+ = +2$, with a semi-circular distribution between the two edges, and, correspondingly, a square-root singularity of the density of eigenvalues close to the edges. This is also the case for Wishart matrices, for which again the density of eigenvalues has square-root singularities close to both edges. As discussed in Section 5.3.2, this is a generic property, with a few notable exceptions. One example is provided by Wishart matrices with parameter $q = 1$, for which the eigenvalue spectrum extends down to $\lambda = 0$ with an inverse square-root singularity there. Another case is that of Wigner matrices constrained to have all eigenvalues positive: the spectrum also has an inverse square-root singularity – see Eq. (5.94). One speaks of a "hard edge" in that case, because the minimum eigenvalue is imposed by a strict constraint. The Wigner semi-circle edge at $\lambda_+ = 2$, on the other hand, is "soft" and appears naturally as a result of the minimization of the energy of a collection of interacting Coulomb charges in an external potential.[1]

Consider for example Wigner matrices of size N. The existence of sharp edges delimiting a region where one expects to see a non-zero density of eigenvalues from a region where there should be none is only true in the asymptotically large size limit $N \to \infty$. For large but finite N, on the other hand, one expects that the probability to find an eigenvalue beyond the Wigner sea is very small but non-zero. The width of the transition region, and the tail of the density of states was investigated a while ago, culminating in the beautiful results by Tracy and Widom on the distribution of the *largest* eigenvalue of a random matrix, which we will describe in the next section. The most important result is that the width of the region around λ_+ within which one expects to observe the largest eigevalue of a Wigner matrix goes down as $N^{-2/3}$.

Hence the largest eigenvalue $\lambda_{\max}$ does not fluctuate very far away from the classical edge λ_+. Take for example $N = 1000$; $\lambda_{\max}$ is within $1000^{-2/3} = 0.01$ away from $\lambda_+ = 2$. In real applications the largest eigenvalue can deviate quite substantially from the classical

[1] Note that there are also cases where the soft edge has a different singularity, see Section 5.3.3, and cases where the eigenvalue spectrum extends up to infinity, for example "Lévy matrices" with IID elements of infinite variance.

edge. The origin of such a large eigenvalue is usually not an improbably large Tracy–Widom fluctuation but rather a true outlier that should be modeled separately. This is the goal of the present chapter. We will see in particular that perturbing a Wigner (or Wishart) matrix with a deterministic, low-rank matrix of sufficient amplitude $a > a_c$ generates "true" outliers, which remain at a distance $O(a)$ from the upper edge. For $a < a_c$ on the other hand, the largest eigenvalue remains at distance $N^{-2/3}$ from λ_+.

14.1 The Tracy–Widom Regime

The Tracy–Widom result characterizes precisely the distance between the largest eigenvalue $\lambda_{\max}$ of Gaussian Wigner or Wishart matrices and the upper edge of the spectrum which we denoted by λ_+. This result can be (formally) stated as follows: the rescaled distribution of $\lambda_{\max} - \lambda_+$ converges, for $N \to \infty$, towards the Tracy–Widom distribution, usually noted F_1:

$$\mathbb{P}\left(\lambda_{\max} \le \lambda_+ + \gamma N^{-2/3} u\right) = F_1(u), \tag{14.1}$$

where γ is a constant that depends on the problem and $F_1(u)$ is the $\beta = 1$ Tracy–Widom distribution. For the Wigner problem, $\lambda_+ = 2$ and $\gamma = 1$, whereas for Wishart matrices, $\lambda_+ = (1 + \sqrt{q})^2$ and $\gamma = \sqrt{q}\lambda_+^{2/3}$. In fact, Eq. (14.1) holds for a much wider large class of $N \times N$ random matrices, for example symmetric random matrices with arbitrary IID elements with a finite fourth moment. The Tracy–Widom distribution for all three values of β is plotted in Figure 14.1. (The case where the fourth moment is infinite is discussed in Section 14.3 below.)

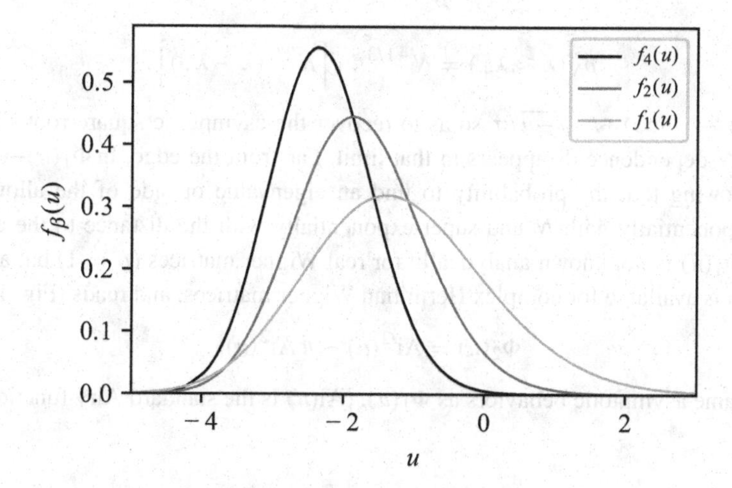

Figure 14.1 Rescaled and shifted probability density of the largest eigenvalue for a large class of random matrices such as Wigner and Wishart: the Tracy–Widom distribution. The distribution depends on the Dyson index (β) and is shown here for $\beta = 1, 2$ and 4.

Everything is known about the Tracy–Widom density $f_1(u) = F_1'(u)$, in particular its left and right far tails:

$$\ln f_1(u) \propto -u^{3/2}, \quad (u \to +\infty); \qquad \ln f_1(u) \propto -|u|^3, \quad (u \to -\infty). \tag{14.2}$$

One notices that the left tail is much thinner than the right tail: pushing the largest eigenvalue inside the Wigner sea implies compressing the whole Coulomb gas of repulsive charges, which is more difficult than pulling one eigenvalue away from λ_+. Using this analogy and the formalism of Section 5.4.2, the large deviation regime of the Tracy–Widom problem (i.e. for $\lambda_{\max} - \lambda_+ = O(1)$) can be obtained. Note that the result is exponentially small in N as the $u^{3/2}$ behavior for $u \to \infty$ combines with $N^{2/3}$ to give a linear in N dependence.

The distribution of the smallest eigenvalue $\lambda_{\min}$ around the lower edge λ_- is also Tracy–Widom, except in the particular case of Wishart matrices with $q = 1$. In this case $\lambda_- = 0$, which is a "hard" edge since all eigenvalues of the empirical matrix must be non-negative.[2]

The behavior of the width of the transition region can be understood using a simple heuristic argument. Suppose that the $N = \infty$ density goes to zero near the upper edge λ_+ as $(\lambda_+ - \lambda)^\theta$ (generically, $\theta = 1/2$ as is the case for the Wigner and the Marčenko–Pastur distributions). For finite N, one expects not to be able to say whether the density is zero or non-zero when the probability to observe an eigenvalue is of order $1/N$, i.e. when the $O(1)$ eigenvalue is within the "blurred" region. This leads to a blurred region of width

$$|\lambda^* - \lambda_+|^{\theta+1} \propto \frac{1}{N} \to \Delta\lambda^* \sim N^{-\frac{1}{1+\theta}}, \tag{14.3}$$

which goes to zero as $N^{-2/3}$ in the generic square-root case $\theta = 1/2$. More precisely, for Gaussian ensembles, the average density of states at a distance $\sim N^{-2/3}$ from the edge behaves as

$$\rho_N(\lambda \approx \lambda_+) = N^{-1/3}\Phi_1\left[N^{2/3}(\lambda - \lambda_+)\right], \tag{14.4}$$

with $\Phi_1(u \to -\infty) \approx \sqrt{-u}/\pi$ so as to recover the asymptotic square-root singularity, since the N dependence disappears in that limit. Far from the edge, $\ln \Phi_1(u \to +\infty) \propto -u^{3/2}$, showing that the probability to find an eigenvalue outside of the allowed band decays exponentially with N and super-exponentially with the distance to the edge. The function $\Phi_1(u)$ is not known analytically for real Wigner matrices ($\beta = 1$) but an explicit expression is available for complex Hermitian Wigner matrices, and reads (Fig. 14.2)

$$\Phi_2(u) = \mathrm{Ai}'^2(u) - u\,\mathrm{Ai}^2(u), \tag{14.5}$$

with the same asymptotic behaviors as $\Phi_1(u)$. ($\mathrm{Ai}(u)$ is the standard Airy function.)

[2] This special case is treated in Péché [2003].

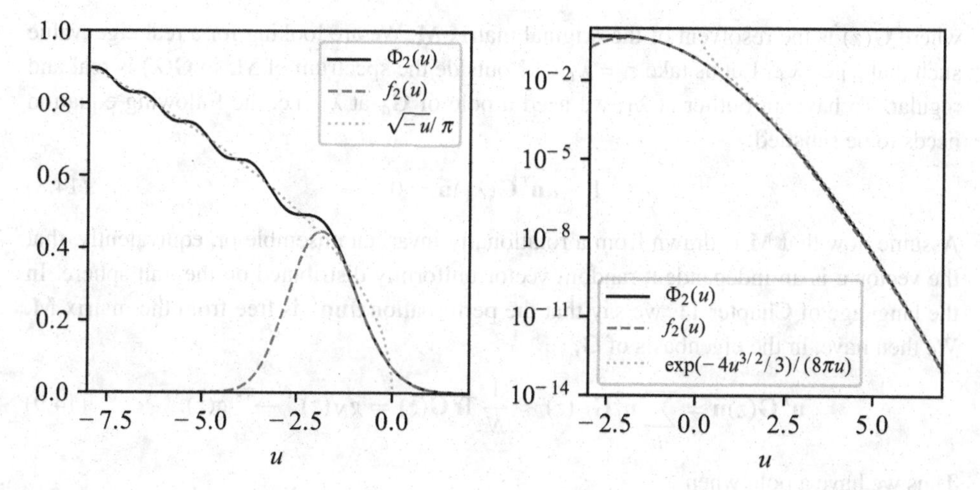

Figure 14.2 Behavior of the density near the edge λ_+ at the scale $N^{-2/3}$ for complex Hermitian Wigner matrices given by Eq. (14.5). For comparison the probability of the largest eigenvalue $f_2(u)$ is also shown. For positive values, the two functions are almost identical and behave as $\exp(-4u^{3/2}/3)/(8\pi u)$ for large u (right). For negative arguments the functions are completely different: $\Phi_2(u)$ behaves as $\sqrt{-u}/\pi$ for large negative u while $f_2(u) \to 0$ as the largest eigenvalue cannot be in the bulk (left).

14.2 Additive Low-Rank Perturbations

14.2.1 Eigenvalues

We will now study the outliers for an additive perturbation to a large random matrix. Take a large symmetric random matrix $\mathbf{M}$ (e.g. Wigner or Wishart) with a well-behaved asymptotic spectrum that has a deterministic right edge λ_+. We would like to know what happens when one adds to $\mathbf{M}$ a low-rank (deterministic) perturbation. For simplicity, we only consider the rank-1 perturbation $a\mathbf{u}\mathbf{u}^T$ with $\|\mathbf{u}\| = 1$ and a of order 1, but the results below easily generalize to the case of a rank-n perturbation with $n \ll N$.

We want to know whether there will be an isolated eigenvalue of $\mathbf{M} + a\mathbf{u}\mathbf{u}^T$ outside the spectrum of $\mathbf{M}$ (i.e. an "outlier") or not. To answer this question, we calculate the matrix resolvent

$$\mathbf{G}_a(z) = \left(z - \mathbf{M} - a\mathbf{u}\mathbf{u}^T\right)^{-1}. \tag{14.6}$$

The matrix $\mathbf{G}_a(z)$ has a pole at every eigenvalue of $\mathbf{M} + a\mathbf{u}\mathbf{u}^T$. An alternative approach would have been to study the zeros of the function $\det\left(z - \mathbf{M} - a\mathbf{u}\mathbf{u}^T\right)$, but the full resolvent $\mathbf{G}_a(z)$ also gives us information about the eigenvectors.

Now we apply the Sherman–Morrison formula (1.28); taking $\mathbf{A} = z - \mathbf{M}$, we get

$$\mathbf{G}_a(z) = \mathbf{G}(z) + a\frac{\mathbf{G}(z)\mathbf{u}\mathbf{u}^T\mathbf{G}(z)}{1 - a\mathbf{u}^T\mathbf{G}(z)\mathbf{u}}, \tag{14.7}$$

where $\mathbf{G}(z)$ is the resolvent of the original matrix $\mathbf{M}$. We are looking for a real eigenvalue such that $\lambda_1 > \lambda_+$. Let us take $z = \lambda_1 \in \mathbb{R}$ outside the spectrum of $\mathbf{M}$, so $\mathbf{G}(\lambda)$ is real and regular. To have an outlier at λ_1, we need a pole of $\mathbf{G}_a$ at λ_1, i.e. the following equation needs to be satisfied:

$$1 - a\mathbf{u}^T\mathbf{G}(\lambda_1)\mathbf{u} = 0. \tag{14.8}$$

Assume now that $\mathbf{M}$ is drawn from a rotationally invariant ensemble or, equivalently, that the vector $\mathbf{u}$ is an independent random vector uniformly distributed on the unit sphere. In the language of Chapter 11, we say that the perturbation $a\mathbf{u}\mathbf{u}^T$ is free from the matrix $\mathbf{M}$. We then have, in the eigenbasis of $\mathbf{G}$,

$$\mathbf{u}^T\mathbf{G}(z)\mathbf{u} = \sum_i \mathbf{u}_i^2\mathbf{G}_{ii}(z) \approx \frac{1}{N}\operatorname{Tr}\mathbf{G}(z) = g_N(z) \overset{N\to\infty}{\to} \mathfrak{g}(z). \tag{14.9}$$

Thus we have a pole when

$$a\mathfrak{g}(\lambda_1) = 1 \Rightarrow \mathfrak{g}(\lambda_1) = 1/a. \tag{14.10}$$

If $\mathfrak{z}(g)$, the inverse function of $\mathfrak{g}(z)$, exists, we arrive at

$$\lambda_1 = \mathfrak{z}\left(\frac{1}{a}\right). \tag{14.11}$$

The condition for the invertibility of $\mathfrak{g}(z)$ happens to be precisely the same as the condition to have an outlier, i.e. $\lambda_1 > \lambda_+$ – see Section 10.4. We have established there that $\lambda_1 = \mathfrak{z}(1/a)$ is monotonically increasing in a, and $\lambda_1 = \lambda_+$ when $a = a^* = 1/\mathfrak{g}(\lambda_+)$, which is the critical value of a for which an outlier first appears. Generically, $g_+ = \mathfrak{g}(\lambda_+)$ is a minimum of $\mathfrak{z}(g)$:

$$\left.\frac{d\mathfrak{z}(g)}{dg}\right|_{g_+} = 0 \quad \text{when} \quad \mathfrak{z}(g_+) = \lambda_+. \tag{14.12}$$

For instance, for Wigner matrices, we have $\mathfrak{z}(g) = \sigma^2 g + g^{-1}$, for which

$$\sigma^2 - g_+^{-2} = 0 \Rightarrow g_+ = \sigma^{-1}, \tag{14.13}$$

and $\lambda_+ = \mathfrak{z}(\sigma^{-1}) = 2\sigma$, which is indeed the right edge of the semi-circle law.

In sum, for $a > a^* = 1/g_+$, there exists a unique outlier eigenvalue that is increasing with a. The smallest value for which we can have an outlier is $a^* = 1/g_+$, corresponding to $\lambda_1 = \lambda_+$. For $a < a^*$ there is no outlier to the right of λ_+.[3]

Using the relation between the inverse function $\mathfrak{z}(g)$ and the R-transform (10.10), we can express the position of the outlier as

$$\lambda_1 = R\left(\frac{1}{a}\right) + a \quad \text{for} \quad a > a^* = \frac{1}{g_+}. \tag{14.14}$$

[3] Outliers such that $\lambda < \lambda_-$ behave similarly, we just need to consider the matrix $-\mathbf{M} - a\mathbf{u}\mathbf{u}^T$ and follow the same logic.

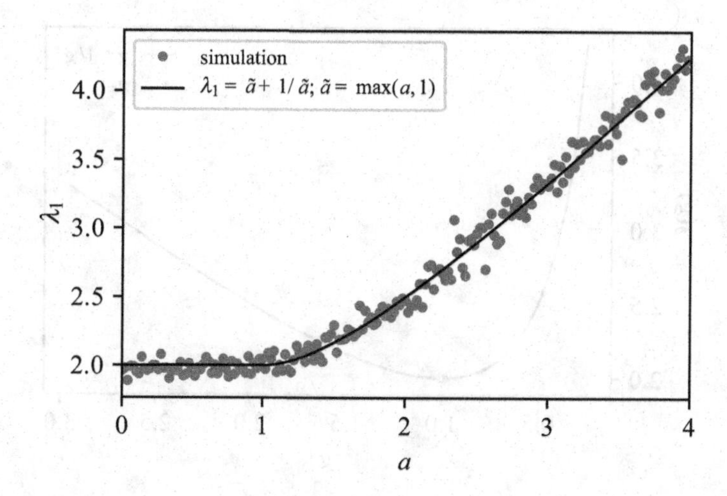

Figure 14.3 Largest eigenvalue of a Gaussian Wigner matrix with $\sigma^2 = 1$ with a rank-1 perturbation of magnitude a. Each dot is the largest eigenvalue of a single random matrix with $N = 200$. Equation (14.16) is plotted as the solid curve. For $a < 1$, the fluctuations follow a Tracy–Widom law with $N^{-2/3}$ scaling, while for $a > 1$ the fluctuations are Gaussian with $N^{-1/2}$ scaling. From the graph, we see fluctuations that are indeed smaller when $a < 1$. They also have a negative mean and positive skewness, in agreement with the Tracy–Widom distribution.

Using the cumulant expansion of the R-transform (11.63), we then get a general expression for large a:

$$\lambda_1 = a + \tau(\mathbf{M}) + \frac{\kappa_2(\mathbf{M})}{a} + O(a^{-2}). \tag{14.15}$$

For Wigner matrices, we actually have for all a (see Fig. 14.3)

$$\lambda_1 = a + \frac{\sigma^2}{a} \quad \text{for} \quad a > a^* = \sigma. \tag{14.16}$$

When $a \to a^*$, on the other hand, one has

$$\left.\frac{\mathrm{d}\lambda_1(a)}{\mathrm{d}a}\right|_{a^*} = \left.\frac{\mathrm{d}\mathfrak{z}(g)}{\mathrm{d}g}\right|_{g_+=1/a^*} = 0. \tag{14.17}$$

Hence, one has, for $a \to a^*$ and for generic square-root singularities,

$$\lambda_1 = \lambda_+ + C(a - a^*)^2 + O\left((a - a^*)^3\right), \tag{14.18}$$

where C is some problem dependent coefficient.

By studying the fluctuations of $\mathbf{u}\mathbf{G}(\lambda)\mathbf{u}^T$ around $\mathfrak{g}(\lambda)$, one can show that the fluctuations of the outlier around $\lambda_1 = R(a^{-1}) + a$ are Gaussian and of order $N^{-1/2}$. This is to be contrasted with the fluctuations of the largest eigenvalue when there are no outliers ($a < g_+$), which are Tracy–Widom and of order $N^{-2/3}$. The transition between the two regimes is called the Baik–Ben Arous–Péché (BBP) transition.

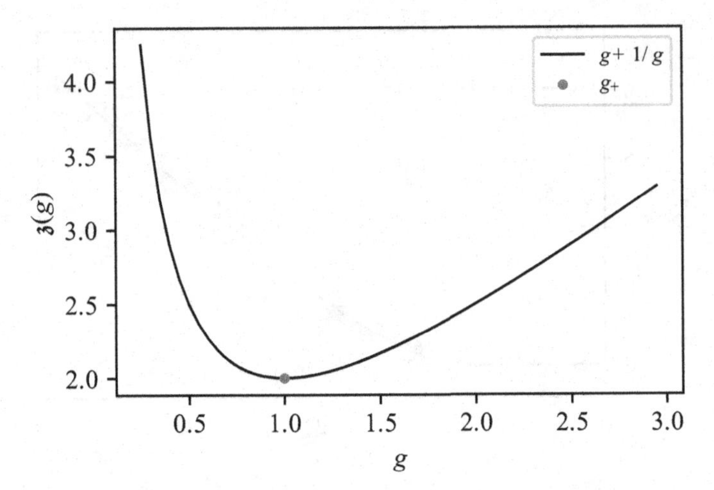

Figure 14.4 Plot of the inverse function $\mathfrak{z}(g) = g + 1/g$ for the unit Wigner function $\mathfrak{g}(z)$. The gray dot indicates the point (g_+, λ_+). The line to the left of this point is the true inverse of $\mathfrak{g}(z)$: $\mathfrak{z}(g)$ is defined on $[0, g_+)$ and is monotonously decreasing in g. The line to the right is a spurious solution introduced by the R-transform. Note that the point $g = g_+$ is a minimum of $\mathfrak{z}(g) = g + 1/g$.

We finish this section with two remarks.

- One concerns the solutions to Eq. (14.14), and the way to find the value a^* beyond which an outlier appears. The point is that, while the function $R(g = 1/a)$ is well defined for $g \in [0, g_+)$, it also often makes sense even beyond g_+ (see the discussion in Section 10.4). In that case, one will find spurious solutions: Figure 14.4 shows a plot of $\mathfrak{z}(g) = R(g) + 1/g$ in the unit Wigner case, which is still well defined for $g > g_+ = 1$ even if this function is no longer the inverse of $\mathfrak{g}(z)$ (Section 10.4). There are two solutions to $\mathfrak{z}(g) = \lambda_1$, one such that $g < g_+$ and the other such that $g > g_+$. As noted above, the point g_+ is a minimum of $\mathfrak{z}(g)$, beyond which the relation between λ_1 and a is monotonically increasing because $\mathfrak{g}(z)$ is monotonically decreasing for $z > \lambda_+$.
- The second concerns the case of a free rank-n perturbation, when $n \ll N$. In this case one cannot use the Sherman–Morrison formula but one can compute the R-transform of the perturbed matrix, and infer the $1/N$ correction to the Stieltjes transform. The poles of this correction term give the possible outliers. To each eigenvalue a_k ($k = 1, \ldots, n$) of the perturbation, one can associate a candidate outlier λ_k given by

$$\lambda_k = R\left(\frac{1}{a_k}\right) + a_k \quad \text{when} \quad a_k > \frac{1}{g_+}. \tag{14.19}$$

14.2.2 Outlier Eigenvectors

The matrix resolvent in Eq. (14.7) can also tell us about the eigenvectors of the perturbed matrix. We expect that, for a very strong rank-1 perturbation $a\mathbf{u}\mathbf{u}^T$, the eigenvector $\mathbf{v}_1$

associated with the outlier λ_1 will be very close to the perturbation vector $\mathbf{u}$. On the other hand, for $\lambda_1 \approx \lambda_+$, the vector $\mathbf{u}$ will strongly mix with bulk eigenvectors of $\mathbf{M}$ so the eigenvector $\mathbf{v}_1$ will not contain much information about $\mathbf{u}$.

To understand this phenomenon quantitatively, we will study the squared overlap $|\mathbf{v}_1^T\mathbf{u}|^2$. With the spectral decomposition of $\mathbf{M} + a\mathbf{u}\mathbf{u}^T$, we can write

$$\mathbf{G}_a(z) = \sum_{i=1}^{N} \frac{\mathbf{v}_i\mathbf{v}_i^T}{z - \lambda_i}, \tag{14.20}$$

where λ_1 denotes the outlier and $\mathbf{v}_1$ its eigenvector, and $\lambda_i, \mathbf{v}_i$, $i > 1$ all other eigenvalues/eigenvectors. Thus we have

$$\lim_{z\to\lambda_1} \mathbf{u}^T\mathbf{G}_a(z)\mathbf{u} \cdot (z - \lambda_1) = |\mathbf{v}_1^T\mathbf{u}|^2. \tag{14.21}$$

Hence, by (14.7) and (14.9), we get

$$\begin{aligned} |\mathbf{v}_1^T\mathbf{u}|^2 &= \lim_{z\to\lambda_1} \left(\mathfrak{g}(z) + a\frac{\mathfrak{g}(z)^2}{1 - a\mathfrak{g}(z)} \right)(z - \lambda_1) \\ &= \lim_{z\to\lambda_1} \mathfrak{g}(z)\frac{z - \lambda_1}{1 - a\mathfrak{g}(z)}. \end{aligned} \tag{14.22}$$

We cannot simply evaluate the fraction above at $z = \lambda_1$, for at that point $\mathfrak{g}(\lambda_1) = a^{-1}$ and we would get 0/0. We can however use l'Hospital's rule[4] and find

$$|\mathbf{v}_1^T\mathbf{u}|^2 = -\frac{\mathfrak{g}(\lambda_1)^2}{\mathfrak{g}'(\lambda_1)}, \tag{14.24}$$

where we have used $a^{-1} = \mathfrak{g}(\lambda_1)$. The right hand side is always positive since $\mathfrak{g}$ is a decreasing function for $\lambda > \lambda_+$.

We can rewrite Eq. (14.24) in terms of the R-transform and get a more useful formula. To compute $g'(z)$, we take the derivative with respect to z of the implicit equation $z = R(\mathfrak{g}(z)) + \mathfrak{g}^{-1}(z)$ and get

$$1 = R'(\mathfrak{g}(z))\mathfrak{g}'(z) - \frac{\mathfrak{g}'(z)}{\mathfrak{g}^2} \Rightarrow \mathfrak{g}'(z) = \frac{1}{R'(\mathfrak{g}(z)) - \mathfrak{g}^{-2}(z)}. \tag{14.25}$$

Hence we have

$$\begin{aligned} |\mathbf{v}_1^T\mathbf{u}|^2 &= 1 - \mathfrak{g}(\lambda_1)^2 R'(\mathfrak{g}(\lambda_1)) \\ &= 1 - a^{-2}R'(a^{-1}). \end{aligned} \tag{14.26}$$

We can now check our intuition about the overlap for large and small perturbations. For a large perturbation $a \to \infty$, Eq. (14.26) gives

[4] L'Hospital's rule states that

$$\lim_{x\to x_0} \frac{f(x)}{g(x)} = \frac{f'(x_0)}{g'(x_0)} \text{ when } f(x_0) = g(x_0) = 0. \tag{14.23}$$

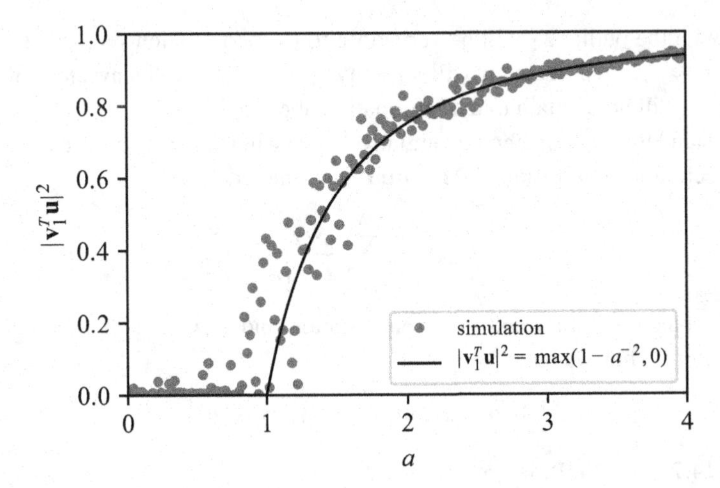

Figure 14.5 Overlap between the largest eigenvector and the perturbation vector of a Gaussian Wigner matrix with $\sigma^2 = 1$ with a rank-1 perturbation of magnitude a. Each dot is the overlap for a single random matrix with $N = 200$. Equation (14.31) is plotted as the solid curve.

$$|\mathbf{v}_1^T\mathbf{u}|^2 = 1 - \frac{\kappa_2(\mathbf{M})}{a^2} + O(a^{-3}) \quad \text{when} \quad a \to \infty. \tag{14.27}$$

As expected, $\mathbf{v}_1 \to \mathbf{u}$ when $a \to \infty$: the angle between the two vectors decreases as $1/a$.

The overlap near the transition $\lambda_1 \to \lambda_+$ can be analyzed as follows. The derivative of $\mathfrak{g}(z)$ can be written as

$$\mathfrak{g}'(z) = -\int_{\lambda_-}^{\lambda_+} \frac{\rho(x)}{(z-x)^2} \mathrm{d}x. \tag{14.28}$$

For a density that vanishes at the edge as $\rho(\lambda) \sim (\lambda_+ - \lambda)^\theta$ with exponent θ between 0 and 1, we have that $\mathfrak{g}(z)$ is finite at $z = \lambda_+$ but $\mathfrak{g}'(z)$ diverges at the same point, as $|z - \lambda_+|^{\theta-1}$. From Eq. (14.24), we have in that case[5]

$$|\mathbf{v}_1^T\mathbf{u}|^2 \propto (\lambda_1 - \lambda_+)^{1-\theta} \quad \text{when} \quad \lambda_1 \to \lambda_+. \tag{14.29}$$

In the generic case, one has $\theta = 1/2$ and, from Eq. (14.18), $\lambda_1 - \lambda_+ \propto (a - a^*)^2$, leading to

$$|\mathbf{v}_1^T\mathbf{u}|^2 \propto a - a^* \quad \text{when} \quad a \to a^*. \tag{14.30}$$

These general results are nicely illustrated by Wigner matrices, for which $R(x) = \sigma^2 x$. The overlap is explicitly given by (see Fig. 14.5)

$$|\mathbf{v}_1^T\mathbf{u}|^2 = 1 - \left(\frac{a^*}{a}\right)^2 \quad \text{when} \quad a > a^* = \sigma. \tag{14.31}$$

[5] In Chapter 5, we encountered a *critical density* where $\rho(\lambda)$ behaves as $(\lambda_+ - \lambda)^\theta$ with an exponent $\theta = \frac{3}{2} > 1$. In this case $\mathfrak{g}'(z)$ does not diverge as $z \to \lambda_+$ and the squared overlap at the edge of the BBP transition does not go to zero (first order transition). For example for the density given by Eq. (5.59) we find $|\mathbf{v}_1^T\mathbf{u}|^2 = \frac{4}{9}$ at the edge $\lambda_1 = 2\sqrt{2}$.

As $a \to a^*$, $\lambda_1 \to 2\sigma$ and $|\mathbf{v}_1^T\mathbf{u}|^2 = 2(a - a^*)/a^* \to 0$: the eigenvector becomes delocalized as the eigenvalue merges with the bulk. For $a < a^*$, one can rigorously show that there is no information left in the eigenvalues of the perturbed matrix that would allow us to reconstruct $\mathbf{u}$.

Note that for $\lambda_1 > \lambda_+$, $|\mathbf{v}_1^T\mathbf{u}|^2$ is of order unity. In Chapter 19 we will see that this is not the case for the overlaps between perturbed and unperturbed eigenvectors in the bulk, which have typical sizes of order N^{-1}.

Exercise 14.2.1 Additive perturbation of a Wishart matrix

Define a modified Wishart matrix $\mathbf{W}_1$ such that every element $(\mathbf{W}_1)_{ij} = \mathbf{W}_{ij} + a/N$, where $\mathbf{W}$ is a standard Wishart matrix and a is a constant of order 1. $\mathbf{W}_1$ is a standard Wishart matrix plus a rank-1 perturbation $\mathbf{W}_1 = \mathbf{W} + a\mathbf{u}\mathbf{u}^T$.

(a) What is the normalized vector $\mathbf{u}$ in this case?
(b) Using Eqs. (14.14) and (10.15) find the value of the outlier and the minimal a in the Wishart case.
(c) The square-overlap between the vector $\mathbf{u}$ and the new eigenvector $\mathbf{v}_1$ is given by Eq. (14.26). Give an explicit expression in the Wishart case.
(d) Generate a large modified Wishart ($q = 1/4$, $N = 1000$) for a few a in the range $[1,5]$. Compute the largest eigenvalue λ_1 and associated eigenvector $\mathbf{v}_1$. Plot λ_1 and $|\mathbf{v}_1^T\mathbf{u}|^2$ as a function of a and compare with the predictions of (b) and (c).

14.3 Fat Tails

The previous section allows us to discuss the very interesting situation of real symmetric random matrices $\mathbf{X}$ with IID elements X_{ij} that have a fat-tailed distribution, but with a finite variance (the case of infinite variance will be alluded to at the end of the section). In order to have eigenvalues of order unity, the random elements must be of typical size $N^{-1/2}$, so we write

$$X_{ij} = \frac{x_{ij}}{\sqrt{N}}, \tag{14.32}$$

with x_{ij} distributed according to some density $P(x)$ of mean zero and variance unity, but that decays as $\mu|x|^{-1-\mu}$ for large x. This means that most elements X_{ij} are small, of order $N^{-1/2}$, with some exceptional elements that are of order unity. The probability that $|X_{ij}| > 1$ is actually given by

$$\mathbb{P}(|X_{ij}| > 1) \approx 2\int_{\sqrt{N}}^{\infty} \mathrm{d}x \frac{\mu}{x^{1+\mu}} = \frac{2}{N^{\mu/2}}. \tag{14.33}$$

Since there are in total $N(N-1)/2 \approx N^2/2$ such random variables, the total number of such variables that exceed unity is given by $N^{2-\mu/2}$. Hence, for $\mu > 4$, this number tends

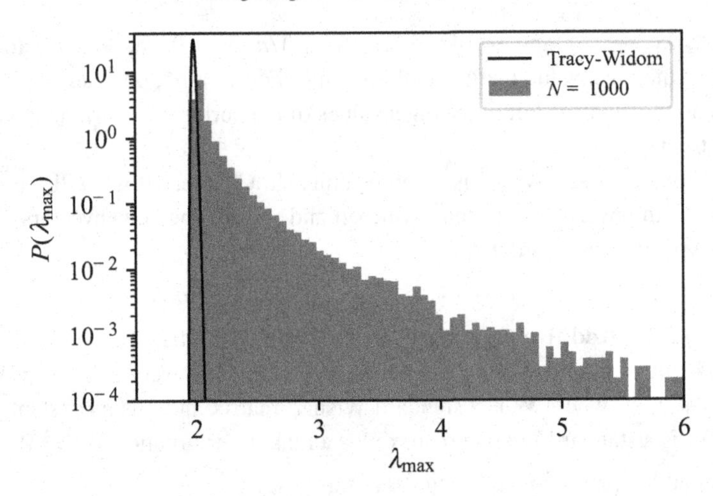

Figure 14.6 Distribution of the largest eigenvalue of $N = 1000$ Wigner matrices with elements drawn from a distribution with $\mu = 5$ compared with the prediction from the Tracy–Widom distribution. Even though as $N \to \infty$ this distribution converges to Tracy–Widom, at $N = 1000$ there is no agreement between the laws as the power law tail $\mathbb{P}(\lambda_{\max} > x) \sim x^{-5}/\sqrt{N}$ still dominates.

to zero with N: there are typically *no* such large elements in the considered random matrix. Since each pair of large entries $X_{ij} = X_{ji}$ can be considered as a rank-2 perturbation of a matrix with all elements of order $N^{-1/2}$, one concludes that for $\mu > 4$ there are no outliers, and the statistics of the largest eigenvalue is given by the Tracy–Widom distribution around $\lambda_+ = 2$. This hand-waving argument can actually be made rigorous: in the large N limit, the finiteness of the fourth moment of the distribution of matrix elements is sufficient to ensure that the largest eigenvalue is given by the Tracy–Widom distribution. However, one should be careful in interpreting this result, because although very large elements appear with vanishing probability, they still dominate the tail of the Tracy–Widom distribution for finite N. The reason is that, whereas the former decreases as $N^{2-\mu/2}$, the latter decreases much faster, as $\exp(-N(\lambda_{\max} - \lambda_+)^{3/2})$ (see Fig. 14.6).

Now consider the case $2 < \mu < 4$. Since $\mu > 2$, the variance of X_{ij} is finite and one knows that the asymptotic distribution of eigenvalues of $\mathbf{X}$ is given by the Wigner semi-circle, with $\lambda_+ = 2$. But now the number of large entries in the matrix $\mathbf{X}$ grows with N as $N^{2-\mu/2}$, which is nevertheless still much smaller than N. Each large pair of entries $X_{ij} = X_{ji} = a$ larger than unity (in absolute value) contributes to two outliers, given by $\lambda = \pm(a + 1/a)$. So there are in total $O(N^{2-\mu/2})$ outliers, the density of which is given by

$$\rho_{\text{out}}(\lambda > 2) = N^{1-\mu/2} \int_1^\infty \mathrm{d}x \frac{\mu}{x^{1+\mu}} \delta\left(\lambda - x - \frac{1}{x}\right). \tag{14.34}$$

This is a rather strange situation where the density of outliers goes to zero as $N \to \infty$ as soon as $\mu > 2$, but at the same time the largest eigenvalue in this problem goes to infinity as

$$\lambda_{\max} \sim N^{\frac{2}{\mu}-\frac{1}{2}}, \qquad 2 < \mu < 4. \tag{14.35}$$

Finally, let us briefly comment on the case $\mu < 2$, for which the variance of the entries of $\mathbf{X}$ diverges (a case called "Lévy matrices" in the literature). For eigenvalues to remain of order unity, one needs to scale the matrix elements differently, as

$$X_{ij} = \frac{x_{ij}}{N^{\frac{1}{\mu}}}, \tag{14.36}$$

with

$$P(x) \underset{|x|\to\infty}{\sim} \frac{\Gamma(1+\mu)\sin(\frac{\pi\mu}{2})}{|x|^{1+\mu}}, \tag{14.37}$$

where the funny factor involving the gamma function is introduced for convenience only. The eigenvalue distribution is no longer given by a semi-circle. In fact, the support of the distribution is in this case unbounded. For completeness, we give here the exact expression of the distribution in terms of Lévy stable laws $L_\mu^{C,\beta}(u)$, where β is called the asymmetry parameter and C the scale parameter.[6] For a given value of λ, one should first solve the following two self-consistent equations for C and β:

$$\begin{aligned} C &= \int_{-\infty}^{+\infty} \mathrm{d}g|g|^{\mu/2-2} L_{\mu/2}^{C,\beta}(\lambda - 1/g), \\ C\beta &= \int_{-\infty}^{+\infty} \mathrm{d}g \,\mathrm{sign}(g)|g|^{\mu/2-2} L_{\mu/2}^{C,\beta}(\lambda - 1/g). \end{aligned} \tag{14.38}$$

Finally, the distribution of eigenvalues $\rho_L(\lambda)$ of Lévy matrices is obtained as

$$\rho_L(\lambda) = L_{\mu/2}^{C(\lambda),\beta(\lambda)}(\lambda). \tag{14.39}$$

One can check in particular that this distribution decays for large λ exactly as $P(x)$ itself. In other words, the tail of the eigenvalue distribution is the same as the tail of the independent entries of the Lévy matrix.

14.4 Multiplicative Perturbation

In data-analysis applications, we often need to understand the largest eigenvalue of a sample covariance matrix. A true covariance with a few isolated eigenvalues can be treated as a matrix $\mathbf{C}_0$ with no isolated eigenvalue plus a low-rank perturbation. The passage from the true covariance to the sample covariance is equivalent to the free product of the true covariance with a white Wishart matrix with appropriate aspect ratio $q = N/T$. To understand such matrices, we will now study outliers for a multiplicative process.

Consider the free product of a certain covariance matrix $\mathbf{C}_0$ with a rank-1 perturbation and another matrix $\mathbf{B}$:

$$\mathbf{E} = \mathbf{B}^{\frac{1}{2}}\mathbf{C}_0^{\frac{1}{2}}\left(\mathbf{1} + a\mathbf{u}\mathbf{u}^T\right)\mathbf{C}_0^{\frac{1}{2}}\mathbf{B}^{\frac{1}{2}}, \tag{14.40}$$

[6] More precisely, $L_\mu^{C,\beta}(x)$ is the Fourier transform of $\exp(-C|k|^\mu(1 + i\beta\mathrm{sign}(k)\tan(\pi\mu/2)))$ for $\mu \neq 1$, and of $\exp(-C|k|(1 + i(2\beta/\pi)\mathrm{sign}(k)\log|k|))$ for $\mu = 1$.

where $\mathbf{u}$ is a normalized eigenvector of $\mathbf{C}_0$ with eigenvalue λ_0, and $\mathbf{B}$ is positive semi-definite, free from $\mathbf{C}_0$, and with $\tau(\mathbf{B}) = 1$. In the special case where $\mathbf{B}$ is a white Wishart, our problem corresponds to a noisy observation of a perturbed covariance matrix, where one of the modes (the one corresponding to $\mathbf{u}$) has a variance boosted by a factor $(1 + a)$.

The matrix $\mathbf{E}_0 := \mathbf{B}^{\frac{1}{2}}\mathbf{C}_0\mathbf{B}^{\frac{1}{2}}$ has an unperturbed spectrum with a lower edge λ_- and an upper edge λ_+. We want to establish, as in the additive case, the condition for the existence of an outlier $\lambda_1 > \lambda_+$ or $\lambda_1 < \lambda_-$, and the exact position of the outlier when it exists.

The eigenvalues of $\mathbf{E}$ are the zeros of its characteristic polynomial, in particular for the largest eigenvalue λ_1 we have

$$\det(\lambda_1\mathbf{1} - \mathbf{E}_0 - a\mathbf{B}^{\frac{1}{2}}\mathbf{C}_0\mathbf{u}\mathbf{u}^T\mathbf{B}^{\frac{1}{2}}) = 0. \tag{14.41}$$

We are looking for an eigenvalue outside the spectrum of $\mathbf{E}_0$, i.e. $\lambda_1 > \lambda_+$ or $\lambda_1 < \lambda_-$. For such a λ_1, the matrix $\lambda_1\mathbf{1} - \mathbf{E}_0$ is invertible and we can use the matrix determinant lemma Eq. (1.30):

$$\det(\mathbf{A} + \mathbf{u}\mathbf{v}^T) = \det\mathbf{A} \cdot \left(1 + \mathbf{v}^T\mathbf{A}^{-1}\mathbf{u}\right), \tag{14.42}$$

with $\mathbf{A} = \lambda_1\mathbf{1} - \mathbf{E}_0$ and $\mathbf{u} = -\mathbf{v} = \sqrt{a}\mathbf{B}^{\frac{1}{2}}\mathbf{C}_0^{\frac{1}{2}}\mathbf{u}$. Equation (14.41) becomes

$$\det(\lambda_1\mathbf{1} - \mathbf{E}_0) \cdot \left(1 - a\mathbf{u}^T\mathbf{C}_0^{\frac{1}{2}}\mathbf{B}^{\frac{1}{2}}\mathbf{G}_0(\lambda_1)\mathbf{B}^{\frac{1}{2}}\mathbf{C}_0^{\frac{1}{2}}\mathbf{u}\right) = 0, \tag{14.43}$$

where we have introduced the matrix resolvent $\mathbf{G}_0(\lambda_1) := (\lambda_1\mathbf{1} - \mathbf{E}_0)^{-1}$. As we said, the matrix $\lambda_1\mathbf{1} - \mathbf{E}_0$ is invertible so its determinant is non-zero. Thus any outlier needs to solve

$$a\lambda_0\mathbf{u}^T\mathbf{B}^{\frac{1}{2}}\mathbf{G}_0(\lambda_1)\mathbf{B}^{\frac{1}{2}}\mathbf{u} = 1. \tag{14.44}$$

Again we assume that $\mathbf{B}$ is a rotationally invariant matrix with respect to $\mathbf{C}_0$. Then we know that in the large N limit $\mathbf{G}_0(z)$ is diagonal in the basis of $\mathbf{B}$ and reads (see, *mutatis mutandis*, Eq. (13.47))

$$\mathbf{G}_0(z) \approx S^*(z)\mathbf{G}_\mathbf{B}(zS^*(z)), \qquad S^*(z) := S_{\mathbf{C}_0}(zg_0(z) - 1). \tag{14.45}$$

Furthermore, since $\mathbf{u}$ and $\mathbf{B}$ are also free,

$$\mathbf{u}^T\mathbf{B}^{\frac{1}{2}}\mathbf{G}_0(\lambda_1)\mathbf{B}^{\frac{1}{2}}\mathbf{u} \approx N^{-1}\operatorname{Tr}\left(\mathbf{B}^{\frac{1}{2}}G_0(\lambda_1)\mathbf{B}^{\frac{1}{2}}\right) \equiv S^*(\lambda_1)\mathrm{t}_\mathbf{B}(\lambda_1 S^*(\lambda_1)), \tag{14.46}$$

where we have recognized the T-transform of the matrix $\mathbf{B}$:

$$\mathrm{t}_\mathbf{B}(z) := \tau\left(\mathbf{B}(z - \mathbf{B})^{-1}\right). \tag{14.47}$$

Thus, the position of the outlier λ_1 is given by the solution of

$$a\lambda_0 S^*(\lambda_1)\mathrm{t}_\mathbf{B}(\lambda_1 S^*(\lambda_1)) = 1. \tag{14.48}$$

In order to keep the calculation simple, we now assume that $\mathbf{C}_0 = \mathbf{1}$. In this case, $S^* = 1$ and $\lambda_0 = 1$, so the equation simplifies to

$$a t_{\mathbf{B}}(\lambda_1) = 1. \tag{14.49}$$

To know whether this equation has a solution we need to know if $t_{\mathbf{B}}(\zeta)$ is invertible. The argument is very similar to the one for $\mathfrak{g}(z)$ in the additive case. In the large N limit, $t_{\mathbf{B}}(\zeta)$ converges to

$$t_{\mathbf{B}}(\zeta) = \int_{\lambda_-}^{\lambda_+} \frac{\rho_{\mathbf{B}}(x)x}{\zeta - x} dx. \tag{14.50}$$

So $t_{\mathbf{B}}(\zeta)$ is monotonically decreasing for $\zeta > \lambda_+$ and is therefore invertible. We then have

$$\lambda_1 = \zeta(a^{-1}) \qquad \text{when} \quad \lambda_1 > \lambda_+, \tag{14.51}$$

where we use the notation $\zeta(t)$ for the inverse of the T-transform of $\mathbf{B}$, in the region where it is invertible.

The inverse function $\zeta(t)$ can be expressed in terms of the S-transform via Eq. (11.92). We get

$$\lambda_1 = \zeta(a^{-1}) = \frac{a+1}{S_{\mathbf{B}}(a^{-1})} \qquad \text{when} \quad a > \frac{1}{t_{\mathbf{B}}(\lambda_+)}. \tag{14.52}$$

Applying the theory to a Wishart matrix $\mathbf{B} = \mathbf{W}$ with

$$S_{\mathbf{W}}(x) = \frac{1}{1+qx}, \qquad \lambda_\pm = (1 \pm \sqrt{q})^2, \tag{14.53}$$

one finds that an outlier appears to the right of λ_+ for $a > \sqrt{q}$, with

$$\lambda_1 = (a+1)\left(1 + \frac{q}{a}\right). \tag{14.54}$$

For large a, we have $\lambda_1 \approx a + 1 + q$, i.e. a large eigenvalue $a+1$ in the covariance matrix $\mathbf{C}$ will appear shifted by q in the eigenvalues of the sample covariance matrix.

Nothing prevents us from considering negative values of a, such that $a > -1$ to preserve the positive definite nature of $\mathbf{C}$. In this case, an outlier appears to the left of λ_- when $a < -\sqrt{q}$. Its position is given by the very same equation (14.54) as above.

Exercise 14.4.1 Transpose version of multiplicative perturbation

Consider a positive definite rotationally invariant random matrix $\mathbf{B}$ and a normalized vector $\mathbf{u}$. In this exercise, we will show that the matrix $\mathbf{F}$ defined by

$$\mathbf{F} = (\mathbf{1} + c\mathbf{u}\mathbf{u}^T)\mathbf{B}(\mathbf{1} + c\mathbf{u}\mathbf{u}^T), \tag{14.55}$$

with $c > 0$ sufficiently large, has an outlier λ_1 given by Eq. (14.52) with $b+1 = (c+1)^2$.

(a) Show that for two positive definite matrices $\mathbf{A}$ and $\mathbf{B}$, $\mathbf{B}^{\frac{1}{2}}\mathbf{A}\mathbf{B}^{\frac{1}{2}}$ has the same eigenvalues as $\mathbf{A}^{\frac{1}{2}}\mathbf{B}\mathbf{A}^{\frac{1}{2}}$.

(b) Show that for a normalized vector $\mathbf{u}$

$$\left(\mathbf{1} + (a-1)\mathbf{u}\mathbf{u}^T\right)^{\frac{1}{2}} = \mathbf{1} + (\sqrt{a}-1)\mathbf{u}\mathbf{u}^T. \tag{14.56}$$

(c) Finish the proof of the above statement.

Exercise 14.4.2 Multiplicative perturbation of an inverse-Wishart matrix

We will see in Section 15.2.3 that the inverse-Wishart matrix $\mathbf{\mathcal{M}}_p$ is defined as

$$\mathbf{\mathcal{M}}_p = (1-q)\mathbf{W}_q^{-1}, \tag{14.57}$$

where $\mathbf{W}_q$ is a Wishart matrix with parameter q and p the variance of the inverse-Wishart is given by $p = \frac{q}{1-q}$. The S-transform of $\mathbf{\mathcal{M}}_p$ is given by

$$S_{\mathbf{\mathcal{M}}_p}(t) = 1 - pt. \tag{14.58}$$

Consider the diagonal matrix $\mathbf{D}$ with $\mathbf{D}_{11} = d$ and all other diagonal entries equal to 1.

(a) $\mathbf{D}$ can be written as $\mathbf{1} + c\mathbf{u}\mathbf{u}^T$. What is the normalized vector $\mathbf{u}$ and the constant c?

(b) Using the result from Exercise 14.4.1, find the value of the largest eigenvalue of the matrix $\mathbf{D}\mathbf{\mathcal{M}}_p\mathbf{D}$ as a function of d. Note that your expression will only be valid for sufficiently large d.

(c) Numerically generate matrices $\mathbf{\mathcal{M}}_p$ with $N = 1000$ and $p = 1/2$ ($q = 1/3$). Find the largest eigenvalue of $\mathbf{D}\mathbf{\mathcal{M}}_p\mathbf{D}$ for various values of d and make a plot of λ_1 vs d. Superimpose your analytical result.

(d) (Harder) Find analytically the minimum value of d to have an outlier λ_1.

14.5 Phase Retrieval and Outliers

Optical detection devices like CCD cameras or photosensitive films measure the photon flux but are blind to the phase of the incoming light. More generally, it is often the case that one can only measure the power spectral density of a signal, which is the magnitude of its Fourier transform. Can one recover the full signal based on this partial information? This problem is called phase retrieval and can be framed mathematically as follows. Let an unknown vector $\mathbf{x} \in \mathbb{R}^N$ be "probed" with T vectors $\mathbf{a}_k$, in the sense that the measurement apparatus gives us $y_k = |\mathbf{a}_k^T\mathbf{x}|^2$ with $k = 1, \ldots, T$.[7] Vectors $\mathbf{x}$ and $\mathbf{a}_k$ are taken to be real but they can easily be made complex.

The phase retrieval problem is

$$\hat{\mathbf{x}} = \underset{\mathbf{x}}{\operatorname{argmin}} \left(\sum_k \left| |\mathbf{a}_k^T\mathbf{x}|^2 - y_k \right|^2 \right). \tag{14.59}$$

[7] We could consider that there is some additional noise in the measurement of y_k but for simplicity we keep here with the noiseless version.

It is a difficult non-convex optimization problem with many local minima. To efficiently find an acceptable solution, we need a starting point $\mathbf{x}_0$ that somehow points in the direction of the true solution $\mathbf{x}$. The problem is that in large dimensions the probability that a random vector $\mathbf{x}_0$ has an overlap $|\mathbf{x}^T\mathbf{x}_0| > \varepsilon$ is exponentially small in N as soon as $\varepsilon > 0$. We will explore here a technique that allows one to find a vector $\mathbf{x}_0$ with non-vanishing overlap with the true $\mathbf{x}$.

The idea is to build some sort of weighted Wishart matrix such that this matrix will have an outlier with non-zero overlap with the unknown true vector $\mathbf{x}$. Consider the following matrix:

$$\mathbf{M} = \frac{1}{T}\sum_{k=1}^{T} f(y_k)\,\mathbf{a}_k\mathbf{a}_k^T, \tag{14.60}$$

where the T vectors $\mathbf{a}_k$ are of size N and $f(y)$ is a function that we will choose later. The function $f(y)$ should be bounded above, otherwise we might have outliers dominated by a few large values of $f(y_k)$. One such function that we will study is the simple threshold $f(y) := \Theta(y-1)$. In large dimensions the results should not depend on the precise statistics of the vectors $\mathbf{a}_k$ provided they are sufficiently random. Here we will assume that all their components are standard IID Gaussian. This assumption makes the problem invariant by rotation. Without loss of generality, we can assume the true vector $\mathbf{x}$ is in the canonical direction $\mathbf{e}_1$. The weights $f(y_k)$ are therefore assumed to be correlated to $|[\mathbf{a}_k]_1|^2 = |\mathbf{a}_k^T\mathbf{e}_1|^2$ and independent of all other components of the vectors $\mathbf{a}_k$.

Given that the first row and column of $\mathbf{M}$ contain the element $[\mathbf{a}_k]_1$, we write the matrix in block form as in Section 1.2.5:

$$\mathbf{M} = \begin{pmatrix} \mathbf{M}_{11} & \mathbf{M}_{12} \\ \mathbf{M}_{21} & \mathbf{M}_{22} \end{pmatrix}, \tag{14.61}$$

with the (11) block of size 1×1 and the (22) block of size $(N-1)\times(N-1)$. To find a potential outlier, we look for the zeros of the Stieltjes transform $g_{\mathbf{M}}(z) = \tau((z\mathbf{1}-\mathbf{M})^{-1})$. Combining Eqs. (1.32) and (1.33),

$$N g_{\mathbf{M}}(z) = \operatorname{Tr}\mathbf{G}_{22}(z) + \frac{1 + \operatorname{Tr}\left[\mathbf{G}_{22}(z)\mathbf{M}_{21}\mathbf{M}_{12}\mathbf{G}_{22}(z)\right]}{z - \mathbf{M}_{11} - \mathbf{M}_{12}\mathbf{G}_{22}(z)\mathbf{M}_{21}}, \tag{14.62}$$

where $\mathbf{G}_{22}(z)$ is the matrix resolvent of the rotationally invariant matrix $\mathbf{M}_{22}$ (i.e. the matrix $\mathbf{M}$ without its first row and column). In the large N limit we expect M_{22} to have a continuous spectrum with an edge λ_+. When the condition for the denominator to vanish, i.e.

$$\lambda_1 - \mathbf{M}_{11} - \mathbf{M}_{12}\mathbf{G}_{22}(\lambda_1)\mathbf{M}_{21} = 0, \tag{14.63}$$

has a solution for $\lambda_1 > \lambda_+$ we can say that the matrix $\mathbf{M}$ has an outlier. The overlap between the corresponding eigenvector $\mathbf{v}_1$ and $\mathbf{x}$ is given by the residue

$$\varrho := \frac{|\mathbf{v}_1^T\mathbf{x}|^2}{|\mathbf{x}|^2} = \left|\mathbf{v}_1^T\mathbf{e}_1\right|^2 = \lim_{z\to\lambda_1} \frac{z-\lambda_1}{z - \mathbf{M}_{11} - \mathbf{M}_{12}\mathbf{G}_{22}(z)\mathbf{M}_{21}}. \tag{14.64}$$

In the large N limit the scalar equation (14.63) becomes self-averaging. We have

$$\mathbf{M}_{11} = \frac{1}{T}\sum_{k=1}^{T} f(y_k)([\mathbf{a}_k]_1)^2 \overset{T\to\infty}{\to} \mathbb{E}\left[f(y)([\mathbf{a}]_1)^2\right]. \tag{14.65}$$

For the second term we have

$$\mathbf{M}_{12}\mathbf{G}_{22}(z)\mathbf{M}_{21} = \sum_{k,\ell=1}^{T} \frac{1}{T^2} f(y_k) f(y_\ell)[\mathbf{a}_k]_1[\mathbf{a}_\ell]_1 \sum_{i,j>1}^{N} [\mathbf{a}_k]_i[\mathbf{G}_{22}(z)]_{ij}[\mathbf{a}_\ell]_j$$

$$\overset{T\to\infty}{\to} q\mathbb{E}\left[f^2(y)([\mathbf{a}]_1)^2\right]h(z), \tag{14.66}$$

where $q = N/T$ and

$$h(z) = \tau\left(\frac{\mathbf{H}\mathbf{H}^T}{T}\mathbf{G}_{22}(z)\right), \quad [\mathbf{H}]_{ik} = [\mathbf{a}_k]_i \quad i > 1. \tag{14.67}$$

We can now put everything together and use l'Hospital's rule to compute the residue. For convenience we define the constants $c_n := \mathbb{E}\left[f^n(y)([\mathbf{a}]_1)^2\right]$. There will be an outlier with overlap

$$\varrho = \frac{1}{1 - qc_2 h'(\lambda_1)} \tag{14.68}$$

when there is a solution $\lambda_1 > \lambda_+$ of

$$\lambda_1 = c_1 + qc_2 h(\lambda_1). \tag{14.69}$$

We will come back later to the computation of $h(z)$. In the $q \to 0$ limit the matrix $\mathbf{M}$ becomes proportional to the identity $\mathbf{M} = \mathbb{E}[f(y)]\mathbf{1} := m_1\mathbf{1}$, so $g_{\mathbf{M}}(z) = 1/(z - m_1)$ and $h(z) = 1/(z - m_1)$. For $q = 0$ we have a solution $\lambda_1 = c_1$ which satisfies $c_1 \geq m_1$. In this limit the overlap tends to one. The linear correction in q is easily obtained as we only need $h(z)$ to order zero. We obtain

$$\lambda_1 = c_1 + q\frac{c_2}{c_1 - m_1} + O(q^2), \quad \varrho = 1 - q\frac{c_2}{(c_1 - m_1)^2} + O(q^2). \tag{14.70}$$

Note that $c_2/(c_1 - m_1)^2$ is always positive so the overlap decreases with q starting from $\varrho = 1$ at $q = 0$. For the unit thresholding function $f(y) = \Theta(y - 1)$ we have $m_1 = \text{erfc}(1/\sqrt{2}) \approx 0.317$ and $c_1 = c_2 = m_1 + \sqrt{2/(e\pi)} \approx 0.801$ (see Fig. 14.7).

Since we have the freedom to choose any bounded function $f(y)$ we should choose the one that gives the largest overlap for the value of q given by our dataset. We will do an easier computation, namely minimize the slope of the linear approximation in q. We want

$$f_{\text{opt}}(y) = \underset{f(y)}{\text{argmin}} \frac{c_2}{(c_1 - m_1)^2} = \underset{f(y)}{\text{argmin}} \frac{\mathbb{E}_a\left[f^2(a^2)a^2\right]}{\mathbb{E}_a\left[f(a^2)(a^2 - 1)\right]^2}, \tag{14.71}$$

where the law of a is $\mathcal{N}(0, 1)$. A variational minimization gives

$$f_{\text{opt}}(y) = 1 - \frac{1}{y}. \tag{14.72}$$

The optimal function is not bounded below and therefore the distribution of eigenvalues is singular with $c_2 \to \infty$ and $m_1 \to -\infty$. One should think of this function as the limit of a series of functions such that $c_2/(c_1 - m_1)^2 \to 0$. In Figure 14.7 we see that numerically this function has indeed an overlap as a function of q with zero slope at the origin. As a consequence it has non-zero overlap for much greater values of q (fewer data T) than the simple thresholding function. It turns out that our small q optimum $f(y) = 1 - 1/y$ is actually the optimal function for all values of q.

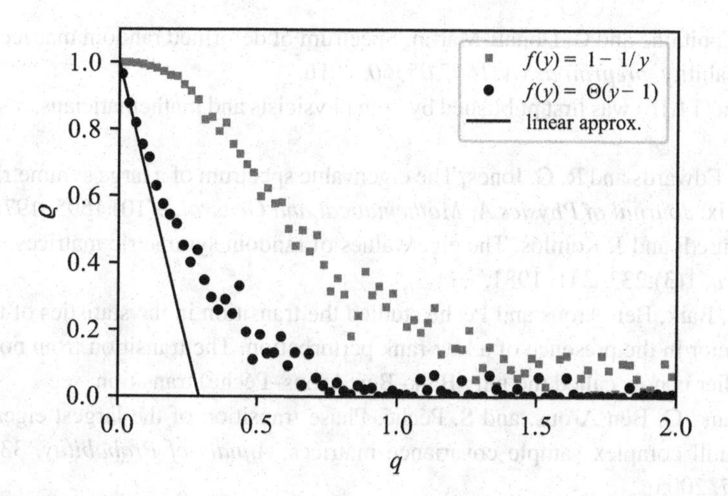

Figure 14.7 Overlap $\varrho := \left|\mathbf{v}_1^T \mathbf{x}\right|^2 / |\mathbf{x}|^2$ between the largest eigenvector and the true signal as a function of $q := N/T$ for two functions: the simple $f(y) = \Theta(y-1)$ and the optimal $f(y) = 1 - 1/y$. Each dot corresponds to a single matrix of aspect ratio q and $NT = 10^6$. The solid line corresponds to the linear q approximation Eq. (14.70) in the thresholding case. For the optimal case, the slope at the origin is zero.

For completeness we show here how to compute the function $h(z)$. We have the following subordination relation (for the (22) block of the relevant matrices):

$$\mathfrak{t}_{\mathbf{M}}(\zeta) = \mathfrak{t}_{\mathbf{W}_q}\left(S_f(q\mathfrak{t}_{\mathbf{M}}(\zeta))\zeta\right), \tag{14.73}$$

where $S_f(t)$ is the S-transform of a diagonal matrix with entries $f(y_k)$. We then have

$$\begin{aligned} h(z) = \tau(\mathbf{W}_q \mathbf{G}_{22}(z)) &= S\tau\left[(Sz\mathbf{1} - \mathbf{M})^{-1}(\mathbf{M} - Sz + Sz)\right] \\ &= S(z\mathfrak{g}_{\mathbf{M}}(z) - 1), \end{aligned} \tag{14.74}$$

with $S := S_f(q(z\mathfrak{g}_{\mathbf{M}}(z) - 1))$. Since

$$S_{\mathbf{M}}(t) = \frac{S_f(qt)}{1+qt} = \frac{t+1}{t\zeta}, \tag{14.75}$$

we have

$$h(z) = \mathfrak{g}_{\mathbf{M}}(z)[1 + q(z\mathfrak{g}_{\mathbf{M}}(z) - 1)], \tag{14.76}$$

where the function $\mathfrak{g}_{\mathbf{M}}(z)$ can be obtained by inverting the relation

$$\mathfrak{z}(g) = \int_{-\infty}^{\infty} \frac{dx}{\sqrt{2\pi}} \frac{f(x^2)e^{-x^2/2}}{1 - qgf(x^2)} + \frac{1}{g}. \tag{14.77}$$

Bibliographical Notes

- For a recent review on the subject of free low-rank additive or multiplicative perturbations, the interested reader should consult

  - M. Capitaine and C. Donati-Martin. Spectrum of deformed random matrices and free probability. *preprint arXiv:1607.05560*, 2016.
- Equation (14.16) was first published by both physicists and mathematicians, respectively, in
  - S. F. Edwards and R. C. Jones. The eigenvalue spectrum of a large symmetric random matrix. *Journal of Physics A: Mathematical and General*, 9(10):1595, 1976,
  - Z. Füredi and J. Komlós. The eigenvalues of random symmetric matrices. *Combinatorica*, 1(3):233–241, 1981.
- In 2005, Baik, Ben Arous and Péché studied the transition in the statistics of the largest eigenvector in the presence of a low-rank perturbation. The transition from no outlier to one outlier is now called the BBP (Baik–Ben Arous–Péché) transition; see
  - J. Baik, G. Ben Arous, and S. Péché. Phase transition of the largest eigenvalue for nonnull complex sample covariance matrices. *Annals of Probability*, 33(5):1643–1697, 2005,
  - D. Féral and S. Péché. The largest eigenvalue of rank one deformation of large Wigner matrices. *Communications in Mathematical Physics*, 272(1):185–228, 2007.
- The case of outliers with a hard edge is treated in
  - S. Péché. *Universality of local eigenvalue statistics for random sample covariance matrices*. PhD thesis, EPFL, 2003.
- The generalization to spiked tensors has recently been studied in
  - G. Ben Arous, S. Mei, A. Montanari, and M. Nica. The landscape of the spiked tensor model. *Communications on Pure and Applied Mathematics*, 72(11):2282–2330, 2019.
- The case of additive or multiplicative perturbation to a general matrix (beyond Wigner and Wishart) was worked out by
  - F. Benaych-Georges and R. R. Nadakuditi. The eigenvalues and eigenvectors of finite, low rank perturbations of large random matrices. *Advances in Mathematics*, 227(1):494–521, 2011.
- For studies about the edge properties of Wigner matrices, see e.g.
  - M. J. Bowick and E. Brzin. Universal scaling of the tail of the density of eigenvalues in random matrix models. *Physics Letters B*, 268(1):21–28, 1991,

  and for the specific problem of the largest eigenvalue, the seminal paper of
  - C. A. Tracy and H. Widom. Level-spacing distributions and the Airy kernel. *Communications in Mathematical Physics*, 159(1):151–174, 1994,

  which has led to a flurry of activity, see e.g.
  - S. Majumdar. Random matrices, the Ulam problem, directed polymers and growth models, and sequence matching. In *Les Houches Lecture Notes "Complex Systems"*, volume 85. Elsevier Science, 2007.
- When the elements of the random matrix are fat tailed, the Tracy–Widom law is modified, see
  - G. Biroli, J.-P. Bouchaud, and M. Potters. On the top eigenvalue of heavy-tailed random matrices. *Europhysics Letters (EPL)*, 78(1):10001, 2007,

– A. Auffinger, G. Ben Arous, and S. Péché. Poisson convergence for the largest eigenvalues of heavy tailed random matrices. *Annales de l'I.H.P. Probabilités et statistiques*, 45(3):589–610, 2009.

- For studies of infinite variance Lévy matrices, see

– P. Cizeau and J. P. Bouchaud. Theory of Lévy matrices. *Physical Review E*, 50:1810–1822, 1994,
– Z. Burda, J. Jurkiewicz, M. A. Nowak, G. Papp, and I. Zahed. Free random Lévy and Wigner-Lévy matrices. *Physical Review E*, 75:051126, 2007,
– G. Ben Arous and A. Guionnet. The spectrum of heavy tailed random matrices. *Communications in Mathematical Physics*, 278(3):715–751, 2008,
– S. Belinschi, A. Dembo, and A. Guionnet. Spectral measure of heavy tailed band and covariance random matrices. *Communications in Mathematical Physics*, 289(3):1023–1055, 2009.

- On the outlier method for the phase recovery problem, the reader is referred to

– Y. M. Lu and G. Li. Phase transitions of spectral initialization for high-dimensional nonconvex estimation. *preprint arXiv:1702.06435*, 2017,
– W. Luo, W. Alghamdi, and Y. M. Lu. Optimal spectral initialization for signal recovery with applications to phase retrieval. *IEEE Transactions on Signal Processing*, 67(9):2347–2356, 2019,

and for a review, see

– A. Stern, editor. *Optical Compressive Imaging*. CRC Press, Boca Raton, Fla., 2017.

Part III

Applications

15

Addition and Multiplication: Recipes and Examples

In the second part of this book, we have built the necessary tools to compute the spectrum of sums and products of free random matrices. In this chapter we will review the results previously obtained and show how they work on concrete, simple examples. More sophisticated examples, and some applications to real world data, will be developed in subsequent chapters.

15.1 Summary

We introduced the concept of freeness which can be summarized by the following intuitive statement: two large matrices are free if their eigenbases are related by a random rotation. In particular a large matrix drawn from a rotationally invariant ensemble is free with respect to any matrix independent of it, for example a deterministic matrix.[1] For example $\mathbf{A}$ and $\mathbf{OBO}^T$ are free when $\mathbf{O}$ is a random rotation matrix (in the large dimension limit). When $\mathbf{A}$ and $\mathbf{B}$ are free, their R- and S-transforms are, respectively, additive and multiplicative:

$$R_{\mathbf{A}+\mathbf{B}}(x) = R_{\mathbf{A}}(x) + R_{\mathbf{B}}(x), \qquad S_{\mathbf{AB}}(t) = S_{\mathbf{A}}(t)S_{\mathbf{B}}(t). \tag{15.1}$$

The free product needs some clarification as $\mathbf{AB}$ is in general not a symmetric matrix, the S-transform $S_{\mathbf{AB}}(t)$ in fact relates to the eigenvalues of the matrix $\sqrt{\mathbf{A}}\mathbf{B}\sqrt{\mathbf{A}}$, which are the same as those of $\sqrt{\mathbf{B}}\mathbf{A}\sqrt{\mathbf{B}}$ when both $\mathbf{A}$ and $\mathbf{B}$ are positive semi-definite (otherwise the square-root is ill defined).

15.1.1 R- and S-Transforms

The R- and S-transforms are defined by the following relations:

$$\mathfrak{g}_{\mathbf{A}}(z) = \tau\left[(z - \mathbf{A})^{-1}\right], \tag{15.2}$$

$$\mathfrak{t}_{\mathbf{A}}(\zeta) = \tau\left[(1 - \zeta^{-1}\mathbf{A})^{-1}\right] - 1 = \zeta\,\mathfrak{g}_{\mathbf{A}}(\zeta) - 1; \tag{15.3}$$

$$R_{\mathbf{A}}(x) = \mathfrak{z}_{\mathbf{A}}(x) - \frac{1}{x}, \quad S_{\mathbf{A}}(t) = \frac{t+1}{t\zeta_{\mathbf{A}}(t)} \text{ if } \tau(\mathbf{A}) \neq 0, \tag{15.4}$$

[1] By *large*, we mean that all normalized moments computed using freeness are correct up to corrections that are $O(1/N)$.

where $\mathfrak{z}_{\mathbf{A}}(x)$ and $\zeta_{\mathbf{A}}(t)$ are the inverse functions of $\mathfrak{g}_{\mathbf{A}}(z)$ and $\mathfrak{t}_{\mathbf{A}}(\zeta)$, respectively.

Under multiplication by a scalar they behave as

$$R_{\alpha\mathbf{A}}(x) = \alpha R_{\mathbf{A}}(\alpha x), \qquad S_{\alpha\mathbf{A}}(t) = \alpha^{-1} S_{\mathbf{A}}(t). \tag{15.5}$$

While the R-transform behaves simply under a shift by a scalar,

$$R_{\mathbf{A}+\alpha\mathbf{1}}(x) = \alpha + R_{\mathbf{A}}(x), \tag{15.6}$$

there is no simple formula for computing the S-transform of a shifted matrix. On the other hand, the S-transform is simple under matrix inversion:

$$S_{\mathbf{A}^{-1}}(x) = \frac{1}{S_{\mathbf{A}}(-x-1)}. \tag{15.7}$$

The two transforms are related by the following equivalent identities:

$$S_{\mathbf{A}}(t) = \frac{1}{R_{\mathbf{A}}(tS_{\mathbf{A}}(t))}, \qquad R_{\mathbf{A}}(x) = \frac{1}{S_{\mathbf{A}}(xR_{\mathbf{A}}(x))}. \tag{15.8}$$

The identity matrix has particularly simple transforms:

$$\mathfrak{g}_{\mathbf{1}}(z) = \frac{1}{z-1} \qquad \mathfrak{t}_{\mathbf{1}}(\zeta) = \frac{1}{\zeta - 1}; \tag{15.9}$$

$$R_{\mathbf{1}}(x) = 1, \qquad S_{\mathbf{1}}(t) = 1. \tag{15.10}$$

The R- and S-transforms have the following Taylor expansion for small arguments:

$$R_{\mathbf{A}}(x) = \kappa_1 + \kappa_2 x + \kappa_3 x^2 + \cdots, \qquad S_{\mathbf{A}}(x) = \frac{1}{\kappa_1} - \frac{\kappa_2}{\kappa_1^3} x + \frac{2\kappa_2^2 - \kappa_1\kappa_3}{\kappa_1^5} x^2 + \cdots, \tag{15.11}$$

where κ_n are the free cumulants of $\mathbf{A}$:

$$\kappa_1 = \tau(\mathbf{A}), \qquad \kappa_2 = \tau(\mathbf{A}^2) - \tau^2(\mathbf{A}), \qquad \kappa_3 = \tau(\mathbf{A}^3) - 3\tau(\mathbf{A})\tau(\mathbf{A}^2) + 2\tau^3(\mathbf{A}). \tag{15.12}$$

Combining Eqs. (15.7) and (15.11), we can obtain the inverse moments of $\mathbf{A}$ from its S-transform. In particular,

$$\tau(\mathbf{A}^{-1}) = S_{\mathbf{A}}(-1), \quad \tau(\mathbf{A}^{-2}) = S_{\mathbf{A}}(-1)\left(S_{\mathbf{A}}(-1) - S'_{\mathbf{A}}(-1)\right). \tag{15.13}$$

15.1.2 Computing the Eigenvalue Density

The R-transform provides a systematic way to obtain the spectrum of the sum $\mathbf{C}$ of two independent matrices $\mathbf{A}$ and $\mathbf{B}$, where at least one of them is rotationally invariant. Here is a simple recipe to compute the eigenvalue density of a free sum of matrices:

1 Find $\mathfrak{g}_{\mathbf{B}}(z)$ and $\mathfrak{g}_{\mathbf{A}}(z)$.
2 Invert $\mathfrak{g}_{\mathbf{B}}(z)$ and $\mathfrak{g}_{\mathbf{A}}(z)$ to get $\mathfrak{z}_{\mathbf{B}}(g)$ and $\mathfrak{z}_{\mathbf{A}}(g)$, and hence $R_{\mathbf{B}}(x)$ and $R_{\mathbf{A}}(x)$.

3 $R_{\mathbf{C}}(x) = R_{\mathbf{B}}(x) + R_{\mathbf{A}}(x)$, which gives $\mathfrak{z}_{\mathbf{C}}(g) = R_{\mathbf{C}}(g) - g^{-1}$.
4 Solve $\mathfrak{z}_{\mathbf{C}}(g) = z$ for $\mathfrak{g}_{\mathbf{C}}(z)$.
5 Use Eq. (2.47) to find the density:

$$\rho_{\mathbf{C}}(\lambda) = \frac{\lim_{\eta \to 0^+} \operatorname{Im} \mathfrak{g}_{\mathbf{C}}(\lambda - i\eta)}{\pi}. \tag{15.14}$$

In the multiplicative case ($\mathbf{C} = \mathbf{A}^{\frac{1}{2}}\mathbf{B}\mathbf{A}^{\frac{1}{2}}$), the recipe is similar:

1 Find $\mathfrak{t}_{\mathbf{B}}(\zeta)$ and $\mathfrak{t}_{\mathbf{A}}(\zeta)$.
2 Invert $\mathfrak{t}_{\mathbf{B}}(\zeta)$ and $\mathfrak{t}_{\mathbf{A}}(\zeta)$ to get $\zeta_{\mathbf{B}}(t)$ and $\zeta_{\mathbf{A}}(t)$, and hence $S_{\mathbf{B}}(t)$ and $S_{\mathbf{A}}(t)$.
3 $S_{\mathbf{C}}(t) = S_{\mathbf{B}}(t)S_{\mathbf{A}}(t)$, which gives $\zeta_{\mathbf{C}}(t)S_{\mathbf{C}}(t)t = t + 1$.
4 Solve $\zeta_{\mathbf{C}}(t) = \zeta$ for $\mathfrak{t}_{\mathbf{C}}(\zeta)$.
5 Equation (2.47) for $\mathfrak{g}_{\mathbf{C}}(z) = (\mathfrak{t}_{\mathbf{C}}(z) + 1)/z$ is equivalent to

$$\rho_{\mathbf{C}}(\lambda) = \frac{\lim_{\eta \to 0^+} \operatorname{Im} \mathfrak{t}_{\mathbf{C}}(\lambda - i\eta)}{\pi\lambda}. \tag{15.15}$$

In some cases, the equation in step 4 is exactly solvable. But it is usually a high order polynomial equation, or worse, a transcendental equation. In these cases numerical solution is still possible. There always exists at least one solution that satisfies

$$\mathfrak{g}(z) = z^{-1} + O(z^{-2}) \tag{15.16}$$

for $z \to \infty$. Since the eigenvalues of $\mathbf{B}$ and $\mathbf{A}$ are real, their R- and S-transforms are real for real arguments. Hence the equation in step 4 is an equation with real coefficients. In order to find a non-zero eigenvalue density we need to find solutions with a strictly positive imaginary part when the parameter η goes to zero. When the equation is quadratic or cubic, complex solutions come in complex conjugated pairs: therefore, at most one solution will have a strictly positive imaginary part. As a numerical trick, $\pi\rho(\lambda)$ can be equated with the maximum of the imaginary part of all two or three solutions (the density will be zero when all solutions are real). For higher order polynomial and transcendental equations, we have to be more careful as there can be spurious complex solutions with positive imaginary part.

Exercise 15.2.1 shows how to do these computations in concrete cases.

15.2 R- and S-Transforms and Moments of Useful Ensembles

15.2.1 Wigner Ensemble

The Wigner ensemble is rotationally invariant, therefore a Wigner matrix is free from any matrix from which it is independent. For a Wigner matrix $\mathbf{X}$ of variance σ^2, the R-transform reads

$$R_{\mathbf{X}}(x) = \sigma^2 x. \tag{15.17}$$

The Wigner matrix is stable under free addition, i.e. the free sum of two Wigner matrices of variance σ_1^2 and σ_2^2 is a Wigner with variance $\sigma_1^2 + \sigma_2^2$.

The Wigner matrix is traceless ($\tau(\mathbf{X}) = 0$), so its S-transform is ill-defined. However, we can shift the mean of the entries of $\mathbf{X}$ by a certain parameter m. We then have $R_{\mathbf{X}+m}(x) = m + \sigma^2 x$. We can use Eq. (11.97) and compute the S-transform:

$$S_{\mathbf{X}+m}(t) = \frac{\sqrt{m^2 + 4\sigma^2 t} - m}{2\sigma^2 t} = \frac{m}{2\sigma^2 t}\left(\sqrt{1 + \frac{4\sigma^2 t}{m^2}} - 1\right). \tag{15.18}$$

It is regular at $t = 0$ whenever $m > 0$ and tends to $(\sigma\sqrt{t})^{-1}$ when $m \to 0$.

Finally, let us recall the formula for the positive moments of Wigner matrices:

$$\tau(\mathbf{X}^{2k}) = \frac{(2k)!}{(k+1)k!^2}\sigma^{2k}; \quad \tau(\mathbf{X}^{2k+1}) = 0. \tag{15.19}$$

The negative moments of $\mathbf{X}$ are all infinite, because the density of zero eigenvalues is positive.

15.2.2 Wishart Ensemble

For a white Wishart matrix $\mathbf{W}_q$ with parameter $q = N/T$, one has (see Section 10.1)

$$R_{\mathbf{W}_q}(x) = \frac{1}{1 - qx}. \tag{15.20}$$

To compute its S-transform we first remember that its Stieltjes transform $\mathfrak{g}(z)$ satisfies Eq. (4.37), which can be written as an equation for $\mathfrak{t}(\zeta)$ or its inverse $\zeta(t)$:

$$\zeta t - (1 + qt)(t + 1) = 0 \quad \Rightarrow \quad S_{\mathbf{W}_q}(t) = \frac{1}{1 + qt}. \tag{15.21}$$

The first few moments of a white Wishart matrix are given by

$$\tau(\mathbf{W}_q) = 1, \quad \tau(\mathbf{W}_q^2) = 1 + q; \quad \tau(\mathbf{W}_q^{-1}) = \frac{1}{1-q}, \quad \tau(\mathbf{W}_q^{-2}) = \frac{1}{(1-q)^3}. \tag{15.22}$$

15.2.3 Inverse-Wishart Ensemble

We take the opportunity of this summary of R- and S-transforms to introduce a very useful ensemble of matrices, namely the inverse-Wishart ensemble. We will call an inverse-Wishart matrix[2] the inverse of a white Wishart matrix, which, we recall, has unit normalized trace.

For a Wishart matrix to be invertible we need to have $q < 1$. Let $\mathbf{W}_q$ be such a matrix. Using Eq. (11.116) we can show that

$$S_{\mathbf{W}_q^{-1}}(t) = 1 - q - qt. \tag{15.23}$$

[2] More generally the inverse of a Wishart matrix with any covariance $\mathbf{C}$ can be called an inverse-Wishart but we will only consider *white* inverse-Wishart matrices.

Since $\tau(\mathbf{W}_q^{-1}) = 1/(1-q)$, we define the (normalized) inverse-Wishart as $\mathbf{M}_p = (1-q)\mathbf{W}_q^{-1}$ and call $p := q/(1-q)$. Rescaling and changing variable we obtain

$$S_{\mathbf{M}_p}(t) = 1 - pt. \tag{15.24}$$

By construction the inverse-Wishart has mean 1 and variance p. Using Eq. (15.11), we find that it has $\kappa_3(\mathbf{M}_p) = 2p^2$, which is higher than the skewness of a white Wishart matrix with the same variance ($\kappa_3(\mathbf{W}_q) = q^2$).

From the S-transform we can find the R-transform using Eq. (15.8):

$$R_{\mathbf{M}_p}(x) = \frac{1 - \sqrt{1-4px}}{2px}. \tag{15.25}$$

To find the Stieltjes transform and the density, it is easier to compute the T-transform from Eq. (11.63) and convert the result into a Stieltjes transform:

$$\mathfrak{g}_{\mathbf{M}_p}(z) = \frac{(1+2p)z - 1 - \sqrt[\oplus]{(z-1)^2 - 4pz}}{2pz^2}. \tag{15.26}$$

We can use Eq. (2.47) to find the density of eigenvalues or do the following change of variable in the white Wishart density (Eq. (4.43)):

$$x = \frac{1-q}{\lambda} \quad \text{and} \quad p = \frac{q}{1-q}. \tag{15.27}$$

Both methods give (see Fig. 15.1)

$$\rho_{\mathbf{M}_p}(x) = \frac{\sqrt{(x_+ - x)(x - x_-)}}{2\pi p x^2}, \quad x_- < x < x_+, \tag{15.28}$$

with the edges of the spectrum given by

$$x_\pm = 2p + 1 \pm 2\sqrt{2(p+1)}. \tag{15.29}$$

From the Stieltjes transform we can obtain

$$\tau(\mathbf{M}_p^{-1}) = -\lim_{z\to 0} \mathfrak{g}_{\mathbf{M}_p}(z) = 1 + p. \tag{15.30}$$

Other low moments of the inverse-Wishart matrix read

$$\tau(\mathbf{M}_p) = 1, \quad \tau(\mathbf{M}_p^2) = 1 + p; \qquad \tau(\mathbf{M}_p^{-1}) = 1 + p, \quad \tau(\mathbf{M}_p^{-2}) = (1+p)(1+2p). \tag{15.31}$$

Finally, the large N inverse-Wishart matrix potential can be obtained from the real part of the Stieltjes transform using (5.38)

$$V_{\mathbf{M}_p}(x) = \frac{1}{px} + \frac{1+2p}{p}\log x. \tag{15.32}$$

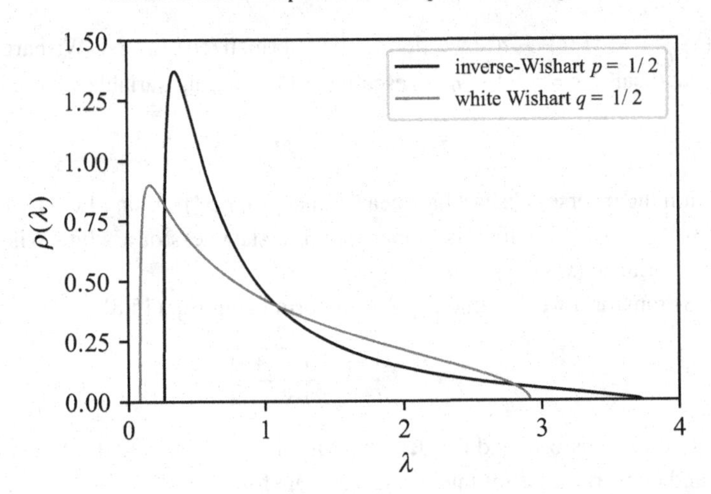

Figure 15.1 Density of eigenvalues for an inverse-Wishart distribution with $p = 1/2$. The white Wishart distribution ($q = 1/2$) is shown for comparison. Both laws are normalized, have unit mean and variance $1/2$.

For completeness we give here the law of the elements of a general (not necessarily white) inverse-Wishart matrix at finite N. We recall the law of a general Wishart matrix Eq. (4.16):

$$P(\mathbf{E}) = \frac{(T/2)^{NT/2}}{\Gamma_N(T/2)} \frac{(\det \mathbf{E})^{(T-N-1)/2}}{(\det \mathbf{C})^{T/2}} \exp\left[-\frac{T}{2}\operatorname{Tr}\left(\mathbf{E}\mathbf{C}^{-1}\right)\right], \tag{15.33}$$

where $\mathbf{E}$ is an $N \times N$ Wishart matrix measured over T time steps with true correlations $\mathbf{C}$ and normalized such that $\mathbb{E}[\mathbf{E}] = \mathbf{C}$. We define the inverse-Wishart as $\mathbf{M} = \mathbf{E}^{-1}$. Note that a finite N Wishart matrix has

$$\mathbb{E}\left[\mathbf{E}^{-1}\right] = \frac{T}{T-N-1}\mathbf{C}^{-1} =: \Sigma, \tag{15.34}$$

where we have defined the matrix Σ such that $\mathbb{E}[\mathbf{M}] = \Sigma$. To do the change of variable $\mathbf{E} \to \mathbf{M}$ in the joint probability density, we need to multiply by the Jacobian $(\det \mathbf{M})^{-N-1}$ (see Eq. (1.41)). Putting everything together we obtain

$$P(\mathbf{M}) = \frac{(T-N-1)^{NT/2}}{2^{NT/2}\Gamma_N(T/2)} \frac{(\det \Sigma)^{T/2}}{(\det \mathbf{M})^{(T+N+1)/2}} \exp\left[-\frac{T-N-1}{2}\operatorname{Tr}\left(\mathbf{M}^{-1}\Sigma\right)\right]. \tag{15.35}$$

In the scalar case $N = 1$, the inverse-Wishart distribution reduces to an inverse-gamma distribution:

$$P(m) = \frac{b^a}{\Gamma(a)} m^{-a-1} \mathrm{e}^{-b/m}, \tag{15.36}$$

with $b = (T-2)\Sigma/2$ and $a = T/2$.

Exercise 15.2.1 Free product of two Wishart matrices

In this exercise, we will compute the eigenvalue distribution of a matrix $\mathbf{E} = (\mathbf{W}_{q_0})^{\frac{1}{2}}\mathbf{W}_q(\mathbf{W}_{q_0})^{\frac{1}{2}}$; as we will see in Section 17.1, this matrix would be the sample covariance matrix of data with true covariance given by a Wishart with q_0 observed over T samples such that $q = N/T$.

(a) Using Eq. (15.21) and the multiplicativity of the S-transform, write the S-transform of $\mathbf{E}$.

(b) Using the definition of the S-transform write an equation for $\mathfrak{t}_\mathbf{E}(z)$. It is a cubic equation in t. If either q_0 or q goes to zero, it reduces to the standard Marčenko–Pastur quadratic equation.

(c) Use Eq. (15.15) and a numerical root finder to plot the eigenvalue density of $\mathbf{E}$ for $q_0 = 1/2$ and $q \in \{1/4, 1/2, 3/4\}$. In practice you can work with $\eta = 0$; of the three roots of your cubic equation, at most one will have a positive imaginary part. When all three solutions are real $\rho_\mathbf{E}(\lambda) = 0$.

(d) Generate numerically two independent Wishart matrices with $q = 1/2$ ($N = 1000$ and $T = 2000$) and compute $\mathbf{E} = (\mathbf{W}_{q_0})^{\frac{1}{2}}\mathbf{W}_q(\mathbf{W}_{q_0})^{\frac{1}{2}}$. Note that the square-root of a matrix is obtained by applying the square-roots to its eigenvalues. Diagonalize your $\mathbf{E}$ and compare its density with your result in (c).

15.3 Worked-Out Examples: Addition

15.3.1 The Arcsine Law

Consider the free sum of two symmetric orthogonal matrices, i.e. matrices with eigenvalues ± 1 with equal weights. Let $\mathbf{M}_1$ and $\mathbf{M}_2$ be two such matrices, their Stieltjes and R-transforms are given by

$$\mathfrak{g}(z) = \frac{z}{z^2 - 1} \quad \text{and} \quad R(g) = \frac{\sqrt{1 + 4g^2} - 1}{2g}, \tag{15.37}$$

from which we can deduce that $\mathbf{M} = \frac{1}{2}(\mathbf{M}_1 + \mathbf{M}_2)$ has an R-transform given by

$$R_\mathbf{M}(g) = \frac{\sqrt{1 + g^2} - 1}{g}, \tag{15.38}$$

where we have used the scaling $R_{\alpha\mathbf{A}}(x) = \alpha R_\mathbf{A}(\alpha x)$ with $\alpha = 1/2$.

The corresponding Stieltjes transform reads

$$\mathfrak{g}_\mathbf{M}(z) = \frac{1}{z\sqrt{1 - 1/z^2}}. \tag{15.39}$$

From this expression we deduce that the density of eigenvalues is given by the centered arcsine law:

$$\rho_{\mathbf{M}}(\lambda) = \frac{1}{\pi}\frac{1}{\sqrt{1-\lambda^2}}, \qquad \lambda \in (-1,1), \tag{15.40}$$

and zero elsewhere. This corresponds to a special case of the Jacobi ensemble that we have encountered in Section 7.1.3.

15.3.2 Sum of Uniform Densities

Suppose now we want to compute the eigenvalue distribution of a matrix $\mathbf{M} = \mathbf{U} + \mathbf{O}\mathbf{U}\mathbf{O}^T$, where $\mathbf{U}$ is a diagonal matrix with entries uniformly distributed between -1 and 1 (e.g. $[\mathbf{U}]_{kk} = 1 + (1-2k)/N$) and $\mathbf{O}$ a random orthogonal matrix. This is the free sum of two matrices with uniform eigenvalue density.

First we need to compute the Stieltjes transform of $\mathbf{U}$. We have

$$\rho_{\mathbf{U}}(\lambda) = \frac{1}{2}, \qquad \lambda \in (-1,1). \tag{15.41}$$

The corresponding Stieltjes transform is[3]

$$\mathfrak{g}_{\mathbf{U}}(z) = \frac{1}{2}\int_{-1}^{1}\frac{d\lambda}{z-\lambda} = \frac{1}{2}\log\left(\frac{z+1}{z-1}\right). \tag{15.42}$$

Note that when $-1 < \lambda < 1$ the argument of the log in $\mathfrak{g}_{\mathbf{U}}(z)$ is negative so $\operatorname{Im}\mathfrak{g}(\lambda - i\eta) = i\pi/2$, consistent with a uniform distribution of eigenvalues. We then compute the R-transform by finding the inverse of $\mathfrak{g}_{\mathbf{U}}(z)$:

$$\mathfrak{z}(g) = \frac{e^{2g}+1}{e^{2g}-1} = \coth(g). \tag{15.43}$$

And so the R-transform of $\mathbf{U}$ is given by

$$R_{\mathbf{U}}(g) = \coth(g) - \frac{1}{g}. \tag{15.44}$$

The R-transform of $\mathbf{M}$ is twice that of $\mathbf{U}$. To find the Stieltjes transform of $\mathbf{U}$ we thus need to solve

$$z = R_{\mathbf{M}}(g) + \frac{1}{g} = 2\coth(g) - \frac{1}{g}, \tag{15.45}$$

for $\mathfrak{g}(z)$. This is a transcendental equation and we need to solve it for complex z near the real axis. Before attempting to do this, it is useful to plot $\mathfrak{z}(g)$ (Fig. 15.2). The region where $z = \mathfrak{z}(g)$ does not have real solutions is where the eigenvalues are. This region is between a local maximum and a local minimum of $\mathfrak{z}(g)$. We should look for complex solutions of Eq. (15.45) near the real axis for $\operatorname{Re}(z)$ between -1.54 and 1.54. We can then put this equation into a complex non-linear solver. The density will be given by $\operatorname{Im}\mathfrak{g}(z)/\pi$ for $\operatorname{Im}(z)$

[3] A more general uniform density between $[m-a, m+a]$ has mean m, variance $a^2/3$ and $\mathfrak{g}_{\mathbf{U}}(z) = \log((z-m+a)/(z-m-a))/(2a)$.

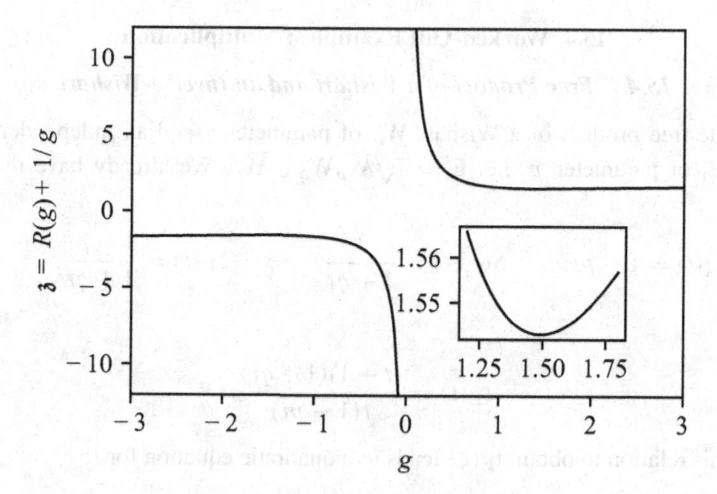

Figure 15.2 The function $\mathfrak{z}(g) = R_{\mathbf{M}}(g) + 1/g$ for the free sum of two flat distributions. Note that there is a region of z near $[-1.5, 1.5]$ when $z = \mathfrak{z}(g)$ does not have real solutions. This is where the eigenvalues lie. The inset shows a zoom of the region near $z = 1.5$, indicating more clearly that $\mathfrak{z}(g)$ has a minimum at $g_+ \approx 1.49$, so $\lambda_+ = \mathfrak{z}(g_+)$. The exact edges of the spectrum are $\lambda_\pm \approx \pm 1.5429$.

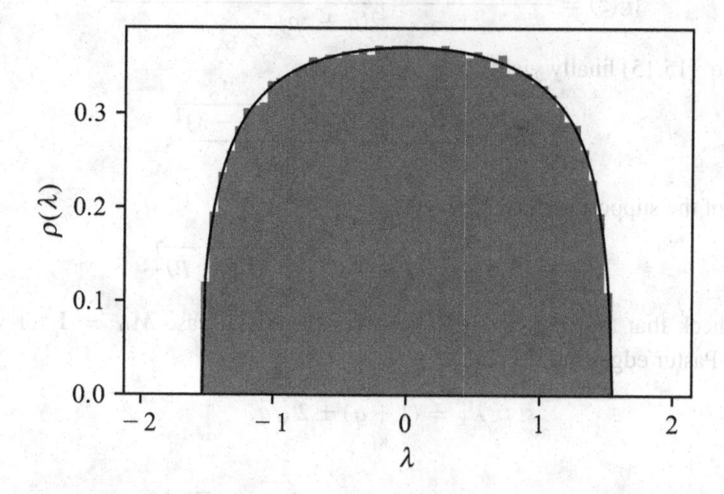

Figure 15.3 Density of eigenvalues for the free sum of two uniform distributions. Continuous curve was computed using a numerical solution of Eq. (15.45). The histogram is a numerical simulation with $N = 5000$.

very small and $\text{Re}(z)$ in the desired range. Note that complex solutions come in conjugated pairs, and it is hard to force the solver to find the correct one. This is not a problem; since their imaginary parts have the same absolute value, we can just use

$$\rho(\lambda) = \frac{|\operatorname{Im} \mathfrak{g}(\lambda - i\eta)|}{\pi} \quad \text{for some small } \eta. \tag{15.46}$$

We have plotted the resulting density in Figure 15.3.

15.4 Worked-Out Examples: Multiplication

15.4.1 Free Product of a Wishart and an Inverse-Wishart

Consider the free product of a Wishart $\mathbf{W}_q$ of parameter q and an independent inverse-Wishart $\mathbf{M}_p$ of parameter p, i.e. $\mathbf{E} = \sqrt{\mathbf{M}_p}\mathbf{W}_q\sqrt{\mathbf{M}_p}$. We already have the building blocks:

$$S_{\mathbf{M}_p}(t) = 1 - pt; \qquad S_{\mathbf{W}_q}(t) = \frac{1}{1+qt} \quad \Rightarrow \quad S_{\mathbf{E}}(t) = \frac{1-pt}{1+qt}, \tag{15.47}$$

leading to

$$\zeta_{\mathbf{E}}(t) = \frac{(t+1)(1+qt)}{t(1-pt)}. \tag{15.48}$$

Inverting this relation to obtain $\mathfrak{t}_{\mathbf{E}}(\zeta)$ leads to a quadratic equation for t:

$$(t+1)(1+qt) = \zeta t(1-pt), \tag{15.49}$$

which can be explicitly solved as

$$\mathfrak{t}_{\mathbf{E}}(z) = \frac{z - q - 1 - \sqrt[\oplus]{(q+1-z)^2 - 4(q+zp)}}{2(q+zp)}. \tag{15.50}$$

Using Eq. (15.15) finally yields

$$\rho_{\mathbf{E}}(\lambda) = \frac{\sqrt{4(p\lambda+q) - (1-q-\lambda)^2}}{2\pi\lambda(p\lambda+q)}. \tag{15.51}$$

The edges of the support are given by

$$\lambda_\pm = \left[1 + q + 2p \pm 2\sqrt{(1+p)(q+p)}\right]. \tag{15.52}$$

One can check that the limit $p \to 0$ recovers the trivial case $\mathbf{M}_0 = \mathbf{1}$ for which the Marčenko–Pastur edges indeed read

$$\lambda_\pm = (1+q) \pm 2\sqrt{q}. \tag{15.53}$$

Exercise 15.4.1 Free product of Wishart and inverse-Wishart

(a) Generate numerically a normalized inverse-Wishart $\mathbf{M}_p$ for $p = 1/4$ and $N = 1000$. Check that $\tau(\mathbf{M}_p) = 1$ and $\tau(\mathbf{M}_p^2) = 1.25$. Plot a normalized histogram of the eigenvalues of $\mathbf{M}_p$ and compare with Eq. (15.28).

(b) Generate an independent Wishart $\mathbf{W}_q$ with $q = 1/4$ and compute $\mathbf{E} = \sqrt{\mathbf{M}_p}\mathbf{W}_q\sqrt{\mathbf{M}_p}$. To compute $\sqrt{\mathbf{M}_p}$, diagonalize $\mathbf{M}_p$, take the square-root of its eigenvalues and reconstruct $\sqrt{\mathbf{M}_q}$. Check that $\tau(\mathbf{E}) = 1$ and $\tau(\mathbf{E}^2) = 1.5$. Plot a normalized histogram of the eigenvalues of $\mathbf{E}$ and compare with Eq. (15.51).

(c) For every eigenvector $\mathbf{v}_k$ of $\mathbf{E}$ compute $\xi_k := \mathbf{v}_k^T \mathbf{M}_p \mathbf{v}_k$, and make a scatter plot of ξ_k vs λ_k, the eigenvalue of $\mathbf{v}_k$. Your scatter plot should show a noisy straight line. We will see in Chapter 19, Eq. (19.49), that this is related to the fact that linear shrinkage is the optimal estimator of the true covariance from the sample covariance when the true covariance is an inverse-Wishart.

15.4.2 Free Product of Projectors

As our last simple example, consider a space of large dimension N, and a projector $\mathbf{P}_1$ on a subspace of dimension $N_1 \leq N$, i.e. a diagonal matrix with N_1 diagonal elements equal to unity, and $N - N_1$ elements equal to zero. We now introduce a second projector $\mathbf{P}_2$ on a subspace of dimension $N_2 \leq N$, and would like to study the eigenvalues of the free product of these two projectors, $\mathbf{P} = \mathbf{P}_1\mathbf{P}_2$. Clearly, all eigenvalues of $\mathbf{P}$ must lie in the interval $(0, 1)$.

As now usual, we first need to compute the T-transform of $\mathbf{P}_1$ and $\mathbf{P}_2$. We define the ratios $q_a = N_a/N$, $a = 1, 2$. Since $\mathbf{P}_a$ has N_a eigenvalues equal to unity and $N - N_a$ eigenvalues equal to zero, one finds

$$\mathfrak{g}_{\mathbf{P}_a}(z) = \frac{1}{N}\left[\frac{N_a}{z-1} + \frac{N - N_a}{z}\right] = \frac{q_a - 1 + z}{z(z-1)} \quad \Rightarrow \quad \mathfrak{t}_{\mathbf{P}_a}(\zeta) = \frac{q_a}{\zeta - 1}. \tag{15.54}$$

Therefore, the inverse of the T-transforms just read $\zeta_{\mathbf{P}_a}(t) = 1 + q_a/t$, and

$$S_{\mathbf{P}_a}(t) = \frac{t+1}{t+q_a} \Longrightarrow \quad S_{\mathbf{P}} = \frac{(t+1)^2}{(t+q_1)(t+q_2)}. \tag{15.55}$$

Now, going backwards,

$$\zeta_{\mathbf{P}}(t) = \frac{(t+q_1)(t+q_2)}{t(t+1)}, \tag{15.56}$$

again leading to a quadratic equation for $\mathfrak{t}_{\mathbf{P}}(\zeta)$:

$$(\zeta - 1)t^2 + (\zeta - q_1 - q_2)t - q_1 q_2 = 0, \tag{15.57}$$

whose solution is

$$\mathfrak{t}(\zeta) = \frac{(q_1 + q_2 - \zeta) + \sqrt[\oplus]{\zeta^2 - 2\zeta(q_1 + q_2 - 2q_1q_2) + (q_1 - q_2)^2}}{2(\zeta - 1)}, \tag{15.58}$$

where the notation $\sqrt[\oplus]{\cdot}$ defined by Eq. (4.56) ensures that we pick the correct root. Note that the argument under the square-root has zeros for

$$\lambda_\pm = q_1 + q_2 - 2q_1q_2 \pm 2\sqrt{q_1q_2(1-q_1)(1-q_2)}. \tag{15.59}$$

One can check that $\lambda_- \geq 0$, the zero bound being reached for $q_1 = q_2$. Note also that $\lambda_+\lambda_- = (q_1 - q_2)^2$.

The Stieltjes transform of $\mathbf{P}$ can thus be written as

$$\mathfrak{g}_{\mathbf{P}}(z) = \frac{1}{z} + \frac{(q_1 + q_2 - z) + \sqrt[\oplus]{z^2 - 2z(q_1 + q_2 - 2q_1q_2) + (q_1 - q_2)^2}}{2z(z-1)}. \quad (15.60)$$

This quantity has poles at $z = 0$ and $z = 1$, and an imaginary part when $z \in (\lambda_+, \lambda_-)$. The spectrum of $\mathbf{P}$ therefore has a continuous part, given by

$$\rho_{\mathbf{P}}(\lambda) = \frac{\sqrt{(\lambda_+ - \lambda)(\lambda - \lambda_-)}}{2\pi\lambda(1-\lambda)}, \quad (15.61)$$

and two delta peaks, $A_0\delta(\lambda)$ and $A_1\delta(\lambda - 1)$. To find the amplitude of the potential poles, we need to compute the numerator of (15.60) at the values $z = 0$ and $z = 1$. Remember that $\sqrt[\oplus]{\cdot}$ equals $-\sqrt{\cdot}$ on the real axis left of the left edge and $\sqrt{\cdot}$ right of the right edge. The amplitude of the $z = 0$ pole of $\mathfrak{g}_{\mathbf{P}}(z)$ is

$$A_0 = 1 - \frac{(q_1 + q_2) - \sqrt{(q_1 - q_2)^2}}{2} = 1 - \min(q_1, q_2), \quad (15.62)$$

while the amplitude of the $z = 1$ pole is

$$\begin{aligned} A_1 &= \frac{(q_1 + q_2 - 1) + \sqrt{1 - 2(q_1 + q_2 - 2q_1q_2) + (q_1 - q_2)^2}}{2} \\ &= \frac{(q_1 + q_2 - 1) + \sqrt{(q_1 + q_2 - 1)^2}}{2} = \max(q_1 + q_2 - 1, 0). \end{aligned} \quad (15.63)$$

This makes a lot of sense geometrically: our product of two projectors can only have a unit eigenvalue if the sum of the dimensions of space spanned by these two projectors exceeds the total dimension N, i.e. when $N_1 + N_2 > N$. Otherwise, there cannot be (generically) any eigenvalue beyond λ_+.

When $q_1 + q_2 < 1$, the density of non-zero eigenvalues (15.61) is the same (up to a normalization) as the density of eigenvalues of a Jacobi matrix (7.20). If we match the edges of the spectrum we find the identification $c_1 = q_{\max}/q_{\min}$ and $c_+ = 1/q_{\min}$. The ratio of normalization $1/c_+ = q_{\min}$ implies that the product of projectors density has a missing mass of $1 - q_{\min}$, which is precisely the Dirac mass at zero. The special case $q_1 = q_2 = 1/2$ was discussed in the context of 2×2 matrices in Exercise 12.5.2. In that case, half of the eigenvalues are zero and the other half are distributed according to the arcsine law: the arcsine law is the limit of a Jacobi matrix with $c_1 \to 1$ and $c_+ \to 2$.

There is an alternative, geometric interpretation of the above calculation that turns out to be useful in many different contexts, see Section 17.4 for an extended discussion. The eigenvectors of projector $\mathbf{P}_1$ form a set of N_1 orthonormal vectors $\mathbf{x}_\alpha$, $\alpha = 1, \ldots, N_1$, from which an $N_1 \times N$ matrix $\mathbf{X}$ of components $(x_\alpha)_i$ can be formed. Similarly, we define an $N_2 \times N$ matrix $\mathbf{Y}$ of components $(y_\beta)_i$, $\beta = 1, \ldots, N_2$. Now, one can write $\mathbf{P}$ as

$$\mathbf{P} = \mathbf{X}^T\mathbf{X}\mathbf{Y}^T\mathbf{Y}. \quad (15.64)$$

The non-zero eigenvalues of $\mathbf{P}$ are the same as the non-zero eigenvalues of $\mathbf{M}^T\mathbf{M}$ (or those of $\mathbf{M}\mathbf{M}^T$), where $\mathbf{M}$ is the $N_1 \times N_2$ matrix of overlaps:

$$\mathbf{M}_{\alpha,\beta} := \sum_{i=1}^{N} (x_\alpha)_i (y_\beta)_i. \tag{15.65}$$

The eigenvalues of $\mathbf{P}$ correspond to the square of the singular values s of $\mathbf{M}$. The geometrical interpretation of these singular values is as follows: the largest singular value corresponds to the maximum overlap between any normalized linear combination of the $\mathbf{x}_\alpha$ on the one hand and of the $\mathbf{y}_\beta$ on the other hand. These two linear combinations define two one-dimensional subspaces of the spaces spanned by $\mathbf{x}_\alpha$ and $\mathbf{y}_\beta$. Once these optimal directions are removed, one can again ask the same question for the remaining $N_1 - 1$ and $N_2 - 1$ dimensional subspaces, defining the second largest singular value of $\mathbf{M}$, and so on.

15.4.3 The Jacobi Ensemble Revisited

We saw in the previous section that the free product of two random projectors has a very simple S-transform and that its non-zero eigenvalues are given by those of a Jacobi matrix. We suspect that the Jacobi ensemble has itself a simple S-transform. Rather than computing its S-transform from its Stieltjes transform (7.18), let us just use the properties of the S-transform to compute it directly from the definition of a Jacobi matrix.

Recall from Chapter 7, an $N \times N$ Jacobi matrix $\mathbf{J}$ is defined as $(\mathbf{1} + \mathbf{E})^{-1}$ where the matrix $\mathbf{E}$ is the free product of the inverse of an unnormalized Wishart matrix $\mathbf{W}_1$ with $T_1 = c_1 N$ and another unnormalized Wishart $\mathbf{W}_2$ with $T_2 = c_2 N$.

The two Wishart matrices have S-transforms given by

$$S_{\mathbf{W}_{1,2}}(t) = T_{1,2}\frac{1}{1 + c_{1,2}^{-1}t} = N\frac{1}{c_{1,2} + t}. \tag{15.66}$$

Using the relation for inverse matrices (15.7), we find

$$S_{\mathbf{W}_1^{-1}}(t) = N^{-1}(c_1 - t - 1). \tag{15.67}$$

The S-transform of $\mathbf{E}$ is just the product

$$S_{\mathbf{E}}(t) = S_{\mathbf{W}_1^{-1}}(t) S_{\mathbf{W}_2}(t) = \frac{c_1 - t - 1}{c_2 + t}. \tag{15.68}$$

The next step is to shift the matrix $\mathbf{E}$ by $\mathbf{1}$. As mentioned earlier, there is no easy rule to compute the S-transform of a shifted matrix. So this will be the hardest part of the computation.

The R-transform behaves simply under shift. The trick is to use one of Eqs. (15.8) to write an equation for $R_{\mathbf{E}}(x)$, shift $\mathbf{E}$ by $\mathbf{1}$ and use the other R-S relation to find back an equation for $S_{\mathbf{E}+\mathbf{1}}(t)$. First we write

$$(c_2 + t)S_{\mathbf{E}} = c_1 - t - 1. \tag{15.69}$$

The second of Eqs. (15.8) can be interpreted as the replacements $S \to 1/R$ and $t \to xR$ and gives

$$(1 + x - c_1 + R_{\mathbf{E}})R_{\mathbf{E}} + c_2 = 0. \tag{15.70}$$

Now $R_{\mathbf{E}}(x) = R_{\mathbf{E}+\mathbf{1}}(x) - 1$, so

$$(x - c_1 + R_{\mathbf{E}+\mathbf{1}})(R_{\mathbf{E}+\mathbf{1}} - 1) + c_2 = 0. \tag{15.71}$$

Following the first of Eqs. (15.8), we make the replacements $R \to 1/S$ and $x \to tS$ and find

$$1 - c_1 + t + (c_2 + c_1 - t - 1)S_{\mathbf{E}+\mathbf{1}} = 0 \quad \Rightarrow \quad S_{\mathbf{E}+\mathbf{1}}(t) = \frac{t + 1 - c_1}{t + 1 - c_2 - c_1}. \tag{15.72}$$

Finally, using the relation for inverse matrices, Eq. (15.7) gives

$$S_{\mathbf{J}}(t) = \frac{t + c_2 + c_1}{t + c_1}. \tag{15.73}$$

We can verify that the T-transform of the Jacobi ensemble

$$\mathfrak{t}_{\mathbf{J}}(\zeta) = \frac{c_1 + 1 - c_+\zeta + \sqrt[\oplus]{c_+^2\zeta^2 - 2(c_1c_+ + c_+ - 2c_1)\zeta + (c_1 - 1)^2}}{2(\zeta - 1)} \tag{15.74}$$

is compatible with our previous result on the Stieltjes transform, Eq. (7.18). We can use the Taylor series of the S-transform (15.11) to find the first few cumulants:

$$\kappa_1 = \frac{c_1}{c_1 + c_2}, \quad \kappa_2 = \frac{c_1c_2}{(c_1 + c_2)^3}, \quad \kappa_3 = \frac{(c_1 - c_2)c_1c_2}{(c_1 + c_2)^5}. \tag{15.75}$$

From the S-transform we can compute the R-transform using Eq. (15.8):

$$R_{\mathbf{J}}(x) = \frac{x - c_1 - c_2 - \sqrt{x^2 + 2(c_1 - c_2)x + (c_1 + c_2)^2}}{2x}. \tag{15.76}$$

Finally, we note that the arcsine law is a Jacobi matrix with $c_1 = c_2 = 1$ and has the following transform:

$$S(t) = \frac{t + 2}{t + 1}, \quad R(x) = \frac{x - 2 - \sqrt{x^2 + 4}}{2x}. \tag{15.77}$$

For the centered arcsine law we have $R_s(t) = 2R(2x) + 1$ and we recover Eq. (15.38).

16

Products of Many Random Matrices

In this chapter we consider an issue of importance in many different fields: that of products of many random matrices. This problem arises, for example, when one considers the transmission of light in a succession of slabs of different optical indices, or the propagation of an electron in a disordered wire, or the way displacements propagate in granular media. It also appears in the context of chaotic systems when one wants to understand how a small difference in initial conditions "propagates" as the dynamics unfolds. In this context, one usually linearizes the dynamics in the vicinity of the unperturbed trajectory. If one takes stroboscopic snapshots of the system, the perturbation is obtained as the product of matrices (corresponding to the linearized dynamics) applied on the initial perturbation (see Chapter 1). If the phase space of the system is large enough, and the dynamics chaotic enough, one may expect that approximating the problem as a product of large, free matrices should be a good starting point.

16.1 Products of Many Free Matrices

The specific problem we will study is therefore the following: consider the symmetrized product of K matrices, defined as

$$\mathbf{M}_K = \mathbf{A}_K\mathbf{A}_{K-1}\dots\mathbf{A}_2\mathbf{A}_1\mathbf{A}_1^T\mathbf{A}_2^T\dots\mathbf{A}_{K-1}^T\mathbf{A}_K^T, \tag{16.1}$$

where all $\mathbf{A}_i$ are identically distributed and mutually free, i.e. randomly rotated with respect to one another. We know now that in such a case the S-transforms simply multiply. Noting as $S_i(z)$ the S-transform of $\mathbf{A}_i\mathbf{A}_i^T$, and $S_{\mathbf{M}_K}(z)$ the S-transform of $\mathbf{M}_K$, one has

$$S_{\mathbf{M}_K}(z) = \prod_{i=1}^{K} S_i(z) \equiv S_1(z)^K. \tag{16.2}$$

Now, it is intuitive that all the eigenvalues of $\mathbf{M}_K$ will behave for large K as μ^K, where μ is itself a random variable which we will characterize below. We take this as an assumption and indeed show that the distribution of μ's tends to a well-defined function $\rho_\infty(\mu)$ as $K \to \infty$. Note here a crucial difference with the case of sums of random matrices. If we assume that the eigenvalues of a sum of K free random matrices behave as $K \times \mu$, one can easily establish that the distribution of μ's collapses for large K to $\delta(\mu - \tau(\mathbf{A}))$, with, once again, $\tau(.) = \mathrm{Tr}(.)/N$. For products of random matrices, on the other hand, the distribution of μ remains non-trivial, as we will find below.

Let us compute $S_{\mathbf{M}_K}(z)$ in the large K limit using our ansatz that the eigenvalues of $\mathbf{M}$ are indeed of the form μ^K. We first compute the function $\mathrm{t}_K(z)$ equal to

$$\mathrm{t}_K(z) := \int \frac{\mu^K}{z - \mu^K} \rho_\infty(\mu)\mathrm{d}\mu = -\int \frac{1}{1 - z\mu^{-K}} \rho_\infty(\mu)\mathrm{d}\mu. \tag{16.3}$$

Setting $z := u^K$, we see that for $K \to \infty$ there is no contribution to this integral from the region $\mu < u$, whereas the region $\mu > u$ simply yields

$$\mathrm{t}_K(z) \approx -P_>(z^{1/K}); \qquad P_>(u) := \int_u^\infty \rho_\infty(\mu)\mathrm{d}\mu. \tag{16.4}$$

The next step to get the S-transform is to compute the functional inverse of $\mathrm{t}_K(z)$. Within the same approximation, this is given by

$$\mathrm{t}_K^{-1}(z) = \left[P_>^{(-1)}(-z)\right]^K, \tag{16.5}$$

where $P_>^{(-1)}$ is the functional inverse of the cumulative distribution function $P_>$. Finally, by definition,

$$S_{\mathbf{M}_K}(z) := \frac{1+z}{z\mathrm{t}_K^{-1}(z)} = S_1(z)^K. \tag{16.6}$$

Hence one finds, in the large K limit where $((1+z)/z)^{1/K} \to 1$,

$$P_>^{(-1)}(-z) = \frac{1}{S_1(z)} \Rightarrow P_>(\mu) = -S_1^{(-1)}\left(\frac{1}{\mu}\right) \tag{16.7}$$

and finally $\rho_\infty(\mu) = -P'_>(\mu)$. The final result is therefore quite simple, and entirely depends on the S-transform of $\mathbf{A}_i\mathbf{A}_i^T$.

A simple case is when $\mathbf{A}_i\mathbf{A}_i^T$ is a large Wishart matrix, with parameter $q \leq 1$. In this case $S_1(z) = (1+qz)^{-1}$, from which one easily works out that $\rho_\infty(\mu) = 1/q$ for $\mu \in (1-q, 1)$ and zero elsewhere (see Fig. 16.1 for an illustration).

In many cases of interest, the eigenvalue spectrum of $\mathbf{A}_i\mathbf{A}_i^T$ has some symmetries, coming from the underlying physical problem one is interested in. For example, when our chaotic system is invariant under time reversal (like the dynamics of a Hamiltonian system), each eigenvalue λ must come with its inverse λ^{-1}. A simple example of a spectrum with such a symmetry is the free log-normal, further discussed in the next section. It is defined from its S-transform, given by

$$S^0_{\mathrm{LN}}(z) = \mathrm{e}^{-a(z+\frac{1}{2})}, \tag{16.8}$$

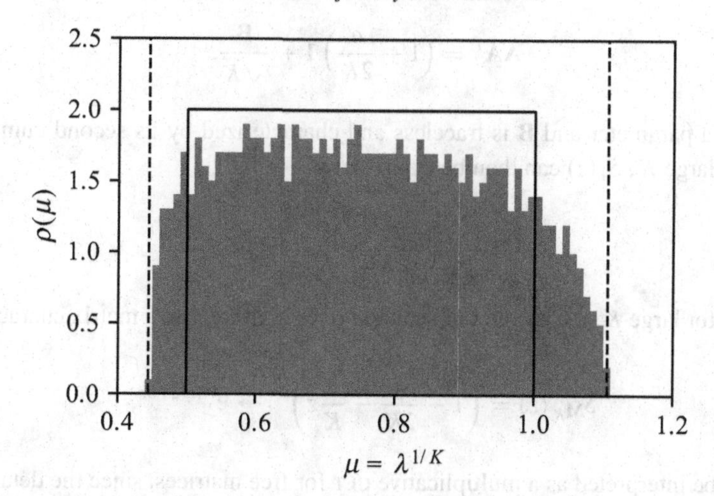

Figure 16.1 Sample density of $\mu = \lambda^{1/K}$ for the free product of $K = 40$ white Wishart matrices with $q = 1/2$ and $N = 1000$. The dark line corresponds to the asymptotic density ($K \to \infty$), which is constant between $1 - q$ and 1 and zero elsewhere. The two dashed vertical lines give the exact positions of the edges of the spectrum ($\mu_- = 0.44$ and $\mu_+ = 1.10$) for $K = 40$, as computed in Exercise 16.1.1.

where the parameter a is related to the trace of the corresponding matrices equal to $e^{a/2}$. Multiplying K such matrices together leads to eigenvalues of the form μ^K, with

$$P_>(\mu) = -\left(S^0_{\text{LN}}\right)^{(-1)}\left(\frac{1}{\mu}\right) = \frac{1}{2} - \frac{\log \mu}{a}, \tag{16.9}$$

corresponding to

$$\rho_\infty(\mu) = -P'_>(\mu) = \frac{1}{a\mu}, \qquad \mu \in (e^{-a/2}, e^{a/2}), \tag{16.10}$$

and zero elsewhere. One can explicitly check that μ^{-1} has the same probability distribution function as μ.

One often describes the eigenvalues of large products of random matrices in terms of the Lyapunov exponents Λ, defined as the eigenvalues of

$$\Lambda = \lim_{K\to\infty} \frac{1}{K} \log \mathbf{M}_K. \tag{16.11}$$

Therefore the Lyapunov exponents are simply related to the μ's above as $\Lambda \equiv \log \mu$. For the free log-normal example, the distribution of Λ is found to be uniform between $-a/2$ and $a/2$.

Let us end this section with an important remark: we have up to now considered products of K matrices with a fixed spectrum, independent of K, which leads to a non-universal distribution of Lyapunov exponents (i.e. a distribution that explicitly depends on the full function $S_1(z)$). Let us now instead assume that these matrices are of the form

$$\mathbf{A}\mathbf{A}^T = \left(1 + \frac{a}{2K}\right)\mathbf{1} + \frac{\mathbf{B}}{\sqrt{K}}, \tag{16.12}$$

where a is a parameter and $\mathbf{B}$ is traceless and characterized by its second cumulant $b = \tau(\mathbf{B}^2)$. For large K, $S_1(z)$ can then be expanded as

$$S_1(z) = 1 - \frac{a}{2K} - \frac{b}{K}z + o(K^{-1}). \tag{16.13}$$

Therefore, for large K, the product of such matrices converges to a matrix characterized by

$$S_{\mathbf{M}_K}(z) = \left(1 - \frac{a}{2K} - \frac{b}{K}z\right)^K \to \mathrm{e}^{-a/2-bz}, \tag{16.14}$$

which can be interpreted as a multiplicative CLT for free matrices, since the detailed statistics of $\mathbf{B}$ has disappeared. The choice $b = a$ corresponds to the free log-normal with inversion symmetry S^0_{LN} (see next section).

Exercise 16.1.1 Edges of the spectrum for the free product of many white Wishart matrices

In this exercise, we will compute the edges of the spectrum of eigenvalues of a matrix $\mathbf{M}$ given by the free product of K large white Wishart matrices with parameter q.

(a) The S-transform of $\mathbf{M}$ is simply given by the S-transform of a white Wishart raised to the power K. Using Eq. (11.92), write an equation for the inverse of the T-transform, $\zeta(t)$, of the matrix $\mathbf{M}$. This is a polynomial equation of order $K+1$.

(b) For odd N, plot $\zeta(t)$ for various K and $0 < q < 1$ and convince yourself that there is always a region of ζ where $\zeta(t) = \zeta$ has no real solution. This region is between a local maximum and a local minimum of $\zeta(t)$. For even N, the argument is more subtle, but the correct branch exists only between the same two extrema.

(c) Differentiate $\zeta(t)$ with respect to t to find an equation for the extrema of $\zeta(t)$. After simplifications and discarding the $t = -1/q$ solution, this equation is quadratic in t^* with two solutions corresponding to the local minimum and maximum. Find the two solutions $t^*_\pm$ and plug these back in your equation for $\zeta(t)$ to find the edges of the spectrum $\lambda_\pm$.

(d) Use your result for $K = 40$ and $q = 0.5$ to verify the edges of the spectrum given in Figure 16.1.

(e) Compute the large K limit of $t^*_\pm$. You should find $t^*_- \to -1$ and $t^*_+ \to (q(K-1))^{-1}$. Show that at large K we have $\lambda_-^{1/K} \to 1 - q$ and $\lambda_+^{1/K} \to 1$.

16.2 The Free Log-Normal

There exists a free version of the log-normal. Its S-transform is given by

$$S_{\mathrm{LN}}(t) = \mathrm{e}^{-a/2-bt}. \tag{16.15}$$

As a two-parameter family, the free log-normal is stable in the sense that the free product of two free log-normals with parameters a_1, b_1 and a_2, b_2 is a free log-normal with parameters $a = a_1 + a_2, b = b_1 + b_2$. The first three free cumulants can be computed from Eq. (16.15):

$$S_{\mathrm{LN}}(t) = \mathrm{e}^{-a/2}\left[1 - bt + \frac{1}{2}b^2t^2\right] + O(t^3). \tag{16.16}$$

Comparing with Eq. (15.11), this leads to

$$\begin{aligned} \kappa_1 &= \mathrm{e}^{a/2}, \\ \kappa_2 &= b\mathrm{e}^{a}, \\ \kappa_3 &= \frac{3b^2}{2}\mathrm{e}^{2a}. \end{aligned} \tag{16.17}$$

In the special case $b = a$, the free log-normal S^0_{LN} has the additional property that its matrix inverse has exactly the same law. Indeed, we have shown in Section 11.4.4 that the following general relation holds:

$$S_{\mathbf{M}^{-1}}(t) = \frac{1}{S_{\mathbf{M}}(-t-1)}, \tag{16.18}$$

or, in the free log-normal case with $a = b$,

$$S_{\mathbf{M}^{-1}}(t) = \mathrm{e}^{a/2-b(1+t)} = S_{\mathbf{M}}(t) \tag{16.19}$$

when $b = a$. This implies that the eigenvalue distribution is invariant under $\lambda \to 1/\lambda$ and therefore that $\mathbf{M}$ has unit determinant. Let us study in more detail the eigenvalue spectrum for the symmetric case $a = b$. By looking for the real extrema of

$$\zeta(t) = \frac{t+1}{t}\mathrm{e}^{a(t+1/2)}, \tag{16.20}$$

we can find the points $t_\pm$ where $t(\zeta)$ ceases to be invertible, which in turn give the edges of the spectrum $\lambda_\pm = \zeta(t_\pm)$:

$$t_\pm = \frac{\pm\sqrt{1+\frac{4}{a}} - 1}{2} \tag{16.21}$$

or

$$\lambda_+ = \frac{1}{\lambda_-} = \left[\sqrt{\frac{a}{4}} + \sqrt{1+\frac{a}{4}}\right]^2 \exp\left(\sqrt{a + \frac{a^2}{4}}\right). \tag{16.22}$$

Note that $\lambda_+ = \lambda_- = 1$ when $a = b = 0$, corresponding to the identity matrix. The eigenvalue distribution is symmetric in $\lambda \to 1/\lambda$ so the density $\rho(\ell)$ of $\ell = \log(\lambda)$ is even. Figure 16.2 shows the density of ℓ for $a = 100$.

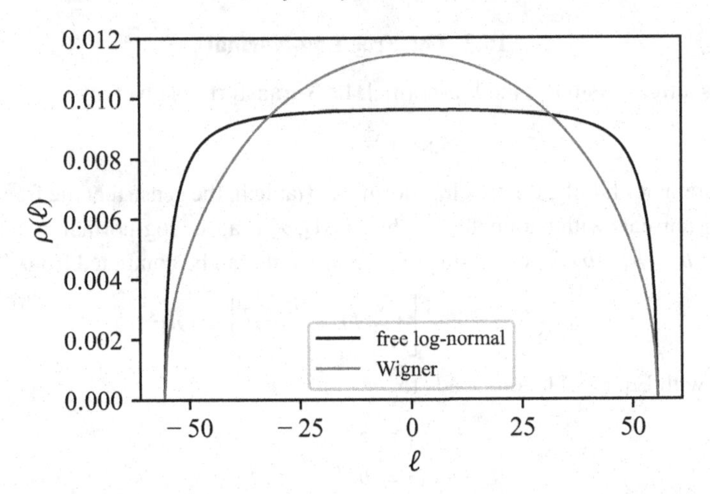

Figure 16.2 Probability density of $\ell = \log(\lambda)$ for a symmetric free log-normal (16.15) with $a = b = 100$ compared with a Wigner semi-circle with the same endpoints. As expected, the distribution is even in ℓ. For $a \lesssim 1$ the density of ℓ is indistinguishable to the eye from a Wigner semi-circle (not shown), whereas for $a \to \infty$ the distribution of ℓ/a tends to a uniform distribution on $[-1/2, 1/2]$.

In the more general case $a \neq b$, the whole distribution of $\ell = \log(\lambda)$ is just shifted by $(a-b)/2$, as expected from the scaling property of the S-transform upon multiplication by a scalar.

16.3 A Multiplicative Dyson Brownian Motion

Let us now consider the problem of multiplying random matrices close to unity from a slightly different angle. Consider the following iterative construction for $N \times N$ matrices:

$$\mathbf{M}_{n+1} = \mathbf{M}_n^{\frac{1}{2}} \left[(1 + \frac{a\varepsilon}{2})\mathbf{1} + \sqrt{\varepsilon}\mathbf{B}_n \right] \mathbf{M}_n^{\frac{1}{2}}, \tag{16.23}$$

where $\mathbf{B}_n$ is a sequence of identical, free, traceless $N \times N$ matrices and $\varepsilon \ll 1$. Using second order perturbation theory, one can deduce an iteration formula for the eigenvalues $\lambda_{i,n}$ of $\mathbf{M}_n$, which reads

$$\lambda_{i,n+1} = \lambda_{i,n} \left(1 + \frac{a\varepsilon}{2} + \sqrt{\varepsilon}\mathbf{v}_{i,n}^T \mathbf{B}_n \mathbf{v}_{i,n} \right) + \varepsilon \sum_{j \neq i} \frac{\lambda_{i,n}\lambda_{j,n}(\mathbf{v}_{i,n}^T \mathbf{B}_n \mathbf{v}_{j,n})^2}{\lambda_{i,n} - \lambda_{j,n}}, \tag{16.24}$$

where $\mathbf{v}_{i,n}$ are the corresponding eigenvectors. Noting that $\mathbf{M}_n$ and $\mathbf{B}_n$ are mutually free and that $\tau(\mathbf{B}_n) = 0$, one has, in the large N limit (using, for example, Eq. (12.8)),

$$\mathbb{E}[\mathbf{v}_{i,n}^T \mathbf{B}_n \mathbf{v}_{i,n}] = 0; \qquad \mathbb{E}[(\mathbf{v}_{i,n}^T \mathbf{B}_n \mathbf{v}_{j,n})^2] = \frac{b}{N}, \tag{16.25}$$

where $b := \tau(\mathbf{B}_n^2)$. Choosing $\varepsilon = \mathrm{d}t$, an infinitesimal time scale, we end up with a multiplicative version of the Dyson Brownian motion in fictitious time t:

$$\frac{\mathrm{d}\lambda_i}{\mathrm{d}t} = \frac{a}{2}\lambda_i + \frac{b}{N}\sum_{j\neq i}\frac{\lambda_i\lambda_j}{\lambda_i - \lambda_j} + \sqrt{\frac{b}{N}}\lambda_i\xi_i, \tag{16.26}$$

where ξ_i is a Langevin noise, independent for each λ_i (compare with Eq. (9.9)).

Now, let us consider the "time" dependent Stieltjes transform, defined as usual as

$$\mathfrak{g}(z,t) = \frac{1}{N}\sum_i \frac{1}{z - \lambda_i(t)}. \tag{16.27}$$

Its evolution is obtained as

$$\frac{\partial \mathfrak{g}}{\partial t} = \frac{1}{N}\sum_i \frac{1}{(z-\lambda_i)^2}\frac{\mathrm{d}\lambda_i}{\mathrm{d}t} = -\frac{1}{N}\frac{\partial}{\partial z}\sum_i \frac{1}{(z-\lambda_i)}\frac{\mathrm{d}\lambda_i}{\mathrm{d}t}. \tag{16.28}$$

After manipulations very similar to those encountered in Section 9.3.1, and retaining only leading terms in N, one finally obtains

$$\frac{\partial \mathfrak{g}}{\partial t} = \frac{1}{2}\frac{\partial}{\partial z}\left[(2b-a)z\mathfrak{g} - bz^2\mathfrak{g}^2\right]. \tag{16.29}$$

Now, introduce the auxiliary function $h(\ell,t) := \mathrm{e}^{\ell}\mathfrak{g}(\mathrm{e}^{\ell},t) + a/2b - 1$, which obeys

$$\frac{\partial h}{\partial t} = -bh\frac{\partial h}{\partial \ell}. \tag{16.30}$$

This is precisely the Burgers' equation (9.37), up to a rescaling of time $t \to bt$. Its solution obeys the following self-consistent equation obtained using the method of characteristics (see Section 10.1):

$$h(\ell,t) = h_0(\ell - bth(\ell,t)); \qquad h_0(\ell) := h(\ell,0) = \frac{1}{1-\mathrm{e}^{-\ell}} + \frac{a}{2b} - 1, \tag{16.31}$$

where we have assumed that at time $t = 0$ the dynamics starts from the identity matrix: $\mathbf{M}_0 = \mathbf{1}$, for which $\mathfrak{g}(z,0) = (z-1)^{-1}$. Hence, with $z = \mathrm{e}^{\ell}$,

$$\mathfrak{g}(z,t) = \frac{1}{z - \mathrm{e}^{t(bz\mathfrak{g}(z,t)+a/2-b)}}. \tag{16.32}$$

Now, let us compare this equation to the one obeyed by the Stieltjes transform of the free log-normal. Injecting $t = z\mathfrak{g}_{\mathrm{LN}} - 1$ in

$$z = \frac{t+1}{tS_{\mathrm{LN}}(t)} \tag{16.33}$$

and using Eq. (16.15), one finds

$$z\mathfrak{g}_{\mathrm{LN}} - 1 = \mathfrak{g}_{\mathrm{LN}}\mathrm{e}^{a/2-b+bz\mathfrak{g}_{\mathrm{LN}}} \to \mathfrak{g}_{\mathrm{LN}} = \frac{1}{z - \mathrm{e}^{bz\mathfrak{g}_{\mathrm{LN}}+a/2-b}}, \tag{16.34}$$

which coincides with Eq. (16.32) for $t = 1$, as it should. For arbitrary times, one finds that the density corresponding to the multiplicative Dyson Brownian motion, Eq. (16.26), is the free log-normal, with parameters ta and tb.

16.4 The Matrix Kesten Problem

The Kesten iteration for scalar random variables appears in many different situations. It is defined by

$$Z_{n+1} = z_n(1 + Z_n), \tag{16.35}$$

where z_n are IID random variables. In the following, we will assume that

$$z_n = 1 + \varepsilon m + \sqrt{\varepsilon}\sigma\eta_n, \tag{16.36}$$

where $\varepsilon \ll 1$ and η_n are IID random variables, of zero mean and unit variance. Setting $Z_n = U_n/\varepsilon$ and expanding to first order in ε, one obtains

$$U_{n+1} = \varepsilon(1 + \varepsilon m + \sqrt{\varepsilon}\sigma\eta_n)\left(1 + \frac{U_n}{\varepsilon}\right) = U_n + \varepsilon m U_n + \sqrt{\varepsilon}\sigma\eta_n U_n + \varepsilon \tag{16.37}$$

or, in the continuous time limit $\mathrm{d}t = \varepsilon$, the following Langevin equation:

$$\frac{\mathrm{d}U}{\mathrm{d}t} = 1 + mU + \sigma\eta U. \tag{16.38}$$

The corresponding Fokker–Planck equation reads

$$\frac{\partial P(U,t)}{\partial t} = -\frac{\partial}{\partial U}[(1 + mU)P] + \frac{\sigma^2}{2}\frac{\partial^2}{\partial U^2}\left[U^2 P\right]. \tag{16.39}$$

This process has a stationary distribution provided the drift m is negative. We will thus write $m = -\hat{m}$ with $\hat{m} > 0$. The corresponding stationary distribution $P_{\text{eq}}(U)$ obeys

$$(1 - \hat{m}U)P_{\text{eq}} = \frac{\sigma^2}{2}\frac{\partial}{\partial U}\left[U^2 P\right], \tag{16.40}$$

which leads to

$$P_{\text{eq}}(U) = \frac{2^\mu}{\Gamma(\mu)\sigma^{2\mu}}\frac{e^{-\frac{2}{\sigma^2 U}}}{U^{1+\mu}}; \qquad \mu := 1 + 2\hat{m}/\sigma^2, \tag{16.41}$$

to wit, the distribution of U is an inverse-gamma, with a power-law tail $U^{-1-\mu}$ with a non-universal exponent $\mu = 1 + 2\hat{m}/\sigma^2$.

Now we can generalize the Kesten iteration for symmetric matrices as[1]

$$\mathbf{U}_{n+1} = \varepsilon\sqrt{\mathbf{1} + \frac{\mathbf{U}_n}{\varepsilon}}\left((1 + m\varepsilon)\mathbf{1} + \sqrt{\varepsilon}\sigma\mathbf{B}\right)\sqrt{\mathbf{1} + \frac{\mathbf{U}_n}{\varepsilon}}, \tag{16.42}$$

or

$$\mathbf{U}_{n+1} - \mathbf{U}_n = \varepsilon\left(\mathbf{1} + m\mathbf{U}_n\right) + \sigma\sqrt{\varepsilon}\sqrt{\mathbf{U}_n}\mathbf{B}\sqrt{\mathbf{U}_n}. \tag{16.43}$$

Following the same steps as in the previous section, we obtain a differential equation for the eigenvalues of $\mathbf{U}$ (where we neglect the noise when $N \to \infty$):

$$\frac{\mathrm{d}\lambda_i}{\mathrm{d}t} = 1 - \hat{m}\lambda_i + \frac{\sigma^2}{N}\sum_{j \neq i}\frac{\lambda_i\lambda_j}{\lambda_i - \lambda_j}, \tag{16.44}$$

where we again assume that $m < 0$ in order to find a stationary state for our process. The corresponding evolution of the Stieltjes transform reads, for large N,

[1] The results of this section have been obtained in collaboration with T. Gautié and P. Le Doussal.

$$\frac{\partial \mathfrak{g}}{\partial t} = \frac{\partial}{\partial z}\left[-\mathfrak{g} + (\sigma^2 + \hat{m})z\mathfrak{g} - \frac{1}{2}\sigma^2 z^2 \mathfrak{g}^2\right]. \tag{16.45}$$

If an equilibrium density exists, its Stieltjes transform must obey

$$\frac{1}{2}\sigma^2 z^2 \mathfrak{g}^2 + (1 - (\sigma^2 + \hat{m})z)\mathfrak{g} + C = 0, \tag{16.46}$$

where C is a constant determined by the fact that $z\mathfrak{g} \to 1$ when $z \to \infty$. Hence,

$$C = \frac{1}{2}\sigma^2 + \hat{m}. \tag{16.47}$$

From the second order equation on $\mathfrak{g}$ one gets

$$\mathfrak{g} = \frac{1}{\sigma^2 z^2}\left[((\sigma^2 + \hat{m})z - 1) - \sqrt[\oplus]{\hat{m}^2 z^2 - 2(\sigma^2 + \hat{m})z + 1}\right]. \tag{16.48}$$

As usual, the density of eigenvalues is non-zero when the square-root becomes imaginary. The edges are thus given by the roots of the second degree polynomial inside the square-root, namely

$$\lambda_{\pm} = \frac{\sigma^2 + \hat{m} \pm \sqrt{\sigma^2(\sigma^2 + 2\hat{m})}}{\hat{m}^2}. \tag{16.49}$$

So only when $\hat{m} \to 0$ can the spectrum extend to infinity, with a power-law decay as $\lambda^{-3/2}$. Otherwise, the power law is truncated beyond $2\sigma^2/\hat{m}^2$. Note that, contrary to the scalar Kesten case, the exponent of the power law is universal, with $\mu = 1/2$.

In fact, if one stares at Eq. (16.48), one realizes that the stationary Kesten matrix **U** is an inverse-Wishart matrix. Indeed, the eigenvalue spectrum given by Eq. (16.48) maps into the Marčenko–Pastur law, Eq. (4.43), provided one makes the following transformation:

$$\lambda \to x = \frac{2}{\sigma^2 + 2\hat{m}}\frac{1}{\lambda}. \tag{16.50}$$

The parameter q of the Marčenko–Pastur law is then given by

$$q = \frac{\sigma^2}{\sigma^2 + 2\hat{m}} = \frac{1}{\mu} < 1. \tag{16.51}$$

Although not trivial, this result is not so surprising: since Wishart matrices are the matrix equivalent of the scalar gamma distribution, the matrix equivalent of the Kesten variable distributed as an inverse-gamma, Eq. (16.41), is an inverse-Wishart.

Bibliographical Notes

- For a general reference on products of random matrices and their applications, see
  - A. Crisanti, G. Paladin, and A. Vulpiani. *Products of Random Matrices in Statistical Physics*. Series in Solid-State Sciences, Vol. 104. Springer Science & Business Media, 2012.
- For the specific case of electrons in disordered media, see
  - C. W. J. Beenakker. Random-matrix theory of quantum transport. *Reviews of Modern Physics*, 69:731–808, 1997,

  and for applications to granular media, see

- L. Yan, J.-P. Bouchaud, and M. Wyart. Edge mode amplification in disordered elastic networks. *Soft Matter*, 13:5795–5801, 2017.

- For the numerical calculation of the spectrum of Lyapunov exponents, see
  - G. Benettin, L. Galgani, A. Giorgilli, and J.-M. Strelcyn. Lyapunov characteristic exponents for smooth dynamical systems and for Hamiltonian systems; a method for computing all of them. Part 1: Theory. *Meccanica*, 15(1):9–20, 1980.
- For the study of the limiting distribution of the spectrum of Lyapunov exponents, see the early work of
  - C. M. Newman. The distribution of Lyapunov exponents: Exact results for random matrices. *Communications in Mathematical Physics*, 103(1):121–126, 1986,

  and, using free random matrix methods,
  - G. Tucci. Asymptotic products of independent gaussian random matrices with correlated entries. *Electronic Communications in Probability*, 16:353–364, 2011.
- About the classical Kesten variable, see
  - H. Kesten. Random difference equations and renewal theory for products of random matrices. *Acta Math.*, 131:207–248, 1973,
  - C. de Calan, J. M. Luck, T. M. Nieuwenhuizen, and D. Petritis. On the distribution of a random variable occurring in 1d disordered systems. *Journal of Physics A: Mathematical and General*, 18(3):501–523, 1985,

  and for the particular equation (16.38),
  - J.-P. Bouchaud, A. Comtet, A. Georges, and P. Le Doussal. Classical diffusion of a particle in a one-dimensional random force field. *Annals of Physics*, 201(2):285–341, 1990,
  - J.-P. Bouchaud and M. Mézard. Wealth condensation in a simple model of economy. *Physica A: Statistical Mechanics and Its Applications*, 282(3):536–545, 2000.

17

Sample Covariance Matrices

In this chapter, we will show how to compute the various transforms ($S(t)$, $\mathfrak{t}(z)$, $\mathfrak{g}(z)$) for sample covariance matrices (SCM) when the data has non-trivial true correlations, i.e. is characterized by a non-diagonal true underlying covariance matrix $\mathbf{C}$ and possibly non-trivial temporal correlations as well. More precisely, N time series of length T are stored in a rectangular $N \times T$ matrix $\mathbf{H}$. The *sample covariance matrix* is defined as

$$\mathbf{E} = \frac{1}{T}\mathbf{H}\mathbf{H}^T. \tag{17.1}$$

If the N time series are stationary, we expect that for $T \gg N$, the SCM $\mathbf{E}$ converges to the "true" covariance matrix $\mathbf{C}$. The non-trivial correlations encoded in the off-diagonal elements of $\mathbf{C}$ are what we henceforth call *spatial* (or cross-sectional) correlations. But the T samples might also be non-independent and we will also model these *temporal* correlations. Of course, the data might have both types of correlations (spatial and temporal).

We will be interested in the eigenvalues $\{\lambda_k\}$ of $\mathbf{E}$ and their density $\rho_{\mathbf{E}}(\lambda)$, which we will compute from the knowledge of its Stieltjes transform $\mathfrak{g}_{\mathbf{E}}(z)$ using Eq. (2.47). We can also compute the singular values $\{s_k\}$ of $\mathbf{H}$; note that these singular values are related to the eigenvalues of $\mathbf{E}$ via $s_k = \sqrt{T\lambda_k}$.

17.1 Spatial Correlations

Consider the case where $\mathbf{H}$ are multivariate Gaussian observations, drawn from $\mathcal{N}(0, \mathbf{C})$. We saw in Section 4.2.4 that $\mathbf{E}$ is then a general Wishart matrix with column covariance $\mathbf{C}$, and can be written as

$$\mathbf{E} = \mathbf{C}^{\frac{1}{2}}\mathbf{W}_q\mathbf{C}^{\frac{1}{2}}. \tag{17.2}$$

We recognize this formula as the free product of the covariance matrix $\mathbf{C}$ and a white Wishart of parameter q, $\mathbf{W}_q$. Note that since the white Wishart is rotationally invariant, it is free from any matrix $\mathbf{C}$. From the multiplicativity of the S-transform and the form of the S-transform of the white Wishart (Eq. (15.21)), we have

$$S_{\mathbf{E}}(t) = \frac{S_{\mathbf{C}}(t)}{1 + qt}. \tag{17.3}$$

We can also use the subordination relation of the free product (Eq. (11.109)) to write this relation in terms of T-transforms:

$$\mathfrak{t}_{\mathbf{E}}(z) = \mathfrak{t}_{\mathbf{C}}\left(Z(z)\right), \qquad Z(z) = \frac{z}{1 + q\mathfrak{t}_{\mathbf{E}}(z)}. \tag{17.4}$$

This last expression can be written in terms of the more familiar Stieltjes transform using $\mathfrak{t}(z) = z\mathfrak{g}(z) - 1$:

$$z\mathfrak{g}_{\mathbf{E}}(z) = Z\mathfrak{g}_{\mathbf{C}}(Z), \quad \text{where} \quad Z = \frac{z}{1 - q + qz\mathfrak{g}_{\mathbf{E}}(z)}. \tag{17.5}$$

This is the central equation that allows one to infer the "true" spectral density of $\mathbf{C}$, $\rho_{\mathbf{C}}(\lambda)$, from the empirically observed spectrum of $\mathbf{E}$. Note that this equation can equivalently be rewritten in terms of the spectral density of $\mathbf{C}$ as

$$\mathfrak{g}_{\mathbf{E}}(z) = \int \frac{\rho_{\mathbf{C}}(\mu)\mathrm{d}\mu}{z - \mu(1 - q + qz\mathfrak{g}_{\mathbf{E}}(z))}. \tag{17.6}$$

We will see in Chapter 20 some real world applications of this formula. One of the most important properties of Eq. (17.5) is its universality: it holds (in the large N limit) much beyond the restricted perimeter of multivariate Gaussian observations $\mathbf{H}$. In fact, as soon as the observations have a finite second moment, the relation between the "true" spectral density $\rho_{\mathbf{C}}$ and the empirical Stieltjes transform $\mathfrak{g}_{\mathbf{E}}(z)$ is given by Eq. (17.5).

Let us discuss some interesting limiting cases. First, when $q \to 0$, i.e. when $T \gg N$, one expects that $\mathbf{E} \approx \mathbf{C}$. This is indeed what one finds since in that limit $Z = z + O(q)$; hence $\mathfrak{g}_{\mathbf{E}}(z) = \mathfrak{g}_{\mathbf{C}}(z)$ and $\rho_{\mathbf{E}} = \rho_{\mathbf{C}}$.

Second, consider the case $\mathbf{C} = \mathbf{1}$, for which $\mathfrak{g}_{\mathbf{C}}(Z) = 1/(Z - 1)$. We thus obtain

$$z\mathfrak{g}_{\mathbf{E}}(z) = \frac{Z}{Z-1} = \frac{z}{(z - 1 + q - qz\mathfrak{g}_{\mathbf{E}}(z))} \to \frac{1}{\mathfrak{g}_{\mathbf{E}}(z)} = z1 + q - qz\mathfrak{g}_{\mathbf{E}}(z), \tag{17.7}$$

which coincides with Eq. (4.37). In the next exercise, we consider the case where $\mathbf{C}$ is an inverse-Wishart matrix, in which case some explicit results can be obtained.

We can also infer some properties of the spectrum of $\mathbf{E}$ using the moment generating function. The T-transform of $\mathbf{E}$ can be expressed as the following power series for $z \to \infty$:

$$\mathfrak{t}_{\mathbf{E}}(z) \underset{z\to\infty}{\longrightarrow} \sum_{k=1}^{\infty} \tau(\mathbf{E}^k)z^{-k}. \tag{17.8}$$

We thus deduce that

$$Z(z) \underset{z\to\infty}{\longrightarrow} \frac{z}{1 + q\sum_{k=1}^{\infty} \tau(\mathbf{E}^k)z^{-k}}.$$

Therefore we have, for $z \to \infty$,

$$\mathfrak{t}_{\mathbf{C}}(Z(z)) \underset{z\to\infty}{\longrightarrow} \sum_{k=1}^{\infty} \frac{\tau(\mathbf{C}^k)}{z^k} \left(1 + q\sum_{\ell=1}^{\infty} \tau(\mathbf{E}^\ell)z^{-\ell}\right)^k. \tag{17.9}$$

Hence, one can thus relate the moments of $\rho_{\mathbf{E}}$ with the moments of $\rho_{\mathbf{C}}$ by taking $z \to \infty$ in Eq. (17.4), namely

$$\sum_{k=1}^{\infty} \frac{\tau(\mathbf{E}^k)}{z^k} = \sum_{k=1}^{\infty} \frac{\tau(\mathbf{C}^k)}{z^k} \left(1 + q \sum_{\ell=1}^{\infty} \tau(\mathbf{E}^\ell) z^{-\ell}\right)^k. \tag{17.10}$$

In particular, Eq. (17.10) yields the first three moments of $\rho_{\mathbf{E}}$:

$$\begin{aligned} \tau(\mathbf{E}) &= \tau(\mathbf{C}), \\ \tau(\mathbf{E}^2) &= \tau(\mathbf{C}^2) + q, \\ \tau(\mathbf{E}^3) &= \tau(\mathbf{C}^3) + 3q\tau(\mathbf{C}^2) + q^2. \end{aligned} \tag{17.11}$$

We thus see that the mean of $\mathbf{E}$ is equal to that of $\mathbf{C}$, whereas the variance of $\mathbf{E}$ is equal to that of $\mathbf{C}$ plus q. As expected, the spectrum of the sample covariance matrix $\mathbf{E}$ is always wider (for $q > 0$) than the spectrum of the population covariance matrix $\mathbf{C}$.

Another interesting expansion concerns the case where $q < 1$, such that $\mathbf{E}$ is invertible. Hence $\mathfrak{g}_{\mathbf{E}}(z)$ for $z \to 0$ is analytic and one readily finds

$$\mathfrak{g}_{\mathbf{E}}(z) \underset{z \to 0}{\longrightarrow} -\sum_{k=1}^{\infty} \tau\left(\mathbf{E}^{-k}\right) z^{k-1}. \tag{17.12}$$

This allows us to study the moments of $\mathbf{E}^{-1}$, which turn out to be important quantities for many applications. Using Eq. (17.5), we can actually relate the moments of the spectrum $\mathbf{E}^{-1}$ to those of $\mathbf{C}^{-1}$. Indeed, for $z \to 0$,

$$Z(z) = \frac{z}{1 - q - q\sum_{k=1}^{\infty} \tau\left(\mathbf{E}^{-k}\right) z^k}.$$

Hence, we obtain the following expansion:

$$\sum_{k=1}^{\infty} \tau\left(\mathbf{E}^{-k}\right) z^k = \sum_{k=1}^{\infty} \tau\left(\mathbf{C}^{-k}\right) \left(\frac{z}{1-q}\right)^k \left(\frac{1}{1 - \frac{q}{1-q}\sum_{\ell=1}^{\infty} \tau\left(\mathbf{E}^{-\ell}\right) z^\ell}\right)^k. \tag{17.13}$$

After a little work, we get

$$\tau\left(\mathbf{E}^{-1}\right) = \frac{\tau(\mathbf{C}^{-1})}{1-q}, \qquad \tau\left(\mathbf{E}^{-2}\right) = \frac{\tau(\mathbf{C}^{-2})}{(1-q)^2} + \frac{q\tau\left(\mathbf{C}^{-1}\right)^2}{(1-q)^3}. \tag{17.14}$$

We will discuss in Section 20.2.1 a direct application of these formulas: $\tau(\mathbf{E}^{-1})$ turns out to be related to the "out-of-sample" risk of an optimized portfolio of financial instruments.

Exercise 17.1.1 The exponential moving average sample covariance matrix (EMA-SCM)

Instead of measuring the sample covariance matrix using a flat average over a fixed time window T, one can compute the average using an exponential weighted moving average. Let us compute the spectrum of such a matrix in the null case of IID data. Imagine we have an infinite time series of vectors of size N $\{\mathbf{x}_t\}$ for t from minus infinity to now. We define the EMA-SCM (on time scale τ_c) as

$$\mathbf{E}(t) = \gamma_c \sum_{t'=-\infty}^{t} (1-\gamma_c)^{t-t'} \mathbf{x}_{t'} \mathbf{x}_{t'}^T, \tag{17.15}$$

where $\gamma_c := 1/\tau_c$. Hence,

$$\mathbf{E}(t) = (1-\gamma_c)\mathbf{E}(t-1) + \gamma_c \mathbf{x}_{t'} \mathbf{x}_{t'}^T. \tag{17.16}$$

The second term on the right hand side can be thought of as a Wishart matrix with $T = 1$ (or $q = N$). Now, both $\mathbf{E}(t)$ and $\mathbf{E}(t-1)$ are equal in law so we write

$$\mathbf{E} \overset{\text{in law}}{=} (1-\gamma_c)\mathbf{E} + \gamma_c \mathbf{W}_{q=N}. \tag{17.17}$$

(a) Given that $\mathbf{E}$ and $\mathbf{W}$ are free, use the properties of the R-transform to get the equation

$$R_{\mathbf{E}}(x) = (1-\gamma_c) R_{\mathbf{E}}((1-\gamma_c)x) + \gamma_c(1 - N\gamma_c x). \tag{17.18}$$

(b) Take the limit $N \to \infty$, $\tau_c \to \infty$ with $q := N/\tau_c$ fixed to get the following differential equation for $R_{\mathbf{E}}(x)$:

$$R_{\mathbf{E}}(x) = -x \frac{\mathrm{d}}{\mathrm{d}x} R_{\mathbf{E}}(x) + \frac{1}{1-qx}. \tag{17.19}$$

(c) The definition of $\mathbf{E}$ is properly normalized, $\tau(\mathbf{E}) = 1$ [show this using Eq. (17.17)], so we have the initial condition $R(0) = 1$. Show that

$$R_{\mathbf{E}}(x) = -\frac{\log(1-qx)}{qx} \tag{17.20}$$

solves your equation with the correct initial condition. Compute the variance $\kappa_2(\mathbf{E})$.

(d) To compute the spectrum of eigenvalues of $\mathbf{E}$, one needs to solve a complex transcendental equation. First write $\mathfrak{z}(g)$, the inverse of $\mathfrak{g}(z)$. For $q = 1/2$ plot $\mathfrak{z}$ as a function of g (for $-4 < g < 2$). You will see that there are values of z that are never attained by $\mathfrak{z}(g)$, in other words $\mathfrak{g}(z)$ has no real solutions for these z. Numerically find complex solutions for $\mathfrak{g}(z)$ in that range. Plot the density of eigenvalues $\rho_{\mathbf{E}}(\lambda)$ given by Eq. (2.47). Plot also the density for a Wishart with the same mean and variance.

(e) Construct numerically the matrix $\mathbf{E}$ as in Eq. (17.15). Use $N = 1000$, $\tau_c = 2000$ and use at least 10 000 values for t'. Plot the eigenvalue distribution of your numerical $\mathbf{E}$ against the distribution found in (d).

17.2 Temporal Correlations

17.2.1 General Case

A common problem in data analysis arises when samples are not independent. Intuitively, correlated samples are somehow redundant and the sample covariance matrix should behave as if we had observed not T samples but an effective number $T^* < T$. Let us analyze more precisely the sample covariance matrix in the presence of correlated samples. We will start with the case when the true spatial correlations are zero, i.e. $\mathbf{C} = \mathbf{1}$. Our data can then be written in a rectangular $N \times T$ matrix $\mathbf{H}$ satisfying

$$\mathbb{E}[\mathbf{H}_{it}\mathbf{H}_{js}] = \delta_{ij}\mathbf{K}_{ts}, \tag{17.21}$$

where $\mathbf{K}$ is the $T \times T$ temporal covariance matrix that we assumed to be normalized as $\tau(\mathbf{K}) = 1$. Following the same arguments as in Section 4.2.4, we can write

$$\mathbf{H} = \mathbf{H}_0\mathbf{K}^{\frac{1}{2}}, \tag{17.22}$$

where $\mathbf{H}_0$ is a white rectangular matrix. So the sample covariance matrix becomes

$$\mathbf{E} = \frac{1}{T}\mathbf{H}\mathbf{H}^T = \frac{1}{T}\mathbf{H}_0\mathbf{K}\mathbf{H}_0^T. \tag{17.23}$$

Now this is not quite the free product of the matrix $\mathbf{K}$ and a white Wishart, but if we define the $(T \times T)$ matrix $\mathbf{F}$ as

$$\mathbf{F} = \frac{1}{N}\mathbf{H}^T\mathbf{H} = \frac{1}{N}\mathbf{K}^{\frac{1}{2}}\mathbf{H}_0^T\mathbf{H}_0\mathbf{K}^{\frac{1}{2}} \equiv \mathbf{K}^{\frac{1}{2}}\mathbf{W}_{1/q}\mathbf{K}^{\frac{1}{2}}, \tag{17.24}$$

then $\mathbf{F}$ is the free product of the matrix $\mathbf{K}$ and a white Wishart matrix with parameter $1/q$. Hence,

$$S_{\mathbf{F}}(t) = \frac{S_{\mathbf{K}}(t)}{1 + t/q}. \tag{17.25}$$

To find the S-transform of $\mathbf{E}$, we go back to Section 4.1.1, where we obtained Eq. (4.5) relating the Stieltjes transforms of $\mathbf{E}$ and $\mathbf{F}$. In terms of the T-transform, the relation is even simpler:

$$\mathfrak{t}_{\mathbf{F}}(z) = q\mathfrak{t}_{\mathbf{E}}(qz) \quad \Rightarrow \quad \zeta_{\mathbf{E}}(t) = q\zeta_{\mathbf{F}}(qt), \tag{17.26}$$

where the functions $\zeta(t)$ are the inverse T-transforms. Using the definition of the S-transform (Eq. (11.92)), we finally get

$$S_{\mathbf{E}}(t) = \frac{S_{\mathbf{K}}(qt)}{1 + qt}, \tag{17.27}$$

which can be expressed as a relation between inverse T-transforms:

$$\zeta_{\mathbf{E}}(t) = q(1 + t)\zeta_{\mathbf{K}}(qt). \tag{17.28}$$

We can also write a subordination relation between the T-transforms:

$$q\mathrm{t}_{\mathbf{E}}(z) = \mathrm{t}_{\mathbf{K}}\left(\frac{z}{q(1+\mathrm{t}_{\mathbf{E}}(z))}\right). \tag{17.29}$$

This is a general formula that we specialize to the case of exponential temporal correlations in the next section. Note that in the limit $z \to 0$, the above equation gives access to $\tau(\mathbf{E}^{-1})$. Using

$$\mathrm{t}_{\mathbf{E}}(z) \underset{z\to 0}{=} -1 - \tau(\mathbf{E}^{-1})z + O(z^2), \tag{17.30}$$

we find

$$\tau(\mathbf{E}^{-1}) = -\frac{1}{q\zeta_{\mathbf{K}}(-q)}. \tag{17.31}$$

17.2.2 Exponential Correlations

The most common form of temporal correlation in experimental data is the decaying exponential, corresponding to a matrix $\mathbf{K}_{ts}$ in Eq. (17.21) given by

$$\mathbf{K}_{ts} := a^{|t-s|}, \tag{17.32}$$

where $1/\log(a)$ defines the temporal span of the correlations.

In Appendix A.3 we explicitly compute the S-transform of $\mathbf{K}$. The result reads

$$S_{\mathbf{K}}(t) = \frac{t+1}{\sqrt{1+(b^2-1)t^2}+bt}, \tag{17.33}$$

where $b := (1+a^2)/(1-a^2)$. From $S_{\mathbf{K}}$ one can also obtain $\zeta_{\mathbf{K}}$ and its inverse $\mathrm{t}_{\mathbf{K}}$, which read

$$\zeta_{\mathbf{K}}(t) = \frac{\sqrt{1+(b^2-1)t^2}}{t} + b, \qquad \mathrm{t}_{\mathbf{K}}(\zeta) = -\frac{1}{\sqrt{\zeta^2-2\zeta b+1}}. \tag{17.34}$$

Combining Eq. (17.27) with Eq. (17.33), we get

$$S_{\mathbf{E}}(t) = \frac{1}{\sqrt{1+(b^2-1)(qt)^2}+bqt}. \tag{17.35}$$

From the S-transform, we find

$$\zeta_{\mathbf{E}}(t) = \frac{1+t}{tS_{\mathbf{E}}(t)} = \frac{1+t}{t}\left(\sqrt{1+(b^2-1)(qt)^2}+bqt\right), \tag{17.36}$$

which when inverted leads to a fourth order equation for $\mathrm{t}_{\mathbf{E}}(z)$ that must be solved numerically, leading to the densities plotted in Fig. 17.1. However, one can obtain some information on $\tau(\mathbf{E}^{-1})$. From Eqs. (17.31) and (17.34), one obtains

$$\tau(\mathbf{E}^{-1}) = \frac{1}{\sqrt{q^2(b^2-1)+1}-bq} := \frac{1}{1-q^*}, \tag{17.37}$$

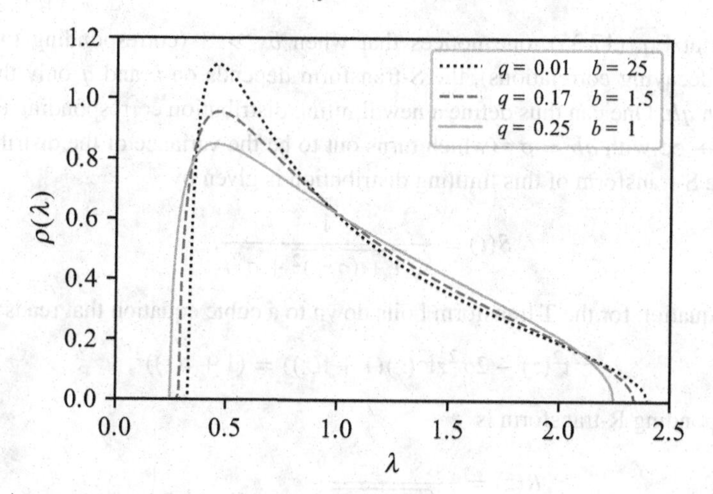

Figure 17.1 Density of eigenvalues for a sample covariance matrix with exponential temporal correlations for three choices of parameters q and b such that $qb = 0.25$. All three densities are normalized, have mean 1 and variance $\sigma_{\mathbf{E}}^2 = qb = 0.25$. The solid light gray one is the Marčenko–Pastur density ($q = 0.25$), the dotted black one is very close to the limiting density for $q \to 0$ with $\sigma^2 = bq$ fixed.

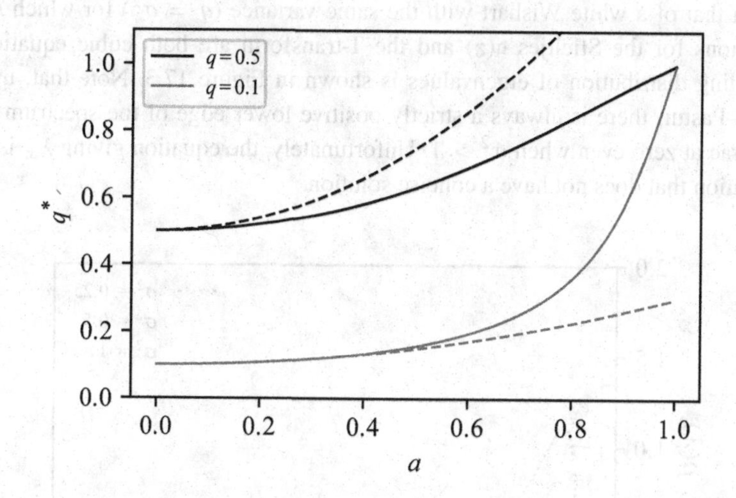

Figure 17.2 Effective value q^* versus the one-lag autocorrelation coefficient a for a sample covariance matrix with exponential temporal correlations shown for two values of q. The dashed lines indicate the approximation (valid at small a) $q^* = q(1 + 2a^2)$. The approximation means that, for 10% autocorrelation, q^* is only 2% greater than q.

where $q^* = N/T^*$ defines the effective length of the time series, reduced by temporal correlations (compare with Eq. (17.14) with $\mathbf{C} = \mathbf{1}$). Figure 17.2 shows q^* as a function of a. As expected, $q^* = q$ for $a = 0$ (no temporal correlations), whereas $q^* \to 1$ when $a \to 1$, i.e. when $\tau_c \to \infty$. In this limit, $\mathbf{E}$ becomes singular.

Looking at Eq. (17.35), one notices that when $b \gg 1$ (corresponding to $a \to 1$, i.e. slowly decaying correlations), the S-transform depends on b and q only through the combination qb. One can thus define a new limiting distribution corresponding to the limit $q \to 0$, $b \to \infty$ with $qb = \sigma^2$ (which turns out to be the variance of the distribution, see below). The S-transform of this limiting distribution is given by

$$S(t) = \frac{1}{\sqrt{1 + (\sigma^2 t)^2} + \sigma^2 t}, \tag{17.38}$$

while the equation for the T-transform boils down to a cubic equation that reads:

$$z^2 \mathfrak{t}^2(z) - 2\sigma^2 z \mathfrak{t}^2(z)(1 + \mathfrak{t}(z)) = (1 + \mathfrak{t}(z))^2. \tag{17.39}$$

The corresponding R-transform is

$$\begin{aligned} R(z) &= \frac{1}{\sqrt{1 - 2\sigma^2 z}} \\ &= 1 + \sigma^2 z + \frac{3}{2}\sigma^4 z^2 + O(z^3). \end{aligned} \tag{17.40}$$

The last equation gives its first three cumulants: its average is equal to one, its variance is σ^2 as announced above, and its skewness is $\kappa_3 = \frac{3}{2}\sigma^4$. We notice that this skewness is larger than that of a white Wishart with the same variance ($q = \sigma^2$) for which $\kappa_3 = \sigma^4$. The equations for the Stieltjes $\mathfrak{g}(z)$ and the T-transform are both cubic equations. The corresponding distribution of eigenvalues is shown in Figure 17.3. Note that, unlike the Marčenko–Pastur, there is always a strictly positive lower edge of the spectrum $\lambda_- > 0$ and no Dirac at zero even when $\sigma^2 > 1$. Unfortunately, the equation giving $\lambda_\pm$ is a fourth order equation that does not have a concise solution.

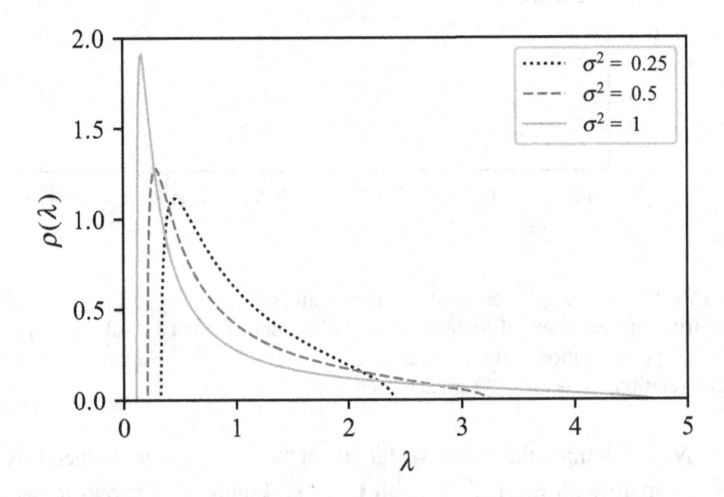

Figure 17.3 Density of eigenvalues for the limiting distribution of sample covariance matrix with exponential temporal correlations $\mathbf{W}_{\sigma^2}$ for three choices of the parameter σ^2: 0.25, 0.5 and 1.

An intuitive way to understand this particular random matrix ensemble is to consider N independent Ornstein–Uhlenbeck processes with the same correlation time τ_c that we record over a long time T. We sample the data at interval Δ, such that the total number of observations is T/Δ. We then construct a sample covariance matrix of the N variables from these observations. If $\Delta \gg \tau_c$, then each sample can be considered independent and the sample covariance matrix will be a Marčenko–Pastur with $q = N\Delta/T$. But if we "oversample" at intervals $\Delta \ll \tau_c$, such that our observations are strongly correlated, then the resulting sample covariance matrix no longer depends on Δ but only on τ_c. The sample covariance matrix converges in this case to our new random matrix characterized by Eq. (17.38), with parameter $\sigma^2 = qb = N\tau_c/T$.

17.2.3 Spatial and Temporal Correlations

In the general case where spatial and temporal correlations exist, the sample covariance matrix can be written as

$$\mathbf{E} = \frac{1}{T}\mathbf{H}\mathbf{H}^T = \frac{1}{T}\mathbf{C}^{\frac{1}{2}}\mathbf{H}_0\mathbf{K}\mathbf{H}_0^T\mathbf{C}^{\frac{1}{2}}, \tag{17.41}$$

using the same notations as above. After similar manipulations, the S-transform of $\mathbf{E}$ is found to be given by

$$S_{\mathbf{E}}(t) = \frac{S_{\mathbf{C}}(t)S_{\mathbf{K}}(qt)}{1+qt}, \tag{17.42}$$

which leads to

$$\zeta_{\mathbf{E}}(t) = qt\zeta_{\mathbf{C}}(t)\zeta_{\mathbf{K}}(qt), \tag{17.43}$$

or, in terms of T-transforms,

$$q\mathfrak{t}_{\mathbf{E}}(z) = \mathfrak{t}_{\mathbf{K}}\left(\frac{z}{q\mathfrak{t}_{\mathbf{E}}(z)\zeta_{\mathbf{C}}(\mathfrak{t}_{\mathbf{E}}(z))}\right). \tag{17.44}$$

When $\mathbf{C} = \mathbf{1}$, $\zeta_{\mathbf{C}}(t) = (1+t)/t$ and one recovers Eq. (17.29). Specializing to the case of exponential correlations in the limit $q \to 0$, $a \to 1$, $qb = \sigma^2$, we obtain the following equation for the T-transform of the limiting distribution, now for an arbitrary covariance matrix $\mathbf{C}$:

$$z^2 - 2\sigma^2 z\mathfrak{t}_{\mathbf{E}}(z)\zeta_{\mathbf{C}}(\mathfrak{t}_{\mathbf{E}}(z)) = \zeta_{\mathbf{C}}^2(\mathfrak{t}_{\mathbf{E}}(z)), \tag{17.45}$$

where we used $\mathfrak{t}_{\mathbf{K}}(z) = -1/\sqrt{z^2 - 2zb + 1}$. When $\mathbf{C} = \mathbf{1}$, one recovers Eq. (17.39). When $\mathbf{C}$ is an inverse-Wishart matrix, $\zeta_{\mathbf{C}}(t) = (t+1)/t(1-pt)$, the equation for $\mathfrak{t}_{\mathbf{E}}(z)$ is of fourth order.

Note finally that Eq. (17.44), in the limit $z \to 0$, yields a simple generalization of Eq. (17.31) that reads

$$\tau(\mathbf{E}^{-1}) = -\frac{\tau(\mathbf{C}^{-1})}{q\zeta_{\mathbf{K}}(-q)}. \tag{17.46}$$

Comparing with Eq. (17.14) allows us to define an effective length of the time series which, interestingly, is *independent* of **C** and reads

$$q^* := \frac{N}{T^*} = 1 + q\zeta_{\mathbf{K}}(-q). \tag{17.47}$$

Exercise 17.2.1 On the futility of oversampling

Consider data consisting of N variables (columns) with true correlation $\mathbf{C}$ and T independent observations (rows). Instead of computing the sample covariance matrix with these T observations, we repeat each one m times and sum over mT columns. Obviously the redundant columns should not change the sample covariance matrix, hence it should have the same spectrum as the one using only the original T observations.

(a) The redundancy of columns can be modeled as a temporal correlation with an $mT \times mT$ covariance matrix $\mathbf{K}$ that is block diagonal with T blocks of size K and all the values within one block equal to 1 and zero outside the blocks. Show that this matrix has T eigenvalues equal to m and $(T-1)m$ zero eigenvalues.

(b) Compute $\mathfrak{t}_{\mathbf{K}}(z)$ for this model.

(c) Show that $S_{\mathbf{K}}(t) = (1+t)/(1+mt)$.

(d) If we include the redundant columns we have a value of $q_m = N/(mT)$, but we need to take temporal correlations into account so $S_{\mathbf{E}}(t) = S_{\mathbf{C}}(t)S_{\mathbf{K}}(q_m t)/(1+q_m t)$. Show that in this case $S_{\mathbf{E}}(t) = S_{\mathbf{C}}(t)/(1+qt)$ with $q = N/T$, which is the result without the redundant columns.

17.3 Time Dependent Variance

Another common and important situation is when the N correlated time series are heteroskedastic, i.e. have a time dependent variance. More precisely, we consider a model where

$$x_i^t = \sigma_t \mathbf{H}_{it}, \tag{17.48}$$

where σ_t is time dependent, and

$$\mathbb{E}[\mathbf{H}_{it}\mathbf{H}_{js}] = \delta_{ts}\mathbf{C}_{ij}, \tag{17.49}$$

i.e. x_i^t is the product of a time dependent factor σ_t and a random variable with a general correlation structure **C** but no time correlations. The SCM **E** can be expressed as

$$\mathbf{E} = \sum_{t=1}^{T} \mathbf{P}_t, \qquad \mathbf{P}_t := \frac{1}{T}\sigma_t^2 \mathbf{H}_t \mathbf{H}_t^T, \tag{17.50}$$

where each $\mathbf{P}_t$ is a rank-1 matrix with a non-zero eigenvalue that converges, when N and T tend to infinity, to $q\sigma_t^2\tau(\mathbf{C})$ with, as always, $q = N/T$.

We will first consider the case $\mathbf{C} = \mathbf{1}$, i.e. a structureless covariance matrix. In this case, the vectors $\mathbf{x}^t$ are rotationally invariant, the matrix $\mathbf{E}$ can be viewed as the free sum of a large number of rank-1 matrices, each with a non-zero eigenvalue equal to $q\sigma_t^2$. Hence,

$$R_{\mathbf{E}}(g) = \sum_{t=1}^{T} R_t(g). \tag{17.51}$$

To compute the R-transform of the matrix $\mathbf{E}$ we need to compute the R-transform of a rank-1 matrix. Note that since there are T terms in the sum, we will need to know $R_t(g)$ including correction of order $1/N$:

$$\mathfrak{g}_t(z) = \frac{1}{N}\left(\frac{N-1}{z} + \frac{1}{z - q\sigma_t^2}\right) = \frac{1}{z} + \frac{1}{N}\frac{q\sigma_t^2}{z(z - q\sigma_t^2)}. \tag{17.52}$$

Inverting to first order in $1/N$ we find

$$\mathfrak{z}_t(g) = \frac{1}{g} + \frac{1}{N}\frac{q\sigma_t^2}{1 - q\sigma_t^2 g}. \tag{17.53}$$

Now, since $R(z) = \mathfrak{z}(g) - 1/z$, we find

$$R_{\mathbf{E}}(g) = \frac{1}{T}\sum_{t=1}^{T}\frac{\sigma_t^2}{1 - q\sigma_t^2 g}. \tag{17.54}$$

The fluctuations of σ_t^2 can be stochastic or deterministic. In the large T limit we can encode them with a probability density $P(s)$ for $s = \sigma^2$ and convert the sum into an integral, leading to[1]

$$R_{\mathbf{E}}(g) = \int_0^{\infty}\frac{sP(s)}{1 - qsg}ds. \tag{17.55}$$

Note that if the variance is always 1 (i.e. $P(s) = \delta(s-1)$), we recover the R-transform of a Wishart matrix of parameter q:

$$R_q(g) = \frac{1}{1 - qg}. \tag{17.56}$$

In the general case, the R-transform of $\mathbf{E}$ is simply related to the T-transform of the distribution of s:

$$R_{\mathbf{E}}(g) = \mathfrak{t}_s\left(\frac{1}{qg}\right). \tag{17.57}$$

[1] When the distribution of s is bounded, the integral (17.55) always converges for small enough g and the R-transform is well defined near zero. For unbounded s, the R-transform can be singular at zero indicating that the distribution of eigenvalues doesn't have an upper edge.

In the more general case where $\mathbf{C}$ is not the identity matrix, one can again write the SCM as $\widetilde{\mathbf{E}} = \mathbf{C}^{\frac{1}{2}}\mathbf{E}\mathbf{C}^{\frac{1}{2}}$, where $\mathbf{E}$ corresponds to the case $\mathbf{C} = \mathbf{1}$ that we just treated. Hence, using the fact that $\mathbf{C}$ and $\mathbf{E}$ are mutually free, the S-transform of $\widetilde{\mathbf{E}}$ is simply given by

$$S_{\widetilde{\mathbf{E}}}(t) = S_{\mathbf{C}}(t)S_{\mathbf{E}}(t). \tag{17.58}$$

Another way to treat the problem is to view the fluctuating variance as a diagonal temporal covariance matrix with entries drawn from $P(s)$. Following Section 17.2.3, we can write

$$S_{\mathbf{E}}(t) = \frac{S_s(qt)}{1+qt}, \quad S_{\widetilde{\mathbf{E}}}(t) = \frac{S_{\mathbf{C}}(t)S_s(qt)}{1+qt}, \tag{17.59}$$

with $S_s(t)$ the S-transform associated with $\mathrm{t}_s(\zeta)$.

A particular case of interest for financial applications is when $P(s)$ is an inverse-gamma distribution. When $\mathbf{x}^t$ is a Gaussian multivariate vector, one obtains for $\sigma_t\mathbf{x}^t$ a *Student multivariate distribution* (see bibliographical notes for more on this topic).

17.4 Empirical Cross-Covariance Matrices

Let us now consider two time series $\mathbf{x}^t$ and $\mathbf{y}^t$, each of length T, but of different dimensions, respectively N_1 and N_2. The empirical cross-covariance matrix is an $N_1 \times N_2$ rectangular matrix defined as

$$\mathbf{E}_{xy} = \frac{1}{T}\sum_{t=1}^{T} \mathbf{x}^t(\mathbf{y}^t)^T. \tag{17.60}$$

Let us assume that the "true" cross-covariance matrix $\mathbb{E}[\mathbf{x}\mathbf{y}^T]$ is zero, i.e. that there are no true cross-correlations between our two sets of variables. What is the singular value spectrum of $\mathbf{E}_{xy}$ in this case?

As with SCM that are described by the Marčenko–Pastur law when $N, T \to \infty$ with a fixed ratio $q = N/T$, we expect that some non-trivial results will appear in the limit $N_1, N_2, T \to \infty$ with $q_1 = N_1/T$ and $q_2 = N_2/T$ finite. A convenient way to perform this analysis is to consider the eigenvalues of the $N_1 \times N_1$ matrix $\mathbf{M}_{xy} = \mathbf{E}_{xy}\mathbf{E}_{xy}^T$, which are equal to the square of the singular values s of $\mathbf{E}_{xy}$.

The matrix $\mathbf{M}_{xy}$ shares the same non-zero eigenvalues as those of $\widehat{\mathbf{E}}_x\widehat{\mathbf{E}}_y$, where $\widehat{\mathbf{E}}_x$ and $\widehat{\mathbf{E}}_y$ are the dual $T \times T$ sample covariance matrices:

$$\widehat{\mathbf{E}}_x = \mathbf{x}^T\mathbf{x}, \qquad \widehat{\mathbf{E}}_y = \mathbf{y}^T\mathbf{y}. \tag{17.61}$$

Hence one can compute the spectral density of $\mathbf{M}_{xy}$ using the free product formalism and infer the spectrum of the product $\widehat{\mathbf{E}}_x\widehat{\mathbf{E}}_y$. However, the result will depend on the "true" covariance matrices of $\mathbf{x}$ and $\mathbf{y}$, which are usually unknown in practical applications.

A way to obtain a universal result is to consider the sample-normalized principal components of $\mathbf{x}$ and of $\mathbf{y}$, which we call $\widetilde{\mathbf{x}}$ and $\widetilde{\mathbf{y}}$, such that the corresponding dual covariance matrix $\widehat{\mathbf{E}}_{\widetilde{x}}$ has N_1 eigenvalues exactly equal to 1 and $T - N_1$ eigenvalues exactly equal to zero, whereas $\widehat{\mathbf{E}}_{\widetilde{y}}$ has N_2 eigenvalues exactly equal to 1 and $T - N_2$ eigenvalues exactly equal to zero. This is precisely the problem studied in Section 15.4.2. The singular value spectrum of $\mathbf{E}_{xy}$ is thus given by

$$\rho(s) = \max(q_1 + q_2 - 1, 0)\delta(s-1) + \mathrm{Re}\,\frac{\sqrt{(s^2-\gamma_-)(\gamma_+ - s^2)}}{\pi s(1-s^2)}, \tag{17.62}$$

where $\gamma_\pm$ are given by

$$\gamma_\pm = q_1 + q_2 - 2q_1q_2 \pm 2\sqrt{q_1q_2(1-q_1)(1-q_2)}, \quad 0 \le \gamma_\pm \le 1. \tag{17.63}$$

The allowed s's are all between 0 and 1, as they should be, since these singular values can be interpreted as correlation coefficients between some linear combination of the **x**'s and some other linear combination of the **y**'s.

In the limit $T \to \infty$ at fixed N_1, N_2, all singular values collapse to zero, as they should since there are no true correlations between **x** and **y**. The allowed band in the limit $q_1, q_2 \to 0$ becomes

$$s \in \left[|\sqrt{q_1} - \sqrt{q_2}|, \sqrt{q_1} + \sqrt{q_2} \right],$$

showing that for fixed N_1, N_2, the order of magnitude of allowed singular values decays as $T^{-\frac{1}{2}}$. The above result allows one to devise precise statistical tests to detect "true" cross-correlations between sets of variables.

Bibliographical Notes

- The subject of this chapter is treated in several books, see e.g.
  - Z. Bai and J. W. Silverstein. *Spectral Analysis of Large Dimensional Random Matrices*. Springer-Verlag, New York, 2010,
  - A. M. Tulino and S. Verdú. *Random Matrix Theory and Wireless Communications*. Now publishers, Hanover, Mass., 2004,
  - L. Pastur and M. Scherbina. *Eigenvalue Distribution of Large Random Matrices*. American Mathematical Society, Providence, Rhode Island, 2010,
  - R. Couillet and M. Debbah. *Random Matrix Methods for Wireless Communications*. Cambridge University Press, Cambridge, 2011.
- The initial "historical" paper is of course
  - V. A. Marchenko and L. A. Pastur. Distribution of eigenvalues for some sets of random matrices. *Matematicheskii Sbornik*, 114(4):507–536, 1967,

  with many posterior rediscoveries – as for example in[2]
  - A. Crisanti and H. Sompolinsky. Dynamics of spin systems with randomly asymmetric bonds: Langevin dynamics and a spherical model. *Physical Review A*, 36:4922–4939, 1987,
  - A. Sengupta and P. P. Mitra. Distributions of singular values for some random matrices. *Physical Review E*, 60(3):3389, 1999.
- The universality of Eq. (17.5) is discussed in
  - J. W. Silverstein and Z. Bai. On the empirical distribution of eigenvalues of a class of large dimensional random matrices. *Journal of Multivariate Analysis*, 54(2):175–192, 1995.
- The case of power-law distributed variables and deviations from Marčenko–Pastur are discussed in

[2] In fact, Crisanti and Sompolinsky [1987] themselves cite an unpublished work of D. Movshovitz and H. Sompolinsky.

- S. Belinschi, A. Dembo, and A. Guionnet. Spectral measure of heavy tailed band and covariance random matrices. *Communications in Mathematical Physics*, 289(3):1023–1055, 2009,
- G. Biroli, J.-P. Bouchaud, and M. Potters. On the top eigenvalue of heavy-tailed random matrices. *Europhysics Letters (EPL)*, 78(1):10001, 2007,
- A. Auffinger, G. Ben Arous, and S. Péché. Poisson convergence for the largest eigenvalues of heavy tailed random matrices. *Annales de l'I.H.P. Probabilités et statistiques*, 45(3):589–610, 2009.

• The case with spatial and temporal correlations was studied in
  - A. Sengupta and P. P. Mitra. Distributions of singular values for some random matrices. *Physical Review E*, 60(3):3389, 1999,
  - Z. Burda, J. Jurkiewicz, and B. Wacław. Spectral moments of correlated Wishart matrices. *Physical Review E*, 71:026111, 2005.

• Multivariate Student (or elliptical) variables and their associated sample covariance matrices are studied in
  - G. Biroli, J.-P. Bouchaud, and M. Potters. The Student ensemble of correlation matrices: eigenvalue spectrum and Kullback-Leibler entropy. *preprint arXiv:0710.0802*, 2007,
  - N. El Karoui et al. Concentration of measure and spectra of random matrices: Applications to correlation matrices, elliptical distributions and beyond. *The Annals of Applied Probability*, 19(6):2362–2405, 2009.

• Large cross-correlation matrices and corresponding null-hypothesis statistical tests are studied in
  - I. M. Johnstone. Multivariate analysis and Jacobi ensembles: Largest eigenvalue, TracyWidom limits and rates of convergence. *The Annals of Statistics*, 36(6):2638–2716, 2008,
  - J.-P. Bouchaud, L. Laloux, M. A. Miceli, and M. Potters. Large dimension forecasting models and random singular value spectra. *The European Physical Journal B*, 55(2):201–207, 2007,
  - Y. Yang and G. Pan. Independence test for high dimensional data based on regularized canonical correlation coefficients. *The Annals of Statistics*, 43(2):467–500, 2015.

• For an early derivation of Eq. (17.62) without the use of free probability methods, see
  - K. W. Wachter. The limiting empirical measure of multiple discriminant ratios. *The Annals of Statistics*, 8(5):937–957, 1980.

18
Bayesian Estimation

In this chapter we will review the subject of Bayesian estimation, with a particular focus on matrix estimation. The general situation one encounters is one where the observed matrix is a noisy version of the "true" matrix one wants to estimate. For example, in the case of additive noise, one observes a matrix $\mathbf{E}$ which is the true matrix $\mathbf{C}$ plus a random matrix $\mathbf{X}$ that plays the role of noise, to wit,

$$\mathbf{E} = \mathbf{C} + \mathbf{X}. \tag{18.1}$$

In the case of multiplicative noise, the observed matrix $\mathbf{E}$ has the form

$$\mathbf{E} = \mathbf{C}^{\frac{1}{2}}\mathbf{W}\mathbf{C}^{\frac{1}{2}}. \tag{18.2}$$

When $\mathbf{W}$ is a white Wishart matrix, this is the problem of sample covariance matrix encountered in Chapter 17.

In general, the true matrix $\mathbf{C}$ is unknown to us. We would like to know the probability of $\mathbf{C}$ given that we have observed $\mathbf{E}$, i.e. compute $P(\mathbf{C}|\mathbf{E})$. This is the general subject of Bayesian estimation, which we introduce and discuss in this chapter.

18.1 Bayesian Estimation

Before doing Bayesian theory on random matrices (see Section 18.3), we first review Bayesian estimation and see it at work on simpler examples.

18.1.1 General Framework

Imagine we have an observable variable y that we would like to infer from the observation of a related variable x. The variables x and y can be scalars, vectors, matrices, higher dimensional objects ... We postulate that we know the random process that generates y given x, i.e. y could be a noisy version of x or more generally y could be drawn from a known distribution with x as a parameter. The generation process of y is encoded in a probability distribution $P(y|x)$, which is called the *sampling distribution* or the *likelihood function.*

Given our knowledge of $P(y|x)$, we would like to write the inference probability $P(x|y)$, also called the *posterior distribution*. To do so, we can use Bayes' rule:

$$P(x|y) = \frac{P(y|x)P_0(x)}{P(y)}. \tag{18.3}$$

To obtain the desired probability, Bayes' rule tells us that we need to know the *prior distribution* $P_0(x)$. In theory $P_0(x)$ is the distribution from which x is drawn and it is in some cases knowable. In many practical applications, however, x is actually not random but simply unknown and $P_0(x)$ encodes our ignorance of x. It should represent our best (probabilistic) guess of x before we observe the data y. The determination (or arbitrariness) of the prior $P_0(x)$ is considered to be one of the weak points of the Bayesian approach. Often $P_0(x)$ is just taken to be constant, i.e. no prior knowledge at all on x. However, note that $P_0(x) = \text{constant}$ is not invariant upon changes of variables, for if $x' = f(x)$ is a non-linear transformation of x, then $P_0(x')$ is no longer constant! In Section 18.1.3, we will see how the arbitrariness in the choice of $P_0(x)$ can be used to simplify modeling.

The other distribution appearing in Bayes' rule $P(y)$ is actually just a normalization factor. Indeed, y is assumed to be known, therefore $P(y)$ is just a fixed number that can be computed by normalizing the posterior distribution. One therefore often simplifies Bayes' rule as

$$P(x|y) = \frac{1}{Z}P(y|x)P_0(x), \qquad Z := \int \mathrm{d}x\, P(y|x)P_0(x), \tag{18.4}$$

where $P(y|x)$ represents the measurement (or noise) process and $P_0(x)$ the (often arbitrary) prior distribution.

From the posterior distribution $P(x|y)$ we can build an estimator of x. The optimal estimator depends on the problem at hand, namely, which quantity are we trying to optimize. The most common Bayesian estimators are

1 MMSE: The posterior mean $\mathbb{E}[x]_y$. It minimizes a quadratic loss function and is hence called the Minimum Mean Square Error estimator.
2 MAVE: The posterior median or Minimum Absolute Value Error estimator.
3 MAP: The Maximum *A Posteriori* estimator, defined as $\hat{x} = \operatorname{argmax}_x P(x|y)$.

18.1.2 A Simple Estimation Problem

Consider the simplest one-dimensional estimation problem:

$$y = x + \varepsilon, \tag{18.5}$$

where x is some signal to be estimated, ε is an independent noise, and y is the observation. Then $P(y|x)$ is simply $P_\varepsilon(.)$ evaluated at $y - x$:

$$P(y|x) = P_\varepsilon(y - x). \tag{18.6}$$

Suppose further that ε is a centered Gaussian noise with variance σ_{n}^2, where the subscript n means "noise". Then we have

$$P(y|x) = \frac{1}{\sqrt{2\pi\sigma_{\mathrm{n}}^2}} \exp\left(-\frac{(y-x)^2}{2\sigma_{\mathrm{n}}^2}\right). \tag{18.7}$$

Then we get that

$$P(x|y) \propto P_0(x) \exp\left(\frac{2xy - x^2}{2\sigma_{\mathrm{n}}^2}\right), \tag{18.8}$$

where $P_0(x)$ is the prior distribution of x and we have dropped x-independent factors. Depending on the choice of $P_0(x)$ we will get different posterior distributions and hence different estimators of x.

Gaussian Prior

Suppose first $P_0(x)$ is a Gaussian with variance σ_{s}^2 (for signal) centered at x_0. Then

$$\begin{aligned} P(x|y) &\propto \exp\left(-\frac{(x-x_0)^2}{2\sigma_{\mathrm{s}}^2} + \frac{2xy - x^2}{2\sigma_{\mathrm{n}}^2}\right) \\ &= \frac{1}{\sqrt{2\pi\sigma^2}} \exp\left(-\frac{(x-\hat{x})^2}{2\sigma^2}\right), \end{aligned} \tag{18.9}$$

with

$$\hat{x} := x_0 + r(y - x_0) = (1-r)x_0 + ry; \qquad \sigma^2 := r\sigma_{\mathrm{n}}^2, \tag{18.10}$$

where the signal-to-noise ratio r is $r = \sigma_{\mathrm{s}}^2/(\sigma_{\mathrm{s}}^2 + \sigma_{\mathrm{n}}^2)$. The posterior distribution is thus a Gaussian centered around $\hat{x}$ and of variance σ^2.

For a Gaussian distribution the mean, median and maximum probability values are all equal to $\hat{x}$, which is therefore the optimal estimator in all three standard procedures, MMSE, MAVE and MAP. This estimator is called the *linear shrinkage estimator* as it is linear in the observed variable y. The linear coefficient of y is the signal-to-noise ratio r, a number smaller than one that *shrinks* the observed value towards the *a priori* mean x_0.

Note that this estimator can also be obtained in a completely different framework: it is the affine estimator that minimizes the mean square error. The estimator is affine by construction and minimization only involves first and second moments; it is therefore not too surprising that we recover Eq. (18.10), see Exercise 18.1.2. As so often in optimization problems, assuming Gaussian fluctuations is equivalent to imposing an affine solution.

Another important property of the linear shrinkage estimator is that it is rather conservative: it is biased towards x_0. By assumption x fluctuates with variance σ_{s}^2 and y fluctuates with variance $\sigma_{\mathrm{s}}^2 + \sigma_{\mathrm{n}}^2$. This allows us to compute the variance of the estimator $\hat{x}(y)$ as

$$\mathbb{V}[\hat{x}(y)] = r^2(\sigma_{\mathrm{s}}^2 + \sigma_{\mathrm{n}}^2) = \frac{\sigma_{\mathrm{s}}^4}{\sigma_{\mathrm{s}}^2 + \sigma_{\mathrm{n}}^2} \le \sigma_{\mathrm{s}}^2. \tag{18.11}$$

So the variance of the estimator[1] is not only smaller than that of the observed variable y it is also smaller than the fluctuations of the true variable x!

Exercise 18.1.1 Optimal affine estimator

Suppose that we observe a variable y that has some non-zero covariance with an unknown variable x that we would like to estimate. We will show that the best affine estimator of x is given by the linear shrinkage estimator (18.10). The variables x and y can be drawn from any distribution with finite variance. We write the general affine estimator

$$\hat{x} = ay + b, \tag{18.12}$$

and choose a and b to minimize the expected mean square error.

(a) Initially assume that x and y have zero mean – we will relax this assumption later. Show that

$$\mathbb{E}\left[(x - \hat{x})^2\right] = a^2\sigma_y^2 + b^2 + \sigma_x^2 - 2a\sigma_{xy}^2, \tag{18.13}$$

where σ_x^2, σ_y^2 and σ_{xy}^2 are the variances of x, y and their covariance.

(b) Show that the optimal estimator has $a = \sigma_{xy}^2/\sigma_y^2$ and $b = 0$.

(c) Compute b in the non-zero mean case by considering $x - x_0$ estimated using $y - y_0$.

(d) Compute σ_y^2 and σ_{xy}^2 when $y = x + \varepsilon$ with ε independent of x.

(e) Show that when $\mathbb{E}[\varepsilon] = 0$ we recover Eq. (18.10).

Bernoulli Prior

When $P_0(x)$ is non-Gaussian, the obtained estimators are in general non-linear. As a second example suppose that $P_0(x)$ is Bernoulli random variable with $P_0(x = 1) = P_0(x = -1) = 1/2$. Then, after a few simple manipulations one obtains

$$P(x|y) = \frac{1}{2}\left(\left(1 + \tanh\left(\frac{y}{\sigma_n^2}\right)\right)\delta_{x,1} + \left(1 - \tanh\left(\frac{y}{\sigma_n^2}\right)\right)\delta_{x,-1}\right). \tag{18.14}$$

The posterior distribution is now a discrete function that takes on only two values, namely ± 1. In this case the maximum probability and the median are such that

$$\hat{x}_{\text{MAP}}(y) = \text{sign}(y). \tag{18.15}$$

[1] One should not confuse the variance of the posterior distribution $r\sigma_n^2$ with the variance of the estimator $r\sigma_s^2$. The first one measures the remaining uncertainty about x once we have observed y while the second measures the variability of $\hat{x}(y)$ when we repeat the experiment multiple times with varying x and noise ε.

It is also easy to calculate the MMSE estimator:

$$\hat{x}_{\text{MMSE}}(y) = \mathbb{E}[x]_y = \tanh\left(\frac{y}{\sigma_n^2}\right). \tag{18.16}$$

It may seem odd that the MMSE estimator takes continuous values between -1 and 1 while we postulated that the true x can only be equal to ± 1. Nevertheless, in order to minimize the variance it is optimal to shoot somewhere in the middle of -1 and 1 as choosing the wrong sign costs a lot in terms of variance. The estimator $\hat{x}(y)$ is biased, i.e. $\mathbb{E}[\hat{x}_{\text{MMSE}}|x] \neq x$. It also has a variance strictly less than 1, whereas the variance of the true x is unity.

Laplace Prior

As a third example, consider a Laplace distribution

$$P_0(x) = \frac{b}{2}\mathrm{e}^{-b|x|} \tag{18.17}$$

for the prior, with variance $2b^{-2}$. In this case the posterior distribution is given by

$$P(x|y) \propto \exp\left(-b|x| + \frac{2xy - x^2}{2\sigma_n^2}\right). \tag{18.18}$$

The MMSE and MAVE estimators can be computed but the results are not very enlightening as they are given by an ugly combination of error functions and even inverse error functions (for MAVE). The MAP estimator is both simpler and more interesting in this case. It is given by

$$\hat{x}_{\text{MAP}}(y) = \begin{cases} 0 & \text{for } |y| < b\sigma_n^2, \\ y - b\sigma_n^2 \operatorname{sign}(y) & \text{otherwise.} \end{cases} \tag{18.19}$$

The MAP estimator is sparse in the sense that in a non-zero fraction of cases it takes the exact value of zero. Note that the true variable x itself is not sparse: it is almost surely non-zero. This example is a toy-model for the "LASSO" regularization that we will study in Section 18.2.2.

Non-Gaussian Noise

The noise in Eq. (18.5) can also be non-Gaussian. When the noise has fat tails, one can even be in the counter-intuitive situation where the estimator is not monotonic in the observed variable, i.e. the best estimate of x decreases as a function of its noisy version y. For example, if x is centered unit Gaussian and ε is a centered unit Cauchy noise, we have

$$P(x|y) \propto \frac{\mathrm{e}^{-x^2/2}}{(y-x)^2 + 1}. \tag{18.20}$$

Whereas the Cauchy noise ε and the observation y do not have a first moment, the posterior distribution of x is regularized by the Gaussian weight and all its moments

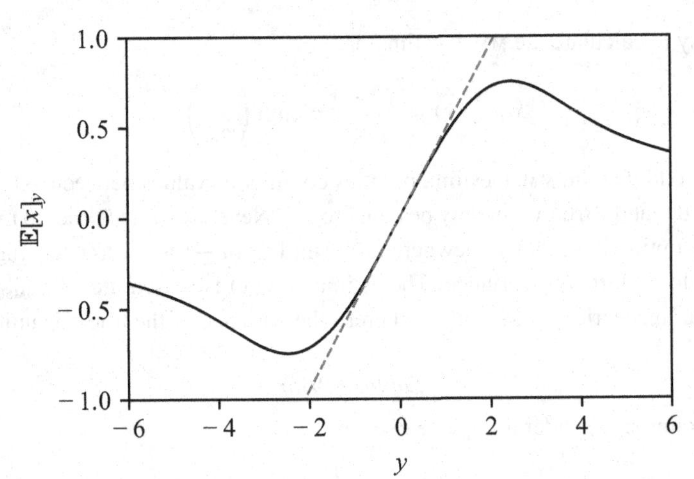

Figure 18.1 A non-monotonic optimal estimator. The MMSE estimator of a Gaussian variable corrupted by Cauchy noise (see Eq. (18.21)). For small absolute observations y, the estimator is almost linear with slope $2 - \sqrt{2/e\pi}/\mathrm{erfc}(1/\sqrt{2}) \approx 0.475$ (dashed line).

are finite. After some tedious calculation we arrive at the conditional mean or MMSE estimator:

$$\mathbb{E}[x]_y = y + \frac{\mathrm{Im}(\Phi)}{\mathrm{Re}(\Phi)}, \quad \text{where} \quad \Phi = e^{iy}\mathrm{erfc}\left(\frac{1+iy}{\sqrt{2}}\right). \tag{18.21}$$

The shape of the estimator as a function of y is not obvious from this expression but it is plotted numerically in Figure 18.1. The interpretation is the following:

- When we observe a small (order 1) value of y, we can assume that it was generated by a moderate x with moderate noise, hence we are in the regime of the linear estimator with a signal-to-noise ratio close to one-half ($\hat{x} \approx 0.475y$).
- On the other hand, when y is much larger than the standard deviation of x it becomes clear that y can only be large because the noise takes extreme values. When the noise is large our knowledge of x decreases, hence the estimator tends to zero as $|y| \to \infty$.

18.1.3 Conjugate Priors

The main weakness of Bayesian estimation is the reliance on a prior distribution for the variable we want to estimate. In many practical applications one does not have a probabilistic or statistical knowledge of $P_0(x)$. The variable x is a fixed quantity that we do not know, so how are we supposed to know about $P_0(x)$? In such cases we are left with making a reasonable practical guess. Since $P_0(x)$ is just a guess, we can at least choose a functional form for $P_0(x)$ that makes computation easy. This is the idea behind "conjugate priors".

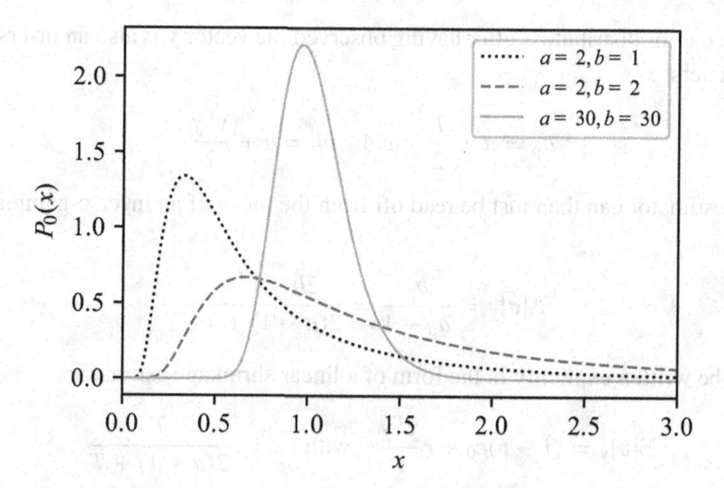

Figure 18.2 The inverse-gamma distribution Eq. (18.24). Its mean is given by $b/(a-1)$ and it becomes increasingly peaked around this mean as both a and b become large.

When we studied the one-dimensional estimation of a variable x corrupted by additive Gaussian noise (Eq. (18.5), with Gaussian ε) we found that choosing a Gaussian prior for x gave us a Gaussian posterior distribution. In many other cases, we can find a family of prior distributions that will similarly keep the posterior distribution in the same family. This concept is better explained in an example.

Imagine we are given a series of T numbers $\{y_i\}$ generated independently from a centered Gaussian distribution of variance c that is unknown to us. We use the variable c rather than σ^2 to avoid the confusion between the estimation of σ and that of $c = \sigma^2$. The joint probability of the y's is given by

$$P(\mathbf{y}|c) = \frac{1}{(2\pi c)^{T/2}} \exp\left(-\frac{\mathbf{y}^T\mathbf{y}}{2c}\right). \tag{18.22}$$

The posterior distribution is thus given by

$$P(c|\mathbf{y}) \propto P_0(c)c^{-T/2} \exp\left(-\frac{\mathbf{y}^T\mathbf{y}}{2c}\right). \tag{18.23}$$

Now if the prior $P_0(c)$ has the form $P_0(c) \propto c^{-a-1}\mathrm{e}^{-b/c}$, the posterior will also be of that form with modified values for a and b. Such a $P_0(c)$ will thus be our conjugate prior. This law is precisely the inverse-gamma distribution (see Fig. 18.2):

$$P_0(c) = \frac{b^a}{\Gamma(a)} c^{-a-1}\mathrm{e}^{-b/c} \qquad (c \geq 0). \tag{18.24}$$

It describes a non-negative variable, as a variance should. It is properly normalized when $a > 0$ and has mean $b/(a-1)$ whenever $a > 1$. If we choose such a law as our variance prior, the posterior distribution after having observed the vector $\mathbf{y}$ is also an inverse-gamma with parameters

$$a_p = a + \frac{T}{2} \quad \text{and} \quad b_p = b + \frac{\mathbf{y}^T\mathbf{y}}{2}. \tag{18.25}$$

The MMSE estimator can then just be read off from the mean of an inverse-gamma distribution:

$$\mathbb{E}[c]_{\mathbf{y}} = \frac{b_p}{a_p - 1} = \frac{2b + \mathbf{y}^T\mathbf{y}}{2(a-1) + T}, \tag{18.26}$$

which can be written explicitly in the form of a linear shrinkage estimator:

$$\mathbb{E}[c]_{\mathbf{y}} = (1-r)c_0 + r\frac{\mathbf{y}^T\mathbf{y}}{T} \quad \text{with} \quad r = \frac{T}{2(a-1)+T}, \tag{18.27}$$

and $c_0 = b/(a-1)$ is the mean of the prior. We see that $r \to 1$ when $T \to \infty$: in this case the prior guess on c_0 disappears and one is left with the naive empirical estimator $\mathbf{y}^T\mathbf{y}/T$.

Exercise 18.1.2 Conjugate prior for the amplitude of a Laplace distribution
Suppose that we observe T variables y_i drawn from a Laplace distribution (18.17) with unknown amplitude b. We would like to estimate b using the Bayesian method with conjugate prior.

(a) Write the joint probability density of elements of the vector $\mathbf{y}$ for a given b. This is the likelihood function $P(\mathbf{y}|b)$.
(b) As a function of b, the likelihood function has the same form as a gamma distribution (4.17). Using a gamma distribution with parameters a_0 and b_0 for the prior on b show that the posterior distribution of b is also a gamma distribution. Find the posterior parameters a_p and b_p.
(c) Given that the mean of a gamma distribution is given by a/b, write the MMSE estimator in this case.
(d) Compute the estimator in the two limiting cases $T = 0$ and $T \to \infty$.
(e) Write your estimator from (c) as a shrinkage estimator interpolating between these two limits. Show that the signal-to-noise ratio r is given by $r = Tm/(Tm + 2b_0)$ where $m = \sum |y_i|/T$. Note that in this case the shrinkage estimator is non-linear in the naive estimate $\hat{b} = 1/(2m)$.

18.2 Estimating a Vector: Ridge and LASSO

A very standard problem for which Bayesian ideas are helpful is linear regression. Assume we want to estimate the parameters a_i of a multi-linear regression, where we assume that an observable y can be written as

$$y = \sum_{i=1}^{N} a_i x_i + \varepsilon, \tag{18.28}$$

where x_i are N observable quantities and ε is noise (not directly observable). We observe a time series of y of length T that we stack into a vector $\mathbf{y}$, whereas the different x_i are stacked into an $N \times T$ data matrix $\mathbf{H}_{it} = x_i^t$, and ε is the corresponding T-dimensional noise vector. We thus write

$$\mathbf{y} = \mathbf{H}^T \mathbf{a} + \varepsilon, \tag{18.29}$$

where $\mathbf{a}$ is an N-dimensional vector of coefficients we want to estimate. We assume the following structure for the random variables x and ε:

$$\frac{1}{T}\mathbb{E}[\varepsilon\varepsilon^T] = \sigma_n^2 \mathbf{1}; \qquad \frac{1}{T}\mathbb{E}[\mathbf{H}\mathbf{H}^T] = \mathbf{C}, \tag{18.30}$$

where $\mathbf{C}$ can be an arbitrary covariance matrix, but we will assume it to be the identity $\mathbf{1}$ in the following, unless otherwise stated.

Classical linear regression would find the coefficient vector $\mathbf{a}$ that minimizes the error $\mathcal{E} = \|\mathbf{y} - \mathbf{H}^T\mathbf{a}\|^2$ on a given dataset. As is well known, the regression coefficients are given by

$$\mathbf{a}_{\text{reg}} = \left(\mathbf{H}\mathbf{H}^T\right)^{-1} \mathbf{H}\mathbf{y}. \tag{18.31}$$

This equation can be derived easily by taking the derivatives of $\mathcal{E}$ with respect to all a_i and setting them to zero. Note that when $q := N/T < 1$, $\mathbf{H}\mathbf{H}^T$ is in general invertible, but when $q \geq 1$ (i.e. when there is not enough data), Eq. (18.31) is *a priori* ill defined.

In a Bayesian estimation framework, we want to write the posterior distribution $P(\mathbf{a}|\mathbf{y})$ and build an estimator of $\mathbf{a}$ from it. We expect that the Bayesian approach will work better than linear regression "out of sample", i.e. on a new independent sample. The reason is that the linear regression method minimizes an "in-sample" error, and is thus devised to fit best the details of the observed dataset, with no regard to overfitting considerations. These concepts will be clarified in Section 18.2.3.

Following the approach of Section 18.1, we write the posterior distribution as

$$P(\mathbf{a}|\mathbf{y}) \propto P_0(\mathbf{a}) \exp\left(-\frac{1}{2\sigma_n^2}\left\|\mathbf{y} - \mathbf{H}^T\mathbf{a}\right\|^2\right), \tag{18.32}$$

where σ_n^2 is the variance of the noise ε. Now, the art is to choose an adequate prior distribution $P_0(\mathbf{a})$.

18.2.1 Ridge Regression

The likelihood function in Eq. (18.32) is a Gaussian function of $\mathbf{a}$, so choosing a Gaussian prior for $P_0(\mathbf{a})$ will give us a Gaussian posterior. To construct a Gaussian distribution for $P_0(\mathbf{a})$ we need to choose a prior mean $\mathbf{a}_0$ and a prior covariance matrix.

Regression coefficients can be positive or negative, so the most natural prior mean is the zero vector $\mathbf{a}_0 = \mathbf{0}$. In the absence of any other information about the direction in which $\mathbf{a}$ may point, we should make a rotationally invariant prior for the covariance matrix.[2] The only rotationally invariant choice is a multiple of the identity $\sigma_s^2\mathbf{1}$ for the prior covariance. Assuming that the coefficients a_i are IID gives the same answer. However, we do not have a good argument to set the scale of the covariance σ_s^2; we will come back to this point later.

The posterior distribution is then written

$$P(\mathbf{a}|\mathbf{y}) \propto \exp\left(-\frac{1}{2\sigma_n^2}\left(\mathbf{a}^T\left(\mathbf{H}\mathbf{H}^T + \frac{\sigma_n^2}{\sigma_s^2}\mathbf{1}\right)\mathbf{a} - 2\mathbf{a}^T\mathbf{H}\mathbf{y}\right)\right). \tag{18.33}$$

As announced, the posterior is a multivariate Gaussian distribution. The MMSE, MAVE and MAP estimator are all equal to the mode of the distribution, given by[3]

$$\mathbb{E}[\mathbf{a}]_{\mathbf{y}} = \left(\frac{\mathbf{H}\mathbf{H}^T}{T} + \zeta\mathbf{1}\right)^{-1}\frac{\mathbf{H}\mathbf{y}}{T}, \qquad \zeta := \frac{\sigma_n^2}{T\sigma_s^2}. \tag{18.34}$$

This is called the "ridge" regression estimator, as it amounts to adding weight on the diagonal of the sample covariance matrix $(\mathbf{H}\mathbf{H}^T)/T$. This can also be seen as a shrinkage of the covariance matrix towards the identity, as we will discuss further in Section 18.3 below.

Another way to understand what ridge regression means is to notice that Eq. (18.31) involves the inverse of the covariance matrix $(\mathbf{H}\mathbf{H}^T)/T$, which can be unstable in large dimensions. This instability can lead to very large coefficients in $\mathbf{a}$. One can thus regularize the regression problem by adding a quadratic (or L^2-norm) penalty for $\mathbf{a}$ so the vector does not become too big:

$$\mathbf{a}_{\text{ridge}} = \underset{\mathbf{a}}{\operatorname{argmin}}\left[\left\|\mathbf{y} - \mathbf{H}^T\mathbf{a}\right\|^2 + T\zeta\left\|\mathbf{a}\right\|^2\right]. \tag{18.35}$$

Setting $\zeta = 0$ we recover the standard regression. The solution of the regularized optimization problem yields exactly Eq. (18.34); it is often called the Tikhonov regularization. Note that the resulting equation for $\mathbf{a}_{\text{ridge}}$ remains well defined even when $q \geq 1$ as long as $\zeta > 0$.

In both approaches (Bayesian and Tikhonov regularization) the result depends on the choice of the parameter $\zeta = \sigma_n^2/(T\sigma_s^2)$ which is hard to estimate *a priori*. The modern way of fixing ζ in practical applications is by using a validation (or cross-validation) method. The idea is to find the value of $\mathbf{a}_{\text{ridge}}$ on part of the data (the "training set") and measure the quality of the regression on another, non-overlapping part of the data (the "validation set"). The value of ζ is then chosen as the one that gives the lowest error on the validation set.

[2] This assumption relies on our hypothesis that the covariance matrix $\mathbf{C}$ of the x's is the identity matrix. Otherwise, the eigenvectors of $\mathbf{C}$ could be used to construct non-rotationally invariant priors.

[3] We have introduced a factor of $1/T$ in the definition of ζ so it parameterizes the shift in the *normalized* covariance matrix $(\mathbf{H}\mathbf{H}^T)/T$. It turns out to be the proper scaling in the large N limit with $q = N/T$ fixed. Note that if the elements of $\mathbf{a}$ and $\mathbf{H}$ are of order one, the variance of the elements of $\mathbf{H}^T\mathbf{a}$ is of order N; for the noise to contribute significantly in the large N limit we must have σ_n^2 of order N and hence ζ of order 1.

In cross-validation, the procedure is repeated with multiple validation sets (always disjoint from the training set) and the error is then averaged over these sets.

18.2.2 LASSO

Another common estimating method for vectors is the "LASSO" method[4] which combines a Laplace prior with the MAP estimator.

In this method, the prior distribution amounts to assuming that the coefficients of **a** are IID Laplace random number with variance $2b^{-2}$. The posterior then becomes

$$P(\mathbf{a}|\mathbf{y}) \propto \exp\left(-b\sum_{i=1}^{N}|a_i| - \frac{1}{2\sigma_n^2}\left\|\mathbf{y}-\mathbf{H}^T\mathbf{a}\right\|^2\right).$$

As in the toy model Eq. (18.18), the MMSE and MAVE estimators look rather ugly, but the MAP one is quite simple. It is given by the maximum of the argument of the above exponential:

$$\mathbf{a}_{\text{LASSO}} = \underset{\mathbf{a}}{\operatorname{argmin}}\left[2b\sigma_n^2\sum_{i=1}^{N}|a_i| + \left\|\mathbf{y}-\mathbf{H}^T\mathbf{a}\right\|^2\right]. \tag{18.36}$$

This minimization amounts to regularizing the standard regression estimation with an absolute value penalty (also called L^1-norm penalty), instead of the quadratic penalty for the ridge regression. Interestingly, the solution to this minimization problem leads to a sparse estimator: the absolute value penalty strongly disfavors small values of $|a_i|$ and prefers to set these values to zero. Only sufficiently relevant coefficients a_i are retained – LASSO automatically selects the salient factors (this is the "so" part in LASSO), which is very useful for interpreting the regression results intuitively.

Note that the true vector **a** is not sparse, as the probability to find a coefficient a_i to be exactly zero is itself zero for the prior Laplace distribution, which does not contain a singular $\delta(a)$ peak. The sparsity of the LASSO estimator $\mathbf{a}_{\text{LASSO}}$ is controlled by the parameter b. When $b\sigma_n^2 \to 0$, the penalty disappears and all the coefficients of the vector **a** are non-zero (barring exceptional cases). When $b\sigma_n^2 \to \infty$, on the other hand, all coefficients are zero. In fact, the number of non-zero coefficients is a monotonic decreasing function of $b\sigma_n^2$. As for the parameter ζ for the ridge regression, it is hard to come up with a good prior value for b, which should be estimated again using validation or cross-validation methods (Figure 18.3). Finally we note that it is sometimes useful to combine the L^1 penalty of LASSO with the L^2 penalty of ridge, the resulting estimator is called an elastic net.

18.2.3 In-Sample and Out-of-Sample Error

Standard linear regression is built to minimize the sum of the squared-residuals on the dataset at hand. We call this error the *in-sample* error. In many cases, we are interested in

[4] LASSO stands for Least Absolute Shrinkage and Selection Operator.

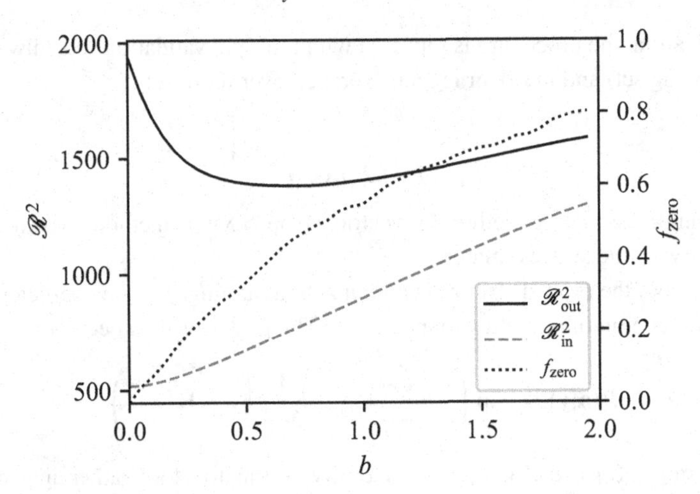

Figure 18.3 Illustration of the validation method in LASSO regularization. We built a linear model with 500 coefficients drawn from a Laplace distribution with $b = 1$ and Gaussian noise $\sigma_n^2 = 1000$. The model is estimated using $T = 1000$ unit Gaussian data and validated using a different set of the same size. The error on the training set is minimal with no regularization and gets worse as b increases (dashed line, left axis). The validation error (full line, left axis) is minimal for b about equal to 1. The dotted line (right axis) shows the fraction of the coefficient estimated to be exactly zero; this number grows from zero (no regularization) to almost 1 (strong regularization).

the predictive power of a linear model and the relevant error is the mean square error on a new independent but statistically equivalent dataset: the *out-of-sample* error. If the number of fitted variables (degree of freedom) is small with respect to the number of samples, the in-sample error is a good estimator of the out-of-sample error and the standard linear regression is also the optimal linear predictive model. The situation changes radically when the number of fitted variables becomes comparable to the number of samples, as we discuss below.

To summarize, linear regression, regularized or not, addresses two types of task:

- **In-sample estimator**: we observe some $\mathbf{H}_1$ and $\mathbf{y}_1$, and estimate $\mathbf{a}$.
- **Out-of-sample prediction**: we observe some other $\mathbf{H}_2$, non-overlapping with $\mathbf{H}_1$, and use them to predict $\mathbf{y}_2$ with the in-sample estimate of $\mathbf{a}$.

The result of the in-sample estimation is given by Eq. (18.31), which we write as

$$\mathbf{a}_{\text{reg}} = \mathbf{E}^{-1}\mathbf{b}; \qquad \mathbf{E} := \frac{1}{T}\mathbf{H}_1\mathbf{H}_1^T, \quad \mathbf{b} := \frac{1}{T}\mathbf{H}_1\mathbf{y}_1. \tag{18.37}$$

This is the best in-sample estimator. However, this is not necessarily the case for the out-of-sample prediction.

Note that both the standard regression and the ridge regression estimator (Eq. (18.34)) are of the form $\hat{\mathbf{a}} = \Xi^{-1}\mathbf{b}$ with $\Xi = \mathbf{E}$ and $\Xi = \mathbf{E} + \zeta\mathbf{1}$, respectively. We will compute in the following the in-sample and out-of-sample estimation error for any estimator of that nature.

Recalling that $\mathbb{E}[\boldsymbol{\varepsilon}\boldsymbol{\varepsilon}^T] = \sigma_n^2 \mathbf{1}$ and after some calculations, one finds that the in-sample ($\mathcal{R}^2_{\text{in}}$) error is given by

$$\mathcal{R}^2_{\text{in}}(\hat{\mathbf{a}}) = \frac{1}{T}\left\{\sigma_n^2\left[T - 2\,\text{Tr}(\Xi^{-1}\mathbf{E}) + \text{Tr}(\Xi^{-1}\mathbf{E}\Xi^{-1}\mathbf{E})\right] + \mathbf{a}^T\left(\mathbf{E} - 2\mathbf{E}\Xi^{-1}\mathbf{E} + \mathbf{E}\Xi^{-1}\mathbf{E}\Xi^{-1}\mathbf{E}\right)\mathbf{a}\right\}. \tag{18.38}$$

In the special case $\Xi = \mathbf{E}$, we have

$$\mathcal{R}^2_{\text{in}}(\mathbf{a}_{\text{reg}}) = \frac{\sigma_n^2}{T}(T - N) = (1-q)\sigma_n^2,$$

which is smaller than the true error, which is simply equal to σ_n^2. In fact, the error goes to zero as $q \to 1$, i.e. when the number of parameters becomes equal to the number of observations. This error reduction is called "overfitting", or in-sample bias: if the task is to find the best model that explains past data, one can do better than the true error. Note that the above result is quite special, in the sense that it actually does not depend on either $\mathbf{E}$ or $\mathbf{a}$.

Next we calculate the expected out-of-sample ($\mathcal{R}^2_{\text{out}}$) error. We draw another matrix $\mathbf{H}_2$ of size $N \times T_2$ and consider another independent noise vector $\boldsymbol{\varepsilon}_2$ of variance σ_n^2 and size T_2 (where T_2 does not need to be equal to T, it can even be equal to 1). We calculate

$$\begin{aligned}\mathcal{R}^2_{\text{out}}(\hat{\mathbf{a}}) &= \frac{1}{T_2}\mathbb{E}_{\mathbf{H}_2,\boldsymbol{\varepsilon}_2}\left[\left\|\mathbf{H}_2^T\mathbf{a} + \boldsymbol{\varepsilon}_2 - \mathbf{H}_2^T\hat{\mathbf{a}}\right\|^2\right] \\ &= \frac{1}{T_2}\mathbb{E}_{\mathbf{H}_2,\boldsymbol{\varepsilon}_2}\left[\left\|\mathbf{H}_2^T\mathbf{a} + \boldsymbol{\varepsilon}_2 - \mathbf{H}_2^T\Xi^{-1}\mathbf{E}_1\mathbf{a} - \mathbf{H}_2^T\Xi^{-1}\frac{1}{T}\mathbf{H}_1\boldsymbol{\varepsilon}_1\right\|^2\right],\end{aligned} \tag{18.39}$$

where we denote $\mathbf{E}_1 := T^{-1}\mathbf{H}_1\mathbf{H}_1^T$. We now assume that $T_2^{-1}\mathbb{E}[\mathbf{H}_2\mathbf{H}_2^T] = \mathbf{C}$ with a general covariance $\mathbf{C}$.

In the standard regression case, $\Xi = \mathbf{E}$ and $\hat{\mathbf{a}} = \mathbf{a}_{\text{reg}}$ and we have

$$\mathcal{R}^2_{\text{out}}(\mathbf{a}_{\text{reg}}) = \frac{1}{T_2}\mathbb{E}_{\mathbf{H}_2,\boldsymbol{\varepsilon}_2}\left[\left\|\boldsymbol{\varepsilon}_2 - \mathbf{H}_2^T\Xi^{-1}\frac{1}{T}\mathbf{H}_1\boldsymbol{\varepsilon}_1\right\|^2\right] = \sigma_n^2 + \frac{\sigma_n^2}{T}\text{Tr}(\mathbf{E}^{-1}\mathbf{C}). \tag{18.40}$$

Now since $\mathbf{E}$ is a sample covariance matrix with true covariance $\mathbf{C}$, we have

$$\text{Tr}(\mathbf{E}^{-1}\mathbf{C}) = \text{Tr}\left(\mathbf{C}^{-\frac{1}{2}}\mathbf{W}_q^{-1}\mathbf{C}^{-\frac{1}{2}}\mathbf{C}\right) = \text{Tr}(\mathbf{W}_q^{-1}) \approx \frac{N}{1-q}, \tag{18.41}$$

where $\mathbf{W}_q$ denotes a standard Wishart matrix. Thus we find

$$\mathcal{R}^2_{\text{out}}(\mathbf{a}_{\text{reg}}) = \sigma_n^2 + \frac{q\sigma_n^2}{1-q} = \frac{\sigma_n^2}{1-q} = \frac{\mathcal{R}^2_{\text{in}}(\mathbf{a}_{\text{reg}})}{(1-q)^2}. \tag{18.42}$$

As an illustration see Figure 18.3 where without regularization ($b = 0$) we have indeed $\mathcal{R}^2_{\text{out}}/\mathcal{R}^2_{\text{in}} \approx (1-q)^{-2} = 4$. Thus, we see that we can make precise statements about the following intuitive inequalities:

$$\text{in-sample error} \leq \text{true error} \leq \text{out-of-sample error}. \tag{18.43}$$

Note that the out-of-sample error tends to ∞ as $N \to T$.

Now, let us compute the expected out-of-sample error ($\mathcal{R}^2_{\text{out}}$) for the ridge predictor $\mathbf{a}_{\text{ridge}}$, parameterized by ζ. The result reads

$$\mathcal{R}^2_{\text{out}}(\mathbf{a}_{\text{ridge}}) = \sigma_{\text{n}}^2 + \frac{\sigma_{\text{n}}^2}{T}\operatorname{Tr}(\mathbf{C}\Xi^{-1}\mathbf{E}\Xi^{-1}) + \zeta^2 \operatorname{Tr}(\mathbf{C}\Xi^{-1}\mathbf{a}\mathbf{a}^T\Xi^{-1}), \tag{18.44}$$

with $\Xi = \mathbf{E} + \zeta\mathbf{1}$. Expanding to linear order for small ζ then leads to

$$\begin{aligned} \mathcal{R}^2_{\text{out}}(\mathbf{a}_{\text{ridge}}) &= \mathcal{R}^2_{\text{out}}(\mathbf{a}_{\text{reg}}) - \frac{2\sigma_{\text{n}}^2}{T}\operatorname{Tr}(\mathbf{C}\mathbf{E}^{-2})\zeta + O(\zeta^2) \\ &= \mathcal{R}^2_{\text{out}}(\mathbf{a}_{\text{reg}}) - \frac{2\sigma_{\text{n}}^2 q}{(1-q)^3}\tau(\mathbf{C}^{-1})\zeta + O(\zeta^2), \end{aligned} \tag{18.45}$$

where $\tau(.) = \operatorname{Tr}(.)/N$ and we have used the fact that $\tau(\mathbf{W}_q^{-2}) = (1-q)^{-3}$. The important point here is that the coefficient in front of ζ is *negative*, i.e. to first order, the ridge estimator has a lower out-of-sample error than the naive regression estimator:

$$\text{ridge estimation error} < \text{naive estimation error}. \tag{18.46}$$

However, the ridge estimator introduces a systematic bias since $\|\mathbf{a}_{\text{ridge}}\|^2 < \|\mathbf{a}_{\text{reg}}\|^2$ when $\zeta > 0$. This gives the third term in Eq. (18.44), which becomes large for larger ζ. So one indeed expects that there should exist an optimal value of ζ (which depends on the specific problem at hand) which minimizes the out-of-sample error. We now show how this optimal out-of-sample error can be elegantly computed in the large N limit.

The Large N Limit

In the large N limit we can recover the fact that the ridge estimator of Section 18.2.1 minimizes the out-of-sample risk without the need of a Gaussian prior on $\mathbf{a}$. We will also find an interesting relation between the Wishart Stieltjes transform and the out-of-sample risk of the ridge estimator.

In the following we will assume that the elements of the out-of-sample data $\mathbf{H}_2$ are IID with unit variance, i.e. that $\mathbf{C} = \mathbf{1}$. Then when $\Xi = \mathbf{E}_1 + \zeta\mathbf{1}$, Eq. (18.44) becomes, in the large N limit,

$$\mathcal{R}^2_{\text{out}}(\mathbf{a}_{\text{ridge}}) = \sigma_{\text{n}}^2\left(1 - q\mathfrak{g}_{\mathbf{W}_q}(-\zeta)\right) + \zeta\left(q\sigma_{\text{n}}^2 - \zeta|\mathbf{a}|^2\right)\mathfrak{g}'_{\mathbf{W}_q}(-\zeta), \tag{18.47}$$

where we have used that $\mathbf{E}_1$ is a Wishart matrix free from $\mathbf{a}\mathbf{a}^T$, and that

$$\tau((z\mathbf{1} - \mathbf{E}_1)^{-1}) = \mathfrak{g}_{\mathbf{W}_q}(z); \qquad \tau((z\mathbf{1} - \mathbf{E}_1)^{-2}) = -\mathfrak{g}'_{\mathbf{W}_q}(z). \tag{18.48}$$

For $\zeta = 0$, we have $\mathfrak{g}_{\mathbf{W}_q}(0) = -1/(1-q)$ and we thus recover Eq. (18.42). In the large N limit, the out-of-sample error for an estimator with $\Xi = \mathbf{E}_1 + \zeta\mathbf{1}$ depends on the vector $\mathbf{a}$ only through its norm $|\mathbf{a}|^2$, regardless of the distribution of its components. The optimal value of ζ must then also only depend on $|\mathbf{a}|^2$.

Now, we know that when $\mathbf{a}$ is drawn from a Gaussian distribution, the value $\zeta_{\text{opt}} = \sigma_n^2/(T\sigma_s^2)$ is optimal. In the large N limit, $|\mathbf{a}|^2$ is self-averaging and equal to $N\sigma_s^2$. So $\zeta_{\text{opt}} = q\sigma_n^2/|\mathbf{a}|^2$. We can check directly that this value is optimal by computing the derivative of Eq. (18.47) with respect to ζ evaluated at ζ_{opt}. Indeed we have

$$\sigma_n^2 q \mathfrak{g}'_{\mathbf{W}_q}(-\zeta_{\text{opt}}) - \zeta_{\text{opt}}|\mathbf{a}|^2 \mathfrak{g}'_{\mathbf{W}_q}(-\zeta_{\text{opt}}) = 0. \tag{18.49}$$

For the optimal value of ζ we also have

$$\mathcal{R}^2_{\text{out}}(\mathbf{a}_{\text{ridge}}) = \sigma_n^2\left(1 - q\mathfrak{g}_{\mathbf{W}_q}(-\zeta_{\text{opt}})\right), \tag{18.50}$$

where $\mathfrak{g}_{\mathbf{W}_q}(z)$ is given by Eq. (4.40). Since $-\mathfrak{g}_{\mathbf{W}_q}(-z)$ is positive and monotonically decreasing for $z > 0$, we recover that the optimal ridge out-of-sample error is smaller than that of the standard regression.

18.3 Bayesian Estimation of the True Covariance Matrix

We now apply the Bayesian estimation method to covariance matrices. From empirical data, we measure the sample covariance matrix $\mathbf{E}$, and want to infer the most reliable information about the "true" underlying covariance matrix $\mathbf{C}$. Hence we write Bayes' equation for conditional probabilities for matrices:

$$P(\mathbf{C}|\mathbf{E}) \propto P(\mathbf{E}|\mathbf{C})P_0(\mathbf{C}). \tag{18.51}$$

We now recall Eq. (4.16) established in Chapter 4 for Gaussian observations:

$$P(\mathbf{E}|\mathbf{C}) \propto (\det \mathbf{C})^{-T/2} \exp\left[-\frac{T}{2}\operatorname{Tr}\left(\mathbf{C}^{-1}\mathbf{E}\right)\right]. \tag{18.52}$$

As explained in Section 18.1.3, in the absence of any meaningful prior information, it is interesting to pick a conjugate prior, which here is of the form

$$P_0(\mathbf{C}) \propto (\det \mathbf{C})^a \exp\left[-b\operatorname{Tr}\left(\mathbf{C}^{-1}\mathbf{X}\right)\right] \tag{18.53}$$

for some matrix $\mathbf{X}$, which turns out to be proportional to the prior mean of $\mathbf{C}$. Indeed, this prior is in fact the probability density of the elements of an inverse-Wishart matrix. Consider an inverse-Wishart matrix $\mathbf{C}$ of size N, T^* degree of freedom and centered at a (positive definite) matrix $\mathbf{X}$. If $T^* > N + 1$, $\mathbf{C}$ has the density (see Eq. (15.35))

$$P(\mathbf{C}) \propto (\det \mathbf{C})^{-(T^*+N+1)/2} \exp\left[-\frac{T^* - N - 1}{2}\operatorname{Tr}\left(\mathbf{C}^{-1}\mathbf{X}\right)\right]. \tag{18.54}$$

Note that here $T^* > N$ is some parameter that is unrelated to the length of the time series T. The chosen normalization is such that $\mathbb{E}_0[\mathbf{C}] = \mathbf{X}$. As $T^* \to \infty$, we have $\mathbf{C} \to \mathbf{X}$.

With this prior we thus obtain

$$P(\mathbf{C}|\mathbf{E}) \propto (\det \mathbf{C})^{-(T+T^*+N+1)/2} \exp\left[-\frac{T}{2}\operatorname{Tr}\left(\mathbf{C}^{-1}\mathbf{E}^*\right)\right], \tag{18.55}$$

where we define

$$\mathbf{E}^* := \mathbf{E} + \frac{T^* - N - 1}{T}\mathbf{X}. \tag{18.56}$$

We now notice that (18.55) is, by construction, also a probability density for an inverse-Wishart with $\bar{T} = T + T^*$, with

$$\mathbb{E}[\mathbf{C}|\mathbf{E}] = \frac{T\mathbf{E}^*}{T + T^* - N - 1} = r\mathbf{E} + (1 - r)\mathbf{X}, \tag{18.57}$$

with

$$r = \frac{T}{T + T^* - N - 1}. \tag{18.58}$$

Hence we recover a linear shrinkage, similar to Eq. (18.10) in the case of a scalar variable with a Gaussian prior. We will recover this shrinkage formula in the context of rotationally invariant estimators in the next chapter, see Eq. (19.49).

We end with the following remarks:

- The linear shrinkage works even for the finite N case, i.e. without the large N hypothesis.
- In general, if one has no idea of what $\mathbf{X}$ should be, one can use the identity matrix, i.e.

$$\mathbb{E}[\mathbf{C}|\mathbf{E}] = r\mathbf{E} + (1 - r)\mathbf{1}. \tag{18.59}$$

  Another simple choice is a covariance matrix $\mathbf{X}$ corresponding to a one-factor model (see Section 20.4.2):

$$\mathbf{X}_{ij} = \sigma_s^2 \left[\delta_{ij} + \rho(1 - \delta_{ij})\right], \tag{18.60}$$

  where ρ is the average pairwise correlation (which can also be learned using validation).
- Note that T^* (or equivalently r) is generally unknown. It may be inferred from the data or learned using validation.
- As we will see in Chapter 20, the linear shrinkage works quite well in financial applications, showing that inverse-Wishart is not a bad prior for the true covariance matrix in that case (see Fig. 15.1).

Bibliographical Notes

- Some general references on Bayesian methods and statistical inference:

  - E. T. Jaynes. *Probability Theory: The Logic of Science*. Cambridge University Press, Cambridge, 2003,
  - G. James, D. Witten, T. Hastie, and R. Tibshirani. *An Introduction to Statistical Learning: with Applications in R*. Springer, New York, 2013.

19
Eigenvector Overlaps and Rotationally Invariant Estimators

19.1 Eigenvector Overlaps

19.1.1 Setting the Stage

We saw in the first two parts of this book how tools from RMT allow one to infer many properties of the eigenvalue distribution, encoded in the trace of the resolvent of the random matrix under scrutiny. As in the previous chapters, random matrices of particular interest are of the form

$$\mathbf{E} = \mathbf{C} + \mathbf{X}, \quad \text{or} \quad \mathbf{E} = \mathbf{C}^{\frac{1}{2}}\mathbf{W}\mathbf{C}^{\frac{1}{2}}, \tag{19.1}$$

where $\mathbf{X}$ and $\mathbf{W}$ represent some "noise", for example $\mathbf{X}$ might be a Wigner matrix in the additive case and $\mathbf{W}$ a white Wishart matrix in the multiplicative case, whereas $\mathbf{C}$ is the "true", uncorrupted matrix that one would like to measure. One often calls $\mathbf{E}$ the *sample* matrix and $\mathbf{C}$ the *population* matrix.

In this section we want to discuss the properties of the eigenvectors of $\mathbf{E}$, and in particular their relation with the eigenvectors of $\mathbf{C}$. There are, at least, two natural questions about the eigenvectors of the sample matrix $\mathbf{E}$:

1 How similar are sample eigenvectors $[\mathbf{v}_i]_{i\in(1,N)}$ of $\mathbf{E}$ and the true ones $[\mathbf{u}_i]_{i\in(1,N)}$ of $\mathbf{C}$?
2 What information can we learn by observing two independent realizations – say $\mathbf{E} = \mathbf{C}^{\frac{1}{2}}\mathbf{W}\mathbf{C}^{\frac{1}{2}}$ and $\mathbf{E}' = \mathbf{C}^{\frac{1}{2}}\mathbf{W}'\mathbf{C}^{\frac{1}{2}}$ in the multiplicative case – that remain correlated through $\mathbf{C}$?

A natural quantity to characterize the similarity between two arbitrary vectors – say χ and ζ – is the scalar product of χ and ζ. More formally, we define the "overlap" as $\chi^T\zeta$. Since the eigenvectors of real symmetric matrices are only defined up to a sign, it is in fact more natural to consider the squared overlaps $(\chi^T\zeta)^2$. In the first problem alluded to above, we want to understand the relation between the eigenvectors of the sample matrix $[\mathbf{v}_i]_{i\in(1,N)}$ and those of the population matrix $[\mathbf{u}_i]_{i\in(1,N)}$. The matrix of squared overlaps is defined as $(\mathbf{v}_i{}^T\mathbf{u}_j)^2$, which actually forms a so-called bi-stochastic matrix (positive elements with the sums over both rows and columns all equal to unity).

In order to study these overlaps, the central tool is again the resolvent matrix (and not its normalized trace as for the Stieltjes transform), which we recall is defined as

$$\mathbf{G}_{\mathbf{A}}(z) = (z\mathbf{1} - \mathbf{A})^{-1}, \tag{19.2}$$

for any arbitrary symmetric matrix $\mathbf{A}$. Now, if we expand $\mathbf{G}_{\mathbf{E}}$ over the eigenvectors $\mathbf{v}$ of $\mathbf{E}$, we obtain that

$$\mathbf{u}^T \mathbf{G}_{\mathbf{E}}(z)\mathbf{u} = \sum_{i=1}^{N} \frac{\left(\mathbf{u}^T \mathbf{v}_i\right)^2}{z - \lambda_i}, \tag{19.3}$$

for any $\mathbf{u}$ in $\mathbb{R}^N$.

We thus see from Eq. (19.3) that each pole of the resolvent defines a projection onto the corresponding sample eigenvectors. This suggests that the techniques we need to apply are very similar to the ones used above to study the density of states. However, one should immediately stress that contrarily to eigenvalues, each eigenvector $\mathbf{v}_i$ for any given i continues to fluctuate when $N \to \infty$ and never reaches a deterministic limit. As a consequence, we will need to introduce some averaging procedure to obtain a well-defined result. We will thus consider the following quantity:

$$\Phi(\lambda_i, \mu_j) \;:=\; N\mathbb{E}[(\mathbf{v}_i^T \mathbf{u}_j)^2], \tag{19.4}$$

where the expectation $\mathbb{E}$ can be interpreted either as an average over different realizations of the randomness or, perhaps more meaningfully for applications, as an average *for a fixed sample* over small intervals of sample eigenvalues, of width $\mathrm{d}\lambda = \eta$. We choose η in the range $1 \gg \eta \gg N^{-1}$ (say $\eta = N^{-1/2}$) such that there are many eigenvalues in the interval $\mathrm{d}\lambda$, while keeping $\mathrm{d}\lambda$ sufficiently small for the spectral density to be approximately constant. Interestingly, the two procedures lead to the same result for large matrices, i.e. the locally smoothed quantity $\Phi(\lambda, \mu)$ is "self-averaging". A way to do this smoothing automatically is, as we explained in Chapter 2, to choose $z = \lambda_i - i\eta$ in Eq. (19.3), leading to

$$\operatorname{Im} \mathbf{u}_j^T \mathbf{G}_{\mathbf{E}}(\lambda_i - i\eta)\mathbf{u}_j \approx \pi \rho_{\mathbf{E}}(\lambda_i) \times \Phi(\lambda_i, \mu_j), \tag{19.5}$$

provided η is in the range $1 \gg \eta \gg N^{-1}$. Note that we have replaced $N\left(\mathbf{u}_j^T \mathbf{v}_i\right)^2$ with $\Phi(\lambda_i, \mu_j)$, to emphasize the fact that we expect typical square overlaps to be of order $1/N$, such that Φ is of order unity when $N \to \infty$. This assumption will indeed be shown to hold below. In fact, when $\mathbf{u}_j$ and $\mathbf{v}_i$ are completely uncorrelated, one finds $\Phi(\lambda_i, \mu_j) = 1$.

For the second question, the main quantity of interest is, similarly, the (mean squared) overlap between the eigenvectors of two independent noisy matrices $\mathbf{E}$ and $\mathbf{E}'$:

$$\Psi(\lambda_i, \lambda'_j) \;:=\; N\mathbb{E}[(\mathbf{v}_i^T \mathbf{v}'_j)^2], \tag{19.6}$$

where $[\lambda'_i]_{i \in (1,N)}$ and $[\mathbf{v}'_i]_{i \in (1,N)}$ are the eigenvalues and eigenvectors of $\mathbf{E}'$, i.e. another sample matrix that is independent from $\mathbf{E}$ but with the same underlying population matrix **C**. In order to get access to $\Phi(\lambda_i, \lambda'_j)$ defined in Eq. (19.5), one should consider the following quantity:

$$\psi(z, z') = \frac{1}{N} \operatorname{Tr}\left[\mathbf{G_E}(z)\mathbf{G_{E'}}(z')\right]. \tag{19.7}$$

After simple manipulations one readily obtains a generalized Sokhotski–Plemelj formula, where η is such that $1 \gg \eta \gg N^{-1}$:

$$\operatorname{Re}\left[\psi(\lambda_i - i\eta, \lambda'_i + i\eta) - \psi(\lambda_i - i\eta, \lambda'_i - i\eta)\right] \approx 2\pi^2 \rho_{\mathbf{E}}(\lambda_i)\rho_{\mathbf{E'}}(\lambda'_i)\Psi(\lambda_i, \lambda'_j). \tag{19.8}$$

This representation allows one to obtain interesting results for the overlaps between the eigenvectors of two independently drawn random matrices, see Eq. (19.14).

19.1.2 Overlaps in the Additive Case

Now, we can use the subordination relation for the resolvent of the sum of two free matrices established in Chapter 13, Eq. (13.44), which in the present case reads

$$\mathbb{E}[\mathbf{G_E}(z)] = \mathbf{G_C}\left(z - R_{\mathbf{X}}(\mathfrak{g}_{\mathbf{E}}(z))\right). \tag{19.9}$$

Since we choose $\mathbf{u}_j$ to be an eigenvector of **C** with eigenvalue μ_j, one finds

$$\mathbf{u}_j^T \mathbf{G_E}(\lambda_i - i\eta)\mathbf{u}_j = \frac{1}{\lambda_i - i\eta - R_{\mathbf{X}}\left(\mathfrak{g}_{\mathbf{E}}(\lambda_i - i\eta)\right) - \mu_j}, \tag{19.10}$$

where we have dropped the expectation value as the left hand side is self-averaging when η is in the correct range. The imaginary part of this quantity, calculated for $\eta \to 0$, gives access to $\Phi(\lambda_i, \mu_j)$. The formula simplifies in the common case where the noise matrix **X** is a Wigner matrix, such that $R_{\mathbf{X}}(z) = \sigma^2 z$. In this case, one finally obtains a Lorentzian shape for the squared overlaps:

$$\Phi(\lambda, \mu) = \frac{\sigma^2}{(\mu - \lambda + \sigma^2 \mathfrak{h}_{\mathbf{E}}(\lambda))^2 + \sigma^4 \pi^2 \rho_{\mathbf{E}}(\lambda)^2}, \tag{19.11}$$

where we have decomposed the Stieltjes transform into its real and imaginary parts as $\mathfrak{g}_{\mathbf{E}}(x) = \mathfrak{h}_{\mathbf{E}}(x) + i\pi\rho_{\mathbf{E}}(x)$; note that $\mathfrak{h}_{\mathbf{E}}(x)$ is equal to π times the Hilbert transform of $\rho_{\mathbf{E}}(x)$.

In Figure 19.1, we illustrate this formula in the case where **C** is a Wigner matrix with parameter $\sigma^2 = 1$. For a fixed λ, the overlap peaks for

$$\mu = \lambda - \sigma^2 \mathfrak{h}_{\mathbf{E}}(\lambda), \tag{19.12}$$

with a width $\sim \sigma^2 \rho_{\mathbf{E}}(\lambda)$. When $\sigma \to 0$, i.e. in the absence of noise, one recovers

$$\Phi(\lambda, \mu) \to \delta(\lambda - \mu), \tag{19.13}$$

as expected since in this case the eigenvectors of **E** are trivially the same as those of **C**. Note that apart from the singular case $\sigma = 0$, $\Phi(\lambda, \mu)$ is found to be of order unity when $N \to \infty$. But because of the factor N in Eq. (19.6), the overlaps between $\mathbf{v}_i$ and $\mathbf{u}_j$ are of order $N^{-1/2}$ as soon as $\sigma > 0$.

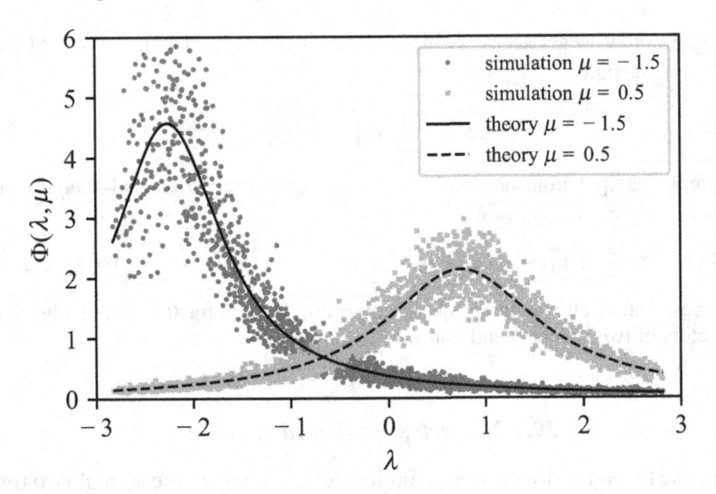

Figure 19.1 Normalized squared-overlap function $\Phi(\lambda, \mu)$ for $\mathbf{E} = \mathbf{X}_1 + \mathbf{X}_2$, the sum of two unit Wigner matrices, compared with a numerical simulation for $\mu = -1.5$ and $\mu = 0.5$. The simulations are for a single sample of size $N = 2000$, each data point corresponds to N times the square overlap between an eigenvector of $\mathbf{E}$ and one of $\mathbf{X}_1$ averaged over eigenvectors with eigenvalues within distance $\eta = 2/\sqrt{N}$ of μ.

Now suppose that $\mathbf{E} = \mathbf{C} + \mathbf{X}$ and $\mathbf{E}' = \mathbf{C} + \mathbf{X}'$, where $\mathbf{X}$ and $\mathbf{X}'$ are two independent Wigner matrices with the same variance σ^2. Using Eq. (19.8), one can compute the expected overlap between the eigenvectors of $\mathbf{E}$ and $\mathbf{E}'$. After a little work, one can establish the following result for $\Psi(\lambda, \lambda)$, i.e. the typical overlap around the same eigenvalues for $\mathbf{E}$ and $\mathbf{E}'$:

$$\Psi(\lambda, \lambda) = \frac{\sigma^2}{2 f_2(\lambda)^2} \frac{\partial_\lambda f_1(\lambda)}{(\partial_\lambda f_1(\lambda))^2 + (\partial_\lambda f_2(\lambda))^2}, \tag{19.14}$$

where

$$f_1(\lambda) = \lambda - \sigma^2 \mathfrak{h}_\mathbf{E}(\lambda); \qquad f_2(\lambda) = \sigma^2 \pi \rho_\mathbf{E}(\lambda); \qquad \mathfrak{h}_\mathbf{E}(\lambda) := \mathrm{Re}[\mathfrak{g}_\mathbf{E}(\lambda)]. \tag{19.15}$$

Note that in the large N limit, $\mathfrak{g}_\mathbf{E}(z) = \mathfrak{g}_{\mathbf{E}'}(z)$.

The formula for $\Psi(\lambda, \lambda')$ is more cumbersome; for a fixed λ', one finds again a humped shaped function with a maximum at $\lambda' \approx \lambda$. The most striking aspect of this formula, however, is that only $\mathfrak{g}_\mathbf{E}(z)$ (which is measurable from data) is needed to compute the expected overlap $\Psi(\lambda, \lambda')$; the knowledge of the "true" matrix $\mathbf{C}$ is not needed to judge whether or not the observed overlap between the eigenvectors of $\mathbf{E}$ and $\mathbf{E}'$ is compatible with the hypothesis that such matrices are both noisy versions of the same unknown $\mathbf{C}$.

19.1.3 Overlaps in the Multiplicative Case

We now repeat the same steps in the case where $\mathbf{E} = \mathbf{C}^{\frac{1}{2}} \mathbf{W}_q \mathbf{C}^{\frac{1}{2}}$, where $\mathbf{W}_q$ is a Wishart matrix of parameter q. We know that in this case the matrix subordination formula reads as (13.47), which can be rewritten as

$$\mathbf{G}_{\mathbf{E}}(z) = \frac{Z(z)}{z}\mathbf{G}_{\mathbf{C}}(Z(z)), \quad \text{with} \quad Z = \frac{z}{1-q+qz\mathfrak{g}_{\mathbf{E}}(z)}. \tag{19.16}$$

This allows us to compute

$$\mathbf{u}_j^T \mathbf{G}_{\mathbf{E}}(\lambda_i - i\eta)\mathbf{u}_j = \frac{Z(\lambda_i - i\eta)}{\lambda_i - i\eta}\frac{1}{Z(\lambda_i - i\eta) - \mu_j}, \tag{19.17}$$

and finally, taking the imaginary part in the limit $\eta \to 0^+$,

$$\Phi(\lambda,\mu) = \frac{q\mu\lambda}{(\mu(1-q) - \lambda + q\mu\lambda\,\mathfrak{h}_{\mathbf{E}}(\lambda))^2 + q^2\mu^2\lambda^2\pi^2\rho_{\mathbf{E}}^2(\lambda)}, \tag{19.18}$$

where, again, $\mathfrak{h}_{\mathbf{E}}$ denotes the real part of the Stieltjes transform $\mathfrak{g}_{\mathbf{E}}$. Note that in the limit $q \to 0$, $\Phi(\lambda,\mu)$ becomes more and more peaked around $\lambda \approx \mu$, with an amplitude that diverges for $q = 0$. Indeed, in this limiting case, one should find that the sample eigenvectors $\mathbf{v}_i$ become equal to the population ones $\mathbf{u}_i$. More generally, $\Phi(\lambda,\mu)$ for a fixed μ has a Lorentzian humped shape as a function of λ, which peaks for $\lambda \approx \mu$.

Now suppose that $\mathbf{E} = \mathbf{C}^{\frac{1}{2}}\mathbf{W}_q\mathbf{C}^{\frac{1}{2}}$ and $\mathbf{E}' = \mathbf{C}^{\frac{1}{2}}\mathbf{W}'_q\mathbf{C}^{\frac{1}{2}}$, where $\mathbf{W}_q$ and $\mathbf{W}'_q$ are two independent Wishart matrices with the same parameter q. Using Eq. (19.8), one can again compute the expected overlap between the eigenvectors of $\mathbf{E}$ and $\mathbf{E}'$. The final formula is however too cumbersome to be reported here, see Bun et al. [2018]. The formula simplifies in the limit where $\mathbf{C}$ is close to the identity matrix, in the sense that $\tau(\mathbf{C}^2) = 1 + \epsilon$, with $\epsilon \to 0$. In this case:

$$\Psi(\lambda,\lambda') = 1 + \epsilon\,[2\,\mathfrak{h}_{\mathbf{E}}(\lambda) - 1]\,[2\,\mathfrak{h}_{\mathbf{E}}(\lambda') - 1] + O(\epsilon^2). \tag{19.19}$$

More generally, the squared overlaps only depend on $\mathfrak{g}_{\mathbf{E}}(z)$ (which is measurable from data). Again, the knowledge of the "true" matrix $\mathbf{C}$ is not needed to judge whether or not the observed overlap between the eigenvectors of $\mathbf{E}$ and $\mathbf{E}'$ is compatible with the hypothesis that such matrices are both noisy versions of the same unknown $\mathbf{C}$. This is particularly important in financial applications, where $\mathbf{E}$ and $\mathbf{E}'$ may correspond to covariance matrices measured on two non-overlapping periods. In such a case, the hypothesis that the true $\mathbf{C}$ is indeed the same in the two periods may not be warranted and can be directly tested using the overlap formula.

19.2 Rotationally Invariant Estimators

19.2.1 Setting the Stage

The results derived above concerning the overlaps between the eigenvectors of sample $\mathbf{E}$ and population (or "true") $\mathbf{C}$ matrices allow one to construct a rotationally invariant estimator of $\mathbf{C}$ knowing $\mathbf{E}$. The idea can be framed within the Bayesian approach of the previous chapter, when the prior knowledge about $\mathbf{C}$ is mute about the possible directions in which the eigenvectors of $\mathbf{C}$ are pointing. More formally, this can be expressed by saying that the prior distribution $P_0(\mathbf{C})$ is rotation invariant, i.e.

$$P_0(\mathbf{C}) = P_0(\mathbf{OCO}^T), \tag{19.20}$$

where $\mathbf{O}$ is an arbitray rotation matrix. Examples of rotationally invariant priors are provided by the orthogonal ensemble introduced in Chapter 5, where $P_0(\mathbf{C})$ only depends on $\mathrm{Tr}(\mathbf{C})$.

Now, since the posterior probability of $\mathbf{C}$ given $\mathbf{E}$ is given by

$$P(\mathbf{C}|\mathbf{E}) \propto (\det \mathbf{C})^{-T/2} \exp\left[-\frac{T}{2}\,\mathrm{Tr}\left(\mathbf{C}^{-1}\mathbf{E}\right)\right] P_0(\mathbf{C}), \tag{19.21}$$

it is easy to verify that the MMSE estimator of $\mathbf{C}$ transforms in the same way as $\mathbf{E}$ under an arbitrary rotation $\mathbf{O}$, i.e.

$$\begin{aligned}\mathbb{E}[\mathbf{C}|\mathbf{O}\mathbf{E}\mathbf{O}^T] &= \int \mathbf{C}\, P(\mathbf{C}|\mathbf{O}\mathbf{E}\mathbf{O}^T) P_0(\mathbf{C}) \mathcal{D}\mathbf{C} \\ &= \mathbf{O}\left[\int \widetilde{\mathbf{C}}\, P(\widetilde{\mathbf{C}}|\mathbf{E}) P_0(\widetilde{\mathbf{C}}) \mathcal{D}\widetilde{\mathbf{C}}\right] \mathbf{O}^T \\ &= \mathbf{O}\mathbb{E}(\mathbf{C}|\mathbf{E})\mathbf{O}^T,\end{aligned} \tag{19.22}$$

using the change of variable $\widetilde{\mathbf{C}} = \mathbf{O}^T\mathbf{C}\mathbf{O}$, and the explicit form of $P(\mathbf{C}|\mathbf{E})$ given in Eq. (19.21).

More generally, if we call $\Xi(\mathbf{E})$ an estimator of $\mathbf{C}$ given $\mathbf{E}$, this estimator is rotationally invariant if and only if

$$\Xi(\mathbf{O}\mathbf{E}\mathbf{O}^T) = \mathbf{O}\Xi(\mathbf{E})\mathbf{O}^T, \tag{19.23}$$

for any orthogonal matrix $\mathbf{O}$. This means in words that, if the SCM $\mathbf{E}$ is rotated by some $\mathbf{O}$, then our estimation of $\mathbf{C}$ must be rotated in the same fashion. Intuitively, this is because we have no prior assumption on the eigenvectors of $\mathbf{C}$, so the only special directions in which $\mathbf{C}$ can point are those singled out by $\mathbf{E}$ itself. Estimators abiding by Eq. (19.23) are called rotationally invariant estimators (RIE).

An alternative interpretation of Eq. (19.23) is that $\Xi(\mathbf{E})$ can be diagonalized in the same basis as $\mathbf{E}$, up to a *fixed* rotation matrix Ω. But consistent with our rotationally invariant prior on $\mathbf{C}$, there is no natural guess for Ω, except the identity matrix $\mathbf{1}$. Hence we conclude that $\Xi(\mathbf{E})$ has the same eigenvectors as those of $\mathbf{E}$, and write

$$\Xi(\mathbf{E}) = \sum_{i=1}^{N} \xi_i \mathbf{v}_i \mathbf{v}_i^T, \tag{19.24}$$

where $\mathbf{v}_i$ are, as above, the eigenvectors of $\mathbf{E}$, and where ξ_i is a function of all empirical eigenvalues $[\lambda_j]_{j\in(1,N)}$. We now show how these ξ_i can be optimally chosen, and operationally computed from data in the limit $N \to \infty$.

19.2.2 The Optimal RIE

Suppose we ask the following question: what is the optimal choice of ξ_i such that $\Xi(\mathbf{E})$, defined by Eq. (19.24), is as close as possible to the true $\mathbf{C}$? If the eigenvectors of $\mathbf{E}$ were equal to those of $\mathbf{C}$, i.e. if $\mathbf{v}_i = \mathbf{u}_i$, $\forall i$, the solution would trivially be $\xi_i = \mu_i$. But in the case where $\mathbf{v}_i \neq \mathbf{u}_i$, the solution is *a priori* non-trivial. So we want to minimize the following least-square error:

$$\mathrm{Tr}(\Xi(\mathbf{E}) - \mathbf{C})^2 = \sum_{i=1}^{N} \mathbf{v}_i^T (\Xi(\mathbf{E}) - \mathbf{C})^2 \mathbf{v}_i = \sum_{i=1}^{N} \left(\xi_i^2 - 2\xi_i \mathbf{v}_i^T \mathbf{C} \mathbf{v}_i + \mathbf{v}_i^T \mathbf{C}^2 \mathbf{v}_i \right). \quad (19.25)$$

Minimizing over ξ_k and noting that the third term in the equation above is independent of the ξ's, it is easy to get the following expression for the optimal ξ_k:

$$\xi_k = \mathbf{v}_k^T \mathbf{C} \mathbf{v}_k. \quad (19.26)$$

This is all very well but seems completely absurd: we assume that we do not know the true $\mathbf{C}$ and want to find the best estimator of $\mathbf{C}$ knowing $\mathbf{E}$, and we find an equation for the ξ that we cannot compute unless we know $\mathbf{C}$.

Because Eq. (19.26) requires in principle knowledge we do not have, it is often called the “oracle” estimator. But as we will see in the next section, the large N limit allows one to actually compute the optimal ξ's from the data alone, without having to know $\mathbf{C}$.

19.2.3 The Large Dimension Miracle

Let us first rewrite Eq. (19.26) in terms of the overlaps introduced in Section 19.1. Expanding over the eigenvectors of $\mathbf{C}$ we find

$$\xi_k = \sum_{j=1}^{N} \mathbf{v}_k^T \mathbf{u}_j \mu_j \mathbf{u}_j^T \mathbf{v}_k = \sum_{j=1}^{N} \mu_j \left(\mathbf{u}_j^T \mathbf{v}_k \right)^2 \quad (19.27)$$

$$\underset{N\to\infty}{\longrightarrow} \int \mathrm{d}\mu\, \rho_{\mathbf{C}}(\mu) \mu \Phi(\lambda_k, \mu), \quad (19.28)$$

where λ_k is the eigenvalue of the sample matrix $\mathbf{E}$ associated with $\mathbf{v}_k$. In other words, ξ_k is an average over the eigenvalues of $\mathbf{C}$, weighted by the square overlaps $\Phi(\lambda_k, \mu)$. Now, using Eq. (19.5), we can also write

$$\xi_k = \sum_{j=1}^{N} \mu_j \left(\mathbf{u}_j^T \mathbf{v}_k \right)^2 = \frac{1}{\pi \rho_{\mathbf{E}}(\lambda_k)} \lim_{\eta \to 0^+} \mathrm{Im} \sum_{j=1}^{N} \mathbf{u}_j^T \mu_j \mathbf{G}_{\mathbf{E}}(\lambda_k - i\eta) \mathbf{u}_j$$

$$= \frac{1}{\pi \rho_{\mathbf{E}}(\lambda_k)} \lim_{\eta \to 0^+} \mathrm{Im}\, \tau \left(\mathbf{C} \mathbf{G}_{\mathbf{E}}(\lambda_k - i\eta) \right). \quad (19.29)$$

We now use the fact that in both the additive and the multiplicative case, the matrices $\mathbf{G}_{\mathbf{E}}$ and $\mathbf{G}_{\mathbf{C}}$ are related by a subordination equation of the type

$$\mathbf{G}_{\mathbf{E}}(z) = Y(z) \mathbf{G}_{\mathbf{C}}(Z(z)), \quad (19.30)$$

with $Y(z) = 1$ in the additive case and $Y(z) = Z(z)/z$ in the multiplicative case. Hence we can write the following series of equalities:

$$\tau\left(\mathbf{C}\mathbf{G}_{\mathbf{E}}(z)\right) = Y\tau\left(\mathbf{C}\mathbf{G}_{\mathbf{C}}(Z)\right) = Y\tau\left((\mathbf{C} - Z\mathbf{1} + Z\mathbf{1})(Z\mathbf{1} - \mathbf{C})^{-1}\right)$$
$$= YZ\mathfrak{g}_{\mathbf{C}}(Z) - Y = Z(z)\mathfrak{g}_{\mathbf{E}}(z) - Y(z). \tag{19.31}$$

But since $Z(z)$ only depend on $\mathfrak{g}_{\mathbf{E}}(z)$, we see that the final formula for ξ_k does not explicitly depend on $\mathbf{C}$ anymore and reads

$$\xi_k = \frac{1}{\pi\rho_{\mathbf{E}}(\lambda_k)} \lim_{\eta\to 0^+} Z(z_k)\mathfrak{g}_{\mathbf{E}}(z_k) - Y(z_k), \qquad z_k := \lambda_k - i\eta. \tag{19.32}$$

Since all the quantities on the right hand side can be estimated from the data alone, this formula will lend itself to real world applications. Let us first explore this formula for two simple cases, for an additive model and for a multiplicative model.

19.2.4 The Additive Case

For a general noise matrix $\mathbf{X}$, one has $Z(z) = z - R_{\mathbf{X}}(\mathfrak{g}_{\mathbf{E}}(z))$, leading to the following mapping between the empirical eigenvalues λ and the RIE eigenvalues ξ:

$$\xi(\lambda) = \lambda - \frac{\lim_{\eta\to 0^+} \operatorname{Im} R_{\mathbf{X}}(\mathfrak{g}_{\mathbf{E}}(z))\mathfrak{g}_{\mathbf{E}}(z)}{\lim_{\eta\to 0^+} \operatorname{Im} \mathfrak{g}_{\mathbf{E}}(z)}, \qquad z = \lambda - i\eta. \tag{19.33}$$

If there is no noise, i.e. $\mathbf{X} = 0$ and hence $R_{\mathbf{X}} = 0$, we find as expected $\xi(\lambda) = \lambda$. If $\mathbf{X}$ is small, then

$$R_{\mathbf{X}}(x) = \epsilon x + \cdots, \tag{19.34}$$

where we have assumed $\tau(\mathbf{X}) = 0$ and $\epsilon = \tau(\mathbf{X}^2)$ is small. Hence we find

$$\xi(\lambda) = \lambda - 2\epsilon\, \mathfrak{h}_{\mathbf{E}}(\lambda) + \cdots. \tag{19.35}$$

A natural case to consider is when $\mathbf{X}$ is Wigner noise, for which $R_{\mathbf{X}}(x) = \sigma_{\mathrm{n}}^2 x$ exactly, such that the equation above is exact with $\epsilon = \sigma_{\mathrm{n}}^2$, for arbitrary values of σ_{n}. When $\mathbf{C}$ is another Wigner matrix with variance σ_{s}^2, then $\mathbf{E}$ is clearly also a Wigner matrix with variance $\sigma^2 = \sigma_{\mathrm{n}}^2 + \sigma_{\mathrm{s}}^2$. In this case, when $-2\sigma < \lambda < 2\sigma$,

$$\mathfrak{h}_{\mathbf{E}}(\lambda) = \frac{\lambda}{2\sigma^2}. \tag{19.36}$$

Hence we obtain, from Eq. (2.38),

$$\xi(\lambda) = \lambda - \frac{\lambda\sigma_{\mathrm{n}}^2}{\sigma^2} = r\lambda, \qquad r := \frac{\sigma_{\mathrm{s}}^2}{\sigma^2}, \tag{19.37}$$

which is the linear shrinkage obtained for Gaussian variables in Chapter 18. In fact, this shrinkage formula is expected elementwise, since all elements are Gaussian random variables:[1]

$$\Xi(\mathbf{E})_{ij} = r\,\mathbf{E}_{ij}, \tag{19.38}$$

see Eq. (18.10) with $x_0 = 0$.

Exercise 19.2.1 Additive RIE for the sum of two matrices from the same distribution

In this exercise we will find a simple form for a RIE estimator when the noise is drawn from the same distribution as the signal, i.e.

$$\mathbf{E} = \mathbf{C} + \mathbf{X}, \tag{19.39}$$

with $\mathbf{X}$ and $\mathbf{C}$ mutually free matrices drawn from the same ensemble.

(a) Write a relationship between $R_{\mathbf{X}}(g)$ and $R_{\mathbf{E}}(g)$.
(b) Given that $\mathfrak{g}_{\mathbf{E}}(z)R_{\mathbf{E}}(\mathfrak{g}_{\mathbf{E}}(z)) = z\mathfrak{g}_{\mathbf{E}}(z) - 1$, what is $\mathfrak{g}_{\mathbf{E}}(z)R_{\mathbf{X}}(\mathfrak{g}_{\mathbf{E}}(z))$?
(c) Use Eq. (19.33) and the fact that z is real in the limit $\eta \to 0^+$ to show that $\xi(\lambda) = \lambda/2$.
(d) Given that $\Xi = \mathbb{E}[\mathbf{C}]_{\mathbf{E}}$ (see Section 19.4), find a simple symmetry argument to show that $\Xi = \mathbf{E}/2$.
(e) Generate numerically two independent symmetric orthogonal matrices $\mathbf{M}_1$ and $\mathbf{M}_2$ with $N = 1000$ (see Exercise 1.2.4). Compute the eigenvalues λ_k and eigenvectors $\mathbf{v}_k$ of the sum of these two matrices.
(f) Plot the normalized histogram of the λ_k's and compare with the arcsine law between -2 and 2 ($\rho(\lambda) = 1/(\pi\sqrt{4-\lambda^2})$).
(g) Make a scatter plot $\mathbf{v}_k^T\mathbf{M}_1\mathbf{v}_k$ vs λ_k and compare with $\lambda/2$.

19.2.5 The Multiplicative Case

We can now tackle the multiplicative case, which includes the important practical problem of estimating the true covariance matrix given a sample covariance matrix. In the multiplicative case, it is more elegant to use the subordination relation Eq. (13.46) for the T-matrix rather than for the resolvent. In the present setting we thus write

$$\mathrm{T}_{\mathbf{E}}(z) = \mathrm{T}_{\mathbf{C}}[z\,S_{\mathbf{W}}(\mathrm{t}_{\mathbf{E}}(z))]. \tag{19.40}$$

[1] Note however that there is a slight subtlety here: the linear shrinkage equation (19.37) only holds in the absence of outliers, i.e. empirical eigenvalues that fall outside the interval $(-2\sigma, 2\sigma)$. For such eigenvalues, shrinkage is non-linear. For a similar situation in the multiplicative case, see Figure 19.2.

In terms of T-transforms, Eq. (19.29) reads

$$\xi(\lambda) = \frac{\lim_{\eta\to 0^+} \operatorname{Im} \tau(\mathbf{C}\mathbf{T}_\mathbf{C}[z\, S_\mathbf{W}(\mathrm{t}_\mathbf{E}(z))])}{\lim_{\eta\to 0^+} \operatorname{Im} \mathrm{t}_\mathbf{E}(z)}; \qquad z = \lambda - i\eta. \tag{19.41}$$

Since $\mathbf{T}_\mathbf{C}(z) = \mathbf{C}(z\mathbf{1} - \mathbf{C})^{-1}$, we have, with $t = \mathrm{t}_\mathbf{E}(z)$ as a shorthand,

$$\begin{aligned}
\tau\left[\mathbf{C}\mathbf{T}_\mathbf{C}(zS_\mathbf{W}(t))\right] &= \tau\left[\mathbf{C}^2(zS_\mathbf{W}(t)\mathbf{1} - \mathbf{C})^{-1}\right] \\
&= \tau\left[\mathbf{C}(\mathbf{C} - zS_\mathbf{W}(t)\mathbf{1} + zS_\mathbf{W}(t)\mathbf{1})(zS_\mathbf{W}(t)\mathbf{1} - \mathbf{C})^{-1}\right] \\
&= \tau(\mathbf{C}) + zS_\mathbf{W}(t)\mathrm{t}_\mathbf{C}(zS_\mathbf{W}(t)) \\
&= \tau(\mathbf{C}) + zS_\mathbf{W}(t)\mathrm{t}_\mathbf{E}(z).
\end{aligned} \tag{19.42}$$

The first term $\tau(\mathbf{C})$ is real and does not contribute to the imaginary part that we have to compute, so we obtain

$$\xi(\lambda) = \lambda \frac{\lim_{\eta\to 0^+} \operatorname{Im} S_\mathbf{W}(\mathrm{t}_\mathbf{E}(z))\mathrm{t}_\mathbf{E}(z)}{\lim_{\eta\to 0^+} \operatorname{Im} \mathrm{t}_\mathbf{E}(z)}, \qquad z = \lambda - i\eta. \tag{19.43}$$

Equation (19.43) is very general. It applies to sample covariance matrices where the noise matrix $\mathbf{W}$ is a white Wishart, but it also applies to more general multiplicative noise processes.

In the special case of sample covariance matrices $\mathbf{E} = \mathbf{C}^{\frac{1}{2}}\mathbf{W}_q\mathbf{C}^{\frac{1}{2}}$ with $N/T = q$, we know that $S_{\mathbf{W}_q}(t) = (1+qt)^{-1}$. In the bulk region $\lambda_- < \lambda < \lambda_+$, $t = \mathrm{t}_\mathbf{E}(z)$ is complex with non-zero imaginary part when $z = \lambda - i\eta$. Hence

$$\xi(\lambda) = \lambda \frac{\lim_{\eta\to 0^+} \operatorname{Im} \frac{t}{1+qt}}{\lim_{\eta\to 0^+} \operatorname{Im} t} = \left.\frac{\lambda}{|1 + q\mathrm{t}_\mathbf{E}(\lambda - i\eta)|^2}\right|_{\eta\to 0}, \tag{19.44}$$

where we have used the fact that

$$\operatorname{Im} \frac{t}{1+qt} = \operatorname{Im} \frac{t(1+qt^\star)}{|1+qt|^2} = \operatorname{Im} \frac{t + q|t|^2}{|1+qt|^2} = \frac{1}{|1+qt|^2} \operatorname{Im} t. \tag{19.45}$$

Equation (19.44) can be interpreted as a form of non-linear shrinkage. A way to see this is to note that below λ_- and above λ_+ (the edges of the sample spectrum) $\mathrm{t}_\mathbf{E}(\lambda)$ is real. From the very definition,

$$\mathrm{t}_\mathbf{E}(\lambda) = \int_{\lambda_-}^{\lambda_+} \mathrm{d}\lambda' \rho_\mathbf{E}(\lambda') \frac{\lambda'}{\lambda - \lambda'} = \lambda g_\mathbf{E}(\lambda) - 1, \tag{19.46}$$

for any λ outside or at the edges of the spectrum. Hence, since $\lambda_- \geq 0$ for covariance matrices, $\mathrm{t}_\mathbf{E}(\lambda_-) < 0$ and $\mathrm{t}_\mathbf{E}(\lambda_+) > 0$. Hence, one directly establishes that the support of the RIE $\Xi(\mathbf{E})$ is narrower than that of $\mathbf{E}$:

$$\xi(\lambda_-) \geq \lambda_-; \qquad \xi(\lambda_+) \leq \lambda_+, \tag{19.47}$$

where the inequalities are saturated for $q = 0$, in which case, as expected $\xi(\lambda) = \lambda$, $\forall \lambda \in (\lambda_-, \lambda_+)$. A more in-depth discussion of the properties of Eq. (19.44) is given in Section 19.3.

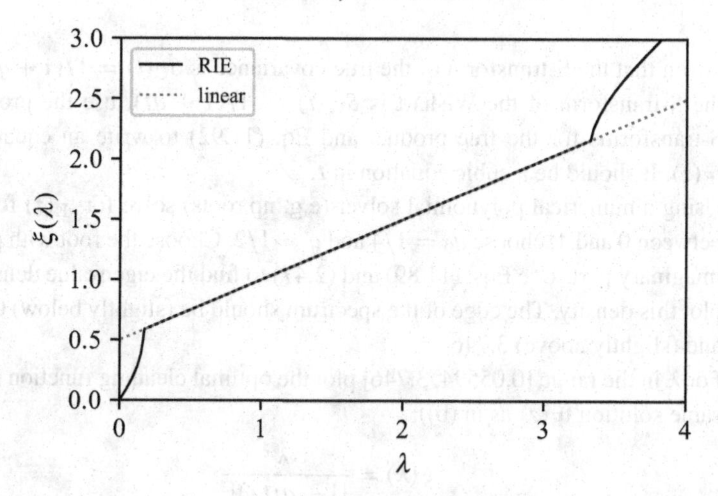

Figure 19.2 The RIE estimator (19.44) for a true covariance matrix given by an inverse-Wishart of variance $p = 0.25$ observed using data with aspect ratio $q = 0.25$. On the support of the sample density ($\lambda \in [0.17, 3.33]$), the RIE matches perfectly the linear shrinkage estimator (19.49) (with $r = 1/2$), but it is different from it outside of the expected spectrum.

Using Eq. (19.46), the shrinkage equation (19.44) can be rewritten as

$$\xi(\lambda) = \left. \frac{\lambda}{|1 - q + q\lambda \mathfrak{g}_{\mathbf{E}}(\lambda - i\eta)|^2} \right|_{\eta \to 0^+}, \tag{19.48}$$

a result first derived in Ledoit and Péché [2011].

Equation (19.44) considerably simplifies in the case where the true covariance matrix $\mathbf{C}$ is an inverse-Wishart matrix of parameter p. Injecting the explicit form of $\mathfrak{t}_{\mathbf{E}}(z)$ given by Eq. (15.50) into Eq. (19.44) leads, after simple manipulations, to

$$\xi(\lambda) = \frac{q + \lambda p}{p + q} = r\lambda + (1 - r); \qquad r := \frac{p}{p + q}, \tag{19.49}$$

i.e. exactly the linear shrinkage result derived in a Bayesian framework in Section 18.3. Note that the result Eq. (19.49) only holds between λ_- and λ_+, given in Eq. (15.52). The full function $\xi(\lambda)$ when $\mathbf{C}$ is an inverse-Wishart matrix is given in Figure 19.2.

Exercise 19.2.2 RIE when the true covariance matrix is Wishart

Assume that the true covariance matrix $\mathbf{C}$ is given by a Wishart matrix with parameter q_0. This case is a tractable model for which the computation can be done semi-analytically (we will get cubic equations!).

We observe a sample covariance matrix $\mathbf{E}$ over $T = qN$ time intervals. $\mathbf{E}$ is the free product of $\mathbf{C}$ and another Wishart matrix of parameter q:

$$\mathbf{E} = \mathbf{C}^{\frac{1}{2}} \mathbf{W} \mathbf{C}^{\frac{1}{2}}. \tag{19.50}$$

(a) Given that the S-transform of the true covariance is $S_{\mathbf{C}}(t) = 1/(1 + q_0 t)$ and the S-transform of the Wishart is $S_{\mathbf{W}}(t) = 1/(1 + qt)$, use the product of S-transforms for the free product and Eq. (11.92) to write an equation for $t_{\mathbf{E}}(z)$. It should be a cubic equation in t.

(b) Using a numerical polynomial solver (e.g. np.roots) solve for $\mathfrak{t}_{\mathbf{E}}(z)$ for z real between 0 and 4, choose $q_0 = 1/4$ and $q = 1/2$. Choose the root with positive imaginary part. Use Eqs. (11.89) and (2.47) to find the eigenvalue density and plot this density. The edge of the spectrum should be (slightly below) 0.05594 and (slightly above) 3.746.

(c) For λ in the range [0.05594, 3.746] plot the optimal cleaning function (use the same solution $\mathfrak{t}_{\mathbf{E}}(z)$ as in (b)):

$$\xi(\lambda) = \frac{\lambda}{|1 + qt(\lambda)|^2}. \tag{19.51}$$

(d) For $N = 1000$ numerically generate $\mathbf{C}$ ($q_0 = 1/4$), two versions of $\mathbf{W}_{1,2}$ ($q = 1/2$) and hence two versions of $\mathbf{E}_{1,2} := \mathbf{C}^{\frac{1}{2}}\mathbf{W}_{1,2}\mathbf{C}^{\frac{1}{2}}$. $\mathbf{E}_1$ will be the "in-sample" matrix and $\mathbf{E}_2$ the "out-of-sample" matrix. Check that $\tau(\mathbf{C}) = \tau(\mathbf{W}_{1,2}) = \tau(\mathbf{E}_{1,2}) = 1$ and that $\tau(\mathbf{C}^2) = 1.25$, $\tau(\mathbf{W}_{1,2}^2) = 1.5$ and $\tau(\mathbf{E}_{1,2}^2) = 1.75$.

(e) Plot the normalized histogram of the eigenvalues of $\mathbf{E}_1$, it should match your plot in (b).

(f) For every eigenvalue, eigenvector pair $(\lambda_k, \mathbf{v}_k)$ of $\mathbf{E}_1$ compute $\xi_{\text{val}}(\lambda_k) := \mathbf{v}_k^T \mathbf{E}_2 \mathbf{v}_k$. Plot $\xi_{\text{val}}(\lambda_k)$ vs λ_k and compare with your answer in (c).

Exercise 19.2.3 Multiplicative RIE when the signal and the noise have the same distribution

(a) Adapt the arguments of Exercise 19.2.1 to the multiplicative case with $\mathbf{C}$ and $\mathbf{W}$ two free matrices drawn from the same ensemble. Show that in this case $\xi(\lambda) = \sqrt{\lambda}$.

(b) Redo Exercise 19.2.2 with $q = q_0 = 1/4$, compare your $\xi(\lambda)$ with $\sqrt{\lambda}$.

19.2.6 RIE for Outliers

So far, we have focused on "cleaning" the bulk eigenvectors. But it turns out that the formulas above are also valid for outliers of $\mathbf{C}$ that appear as outliers of $\mathbf{E}$. One can show that, outside the bulk, $\mathfrak{g}_{\mathbf{E}}(z)$ and $\mathfrak{t}_{\mathbf{E}}(z)$ are analytic on the real axis and thus, for small η,

$$\operatorname{Im} \mathfrak{g}_{\mathbf{E}}(\lambda - i\eta) = -\eta \mathfrak{g}'_{\mathbf{E}}(\lambda), \qquad \operatorname{Im} \mathfrak{t}_{\mathbf{E}}(\lambda - i\eta) = -\eta \mathfrak{t}'_{\mathbf{E}}(\lambda). \tag{19.52}$$

Then Eqs. (19.33) and (19.43) simplify to

$$\xi(\lambda) = \lambda - \frac{\mathrm{d}}{\mathrm{d}g}\left[g R_{\mathbf{X}}(g)\right], \qquad g = \mathfrak{g}_{\mathbf{E}}(\lambda),$$
$$\xi(\lambda) = \lambda \frac{\mathrm{d}}{\mathrm{d}t}\left[t S_{\mathbf{W}}(t)\right], \qquad t = \mathfrak{t}_{\mathbf{E}}(\lambda), \tag{19.53}$$

respectively for the additive and multiplicative cases.

19.3 Properties of the Optimal RIE for Covariance Matrices

Even though the optimal non-linear shrinkage function (19.44), (19.48) seems relatively simple, it is not immediately clear what is the effect induced by the transformation $\lambda_i \to \xi(\lambda_i)$. In this section, we thus give some quantitative properties of the optimal estimator Ξ to understand the impact of the optimal non-linear shrinkage function.

First let us consider the moments of the spectrum of Ξ. From Eqs. (19.24) and (19.26) we immediately derive that

$$\operatorname{Tr}\Xi = \sum_{j=1} \mu_j \mathbf{u}_j^T \left(\sum_{i=1} \mathbf{v}_i \mathbf{v}_i^T \right) \mathbf{u}_j = \operatorname{Tr}\mathbf{C}, \tag{19.54}$$

meaning that the cleaning operation preserves the trace of the population matrix $\mathbf{C}$, as it should do. For the second moment, we have

$$\operatorname{Tr}\Xi^2 = \sum_{j,k=1}^{N} \mu_j \mu_k \sum_{i=1}^{N} (\mathbf{v}_i^T \mathbf{u}_j)^2 (\mathbf{v}_i^T \mathbf{u}_k)^2.$$

Now, if we define the matrix $\mathbf{A}_{jk}$ as $\sum_{i=1}^{N} (\mathbf{v}_i^T \mathbf{u}_j)^2 (\mathbf{v}_i^T \mathbf{u}_k)^2$ for $j,k = 1,N$, it is not hard to see that it is a matrix with non-negative entries and whose rows all sum to unity (remember that all $\mathbf{v}_i$'s are normalized to unity). The matrix $\mathbf{A}$ is therefore a (bi-)stochastic matrix and the Perron–Frobenius theorem tells us that its largest eigenvalue is equal to unity (see Section 1.2.2). Hence, we deduce the following general inequality:

$$\sum_{j,k=1}^{N} \mathbf{A}_{j,k}\, \mu_j \mu_k \le \sum_{j=1}^{N} \mu_j^2,$$

which implies that

$$\operatorname{Tr}\Xi^2 \le \operatorname{Tr}\mathbf{C}^2 \le \operatorname{Tr}\mathbf{E}^2, \tag{19.55}$$

where the last inequality comes from Eq. (17.11). In words, this result states that the spectrum of Ξ is narrower than the spectrum of $\mathbf{C}$, which is itself narrower than the spectrum of $\mathbf{E}$. The optimal RIE therefore tells us that we had better be even more cautious than simply bringing back the sample eigenvalues to their estimated true locations. This is because we have only partial information about the true eigenbasis of $\mathbf{C}$. In particular, one should always shrink downward (resp. upward) the small (resp. top) eigenvalues compared to their true locations μ_i for any $i \in (1,N)$, except for the trivial case $\mathbf{C} = \mathbf{1}$.

Next, we consider the asymptotic behavior of the optimal non-linear shrinkage function (19.44), (19.48). Throughout the following, suppose that we have an outlier at the left of the lower bound of $\rho_{\mathbf{E}}$ and let us assume $q < 1$ so that $\mathbf{E}$ has no exact zero mode. We know from Section 19.2.6 that the estimator (19.44) holds for outliers. Moreover, we have that $\lambda g_{\mathbf{E}}(\lambda) = O(\lambda)$ for $\lambda \to 0$. This allows us to conclude from Eq. (19.26) that, for outliers very close to zero,

$$\xi(\lambda) = \frac{\lambda}{(1-q)^2} + O(\lambda^2), \tag{19.56}$$

which is in agreement with Eq. (19.55): small eigenvalues must be pushed upwards for $q > 0$.

The other asymptotic limit $\lambda \to \infty$ is also useful since it gives us the behavior of the non-linear shrinkage function ξ for large outliers. In that case, we know from Eq. (17.8) that $\lim_{\lambda\to\infty} \lambda t_{\mathbf{E}}(\lambda) \sim \lambda^{-1}\tau(\mathbf{E})$. Therefore, we conclude that

$$\xi(\lambda) \approx \frac{\lambda}{\left(1 + q\lambda^{-1}\tau(\mathbf{E}) + O(\lambda^{-2})\right)^2} \approx \lambda - 2q\tau(\mathbf{E}) + O(\lambda^{-1}). \tag{19.57}$$

If all variances are normalized to unity such that $\tau(\mathbf{E}) = \tau(\mathbf{C}) = 1$, then we simply obtain

$$\xi(\lambda) \approx \lambda - 2q + O(\lambda^{-1}). \tag{19.58}$$

It is interesting to compare this with Eq. (14.54) for large rank-1 perturbations, which gives $\lambda \approx \mu + q$ for $\lambda \to \infty$. As a result, we deduce from Eq. (19.58) that $\xi(\lambda) \approx \mu - q$ and we therefore find the following ordering relation:

$$\xi(\lambda) < \mu < \lambda, \tag{19.59}$$

for isolated and large eigenvalues λ and for $q > 0$. Again, this result is in agreement with Eq. (19.55): large eigenvalues should be reduced downward for any $q > 0$, even below the "true" value of the outlier μ. More generally, the non-linear shrinkage function ξ interpolates smoothly between $\lambda/(1-q)^2$ for small λ to $\lambda - 2q$ for large λ.

19.4 Conditional Average in Free Probability

In this section we give an alternative derivation of the RIE formula, Eq. (19.29). This derivation is more elegant, albeit more abstract. In particular, it does not rely on the computation of eigenvector overlap, so by itself it misses the important link between the RIE and the computation of overlaps.

In the context of free probability, we work with abstract objects ($\mathbf{E}$, $\mathbf{C}$, etc.) that satisfy the axioms of Chapter 11. We can think of them as infinite-dimensional matrices. We are given the matrix $\mathbf{E}$ that was obtained by free operations from an unknown matrix $\mathbf{C}$. For instance it could be given by a combination of free product and free sum.

The matrix $\mathbf{E}$ is generated from the matrix $\mathbf{C}$; in this sense, $\mathbf{E}$ depends on $\mathbf{C}$. We would like to find the best estimator (in the least-square sense) of $\mathbf{C}$ given $\mathbf{E}$. It is given by the conditional average

$$\Xi = \mathbb{E}[\mathbf{C}]_{\mathbf{E}}. \tag{19.60}$$

In this abstract context, the only object we know is $\mathbf{E}$ so Ξ must be a function of $\mathbf{E}$. Let us call this function $\Xi(\mathbf{E})$. The fact that Ξ is a function of $\mathbf{E}$ only imposes that Ξ commutes with $\mathbf{E}$, i.e. that Ξ is diagonal in the eigenbasis of $\mathbf{E}$. One way to determine the function $\Xi(\mathbf{E})$ is to compute all possible moments of the form $m_k = \tau[\Xi(\mathbf{E})\mathbf{E}^k]$. They can be combined in the function

$$F(z) := \tau\left[\xi(\mathbf{E})(z\mathbf{1} - \mathbf{E})^{-1}\right] \tag{19.61}$$

via its Taylor series at $z \to \infty$. Using Eq. (19.60), we write

$$F(z) = \tau\left[\mathbb{E}[\mathbf{C}]|_{\mathbf{E}}\,(z\mathbf{1} - \mathbf{E})^{-1}\right]. \tag{19.62}$$

But the operator $\tau[.]$ contains the expectation value over all variables, both trace and randomness. So by the law of total expectation, $\tau(\mathbb{E}[.]) = \tau(.)$ and

$$F(z) = \tau\left[\mathbf{C}(z\mathbf{1} - \mathbf{E})^{-1}\right]. \tag{19.63}$$

To recover the function $\xi(\lambda)$ from $F(z)$ we use a spectral decomposition of $\mathbf{E}$:

$$F(z) = \int \rho_{\mathbf{E}}(\lambda)\frac{\xi(\lambda)}{z - \lambda}\mathrm{d}\lambda, \tag{19.64}$$

so

$$\lim_{\eta \to 0^+} \operatorname{Im} F(\lambda - i\eta) = \pi\rho_{\mathbf{E}}(\lambda)\xi(\lambda), \tag{19.65}$$

which is equivalent to

$$\xi(\lambda) = \lim_{\eta \to 0^+} \frac{\operatorname{Im} F(\lambda - i\eta)}{\operatorname{Im} \mathfrak{g}_{\mathbf{E}}(\lambda - i\eta)}, \tag{19.66}$$

itself equivalent to Eq. (19.29).

19.5 Real Data

As stated above, the good news about the RIE estimator is that it only depends on transforms of the observable matrix $\mathbf{E}$, such as $\mathfrak{g}_{\mathbf{E}}(z)$ and $\mathfrak{t}_{\mathbf{E}}(z)$ and the R- or S-transform of the noise process. One may think that real world applications should be relatively straightforward. However, we need to know the behavior of the *limiting* transforms on the real axis, precisely where the discrete N transforms $g_N(z)$ and $\mathfrak{t}_N(z)$ fail to converge.

We will discuss here how to compute these transforms using either a parametric fit or a non-parametric approximation on the sample eigenvalues. In both cases we will tackle the multiplicative case with a Wishart noise but the discussion can be adapted to cover the additive case or any other type of noise. In Section 19.6 we will discuss an alternative approach using two datasets (or disjoint subsets of the original data).

19.5.1 Parametric Approach

Ansatz on $\rho_{\mathbf{C}}$ or $S_{\mathbf{C}}$

One can postulate a convenient functional form for $\rho_{\mathbf{C}}(\lambda)$ and fit the associated parameters on the data. This allows one to obtain analytical formulas for all the relevant transforms, from which one can extract the exact behavior on the real axis.

The simplest (most tractable) choice for $\rho_{\mathbf{C}}(\lambda)$ is the inverse-Wishart distribution. In this case $\rho_{\mathbf{E}}(\lambda)$ can be computed exactly (see Eq. (15.51)) and the optimal estimator is linear within the bulk of the spectrum, cf. Eq. (19.49). When the sample covariance matrix is normalized such that $\tau(\mathbf{E}) = 1$, the inverse-Wishart has a single parameter p that needs to be estimated from the data. As an estimate, one can use for example the second moment of $\mathbf{E}$:

$$\tau\left(\mathbf{E}^2\right) = 1 + p + q, \tag{19.67}$$

or its first inverse moment:

$$\tau\left(\mathbf{E}^{-1}\right) = \frac{1+p}{1-q}, \tag{19.68}$$

which is obtained using Eq. (15.13) with $S_{\mathbf{E}}(t) = (1-pt)/(1+qt)$, or simply by noting that $\tau(\mathbf{W}_q^{-1}\mathbf{M}_p^{-1}) = \tau(\mathbf{W}_q^{-1})\tau(\mathbf{M}_p^{-1})$ for free matrices, and using the results of Sections 15.2.2 and 15.2.3.

When the distribution of sample eigenvalues appears to be bounded from above and below, one can use a more complicated but still relatively tractable ansatz for $\rho_{\mathbf{C}}(\lambda)$, by postulating a simple form for its S-transform. For example using

$$S_{\mathbf{C}}(t) = \frac{(1-p_1 t)(1-p_2 t)}{1+q_1 t} \quad \Leftrightarrow \quad S_{\mathbf{E}}(t) = \frac{(1-p_1 t)(1-p_2 t)}{(1+qt)(1+q_1 t)}, \tag{19.69}$$

one finds that $\mathfrak{t}_{\mathbf{E}}(\zeta)$ (and hence $\rho_{\mathbf{E}}(\lambda)$) is the solution of a cubic equation. Higher order terms in t in the numerator or the denominator will give higher order equations for $\mathfrak{t}_{\mathbf{E}}(\zeta)$. The parameters p_1, p_2, q_1, etc. can be evaluated from the first few moments and inverse moments of $\mathbf{E}$ or by fitting the observed density of eigenvalues. However, the particularly convenient choice Eq. (19.69) does not work when the observed distribution of eigenvalues does not have enough skewness, as in the example shown in Figure 19.3.

Parametric Fit of $\rho_{\mathbf{E}}$

Another approach consists of postulating a form for the density of sample eigenvalues and fitting its parameters. For example, one can postulate that

$$\rho_{\mathbf{E}}(\lambda) = Z^{-1} \frac{(1 + a_1\lambda + a_2\lambda^2)\sqrt{(\lambda-\lambda_-)(\lambda_+-\lambda)}}{1 + b_1\lambda + b_2\lambda^2}, \tag{19.70}$$

where $\lambda_\pm$ are fixed to the smallest/largest observed eigenvalues, and a_1, a_2, b_1 and b_2 are fitted on the data by minimizing the square error on the cumulative distribution. The normalization factor Z can be computed during the fitting procedure. This particular form fits very well for sample data generated numerically (see Fig. 19.3 left). To find the optimal shrinkage function (19.48), we then reconstruct the complex Stieltjes transform $\mathfrak{g}_{\mathbf{E}}(x - i0^+)$ numerically, by using the fitted $\rho(\lambda)$ and computing its Hilbert transform. The issue with such an approach is that even when Eq. (19.70) is a good fit to the sample density of eigenvalues, it cannot be obtained as the result of the free product of a Wishart and some

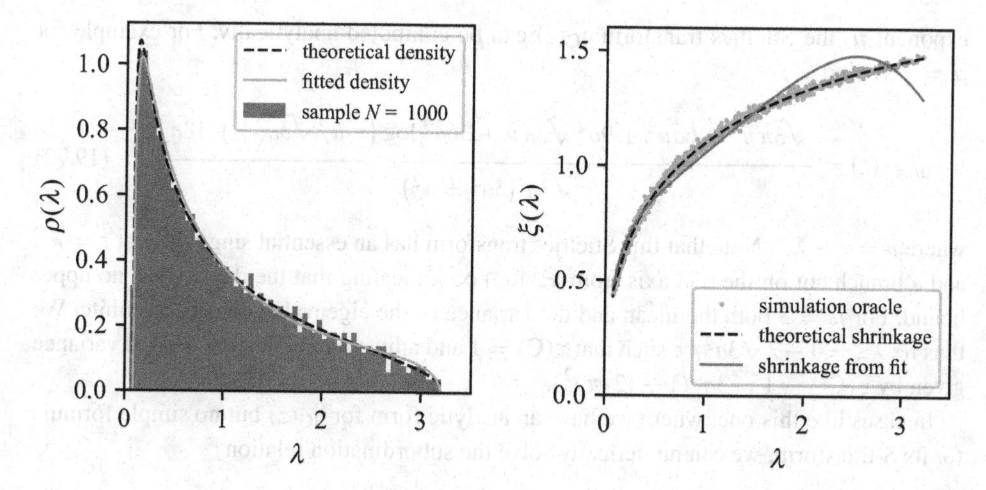

Figure 19.3 Parametric fit illustrated on an example where the true covariance has a uniform density of eigenvalues with mean 1 and variance 0.2 (see Eq. (15.42)). A single sample covariance matrix with $N = 1000$ and $q = 0.4$ was generated, and the ad-hoc distribution (19.70) was fitted to the eigenvalue CDF. The left-hand figure shows a histogram of the sample eigenvalues compared with the theoretical distribution and the ad-hoc fit. The right-hand figure shows the theoretical optimal shrinkage and the one obtained from the fit. The agreement is barely satisfactory, in particular the shrinkage from the fit is non-monotonic. The dots show the oracle estimator $\xi_k = \mathbf{v}_k^T \mathbf{C} \mathbf{v}_k$ computed within the same simulation.

given density. As a consequence the approximate estimator generated by such an ansatz is typically non-monotonic, whereas the exact shrinkage function should be.[2]

The Case of an Unbounded Support

On some real datasets, such as financial time series, it is hard to detect a clear boundary between bulk eigenvalues and the large outliers. In this case one may suspect that the distribution of eigenvalues of the true covariance matrix $\mathbf{C}$ is itself unbounded. In that case, one may try a parametric fit for which the density of $\mathbf{C}$ extends to infinity. For example, if we suspect that the true distribution has a sharp left edge but a power-law right tail, we may choose to model $\rho_{\mathbf{C}}(\lambda)$ as a shifted half Student's t-distribution, i.e.

$$\rho_{\mathbf{C}}(\lambda) = \Theta(\lambda - \lambda_-) \frac{2}{a\sqrt{\pi\mu}} \frac{\Gamma\left(\frac{1+\mu}{2}\right)}{\Gamma\left(\frac{\mu}{2}\right)} \left(1 + \frac{(\lambda - \lambda_-)^2}{a^2 \mu}\right)^{-\frac{1+\mu}{2}}, \tag{19.71}$$

where $\Theta(\lambda - \lambda_-)$ indicates that the density is non-zero only for $\lambda > \lambda_-$, chosen to be the center of the Student's t-distribution. These densities do not have an upper edge, instead they fall off as $\rho(\lambda) \sim \lambda^{-\mu-1}$ for large λ. For integer values of the tail

[2] Although we have not been able to find a simple proof of this property, we strongly believe that it holds in full generality.

exponent μ, the Stieltjes transform $\mathfrak{g}_{\mathbf{C}}(z)$ can be computed analytically. For example for $\mu = 3$ we find

$$\mathfrak{g}_{\mu=3}(z) = \frac{\sqrt{3}\pi u^3 + 6au^2 + 9a^2\sqrt{3}\pi u + 36a^3 \log\left(-u/\sqrt{3a^2}\right) + 18a^3}{\sqrt{3}\pi\left(3a^2 + u^2\right)^2}, \tag{19.72}$$

where $u = z - \lambda_-$. Note that this Stieltjes transform has an essential singularity at $z = \lambda_-$ and a branch cut on the real axis from λ_- to $+\infty$ indicating that the density has no upper bound. For $\mu = 3$ both the mean and the variance of the eigenvalue density are finite. We thus fix $\lambda_- = 1 - 2\sqrt{3a^2}/\pi$ such that $\tau(\mathbf{C}) = 1$ and adjust a to obtain the desired variance given by $\tau(\mathbf{C}^2) - 1 = 3a^2(1 - (2/\pi)^2)$.

In cases like this one, where we have an analytic form for $\mathfrak{g}_{\mathbf{C}}(z)$ but no simple formula for its S-transform, we can numerically solve the subordination relation

$$\mathfrak{t}_{\mathbf{E}}(\zeta) = \mathfrak{t}_{\mathbf{C}}\left(\frac{\zeta}{1 + q\mathfrak{t}_{\mathbf{E}}(\zeta)}\right), \tag{19.73}$$

with $\mathfrak{t}_{\mathbf{C}}(\zeta) = \zeta\mathfrak{g}_{\mathbf{C}}(\zeta) - 1$, using an efficient numerical fixed point equation solver. Most of the time a simple iteration would find the fixed point, but for some values of ζ and q it is sometimes difficult to find an initial condition for the iteration to converge so it is better to use a robust fixed point solver.

Let us end on a technical remark: for unbounded densities, $\mathfrak{g}(z)$ is not analytic at $z = \infty$, which does not conform to some hypotheses made throughout the book. Intuitively, there is no longer any clear distinction between bulk eigenvalues and outliers. For a fixed value of N, and for sufficiently large λ, the distance between two successive eigenvalues will at some point become much larger than $1/N$. Fortunately, the very same RIE formula holds both for bulk and for outlier eigenvalues, so we can close our eyes and safely apply Eq. (19.27) for unbounded densities as well.

19.5.2 Kernel Methods

Another approach to compute the Stieltjes and/or the T-transform on the real axis is to work directly with the discrete eigenvalues λ_k of $\mathbf{E}$. As stated earlier we cannot simply evaluate the discrete $g_N(z)$ at a point $z = \lambda_k$ because $g_N(z)$ is infinite precisely at the points $z \in \{\lambda_k\}$; this is the reason why $g_N(z)$ does not converge to the limiting $\mathfrak{g}_{\mathbf{E}}(z)$ on the support of $\rho(\lambda)$.

The idea here is to generalize the standard kernel method to estimate continuous densities from discrete data. Having observed a set of N eigenvalues $[\lambda_k]_{k\in(1,N)}$, a smooth estimator of the density is constructed as

$$\rho_s(x) := \frac{1}{N}\sum_{k=1}^{N} K_{\eta_k}(x - \lambda_k), \tag{19.74}$$

where K_η is some adequately chosen kernel of width η (possibly k-dependent), normalized such that

$$\int_{-\infty}^{+\infty} \mathrm{d}u\, K_\eta(u) = 1, \tag{19.75}$$

such that

$$\int_{-\infty}^{+\infty} \mathrm{d}x\, \rho_{\mathrm{s}}(x) = 1. \tag{19.76}$$

A standard choice for K is a Gaussian distribution, but we will discuss more appropriate choices for the Stieltjes transform below.

Now, let us similarly define a smoothed Stieltjes transform as

$$\mathfrak{g}_{\mathrm{s}}(z) := \frac{1}{N}\sum_{k=1}^{N} \mathfrak{g}_{K,\eta_k}(z-\lambda_k), \tag{19.77}$$

where $\mathfrak{g}_{K,\eta}$ is the Stieltjes transform of the *kernel* K_η, treated as a density:

$$\mathfrak{g}_{K,\eta}(z) := \int_{-\infty}^{+\infty} \mathrm{d}u \frac{K_\eta(u)}{z-u}; \qquad \mathrm{Im}(z) \neq 0. \tag{19.78}$$

Note that since $\mathrm{Im}\, \mathfrak{g}_{K,\eta}(x - \mathrm{i}0^+) = \mathrm{i}\pi K_\eta(x)$, one immediately concludes that

$$\mathrm{Im}\, \mathfrak{g}_{\mathrm{s}}(x - \mathrm{i}0^+) = \mathrm{i}\pi \rho_{\mathrm{s}}(x) \tag{19.79}$$

for any smoothing kernel K_η. Hence $\mathfrak{g}_{\mathrm{s}}(z)$ is the natural generalization of smoothed densities for Stieltjes transforms. Correspondingly, the real part of the smoothed Stieltjes is the Hilbert transform (up to a π factor) of the smoothed density, i.e.

$$\mathfrak{h}_{\mathrm{s}}(x) := \mathrm{Re}\, \mathfrak{g}_{\mathrm{s}}(x - \mathrm{i}0^+) = ⨍_{-\infty}^{\infty} \mathrm{d}\lambda \frac{\rho_{\mathrm{s}}(\lambda)}{x-\lambda}. \tag{19.80}$$

Two choices for the kernel K_η are specially interesting. One is the Cauchy kernel:

$$K_\eta^{\mathrm{C}}(u) := \frac{1}{\pi}\frac{\eta}{u^2+\eta^2}, \tag{19.81}$$

from which one gets

$$\mathfrak{g}_{K^{\mathrm{C}},\eta}(z) = \frac{1}{z \pm \mathrm{i}\eta}, \qquad \pm = \mathrm{sign}\,(\mathrm{Im}(z)). \tag{19.82}$$

Hence, in this case, we find that the smoothed Stieltjes transform we are looking for is nothing but the discrete Stieltjes transform computed with a k-dependent width η_k:

$$\mathfrak{g}_{\mathrm{s}}^{\mathrm{C}}(z) := \frac{1}{N}\sum_{k=1}^{N} \frac{1}{z-\lambda_k - \mathrm{i}\eta_k}, \qquad \mathrm{Im}(z) < 0, \tag{19.83}$$

which we can now safely compute numerically on the real axis, i.e. when $z = x - \mathrm{i}0^+$, and plug in the corresponding formulas for the RIE estimator $\xi(\lambda)$.

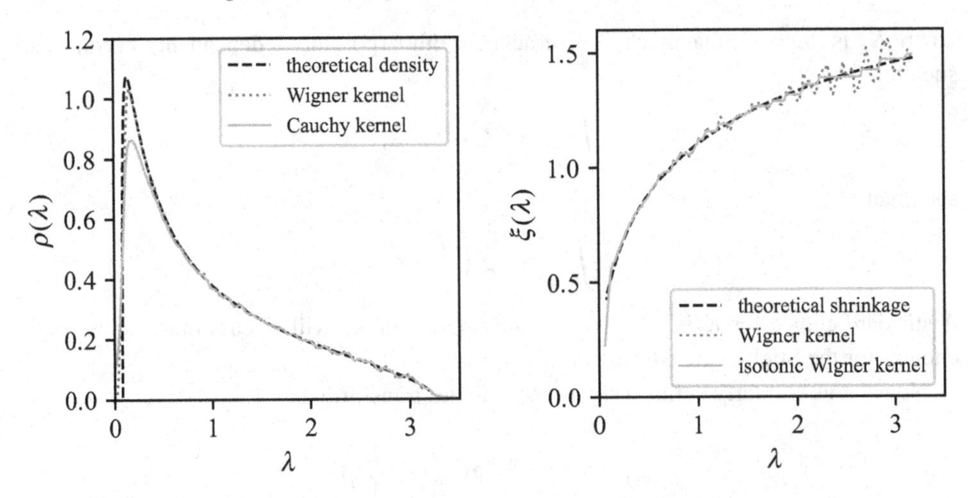

Figure 19.4 Non-parametric kernel methods applied to the same problem as in Figure 19.3. An approximation of $\mathfrak{g}_{\mathbf{E}}(\lambda - \mathrm{i}0^+)$ is computed with the Cauchy kernel (19.82) and the Wigner kernel (19.85) both with $\eta_k = \eta = N^{-1/2}$. (left) We compare the two smoothed densities with the theoretical one. Both are quite good but the Wigner kernel is better where the density changes rapidly. (right) From the smoothed Stieltjes transforms we compute the shrinkage function for both methods. Only the result of the Wigner kernel is shown (the Cauchy kernel is comparable albeit slightly worse). Kernel methods give non-monotonic shrinkage functions which can be easily rectified using an isotonic regression (19.86), which improves the agreement with the theoretical curve.

Another interesting choice for numerical applications is the semi-circle "Wigner kernel", which has sharp edges. To wit,

$$K_\eta^{\mathrm{W}}(u) = \frac{\sqrt{4\eta^2 - u^2}}{2\pi\eta^2} \quad \text{for} \quad -2\eta \le u \le 2\eta, \tag{19.84}$$

and 0 when $|u| > 2\eta$. In this case, we obtain

$$\mathfrak{g}_{\mathrm{s}}^{\mathrm{W}}(z) := \frac{1}{N}\sum_{k=1}^{N} \frac{z-\lambda_k}{2\eta_k^2}\left(1 - \sqrt{1 - \frac{4\eta_k^2}{(z-\lambda_k)^2}}\right). \tag{19.85}$$

Figure 19.4 gives an illustration of the kernel method using both the Cauchy kernel and the Wigner kernel, with $\eta_k = \eta = N^{-1/2}$.

We end this section with two practical implementation points regarding the kernel methods.

1 Since the optimal RIE estimator $\xi(\lambda)$ should be monotonic in λ, one should rectify possibly non-monotonic numerical estimators using an isotonic regression. The isotonic regression $\hat{y}_k$ of some data y_k is given by

$$\hat{y}_k = \operatorname{argmin} \sum_{k=1}^{T} \left(\hat{y}_k - y_k\right)^2 \quad \text{with} \quad \hat{y}_1 \le \hat{y}_2 \le \cdots \le \hat{y}_{T-1} \le \hat{y}_T. \tag{19.86}$$

It is the monotonic sequence that is the closest (in the least-square sense) to the original data.

2 In most situations, we are interested in reconstructing the optimal RIE matrix $\Xi = \sum \xi(\lambda_k)\mathbf{v}_k\mathbf{v}_k^T$ and hence we need to evaluate the shrinkage function $\xi(\lambda)$ precisely at the sample eigenvalues $\{\lambda_k\}$. We have found empirically that excluding the point λ_k itself from the kernel estimator consistently gives better results than including it. For example, in the Cauchy case, one should compute

$$\mathrm{g}_{\mathrm{s}}^{\mathrm{C}}(\lambda_\ell - \mathrm{i}0^+) \approx \frac{1}{N-1}\sum_{\substack{k=1\\k\neq\ell}}^{N}\frac{1}{\lambda_\ell - \lambda_k - \mathrm{i}\eta_k}, \tag{19.87}$$

when estimating Eq. (19.48).

19.6 Validation and RIE

The idea of validation to determine the RIE is to compute the eigenvectors $\mathbf{v}_i$ of $\mathbf{E}$ the SCM of a *training set* and compute their unbiased variance of a different dataset: the *validation set*. More formally, this is written

$$\xi_\times(\lambda_i) := \mathbf{v}_i{}^T\mathbf{E}'\mathbf{v}_i, \tag{19.88}$$

where $\mathbf{E}'$ is the validation SCM. The training set is also called the *in-sample* data and the validation set the *out-of-sample* data.

In practical applications, we have typically a single dataset that needs to be split into a training and a validation set. If we are not too worried about temporal order, any block of the data can serve as the validation set. In *K-fold cross-validation*, the data is split into K blocks, one block is the validation set and the union of the $K-1$ others serves as the training set. The procedure is then repeated successively choosing the K possible validation sets, see Figure 19.5.

In the following, we will assume that the true covariance matrix $\mathbf{C}$ is the same on both datasets so that $\mathbf{E} = \mathbf{C}^{\frac{1}{2}}\mathbf{W}\mathbf{C}^{\frac{1}{2}}$ and $\mathbf{E}' = \mathbf{C}^{\frac{1}{2}}\mathbf{W}'\mathbf{C}^{\frac{1}{2}}$, where $\mathbf{W}'$ is independent from $\mathbf{W}$.

Expanding over the eigenvectors $\mathbf{v}'_j$ of $\mathbf{E}'$, we get

$$\xi_\times(\lambda_i) = \sum_{k=1}^{N}\left(\mathbf{v}_i{}^T\mathbf{v}'_k\right)^2\lambda'_k \tag{19.89}$$

or, in the large N limit and using the definition of Ψ given in Eq. (19.6),

$$\xi_\times(\lambda) \underset{N\to\infty}{\longrightarrow} \int \rho_{\mathbf{E}'}(\lambda')\Psi(\lambda,\lambda')\lambda'\,d\lambda'. \tag{19.90}$$

Now, there is an *exact* relation between Ψ and Φ, which reads

$$\Psi(\lambda,\lambda') = \frac{1}{N}\sum_{j=1}^{N}\Phi(\lambda,\mu_j)\Phi(\lambda',\mu_j) \tag{19.91}$$

or, in the continuum limit,

$$\Psi(\lambda,\lambda') = \int \rho_{\mathbf{C}}(\mu)\Phi(\lambda,\mu)\Phi(\lambda',\mu)\,\mathrm{d}\mu. \tag{19.92}$$

Intuitively, this relation can be understood as follows: we expect that, from the very definition of Φ, the eigenvectors of $\mathbf{E}$ and $\mathbf{E}'$ can be written as

$$\mathbf{v}_i = \frac{1}{\sqrt{N}}\sum_{j=1}^{N} \varepsilon_{ij}\sqrt{\Phi(\lambda_i,\mu_j)}\,\mathbf{u}_j; \qquad \mathbf{v}'_k = \frac{1}{\sqrt{N}}\sum_{\ell=1}^{N} \varepsilon'_{k\ell}\sqrt{\Phi(\lambda'_k,\mu_\ell)}\,\mathbf{u}_\ell, \tag{19.93}$$

where $\mathbf{u}_j$ are the eigenvectors of $\mathbf{C}$ and ε_{ij} are independent random variables of mean zero and variance one, such that

$$\mathbb{E}[\varepsilon_{ij}\varepsilon_{k\ell}] = \delta_{ik}\delta_{j\ell}, \qquad \mathbb{E}[\varepsilon'_{ij}\varepsilon'_{k\ell}] = \delta_{ik}\delta_{j\ell}, \qquad \mathbb{E}[\varepsilon_{ij}\varepsilon'_{k\ell}] = 0. \tag{19.94}$$

This so-called ergodic assumption can be justified from considerations about the Dyson Brownian motion of eigenvectors, see Eq. (9.10), but this goes beyond the scope of this book. In any case, if we now compute $\mathbb{E}[(\mathbf{v}_i^T\mathbf{v}'_k)^2]$ using the ergodic assumption and remembering that $\mathbf{u}_j^T\mathbf{u}_\ell = \delta_{j\ell}$, we find

$$N\mathbb{E}[(\mathbf{v}_i^T\mathbf{v}'_k)^2] = \frac{1}{N}\sum_{j=1}^{N}\Phi(\lambda_i,\mu_j)\Phi(\lambda'_k,\mu_j), \tag{19.95}$$

which is precisely Eq. (19.91).

Injecting Eq. (19.91) into Eq. (19.89), we thus find

$$\begin{aligned}\xi_\times(\lambda) &= \frac{1}{N^2}\sum_{k=1}^{N}\left(\sum_{j=1}^{N}\Phi(\lambda,\mu_j)\Phi(\lambda'_k,\mu_j)\right)\lambda'_k \\ &= \frac{1}{N^2}\sum_{j=1}^{N}\Phi(\lambda,\mu_j)\left(\sum_{k=1}^{N}\Phi(\lambda'_k,\mu_j)\lambda'_k\right).\end{aligned} \tag{19.96}$$

The last term in parenthesis can be computed by using the very definition of $\mathbf{E}'$:

$$\begin{aligned}\sum_{k=1}^{N}\Phi(\lambda'_k,\mu_j)\lambda'_k &\equiv N\mathbf{u}_j{}^T\mathbf{E}'\mathbf{u}_j \\ &= N\mathbf{C}^{\frac{1}{2}}\mathbf{u}_j^T\mathbf{W}'\mathbf{C}^{\frac{1}{2}}\mathbf{u}_j = N\mu_j\mathbf{u}_j^T\mathbf{W}'\mathbf{u}_j.\end{aligned} \tag{19.97}$$

Now, the idea is that since $\mathbf{W}'$ is independent of $\mathbf{C}$, averaging over any small interval of μ_j will amount to replacing $\mathbf{u}_j{}^T\mathbf{W}'\mathbf{u}_j$ by its average over randomly oriented vectors $\mathbf{u}$, which is equal to unity:

$$\mathbb{E}[\mathbf{u}^T\mathbf{W}'\mathbf{u}] = \tau(\mathbf{W}'\mathbf{u}\mathbf{u}^T) = \tau(\mathbf{W}')\tau(\mathbf{u}\mathbf{u}^T) = 1. \tag{19.98}$$

Hence, from Eq. (19.96) we finally obtain

$$\xi_\times(\lambda) = \frac{1}{N}\sum_{j=1}^{N}\Phi(\lambda,\mu_j)\mu_j \underset{N\to\infty}{\longrightarrow} \int \rho_{\mathbf{C}}(\mu)\,\Phi(\lambda,\mu)\,\mu\,\mathrm{d}\mu, \tag{19.99}$$

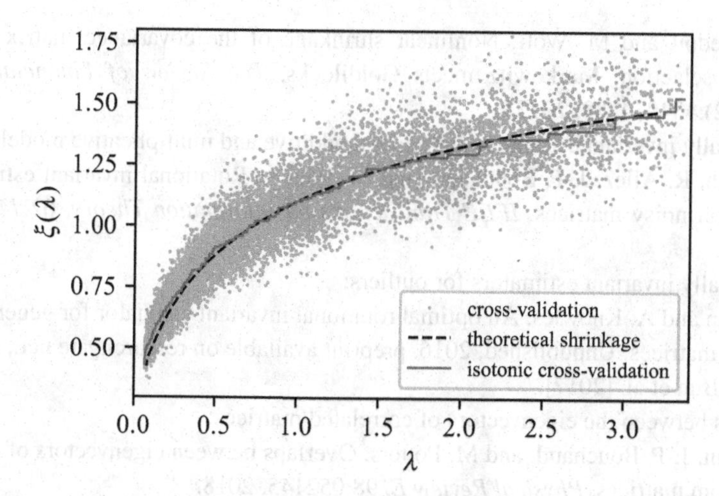

Figure 19.5 Shrinkage function $\xi(\lambda)$ computed for the same problem as in Figure 19.3, now using cross-validation. The dataset is divided into $K = 10$ blocks of equal length. For each block, we compute the $N = 1000$ eigenvalues λ_i^b and eigenvectors $\mathbf{v}_i^b$ of the sample covariance matrix using the rest of the data (of new length $9T/10$), and compute $\xi_\times(\lambda_i^b) := {\mathbf{v}_i^b}^T \mathbf{E}' \mathbf{v}_i^b$, with $\mathbf{E}'$ the sample covariance matrix of the considered block. The dots correspond to the 10×1000 pairs $(\lambda_i^b, \xi_\times(\lambda_i^b))$. The full line is an isotonic regression through the dots. The procedure has a slight bias as we in fact compute the optimal shrinkage for a value of q equal to $q_\times = 10N/9T$, but otherwise the agreement with the optimal curve is quite good.

which precisely coincides with the definition of the optimal non-linear shrinkage function $\xi(\lambda)$, see Eq. (19.27).

This result is very interesting and indicates that one can approximate $\xi(\lambda)$ by considering the quadratic form between the eigenvectors of a given realization of $\mathbf{C}$ – say $\mathbf{E}$ – and another realization of $\mathbf{C}$ – say $\mathbf{E}'$ – even if the two empirical matrices are characterized by different values of the ratio N/T. This method is illustrated in Figure 19.5.

Bibliographical Notes

- For a recent review on the subject of cleaning noisy covariance matrices and RIE, see
  - J. Bun, J.-P. Bouchaud, and M. Potters. Cleaning correlation matrices. *Risk magazine*, 2016,
  - J. Bun, J.-P. Bouchaud, and M. Potters. Cleaning large correlation matrices: Tools from random matrix theory. *Physics Reports*, 666:1–109, 2017.
- The original work of Ledoit and Péché and its operational implementation, see
  - O. Ledoit and S. Péché. Eigenvectors of some large sample covariance matrix ensembles. *Probability Theory and Related Fields*, 151(1-2):233–264, 2011,
  - O. Ledoit and M. Wolf. Nonlinear shrinkage estimation of large-dimensional covariance matrices. *The Annals of Statistics*, 40(2):1024–1060, 2012,

  - O. Ledoit and M. Wolf. Nonlinear shrinkage of the covariance matrix for portfolio selection: Markowitz meets Goldilocks. *The Review of Financial Studies*, 30(12):4349–4388, 2017.
- Rotationally invariant estimators for general additive and multiplicative models:
  - J. Bun, R. Allez, J.-P. Bouchaud, and M. Potters. Rotational invariant estimator for general noisy matrices. *IEEE Transactions on Information Theory*, 62:7475–7490, 2016.
- Rotationally invariant estimators for outliers:
  - J. Bun and A. Knowles. An optimal rotational invariant estimator for general covariance matrices. Unpublished, 2016. preprint available on researchgate.net,,

  see also Bun et al. [2017].
- Overlaps between the eigenvectors of correlated matrices:
  - J. Bun, J.-P. Bouchaud, and M. Potters. Overlaps between eigenvectors of correlated random matrices. *Physical Review E*, 98:052145, 2018.
- Rotationally invariant estimators for cross-correlation matrices:
  - F. Benaych-Georges, J.-P. Bouchaud, and M. Potters. Optimal cleaning for singular values of cross-covariance matrices. *preprint arXiv:1901.05543*, 2019.

20

Applications to Finance

20.1 Portfolio Theory

One of the arch-problems in quantitative finance is portfolio construction. For example, one may consider an investment universe made of N stocks (or more generally N risky assets) that one should bundle up in a portfolio to achieve optimal performance, according to some quality measure that we will discuss below.

20.1.1 Returns and Risk Free Rate

We call $p_{i,t}$ the price of stock i at time t and define the returns over some elementary time scale (say one day) as

$$r_{i,t} := \frac{p_{i,t} - p_{i,t-1}}{p_{i,t-1}}. \tag{20.1}$$

The portfolio weight π_i is the dollar amount invested on asset i, which can be positive (corresponding to buys) or negative (corresponding to short sales). The total capital to be invested is C. Naively, one should have $C = \sum_i \pi_i$. But one can borrow cash, so that $\sum_i \pi_i > C$ and pay the risk free rate r_0 on the borrowed amount, or conversely under-invest in stocks ($\sum_i \pi_i < C$) and invest the remaining capital at the risk free rate, assumed to be the same r_0.[1] Then the total return of the portfolio (in dollar terms) is

$$R_t = \sum_i \pi_i r_{i,t} + (C - \sum_i \pi_i) r_0, \tag{20.2}$$

so that the *excess return* (over the risk free rate) is

$$R_t - C r_0 := \sum_i \pi_i (r_{i,t} - r_0), \tag{20.3}$$

where $r_{i,t} - r_0$ is the excess return of asset i. From now on, we will denote $r_{i,t} - r_0$ by $r_{i,t}$. We will assume that these excess returns are characterized by some expected gains g_i and a covariance matrix $\mathbf{C}$, with

[1] In general, the risk free rate to borrow is different from the one to lend, but we will neglect the difference here.

$$\mathbf{C}_{ij} = \text{Cov}[r_i r_j]. \tag{20.4}$$

The problem, of course, is that both the vector of expected gains $\mathbf{g}$ and the covariance matrix $\mathbf{C}$ are unknown to the investor, who must come up with his/her best guess for these quantities. Forming expectations of future returns is the job of the investor, based on his/her information, anticipations, and hunch. We will not attempt to model the sophisticated process at work in the mind of investors, and simply assume that $\mathbf{g}$ is known. In the simplest case, the investor has no preferences and $\mathbf{g} = g\mathbf{1}$, corresponding to the same expected return for all assets. Another possibility is to assume that $\mathbf{g}$ is, for all practical purposes, a random vector in $\mathbb{R}^N$.

As far as $\mathbf{C}$ is concerned, the most natural choice is to use the sample covariance matrix $\mathbf{E}$, determined using a series of past returns of length T. However, as we already know from Chapter 17, the eigenvalues of $\mathbf{E}$ can be quite far from those of $\mathbf{C}$ when $q = N/T$ is not very small. On the other hand, T cannot be as large as one could wish, the most important reason being that the (financial) world is non-stationary. For a start, many large firms that exist in 2019 did not exist 25 years ago. More generally, it is far from clear that the parameters of the underlying statistical process (if such a thing exists) can be considered as constant in time, so mixing different epochs is in general not warranted. On the other hand, due to experimental constraints, the limitation of data points can be a problem even in a stationary world.

20.1.2 Portfolio Risk

The risk of a portfolio is traditionally measured as the variance of its returns, namely

$$\mathcal{R}^2 := \mathbb{V}[R] = \sum_{i,j} \pi_i \pi_j \text{Cov}[r_i r_j] = \boldsymbol{\pi}^T \mathbf{C} \boldsymbol{\pi}. \tag{20.5}$$

Other measures of risk can however be considered, such as the *expected shortfall* S_p (or conditional value at risk), defined as

$$S_p = -\frac{1}{p} \int_{-\infty}^{R_p} \mathrm{d}R \, R \, P(R), \tag{20.6}$$

where R_p is the p-quantile, with for example $p = 0.01$ for the 1% negative tail events:

$$p = \int_{-\infty}^{R_p} \mathrm{d}R \, P(R). \tag{20.7}$$

If $P(R)$ is Gaussian, then all risk measures are equivalent and subsumed in $\mathbb{V}[R]$.

20.1.3 Markowitz Optimal Portfolio Theory

For the reader not familiar with Markowitz's optimal portfolio theory, we recall in this section some of the most important results. Suppose that an investor wants to invest in a portfolio containing N different assets, with optimal weights $\boldsymbol{\pi}$ to be determined. An intuitive strategy is the so-called mean-variance optimization: the investor seeks an allocation such that the variance of the portfolio is minimized given an expected return target. It is not hard to see that this mean-variance optimization can be translated into a simple quadratic optimization program with a linear constraint. Markowitz's optimal portfolio amounts to solving the following quadratic optimization problem:

$$\begin{cases} \min_{\boldsymbol{\pi}\in\mathbb{R}^N} \frac{1}{2}\boldsymbol{\pi}^T\mathbf{C}\boldsymbol{\pi} \\ \text{subject to } \boldsymbol{\pi}^T\mathbf{g} \geq \mathcal{G} \end{cases} \tag{20.8}$$

where $\mathcal{G}$ is the desired (or should we say hoped for) gain. Without further constraints – such as the positivity of all weights necessary if short positions are not allowed – this problem can be easily solved by introducing a Lagrangian multiplier γ and writing

$$\min_{\boldsymbol{\pi}\in\mathbb{R}^N} \frac{1}{2}\boldsymbol{\pi}^T\mathbf{C}\boldsymbol{\pi} - \gamma\boldsymbol{\pi}^T\mathbf{g}. \tag{20.9}$$

Assuming that $\mathbf{C}$ is invertible, it is not hard to find the optimal solution and the value of γ such that overall expected return is exactly $\mathcal{G}$. It is given by

$$\boldsymbol{\pi}_{\mathbf{C}} = \mathcal{G}\frac{\mathbf{C}^{-1}\mathbf{g}}{\mathbf{g}^T\mathbf{C}^{-1}\mathbf{g}}, \tag{20.10}$$

which, as noted above, requires the knowledge of both $\mathbf{C}$ and $\mathbf{g}$, which are *a priori* unknown. Note that even if the predictions g_i of our investor are completely wrong, it still makes sense to look for the minimum risk portfolio consistent with his/her expectations. But we are left with the problem of estimating $\mathbf{C}$, or maybe $\mathbf{C}^{-1}$ before applying Markowitz's formula, Eq. (20.10). We will see below why one should actually find the best estimator of $\mathbf{C}$ itself before inverting it and determining the weights.

What is the risk associated with this optimal allocation strategy, measured as the variance of the returns of the portfolio? If one knew the population correlation matrix $\mathbf{C}$, the *true* optimal risk associated with $\boldsymbol{\pi}_{\mathbf{C}}$ would be given by

$$\mathcal{R}^2_{\text{true}} := \boldsymbol{\pi}_{\mathbf{C}}^T\mathbf{C}\boldsymbol{\pi}_{\mathbf{C}} = \frac{\mathcal{G}^2}{\mathbf{g}^T\mathbf{C}^{-1}\mathbf{g}}. \tag{20.11}$$

However, the optimal strategy (20.10) is not attainable in practice as the matrix $\mathbf{C}$ is unknown. What can one do then, and how poorly is the realized risk of the portfolio estimated?

20.1.4 Predicted and Realized Risk

One obvious – but far too naive – way to use the Markowitz optimal portfolio is to apply (20.10) using the SCM $\mathbf{E}$ as is, instead of $\mathbf{C}$. Recalling the results of Chapter 17, it is not hard to see that this strategy will suffer from strong biases whenever T is not sufficiently large compared to N.

Notwithstanding, the optimal investment weights using the SCM $\mathbf{E}$ read

$$\boldsymbol{\pi}_{\mathbf{E}} = \mathcal{G}\frac{\mathbf{E}^{-1}\mathbf{g}}{\mathbf{g}^T\mathbf{E}^{-1}\mathbf{g}}, \tag{20.12}$$

and the minimum risk associated with this portfolio is thus given by

$$\mathcal{R}_{\text{in}}^2 = \boldsymbol{\pi}_{\mathbf{E}}^T \mathbf{E}\, \boldsymbol{\pi}_{\mathbf{E}} = \frac{\mathcal{G}^2}{\mathbf{g}^T\mathbf{E}^{-1}\mathbf{g}}, \tag{20.13}$$

which is known as the "in-sample" risk, or the *predicted* risk. It is "in-sample" because it is entirely constructed using the available data. The realized risk in the next period, with fresh data, is correspondingly called *out-of-sample*.

Using the convexity with respect to $\mathbf{E}$ of $\mathbf{g}^T\mathbf{E}^{-1}\mathbf{g}$, we find from the Jensen inequality that, for fixed predicted gains $\mathbf{g}$,

$$\mathbb{E}[\mathbf{g}^T\mathbf{E}^{-1}\mathbf{g}] \geq \mathbf{g}^T\mathbb{E}[\mathbf{E}]^{-1}\mathbf{g} = \mathbf{g}^T\mathbf{C}^{-1}\mathbf{g}, \tag{20.14}$$

where the last equality holds because $\mathbf{E}$ is an unbiased estimator of $\mathbf{C}$. Hence, we conclude that the in-sample risk is lower than the "true" risk and therefore the optimal portfolio $\boldsymbol{\pi}_{\mathbf{E}}$ suffers from an in-sample bias: its predicted risk underestimates the true optimal risk $\mathcal{R}_{\text{true}}$. Intuitively this comes from the fact that $\boldsymbol{\pi}_{\mathbf{E}}$ attempts to exploit all the idiosyncracies that happened during the in-sample period, and therefore manages to reduce the risk below the true optimal risk. But the situation is even worse, because the future out-of-sample or *realized* risk, turns out to be larger than the true risk. Indeed, let us denote by $\mathbf{E}'$ the SCM of this out-of-sample period; the *out-of-sample* risk is then naturally defined by

$$\mathcal{R}_{\text{out}}^2 = \boldsymbol{\pi}_{\mathbf{E}}^T \mathbf{E}'\, \boldsymbol{\pi}_{\mathbf{E}} = \frac{\mathcal{G}^2\mathbf{g}^T\mathbf{E}^{-1}\mathbf{E}'\mathbf{E}^{-1}\mathbf{g}}{(\mathbf{g}^T\mathbf{E}^{-1}\mathbf{g})^2}. \tag{20.15}$$

For large matrices, we expect the result to be self-averaging and given by its expectation value (over the measurement noise). But if the measurement noise in the in-sample period (contained in $\boldsymbol{\pi}_{\mathbf{E}}$) can be assumed to be independent from that of the out-of-sample period, then $\boldsymbol{\pi}_{\mathbf{E}}$ and $\mathbf{E}'$ are uncorrelated and we get, for $N \to \infty$,

$$\boldsymbol{\pi}_{\mathbf{E}}^T \mathbf{E}' \boldsymbol{\pi}_{\mathbf{E}} = \boldsymbol{\pi}_{\mathbf{E}}^T \mathbf{C} \boldsymbol{\pi}_{\mathbf{E}}. \tag{20.16}$$

Now, from the optimality of $\boldsymbol{\pi}_{\mathbf{C}}$, we also know that

$$\boldsymbol{\pi}_{\mathbf{C}}^T \mathbf{C} \boldsymbol{\pi}_{\mathbf{C}} \leq \boldsymbol{\pi}_{\mathbf{E}}^T \mathbf{C} \boldsymbol{\pi}_{\mathbf{E}}, \tag{20.17}$$

so we readily obtain the following general inequalities:

$$\mathcal{R}_{\text{in}}^2 \leq \mathcal{R}_{\text{true}}^2 \leq \mathcal{R}_{\text{out}}^2. \tag{20.18}$$

We plot in Figure 20.1 an illustration of these inequalities. One can see how using $\boldsymbol{\pi}_{\mathbf{E}}$ is clearly overoptimistic and can potentially lead to disastrous results in practice.

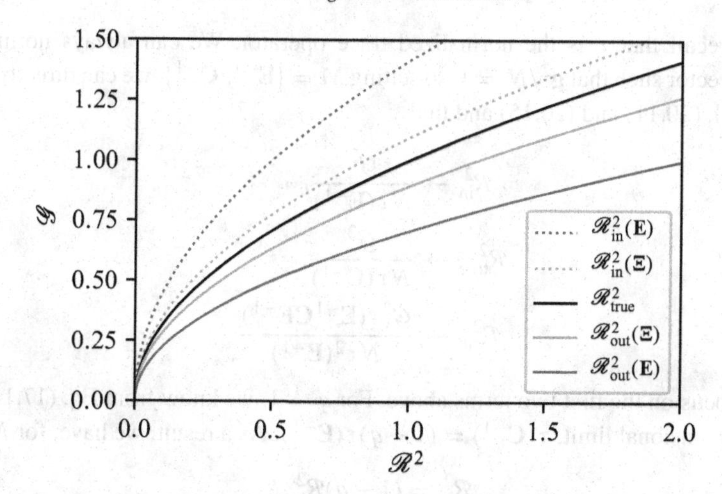

Figure 20.1 Efficient frontier associated with the mean-variance optimal portfolio (20.10) for $\mathbf{g} = \mathbf{1}$ and $\mathbf{C}$ an inverse-Wishart matrix with $p = 0.5$, for $q = 0.5$. The black line depicts the expected gain as a function of the *true* optimal risk (20.11). The gray lines correspond to the realized (out-of-sample) risk using either the SCM $\mathbf{E}$ or its RIE version Ξ. Both estimates are above the true risk, but less so for RIE. Finally, the dashed lines represent the predicted (in-sample) risk, again using either the SCM $\mathbf{E}$ or its RIE version Ξ. $\mathcal{R}$ and $\mathcal{G}$ in arbitrary units, such that $\mathcal{R}_{\text{true}} = 1$ for $\mathcal{G} = 1$.

This conclusion in fact holds for different risk measures, such as the expected shortfall measure mentioned in Section 20.1.2.

20.2 The High-Dimensional Limit

20.2.1 In-Sample vs Out-of-Sample Risk: Exact Results

In the limit of large matrices and with some assumptions on the structure $\mathbf{g}$, we can make the general inequalities Eq. (20.18) more precise using the random matrix theory tools from the previous chapters. Let us suppose that the vector of predictors $\mathbf{g}$ points in a random direction, in the sense that the covariance matrix $\mathbf{C}$ and $\mathbf{g}\mathbf{g}^T$ are mutually free. This is not necessarily a natural assumption. For example, the simplest "agnostic" prediction $\mathbf{g} = \mathbf{1}$ is often nearly collinear with the top eigenvector of $\mathbf{C}$ (see Section 20.4). So we rather think here of market neutral, sector neutral predictors that attempt to capture very idiosyncratic characteristics of firms.

Now, if $\mathbf{M}$ is a positive definite matrix that is free from $\mathbf{g}\mathbf{g}^T$, then in the large N limit:

$$\frac{\mathbf{g}^T\mathbf{M}\mathbf{g}}{N} = \frac{1}{N}\operatorname{Tr}[\mathbf{g}\mathbf{g}^T\mathbf{M}] \underset{\text{freeness}}{=} \frac{\mathbf{g}^2}{N}\tau(\mathbf{M}), \tag{20.19}$$

where we recall that τ is the normalized trace operator. We can always normalize the prediction vector such that $\mathbf{g}^2/N = 1$, so setting $\mathbf{M} = \{\mathbf{E}^{-1}, \mathbf{C}^{-1}\}$, we can directly estimate Eqs. (20.13), (20.11) and (20.15) and find

$$\begin{aligned} \mathcal{R}^2_{\text{in}} &\to \frac{\mathcal{G}^2}{N\tau(\mathbf{E}^{-1})}, \\ \mathcal{R}^2_{\text{true}} &\to \frac{\mathcal{G}^2}{N\tau(\mathbf{C}^{-1})}, \\ \mathcal{R}^2_{\text{out}} &\to \frac{\mathcal{G}^2\tau(\mathbf{E}^{-1}\mathbf{C}\mathbf{E}^{-1})}{N\tau^2(\mathbf{E}^{-1})}. \end{aligned} \tag{20.20}$$

Let us focus on the first two terms above. For $q < 1$, we know from Eq. (17.14) that, in the high-dimensional limit, $\tau(\mathbf{C}^{-1}) = (1-q)\tau(\mathbf{E}^{-1})$. As a result, we have, for $N \to \infty$,

$$\mathcal{R}^2_{\text{in}} = (1-q)\mathcal{R}^2_{\text{true}}. \tag{20.21}$$

Hence, for any $q \in (0,1)$, we see that the in-sample risk associated with $\boldsymbol{\pi}_{\mathbf{E}}$ always provides an overoptimistic estimator. Even better, we are able to quantify precisely the risk underestimation factor thanks to Eq. (20.21).

Next we would like to find the same type of relation for the "out-of-sample" risk. In order to do so, we write $\mathbf{E} = \mathbf{C}^{\frac{1}{2}}\mathbf{W}_q\mathbf{C}^{\frac{1}{2}}$ where $\mathbf{W}_q$ is a white Wishart matrix of parameter q, independent from $\mathbf{C}$. Plugging this representation into Eq. (20.15), we find that the out-of-sample risk can be expressed as

$$\mathcal{R}^2_{\text{out}} = \frac{\mathcal{G}^2\tau(\mathbf{C}^{-1}\mathbf{W}_q^{-2})}{N\tau^2(\mathbf{E}^{-1})}$$

when $N \to \infty$. Now, since $\mathbf{W}_q$ and $\mathbf{C}$ are asymptotically free, we also have

$$\tau(\mathbf{C}^{-1}\mathbf{W}_q^{-2}) = \tau(\mathbf{C}^{-1})\,\tau(\mathbf{W}_q^{-2}). \tag{20.22}$$

Hence, using again Eq. (17.14) yields

$$\mathcal{R}^2_{\text{out}} = \mathcal{G}^2(1-q)^2\frac{\tau(\mathbf{W}_q^{-2})}{N\tau(\mathbf{C}^{-1})}. \tag{20.23}$$

Finally, we know from Eq. (15.22) that $\tau(\mathbf{W}_q^{-2}) = (1-q)^{-3}$ for $q < 1$. Hence, we finally get

$$\mathcal{R}^2_{\text{out}} = \frac{\mathcal{R}^2_{\text{true}}}{1-q}. \tag{20.24}$$

All in all, we have obtained the following asymptotic relation:

$$\frac{\mathcal{R}^2_{\text{in}}}{1-q} = \mathcal{R}^2_{\text{true}} = (1-q)\mathcal{R}^2_{\text{out}}, \tag{20.25}$$

which holds for a completely general $\mathbf{C}$.

Hence, if one invests with the "naive" weights $\pi_{\mathbf{E}}$, it turns out that the predicted risk $\mathcal{R}_{\text{in}}$ underestimates the realized risk $\mathcal{R}_{\text{out}}$ by a factor $(1-q)$, and in the extreme case $N = T$ or $q = 1$, the in-sample risk is equal to zero while the out-of-sample risk diverges (for $N \to \infty$). We thus conclude that, as announced, the use of the SCM $\mathbf{E}$ for the Markowitz optimization problem can lead to disastrous results. This suggests that we should use a more reliable estimator of $\mathbf{C}$ in order to control the out-of-sample risk.

20.2.2 *Out-of-Sample Risk Minimization*

We insisted throughout the last section that the right quantity to control in portfolio management is the realized, out-of-sample risk. It is also clear from Eq. (20.25) that using the sample estimate $\mathbf{E}$ is a very bad idea, and hence it is natural to wonder which estimator of $\mathbf{C}$ one should use to minimize this out-of-sample risk? The Markowitz formula (20.10) naively suggests that one should look for a faithful estimator of the so-called precision matrix $\mathbf{C}^{-1}$. But in fact, since the expected out-of-sample risk involves the matrix $\mathbf{C}$ linearly, it is that matrix that should be estimated.

Let us show this using another route, in the context of rotationally invariant estimators, which we considered in Chapter 19. Let us define our RIE as

$$\Xi = \sum_{i=1}^{N} \xi(\lambda_i)\mathbf{v}_i\mathbf{v}_i^T, \tag{20.26}$$

where we recall that $\mathbf{v}_i$ are the sample eigenvectors of $\mathbf{E}$ and $\xi(\cdot)$ is a function that has to be determined using some optimality criterion.

Suppose that we construct a Markowitz optimal portfolio π_Ξ using this RIE. Again, we assume that the vector $\mathbf{g}$ is random, and independent from Ξ. Consequently, the estimate (20.19) is still valid, such that the realized risk associated with the portfolio π_Ξ reads, for $N \to \infty$,

$$\mathcal{R}^2_{\text{out}}(\Xi) = \mathcal{G}^2 \frac{\operatorname{Tr}\left(\Xi^{-1}\mathbf{C}\Xi^{-1}\right)}{\left(\operatorname{Tr}\Xi^{-1}\right)^2}. \tag{20.27}$$

Using the decomposition (20.26) of Ξ, we can rewrite the numerator as

$$\operatorname{Tr}\left(\Xi^{-1}\mathbf{C}\Xi^{-1}\right) = \sum_{i=1}^{N} \frac{\mathbf{v}_i{}^T\mathbf{C}\mathbf{v}_i}{\xi^2(\lambda_i)}, \tag{20.28}$$

while the denominator of Eq. (20.27) is

$$\left(\operatorname{Tr}\Xi^{-1}\right)^2 = \left(\sum_{i=1}^{N} \frac{1}{\xi(\lambda_i)}\right)^2. \tag{20.29}$$

Regrouping these last two equations allows us to express Eq. (20.27) as

$$\mathcal{R}^2_{\text{out}}(\Xi) = \mathcal{G}^2 \sum_{i=1}^{N} \frac{\mathbf{v}_i{}^T \mathbf{C} \mathbf{v}_i}{\xi^2(\lambda_i)} \left(\sum_{i=1}^{N} \frac{1}{\xi(\lambda_i)} \right)^{-2}. \tag{20.30}$$

Our aim is to find the optimal shrinkage function $\xi(\lambda_j)$ associated with the sample eigenvalues $[\lambda_j]_{j=1}^N$, such that the out-of-sample risk is minimized. This can be done by solving, for a given j, the following first order condition:

$$\frac{\partial \mathcal{R}^2_{\text{out}}(\Xi)}{\partial \xi(\lambda_j)} = 0, \quad \forall j = 1, \ldots, N. \tag{20.31}$$

By performing the derivative with respect to $\xi(\lambda_j)$ in (20.30), one obtains

$$-2\frac{\mathbf{v}_j{}^T \mathbf{C} \mathbf{v}_j}{\xi^3(\lambda_j)} \left(\sum_{i=1}^{N} \frac{1}{\xi(\lambda_i)} \right)^{-2} + \frac{2}{\xi^2(\lambda_j)} \left(\sum_{i=1}^{N} \frac{\mathbf{v}_i{}^T \mathbf{C} \mathbf{v}_i}{\xi^2(\lambda_i)} \right) \left(\sum_{i=1}^{N} \frac{1}{\xi(\lambda_i)} \right)^{-3} = 0. \tag{20.32}$$

The solution to this equation is given by

$$\xi(\lambda_j) = A \mathbf{v}_j{}^T \mathbf{C} \mathbf{v}_j, \tag{20.33}$$

where A is an arbitrary constant at this stage. But since the trace of the RIE must match that of $\mathbf{C}$, this constant A must be equal to 1. Hence we recover precisely the oracle estimator that we have studied in Chapter 19.

As a conclusion, the optimal RIE (19.26) actually minimizes the out-of-sample risk within the class of rotationally invariant estimators. Moreover, the corresponding "optimal" realized risk is given by

$$\mathcal{R}^2_{\text{out}}(\Xi) = \frac{\mathcal{G}^2}{\text{Tr}\left[(\Xi)^{-1}\right]}, \tag{20.34}$$

where we used the notable property that, for any $n \in \mathbb{Z}$,

$$\text{Tr}[(\Xi)^n \mathbf{C}] = \sum_{i=1}^{N} \xi(\lambda_i)^n \, \text{Tr}[\mathbf{v}_i \mathbf{v}_i^T \mathbf{C}] = \sum_{i=1}^{N} \xi(\lambda_i)^n \mathbf{v}_i^T \mathbf{C} \mathbf{v}_i \equiv \text{Tr}[(\Xi)^{n+1}]. \tag{20.35}$$

20.2.3 The Inverse-Wishart Model: Explicit Results

In this section, we specialize the result (20.34) to the case when $\mathbf{C}$ is an inverse-Wishart matrix with parameter $p > 0$, corresponding to the simple linear shrinkage optimal estimator. First, we read from Eq. (15.30) that

$$\tau\left(\mathbf{C}^{-1}\right) = -\mathfrak{g}_{\mathbf{C}}(0) = 1 + p, \tag{20.36}$$

so that we get from Eq. (20.20) that, in the large N limit,

$$\mathcal{R}^2_{\text{true}} = \frac{\mathcal{G}^2}{N} \frac{1}{1+p}. \tag{20.37}$$

Next, we see from Eq. (20.34) that the optimal out-of-sample risk requires the computation of $\tau((\Xi)^{-1})$. In general, the computation of this quantity is highly non-trivial but some simplifications appear when $\mathbf{C}$ is an inverse-Wishart matrix. In the large-dimension limit, the final result reads

$$\tau((\Xi)^{-1}) = -\left(1+\frac{q}{p}\right)\mathfrak{g}_{\mathbf{E}}\left(-\frac{q}{p}\right) = 1+\frac{p^2}{p+q+pq}, \tag{20.38}$$

and therefore we have from Eq. (20.34)

$$\mathcal{R}^2_{\text{out}}(\Xi) = \frac{\mathcal{G}^2}{N}\frac{p+q+pq}{(p+q)(1+p)}, \tag{20.39}$$

and so it is clear from Eqs. (20.39) and (20.37) that, for any $p > 0$,

$$\frac{\mathcal{R}^2_{\text{out}}(\Xi)}{\mathcal{R}^2_{\text{true}}} = 1+q\frac{pq}{(p+q)(1+p)} \geq 1, \tag{20.40}$$

where the last inequality becomes an equality only when $q = 0$, as it should.

It is also interesting to evaluate the in-sample risk associated with the optimal RIE. It is defined by

$$\mathcal{R}^2_{\text{in}}(\Xi) = \mathcal{G}^2\frac{\text{Tr}[(\Xi)^{-1}\mathbf{E}(\Xi)^{-1}]}{N\tau^2((\Xi)^{-1})}, \tag{20.41}$$

where the most challenging term is the numerator. Using the fact that the eigenvalues of Ξ are given by the linear shrinkage formula (19.49), one can once again find a closed formula.[2] The final result is written

$$\mathcal{R}^2_{\text{in}}(\Xi) = \frac{\mathcal{G}^2}{N}\frac{p+q}{(1+p)(p+q(p+1))}, \tag{20.42}$$

and we therefore deduce with Eq. (20.37) that, for any $p > 0$,

$$\frac{\mathcal{R}^2_{\text{in}}(\Xi)}{\mathcal{R}^2_{\text{true}}} = 1-\frac{pq}{p+q(1+p)} \leq 1, \tag{20.43}$$

where the inequality becomes an equality for $q = 0$ as above.

Finally, one may easily check from Eqs. (20.25), (20.40) and (20.43) that

$$\mathcal{R}^2_{\text{in}}(\Xi) - \mathcal{R}^2_{\text{in}}(\mathbf{E}) \geq 0, \qquad \mathcal{R}^2_{\text{out}}(\Xi) - \mathcal{R}^2_{\text{out}}(\mathbf{E}) \leq 0, \tag{20.44}$$

showing explicitly that we indeed reduce the overfitting by using the oracle estimator instead of the SCM in the high-dimensional framework: both the in-sample and out-of-sample risks computed using Ξ are closer to the true risk than when computed with the raw empirical matrix $\mathbf{E}$. The results shown in Figure 20.1 correspond to the inverse-Wishart case with $p = q = \frac{1}{2}$.

Exercise 20.2.1 Optimal portfolio when the true covariance matrix is Wishart

In this exercise we continue the analysis of Exercise 19.2.2 assuming that we measure an SCM from data with a true covariance given by a Wishart with parameter q_0.

[2] Details of this computation can be found in Bun et al. [2017].

(a) The minimum risk portfolio with expected gain G is given by

$$\pi = G\frac{\mathbf{C}^{-1}\mathbf{g}}{\mathbf{g}^T\mathbf{C}^{-1}\mathbf{g}}, \tag{20.45}$$

where $\mathbf{C}$ is the covariance matrix (or an estimator of it) and $\mathbf{g}$ is the vector of expected gains. Compute the matrix Ξ by taking the matrix $\mathbf{E}_1$ and replacing its eigenvalues λ_k by $\xi(\lambda_k)$ and keeping the same eigenvectors. Use the result of Exercise 19.2.2(c) for $\xi(\lambda)$, if some λ_k are below 0.05594 or above 3.746 replace them by 0.05594 and 3.746 respectively. This is so that you do not have to worry about finding the correct solution $\mathfrak{t}_{\mathbf{E}}(z)$ for z outside of the bulk.

(b) Build the three portfolios $\pi_{\mathbf{C}}$, $\pi_{\mathbf{E}}$ and π_{Ξ} by computing Eq. (20.45) for the three matrices $\mathbf{C}$, $\mathbf{E}_1$ and Ξ using $G = 1$ and $\mathbf{g} = \mathbf{e}_1$, the vector with 1 in the first component and 0 everywhere else. These three portfolios correspond to the true optimal, the naive optimal and the cleaned optimal. The true optimal is in general unobtainable. For these three portfolios compute the in-sample risk $R_{\text{in}} := \pi^T\mathbf{E}_1\pi$, the true risk $R_{\text{true}} := \pi^T\mathbf{C}\pi$ and the out-of-sample risk $R_{\text{out}} := \pi^T\mathbf{E}_2\pi$.

(c) Comment on these nine values. For $\pi_{\mathbf{C}}$ and $\pi_{\mathbf{E}}$ you should find exact theoretical values. The out-of-sample risk for π_{Ξ} should better than for $\pi_{\mathbf{E}}$ but worse than for $\pi_{\mathbf{C}}$.

20.3 The Statistics of Price Changes: A Short Overview

20.3.1 Bachelier's First Law

The simplest property of financial prices, dating back to Bachelier's thesis, states that typical price variations grow like the square-root of time. More formally, under the assumption that price returns have zero mean (which is usually a good approximation on short time scales), then the *price variogram*

$$\mathcal{V}(\tau) := \mathbb{E}[(\log p_{t+\tau} - \log p_t)^2] \tag{20.46}$$

grows linearly with time lag τ, such that $\mathcal{V}(\tau) = \sigma^2\tau$.

20.3.2 Signature Plots

Assume now that a price series is described by

$$\log p_t = \log p_0 + \sum_{t'=1}^{t} r_{t'}, \tag{20.47}$$

where the return series r_t is covariance-stationary with zero mean and covariance

$$\text{Cov}\,(r_{t'}, r_{t''}) = \sigma^2 C_r(|t' - t''|). \tag{20.48}$$

The case of a random walk with uncorrelated price returns corresponds to $C_r(u) = \delta_{u,0}$, where $\delta_{u,0}$ is the Kronecker delta function. A trending random walk has $C_r(u) > 0$ and a mean-reverting random walk has $C_r(u) < 0$. How does this affect Bachelier's first law?

One important implication is that the volatility observed by sampling price series on a given time scale τ is itself dependent on that time scale. More precisely, the volatility at scale τ is given by

$$\sigma^2(\tau) := \frac{\mathcal{V}(\tau)}{\tau} = \sigma^2(1)\left[1 + 2\sum_{u=1}^{\tau}\left(1 - \frac{u}{\tau}\right)C_r(u)\right]. \tag{20.49}$$

A plot of $\sigma(\tau)$ versus τ is called a *volatility signature plot*. The case of an uncorrelated random walk leads to a flat signature plot. Positive correlations (which correspond to trends) lead to an increase in $\sigma(\tau)$ with increasing τ. Negative correlations (which correspond to mean reversion) lead to a decrease in $\sigma(\tau)$ with increasing τ.

20.3.3 Volatility Signature Plots for Real Price Series

Quite remarkably, the volatility signature plots of most liquid assets (stocks, futures, FX, ...) are nowadays almost flat for values of τ ranging from a few seconds to a few months (beyond which it becomes dubious whether the statistical assumption of stationarity still holds). For example, for the S&P500 E-mini futures contract, which is one of the most liquid contracts in the world, $\sigma(\tau)$ only decreases by about 20% from short time scales (seconds) to long time scales (weeks). For single stocks, however, some interesting deviations from a flat horizontal line can be detected, see Figure 20.2. The exact form of a volatility signature plot depends on the microstructural details of the underlying asset, but most liquid contracts in this market have a similar volatility signature plot.

20.3.4 Heavy Tails

An overwhelming body of empirical evidence from a vast array of financial instruments (including stocks, currencies, interest rates, commodities, and even implied volatility) shows that unconditional distributions of returns have *fat tails*, which decay as a power law for large arguments and are much heavier than the tails of a Gaussian distribution.

On short time scales (between about a minute and a few hours), the empirical density function of returns r can be fit reasonably well by a Student's t-distribution, see Figure 20.3. Student's t-distributions read

$$P(r) = \frac{1}{a\sqrt{\pi\mu}} \frac{\Gamma\left(\frac{1+\mu}{2}\right)}{\Gamma\left(\frac{\mu}{2}\right)} \left(1 + \frac{r^2}{a^2\mu}\right)^{-\frac{1+\mu}{2}}, \tag{20.50}$$

where a is a parameter fixing the scale of r. Student's t is such that $P(r)$ decays for large r as $|r|^{-1-\mu}$, where μ is the *tail exponent*. Empirically, the tail parameter μ is consistently found to be around 3 for a wide variety of different markets (see Fig. 20.3), which suggests

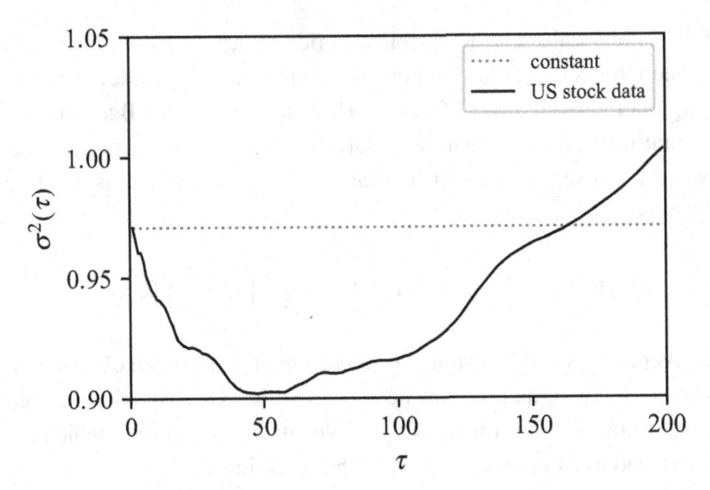

Figure 20.2 Average signature plot for the normalized returns of US stocks, where the x-axis is in days. The data consists of the returns of 1725 US companies over the period 2012–2019 (2000 business days), returns are normalized by a one-year exponential estimate of their past volatility. To a first approximation $\sigma^2(\tau)$ is independent of τ. The signature plot allows us to see deviations from this pure random walk behavior. One can see that stocks tend to mean-revert slightly at short times ($\tau < 50$ days) and trend at longer times. The effect is stronger on the many low liquidity stocks included in this dataset.

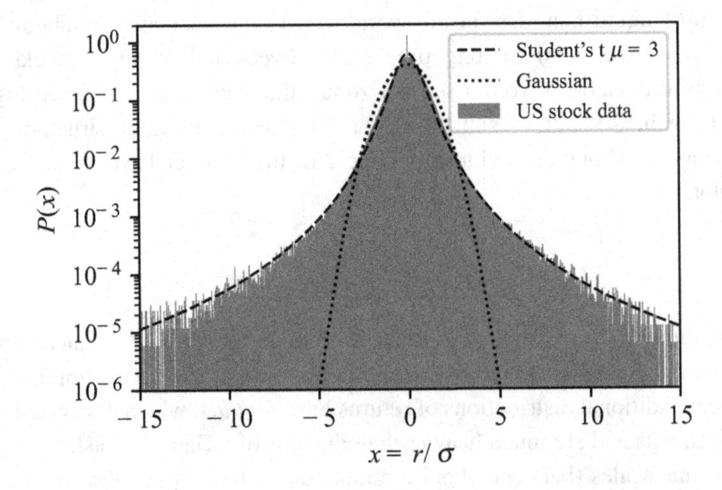

Figure 20.3 Empirical distribution of normalized daily stock returns compared with a Gaussian and a Student's t-distribution with $\mu = 3$ and the same variance. Same data as in Figure 20.2.

some kind of universality in the mechanism leading to extreme returns. This universality hints at the fact that fundamental factors are probably unimportant in determining the amplitude of most large price jumps. Interestingly, many studies indeed suggest that large price moves are often not associated with an identifiable piece of news that would rationally explain wild valuation swings.

20.3.5 Volatility Clustering

Although considering the unconditional distribution of returns is informative, it is also somewhat misleading. Returns are in fact very far from being IID random variables – although they are indeed nearly uncorrelated, as their (almost) flat signature plots demonstrate. Therefore, returns are not simply independent random variables drawn from the Student's t-distribution. Such an IID model would predict that upon time aggregation the distribution of returns would quickly converge to a Gaussian distribution on longer time scales. Empirical data indicates that this is not the case, and that returns remain substantially non-Gaussian on time scales up to weeks or even months.

The dynamics of financial markets is in fact highly intermittent, with periods of intense activity intertwined with periods of relative calm. In intuitive terms, the volatility of financial returns is itself a dynamic variable that changes over time with a broad distribution of characteristic frequencies, a phenomenon called heteroskedasticity. In more formal terms, returns can be represented by the product of a time dependent volatility component σ_t and an IID directional component ε_t,

$$r_t := \sigma_t \varepsilon_t. \tag{20.51}$$

In this representation, ε_t are IID (but not necessarily Gaussian) random variables of unit variance and σ_t are positive random variables with long memory. This is illustrated in Figure 20.4 where we show the autocorrelation of the squared returns, which gives access to $\mathbb{E}[\sigma_t^2 \sigma_{t+\tau}^2]$.

It is worth pointing out that volatilities σ and scaled returns ε are in fact not independent random variables. It is well documented that positive past returns tend to decrease

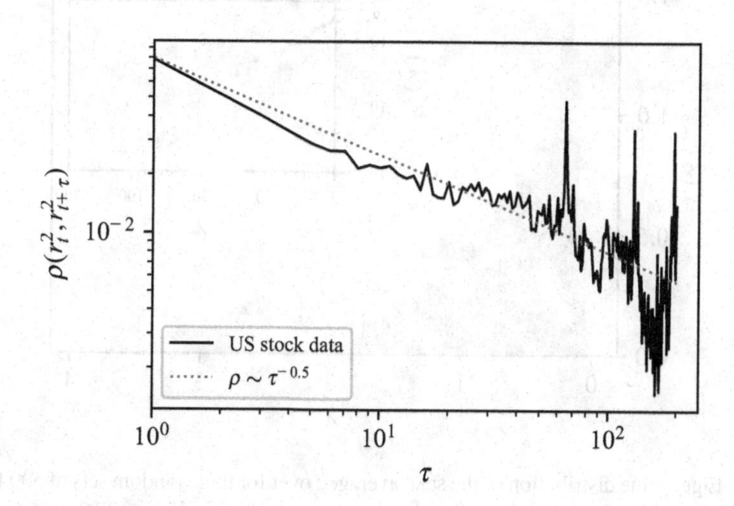

Figure 20.4 Average autocorrelation function of squared daily returns for the US stock data described in Figure 20.3. The autocorrelation decays very slowly with the time difference τ. A power law $\tau^{-\gamma}$ with $\gamma = 0.5$ is plotted to guide the eye. Note the three peaks at $\tau = 65$, 130 and 195 business days correspond to the periodicity of highly volatile earning announcements.

future volatilities and that negative past returns tend to increase future volatilities (i.e. $\mathbb{E}[\varepsilon_t \sigma_{t+\tau}] < 0$ for $\tau > 0$). This is called the *leverage effect*. Importantly, however, past volatilities do not give much information on the sign of future returns (i.e. $\mathbb{E}[\varepsilon_t \sigma_{t+\tau}] \approx 0$ for $\tau < 0$).

20.4 Empirical Covariance Matrices

20.4.1 Empirical Eigenvalue Spectrum

We are now in position to investigate the empirical covariance matrix $\mathbf{E}$ of a collection of stocks. For definiteness, we choose $q = 0.25$ by selecting $N = 500$ stocks observed at the daily time scale, with time series of length $T = 2000$ days. The distribution of eigenvalues of $\mathbf{E}$ is shown in Figure 20.5. We observe a rather broad distribution centered around 1, but with a slowly decaying tail and a top eigenvalue λ_1 found to be $\sim$100 times larger than the mean. The top eigenvector $\mathbf{v}$ corresponds to the dominant risk factor. It is closely aligned with the uniform mode $[\mathbf{e}]_i = 1/\sqrt{N}$, i.e. all stocks moving in sync – hence the name "market mode" to describe the top eigenvector. Numerically, one finds $|\mathbf{v}^T\mathbf{e}| \approx 0.95$.

20.4.2 A One-Factor Model

The simplest model aimed at describing the co-movement of different stocks is to assume that the return $r_{i,t}$ of stock i at time t can be decomposed into a common factor f_t and IID idiosyncratic components $\varepsilon_{i,t}$, to wit,

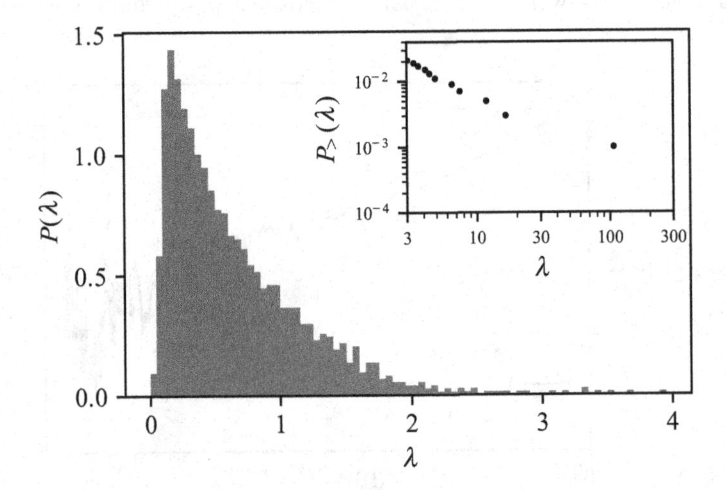

Figure 20.5 Eigenvalue distribution of the SCM, averaged over for three random sets of 500 US stocks, each measured on 2000 business days. Returns are normalized as in Figure 20.3, corresponding to $\bar{\lambda} = 0.97$. The inset shows the complementary cumulative distribution for the largest eigenvalues indicating a power-law behavior for large λ, as $P_>(\lambda) \approx \lambda^{-4/3}$. Note the largest eigenvalue $\lambda_1 \approx 0.2N$, which corresponds to the "market mode", i.e. the risk factor where all stocks move in the same direction.

$$r_{i,t} = \beta_i f_t + \varepsilon_{i,t}, \tag{20.52}$$

where f_t is often thought of as the "market factor". Assuming further that $f_t, \varepsilon_{i,t}$ are uncorrelated random variables of mean zero and variance, respectively, σ_f^2 and σ_ε^2, the covariance matrix $\mathbf{C}_{ij}$ is simply given by

$$\mathbf{C}_{ij} = \beta_i \beta_j \sigma_f^2 + \delta_{ij} \sigma_\varepsilon^2. \tag{20.53}$$

Hence, the matrix $\mathbf{C}$ is equal to $\sigma_\varepsilon^2 \mathbf{1}$ plus a rank-1 perturbation $\sigma_f^2 \boldsymbol{\beta}\boldsymbol{\beta}^T$. The eigenvalues are all equal to σ_ε^2, save the largest one, equal to $\sigma_\varepsilon^2 + \sigma_f^2 |\boldsymbol{\beta}|^2$. The corresponding top eigenvector $\mathbf{u}_f$ is parallel to $\boldsymbol{\beta}$. When the β_i's are not too far from one another, this top eigenvector is aligned with the uniform vector $\mathbf{e}$, as found empirically.

From the analysis conducted in Chapter 14, we know that the eigenvalues of the empirical matrix corresponding to such a model are composed of a Marčenko–Pastur "sea" between $\lambda_- = \sigma_\varepsilon^2(1 - \sqrt{q})^2$ and $\lambda_+ = \sigma_\varepsilon^2(1 + \sqrt{q})^2$, and an outlier located at

$$\lambda_1 = \sigma_\varepsilon^2 (1 + a)\left(1 + \frac{q}{a}\right), \tag{20.54}$$

with $a = \sigma_f^2 |\boldsymbol{\beta}|^2 / \sigma_\varepsilon^2$, and provided $a > \sqrt{q}$ (see Section 14.4). Since $|\boldsymbol{\beta}|^2 = O(N)$, the last condition is easily satisfied for large portfolios. When $a \gg 1$, one thus finds

$$\lambda_1 \approx \sigma_f^2 |\boldsymbol{\beta}|^2. \tag{20.55}$$

Since empirically $\lambda_1 \approx 0.2N$ for the correlation matrix for which $\operatorname{Tr} \mathbf{C} = N$, we deduce that for that normalization $\sigma_\varepsilon^2 \approx 0.8$. The Marčenko–Pastur sea for the value of $q = 1/4$ used in Figure 20.5 should thus extend between $\lambda_- \approx 0.2$ and $\lambda_+ \approx 1.8$. Figure 20.5 however reveals that ~20 eigenvalues lie beyond λ_+, a clear sign that more factors are needed to describe the co-movement of stocks. This is expected: the industrial sector (energy, financial, technology, etc.) to which a given stock belongs is bound to have some influence on its returns as well.

20.4.3 The Rotationally Invariant Estimator for Stocks

We now determine the optimal RIE corresponding to the empirical spectrum shown in Figure 20.5. As explained in Chapter 19, there are two possible ways to do this. One is to use Eq. (19.44) with an appropriately regularized empirical Stieltjes transform – for example by adding a small imaginary part to λ equal to $N^{-1/2}$. The second is to use a cross-validation method, see Eq. (19.88), which is theoretically equivalent as we have shown in Section 19.6. The two methods are compared in Figure 20.6, and agree quite well provided one chooses a slightly higher, effective value q^*, so as to mimic the effect of temporal correlations and fluctuating variance that lead to an effective reduction of the size of the sample (see the discussion in Section 17.2.3).

The shape of the non-linear function $\xi(\lambda)$ is interesting. It is broadly in line with the inverse-Wishart toy model shown in Figure 19.2: $\xi(\lambda)$ is concave for small λ, becomes approximately linear within the Marčenko–Pastur region, and becomes convex for larger λ.

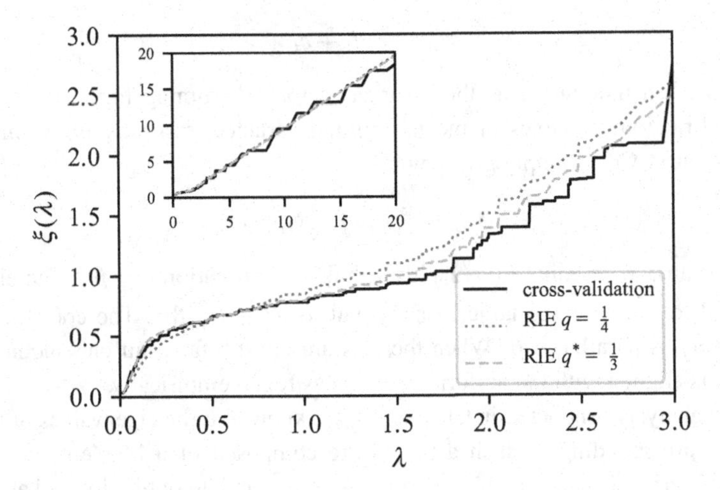

Figure 20.6 Non-linear shrinkage function $\xi(\lambda)$ computed using cross-validation and RIE averaged over three datasets. Each dataset consists of 500 US stocks measured over 2000 business days. Cross-validation is computed by removing a block of 100 days (20 times) to compute the out-of-sample variance of each eigenvector (see Eq. (19.88)). RIE is computed using the sample Stieltjes transform evaluated with an imaginary part $\eta = N^{-1/2}$. Results are shown for $q = N/T = 1/4$ and also for $q^* = 1/3$, chosen to mimic the effects of temporal correlations and fluctuating variance that lead to an effective reduction of the size of the sample (cf. Section 17.2.3). All three curves have been regularized through an isotonic fit, i.e. a fit that respects the monotonicity of the function.

For very large λ, however, $\xi(\lambda)$ becomes linear again (not shown), in a way compatible with the general formula Eq. (19.57). The use of RIE for portfolio construction in real applications is discussed in the references below.

Bibliographical Notes

- For a general introduction on Markowitz portfolio construction, see
  - E. J. Elton and M. J. Gruber. Modern portfolio theory, 1950 to date. *Journal of Banking & Finance*, 21(11):1743–1759, 1997,
  - M. Rubinstein. Markowitz's portfolio selection: A fifty-year retrospective. *The Journal of Finance*, 57(3):1041–1045, 2002,
  - P. N. Kolm, R. Ttnc, and F. J. Fabozzi. 60 years of portfolio optimization: Practical challenges and current trends. *European Journal of Operational Research*, 234(2):356–371, 2014.
- For an application of the RIE technique to portfolio construction, see
  - J. Bun, J.-P. Bouchaud, and M. Potters. Cleaning correlation matrices. *Risk magazine*, 2016,
  - J. Bun, J.-P. Bouchaud, and M. Potters. Cleaning large correlation matrices: Tools from random matrix theory. *Physics Reports*, 666:1–109, 2017,

– P.-A. Reigneron, V. Nguyen, S. Ciliberti, P. Seager, and J.-P. Bouchaud. Agnostic allocation portfolios: A sweet spot in the risk-based jungle? *The Journal of Portfolio Management*, 46 (4), 22–38, 2020,

and references therein.

- For reviews on the main stylized facts of asset returns, see, e.g.

– R. Cont. Empirical properties of asset returns: Stylized facts and statistical issues. *Quantitative Finance*, 1(2):223–236, 2001,
– J.-P. Bouchaud and M. Potters. *Theory of Financial Risk and Derivative Pricing: From Statistical Physics to Risk Management*. Cambridge University Press, Cambridge, 2nd edition, 2003,
– A. Chakraborti, I. M. Toke, M. Patriarca, and F. Abergel. Econophysics review: I. Empirical facts. *Quantitative Finance*, 11(7):991–1012, 2011,
– J.-P. Bouchaud, J. Bonart, J. Donier, and M. Gould. *Trades, Quotes and Prices*. Cambridge University Press, Cambridge, 2nd edition, 2018.

- For early work on the comparison between the Marčenko–Pastur spectrum and the eigenvalues of financial covariance matrices, see

– L. Laloux, P. Cizeau, J.-P. Bouchaud, and M. Potters. Noise dressing of financial correlation matrices. *Physical Review Letters*, 83:1467–1470, Aug 1999,
– V. Plerou, P. Gopikrishnan, B. Rosenow, L. A. N. Amaral, T. Guhr, and H. E. Stanley. Random matrix approach to cross correlations in financial data. *Physical Review E*, 65(6):066126, 2002.

- For a study of the dependence of the instantaneous covariance matrix on some dynamical indicators, and the use of RMT in this case, see

– P.-A. Reigneron, R. Allez, and J.-P. Bouchaud. Principal regression analysis and the index leverage effect. *Physica A: Statistical Mechanics and its Applications*, 390(17):3026–3035, 2011,
– A. Karami, R. Benichou, M. Benzaquen, and J.-P. Bouchaud. Conditional correlations and principal regression analysis for futures. *preprint arXiv:1912.12354*, 2019.

Appendix

Mathematical Tools

A.1 Saddle Point Method

In this appendix, we briefly review the saddle point method (sometimes also called the Laplace method, the steepest descent or the stationary phase approximation). Consider the integral

$$I = \int_{-\infty}^{+\infty} e^{tF(x)} dx. \tag{A.1}$$

We want to find an approximation for this integral when $t \to \infty$. First consider the case where $F(x)$ is real. The key idea of the Laplace method is that when t is large I is dominated by the maximum of $F(x)$ plus Gaussian fluctuations around it. Suppose F reaches its maximum at a unique point x^*, then around x^* we have

$$F(x) = F(x^*) + \frac{F''(x^*)}{2}(x - x^*)^2 + O(|x - x^*|^3), \tag{A.2}$$

where $F''(x^*) < 0$. Thus for large t, we have, after a Gaussian integral over $x - x^*$,

$$I \sim \sqrt{\frac{2\pi}{-F''(x^*)t}} e^{tF(x^*)}, \tag{A.3}$$

where the symbol $\sim$ means that the ratio of both sides of the equation tends to 1 as $t \to \infty$. Often we are only interested in

$$\lim_{t\to\infty} \frac{1}{t} \log I = F(x^*), \tag{A.4}$$

in which case we do not need to compute the prefactor.

Things are more subtle when $F(x)$ is a complex analytic function of $z = x + iy$. One could be tempted to think that one can just apply the Laplace method to $\mathrm{Re}(F(x))$. But this is grossly wrong, because $\exp(it\,\mathrm{Im}(F(x)))$ oscillates so fast that the contribution of any small neighborhood of x^* is killed.

The idea of the steepest descent or the stationary phase approximation relies on the fact that, for any analytic function, the Cauchy–Riemann condition ensures that

$$\vec{\nabla}\,\mathrm{Re}(F(z)) \cdot \vec{\nabla}\,\mathrm{Im}(F(z)) = 0, \tag{A.5}$$

where the gradient is in the two-dimensional complex plane. This means that the contour lines of $\mathrm{Re}(F(z))$ are everywhere orthogonal to the contour lines of $\mathrm{Im}(F(z))$, or alternatively that along the lines of steepest ascent (or descent) of $\mathrm{Re}(F(z))$, the imaginary part $\mathrm{Im}(F(z))$ is constant.

Now, one can always deform the integration contour from the real line in Eq. (A.1) to any curve in the complex plane that starts at $z = -\infty + \mathrm{i}0$ and ends at $z = +\infty + \mathrm{i}0$. Since $\exp(tF(z))$ has no poles, this (by Cauchy's theorem) does not change the value of the integral. But again because of the Cauchy–Riemann condition, $\nabla^2 \mathrm{Re}(F(z)) = 0$, which means that $\mathrm{Re}(F(z))$ has no maximum or minimum in the complex plane, but may have a saddle point z^* where $\vec{\nabla}\,\mathrm{Re}(F(z)) = 0$, increasing in one direction and decreasing in another (see Fig. A.1). Choosing the contour that crosses the saddle point z^* by following a path that is locally orthogonal to the contour lines of $\mathrm{Re}(F(z))$ allows the phase $\mathrm{Im}(F(z))$ to be stationary, hence avoiding the nullification of the integral by rapid oscillations. The final result is then

$$I \sim \sqrt{\frac{2\pi}{-tF''(z^*)}}\, \mathrm{e}^{tF(z^*)}. \tag{A.6}$$

The method is illustrated in Figure A.1 for the example of the Airy function, defined as

$$\mathrm{Ai}(t) = \frac{1}{2\pi}\int_{-\infty}^{+\infty} \mathrm{d}z \mathrm{e}^{tF(z,t)}, \qquad F(z,t) := \mathrm{i}\left(z + \frac{z^3}{3t}\right). \tag{A.7}$$

For a fixed, large positive t, the points for which $F'(z,t) = 0$ are given by $z_\pm = \pm \mathrm{i}\sqrt{t}$. The contour lines of $\mathrm{Re}(F(z,t))$ are plotted in Figure A.1. For $\mathrm{Im}(z) < 0$, $\mathrm{Re}(F(z,t)) \to +\infty$

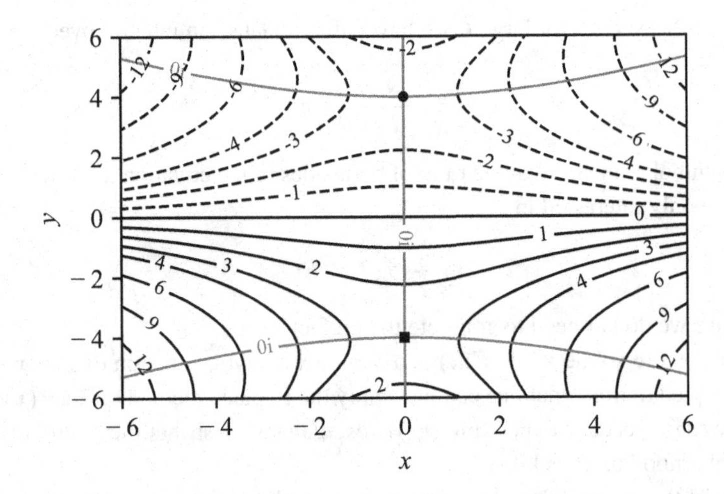

Figure A.1 Contour lines of the real part of $F(z,t) = \mathrm{i}\left(z + z^3/3t\right)$ for $t = 16$ (black lines). The iso-phase line $\mathrm{Im}\, F(z,t) = 0$ is also shown (gray lines). The black circle and square are the two solutions $z_\pm$ of $F'(z,t) = 0$. The relevant saddle is $z^* = +4\mathrm{i}$, for which $F(z^*,t) = -8/3$.

when $\mathrm{Re}(z) \to \pm\infty$, so one cannot deform the contour in that direction without introducing enormous uncontrollable contributions. When $\mathrm{Im}(z) > 0$, on the contrary, $\mathrm{Re}(F(z,t)) \to -\infty$ when $\mathrm{Re}(z) \to \pm\infty$, so one can start at $z = -\infty + \mathrm{i}0$, and travel upwards in the complex plane to the line where $\mathrm{Im}(F(z,t)) = 0$ in a region where $\mathrm{Re}(F(z,t))$ is so negative that there is no contribution to the integral from this part. Then one stays on the iso-phase line $\mathrm{Im}(F(z,t)) = 0$ and climbs upwards to the saddle point $z^* = z_+ = +\mathrm{i}\sqrt{t}$, for which $F(z^*,t) = -2\sqrt{t}/3$. One then continues on the same iso-phase line downwards, and closes the contour towards $z = +\infty + \mathrm{i}0$. Hence, we conclude that

$$\lim_{t\to\infty} \log \mathrm{Ai}(t) = tF(z^*,t) = -\frac{2}{3}t^{3/2}. \tag{A.8}$$

A more precise expression, including the prefactor in Eq. (A.6), is written

$$\mathrm{Ai}(t) \sim \frac{1}{2\sqrt{\pi}t^{1/4}}\mathrm{e}^{-\frac{2}{3}t^{3/2}}. \tag{A.9}$$

Exercise A.1.1 Saddle point method for the factorial function: Stirling's approximation

We are going to estimate the factorial function for large arguments using an integral representation and the saddle point approximation:

$$n! = \Gamma[n+1] = \int_0^\infty x^n \mathrm{e}^{-x}\mathrm{d}x. \tag{A.10}$$

(a) Write $n!$ in the form Eq. (A.1) for some function $F(x)$.

(b) Show that $x^* = n$ is the solution to $F'(x) = 0$.

(c) Let $I_0(n) = nF(x^*(n))$ be an approximation of $\log(n!)$. Compare this approximation to the exact value for $n = 10$ and 100.

(d) Include the Gaussian corrections to the saddle point: Let $I_1(n) = \log(I)$ where I is given by Eq. (A.3) for your function $F(x)$. Show that

$$I_1(n) = n\log(n) - n + \frac{1}{2}\log(2\pi n). \tag{A.11}$$

(e) Compare $I_1(n)$ and $\log(n!)$ for $n = 10$ and 100.

A.2 Tricomi's Formula

Suppose one has the following integral equation to solve for $\rho(x)$:

$$f(x) = \fint \mathrm{d}x' \frac{\rho(x')}{x - x'}, \tag{A.12}$$

where $f(x)$ is an arbitrary function. This problem arises in many contexts, in particular when one studies the equilibrium density of the eigenvalues of some random matrices as in Section 5.5, or the equilibrium density of dislocations in solids, see Landau et al. [2012], Chapter 30, from which the material of the present appendix is derived.

We will consider the case where some "hard walls" are present at $x = a$ and $x = b$, confining the density to an interval smaller than its "natural" extension in the absence of these walls. By continuity, we will also get the solution when these walls are exactly located at these natural boundaries of the spectrum.[1]

The general solution of Eq. (A.12) is given by Tricomi's formula which reads

$$\rho(x) = -\frac{1}{\pi^2 \sqrt{(x-a)(b-x)}} \left(\fint_a^b \mathrm{d}x' \sqrt{(x'-a)(b-x')} \frac{f(x')}{x-x'} + C \right), \tag{A.13}$$

where C is a constant to be determined by some conditions that $\rho(x)$ must fulfill.

The simplest case corresponds to $f(x) = 0$, i.e. no confining potential apart from the walls. The solution then reads

$$\rho(x) = -\frac{C}{\pi^2 \sqrt{(x-a)(b-x)}}. \tag{A.14}$$

The normalization of $\rho(x)$ then yields $C = -\pi$ and one recovers the arcsine law encountered in Sections 5.5, 7.2 and 15.3.1:

$$\rho(x) = \frac{1}{\pi \sqrt{(x-a)(b-x)}}. \tag{A.15}$$

The canonical Wigner case corresponds to $f(x) = x/2\sigma^2$. We look for the values of a, b that are precisely such that the density vanishes at these points, so that the confining walls no longer have any effect. By symmetry one must have $a = -b$. One then obtains

$$\rho(x) = -\frac{1}{2\pi^2\sigma^2 \sqrt{(x+b)(b-x)}} \left(\fint_{-b}^b \mathrm{d}x' \sqrt{(x'+b)(b-x')} \frac{x'-x+x}{x-x'} + C \right). \tag{A.16}$$

Using

$$x \fint_{-b}^b \mathrm{d}x' \frac{\sqrt{(x'+b)(b-x')}}{x-x'} = \pi x^2 \tag{A.17}$$

one has

$$\rho(x) = -\frac{1}{2\pi^2\sigma^2 \sqrt{(x+b)(b-x)}} \left(\pi x^2 + C' \right), \tag{A.18}$$

where $C' = C - \int_{-b}^b \mathrm{d}x' \sqrt{(x'+b)(b-x')} = C - \pi b^2/2$.

For $\rho(x)$ to vanish at the edges, we need to choose $C' = -\pi b^2$, finally leading to

$$\rho(x) = \frac{1}{2\pi\sigma^2} \sqrt{b^2 - x^2}. \tag{A.19}$$

Finally, for this distribution to be normalized one must choose $b = 2\sigma$, as it should be.

[1] Formulas directly adapted to two free boundaries, or one free boundary and one hard wall, can be found in Landau et al. [2012], Chapter 30.

The case studied by Dean and Majumdar, i.e. when all eigenvalues of a Wigner matrix are contrained to be positive (Eq. (5.93)), corresponds to Eq. (A.13) with $a = 0$, $b > 0$ and $\sigma = 1$ (i.e. $f(x) = x/2$), determined again such that $\rho(b) = 0$. The solution is $b^* = 4/\sqrt{3}$ and

$$\rho(x) = \frac{1}{4\pi}\sqrt{\frac{b^* - x}{x}}(b^* + 2x). \tag{A.20}$$

Note that very generically the density of eigenvalues (or dislocations) diverges as $1/\sqrt{d}$ near a hard wall, where d is the distance to that wall. When the boundary is free, on the other hand, the density of eigenvalues (or dislocations) vanishes as $\sqrt{d}$.

A.3 Toeplitz and Circulant Matrices

A Toeplitz matrix is such that its elements $\mathbf{K}_{ij}$ only depend on the "distance" between indices $|i - j|$. An example of a Toeplitz matrix is provided by the covariance of the elements of a stationary time series. For instance consider an AR(1) process x_t in the steady state, defined by

$$x_t = a x_{t-1} + \varepsilon_t, \tag{A.21}$$

where ε_t are IID centered random numbers with variance $1/(1 - a^2)$ such that x_t has unit variance in the steady state. Then

$$\mathbf{K}_{ts} = \mathbb{E}[x_t x_s] = a^{-|t-s|} \text{ with } 0 < a < 1, \tag{A.22}$$

describing decaying exponential correlations. The parameter a measures the decay of the correlation; we can define a correlation time $\tau_c := 1/(1 - a)$. This time is always ≥ 1 (equal to 1 when $\mathbf{K} = \mathbf{1}$) and tends to infinity as $a \to 1$. More explicitly, $\mathbf{K}$ is a $T \times T$ matrix that reads

$$\mathbf{K} = \begin{pmatrix} 1 & a & a^2 & \dots & a^{T-2} & a^{T-1} \\ a & 1 & a & \dots & a^{T-3} & a^{T-2} \\ a^2 & a & 1 & \dots & a^{T-4} & a^{T-3} \\ \vdots & \vdots & \vdots & \ddots & \vdots & \vdots \\ a^{T-2} & a^{T-3} & a^{T-4} & \dots & 1 & a \\ a^{T-1} & a^{T-2} & a^{T-3} & \dots & a & 1 \end{pmatrix}. \tag{A.23}$$

In an infinite system, it can be diagonalized by plane waves (Fourier transform), since

$$\sum_{s=-\infty}^{+\infty} \mathbf{K}_{ts} e^{2\pi i x s} = e^{2\pi i x t} \sum_{s=-\infty}^{+\infty} \mathbf{K}_{ts} e^{2\pi i x (s-t)} = e^{2\pi i x t} \sum_{\ell=-\infty}^{+\infty} a^{|\ell|} e^{2\pi i x \ell}, \tag{A.24}$$

showing that $e^{2\pi i x s}$ are eigenvectors of $\mathbf{K}$ as soon as its elements only depend on $|s - t|$.

For finite T, however, this diagonalization is only approximate as there are "boundary effects" at the edge of the matrix. If the correlation time τ_c is not too large (i.e. if the matrix elements $\mathbf{K}_{ts}$ decay sufficiently fast with $|t-s|$) these boundary effects should be negligible. One way to make this diagonalization exact is to modify the matrix so as to have the distance $|t-s|$ defined on a circle.[2] We define a new matrix $\tilde{\mathbf{K}}$ by

$$\tilde{\mathbf{K}}_{ts} = a^{-\min(|t-s|,\,|t-s+T|,\,|t-s-T|)}, \tag{A.25}$$

$$\tilde{\mathbf{K}} = \begin{pmatrix} 1 & a & a^2 & \dots & a^2 & a \\ a & 1 & a & \dots & a^3 & a^2 \\ a^2 & a & 1 & \dots & a^4 & a^3 \\ \vdots & \vdots & \vdots & \ddots & \vdots & \vdots \\ a^2 & a^3 & a^4 & \dots & 1 & a \\ a & a^2 & a^3 & \dots & a & 1 \end{pmatrix}. \tag{A.26}$$

It may seem that we have greatly modified matrix $\mathbf{K}$ as we have changed about half of its elements, but if $\tau_c \ll T$ most of the elements we have changed were essentially zero and remain essentially zero. Only a finite number ($\approx 2\tau_c^2$) of elements in the bottom right and top left corners have really changed. Changing a finite number of off-diagonal elements in an asymptotically large matrix should not change its spectrum. The matrix $\tilde{\mathbf{K}}$, which we will call $\mathbf{K}$ again, is called a *circulant matrix* and can be exactly diagonalized by Fourier transform for finite T. More precisely, its eigenvectors are

$$[\mathbf{v}_k]_\ell = \mathrm{e}^{2\pi \mathrm{i} k\ell/T} \text{ for } 0 \le k \le T/2. \tag{A.27}$$

Note that to each $\mathbf{v}_k$ correspond two eigenvectors, namely its real and imaginary parts, except for $\mathbf{v}_0$ and $\mathbf{v}_{T/2}$ which are real and have multiplicity 1. The eigenvalues associated with $k=0$ and $k=T/2$ are, respectively, the largest (λ_+) and smallest (λ_-) and are given by

$$\begin{aligned} \lambda_+ &= 1 + 2\sum_{k=1}^{T/2-1} a^k + a^{T/2} \approx \frac{1+a}{1-a}, \\ \lambda_- &= 1 + 2\sum_{k=1}^{T/2-1} (-a)^k + (-a)^{T/2} \approx \frac{1-a}{1+a} = \frac{1}{\lambda_+}. \end{aligned} \tag{A.28}$$

In terms of the correlation time: $\lambda_+ = 2\tau_c - 1$. We label the eigenvalues of $\mathbf{K}$ by an index $x_k = 2k/T$ so that $0 \le x_k \le 1$. As $T \to \infty$, x_k becomes a continuous parameter x and the different multiplicity of the first and the last eigenvalues does not matter. The eigenvalues can be written

$$\lambda(x) = \frac{1-a^2}{1+a^2-2a\cos(\pi x)} \text{ for } 0 \le x \le 1. \tag{A.29}$$

[2] The Toeplitz matrix $\mathbf{K}$ can in fact be diagonalized exactly, see O. Narayan, B. Sriram Shastry, arXiv:2006.15436v2.

For a more general form $\mathbf{K}_{ts} = K(|t-s|)$, the eigenvalues read

$$\lambda(x) = 1 + 2\sum_{\ell=1}^{\infty} K(\ell)\cos(\pi x \ell). \tag{A.30}$$

The T-transform of $\mathbf{K}$ can then be computed as

$$\mathrm{t}_{\mathbf{K}}(z) = \int_0^1 \frac{1-a^2}{z(1+a^2-2a\cos(\pi x)) - (1-a^2)}\mathrm{d}x. \tag{A.31}$$

Using

$$\int_0^1 \frac{\mathrm{d}x}{c - d\cos(\pi x)} = \frac{1}{\sqrt{c-d}\sqrt{c+d}}, \tag{A.32}$$

and after some manipulations we find

$$\mathrm{t}_{\mathbf{K}}(z) = \frac{1}{\sqrt{z-\lambda_-}\sqrt{z-\lambda_+}} \quad \text{with } \lambda_\pm = \frac{1\pm a}{1\mp a}. \tag{A.33}$$

We can also deduce the density of eigenvalues (see Fig. A.2):

$$\rho_{\mathbf{K}}(\lambda) = \frac{1}{\pi\lambda\sqrt{(\lambda-\lambda_-)(\lambda_+-\lambda)}} \quad \text{for } \lambda_- < \lambda < \lambda_+. \tag{A.34}$$

This density has integrable singularities at $\lambda = \lambda_\pm$. It is normalized and its mean is 1. We can also invert $\mathrm{t}_{\mathbf{K}}(z)$ with the equation

$$t^2\zeta_{\mathbf{K}}^2 - 2bt^2\zeta_{\mathbf{K}} + t^2 - 1 = 0, \text{ where } b = \frac{1+a^2}{1-a^2}, \tag{A.35}$$

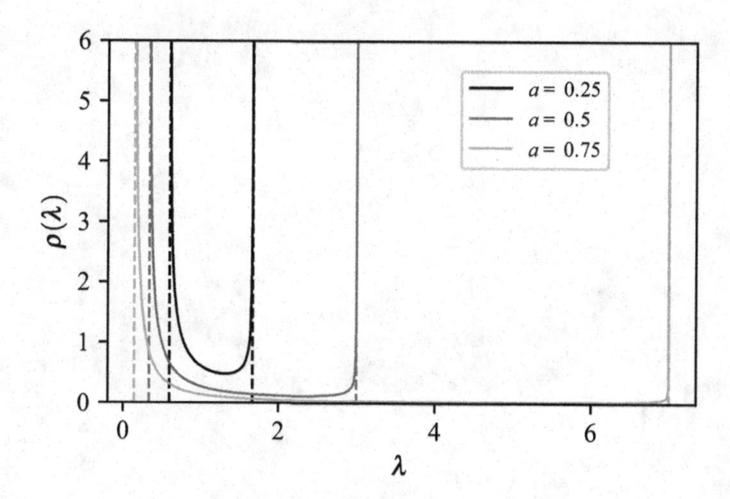

Figure A.2 Density of eigenvalues for the decaying exponential covariance matrix $\mathbf{K}$ for three values of a: 0.25, 0.5 and 0.75.

and get

$$\zeta_{\mathbf{K}}(t) = \frac{bt^2 + \sqrt{(b^2-1)t^4 + t^2}}{t^2}, \tag{A.36}$$

so the S-transform is given by

$$\begin{aligned} S_{\mathbf{K}}(t) &= \frac{t+1}{\sqrt{1+(b^2-1)t^2} + bt} \\ &= 1 - (b-1)t + O(t^2), \end{aligned} \tag{A.37}$$

where the last equality tells us that the matrix $\mathbf{K}$ has mean 1 and variance $\sigma_{\mathbf{K}}^2 = b - 1 = a^2/(1-a^2)$.

Bibliographical Notes

- On the saddle point method, see the particularly clear exposition in
  - J. Hinch. *Perturbation Methods*. Cambridge Texts in Applied Mathematics. Cambridge University Press, Cambridge, 1991.
- About the Tricomi method and the role of boundaries, Chapter 30 of Landau and Lifshitz's *Theory of Elasticity* is enlightening:
  - L. D. Landau, L. P. Pitaevskii, A. M. Kosevich, and E. M. Lifshitz. *Theory of Elasticity*. Butterworth-Heinemann, Oxford, 3rd edition, 2012.

  For its use in the context of random matrix theory, see
  - D. S. Dean and S. N. Majumdar. Extreme value statistics of eigenvalues of Gaussian random matrices. *Physical Review E*, 77:041108, 2008.

Index

Printed in the United States
By Bookmasters